Sequential Logic
and Verilog HDL
Fundamentals

Sequential Logic
and Verilog HDL
Fundamentals

Joseph Cavanagh

CRC Press
Taylor & Francis Group
Boca Raton London New York

CRC Press is an imprint of the
Taylor & Francis Group, an **informa** business

CRC Press
Taylor & Francis Group
6000 Broken Sound Parkway NW, Suite 300
Boca Raton, FL 33487-2742

First issued in hardback 2019

© 2016 by Taylor & Francis Group, LLC
CRC Press is an imprint of Taylor & Francis Group, an Informa business

No claim to original U.S. Government works

ISBN-13: 978-1-4987-3822-4 (hbk)

Visit the Taylor & Francis Web site at
http://www.taylorandfrancis.com

and the CRC Press Web site at
http://www.crcpress.com

In memory of Ivan Pesic and for his wife Kathy and son Illya
Founders and owners of Silvaco, Inc.
for generously providing the SILOS Simulation Environment software
for all of my books that use Verilog HDL and for their continued support

CONTENTS

Preface xi

Chapter 1   Introduction to Verilog HDL   1

1.1   Built-In Primitives   2
1.2   User-Defined Primitives   18
    1.2.1   Defining a User-Defined Primitive   18
    1.2.2   Combinational User-Defined Primitives   19
1.3   Dataflow Modeling   35
    1.3.1   Continuous Assignment   35
    1.3.2   Reduction Operators   41
    1.3.3   Conditional Operator   45
    1.3.4   Relational Operators   48
    1.3.5   Logical Operators   50
    1.3.6   Bitwise Operators   53
    1.3.7   Shift Operators   59
1.4   Behavioral Modeling   63
    1.4.1   Initial Statement   64
    1.4.2   Always Statement   66
    1.4.3   Intrastatement Delay   70
    1.4.4   Interstatement Delay   73
    1.4.5   Blocking Assignments   75
    1.4.6   Nonblocking Assignments   78
    1.4.7   Conditional Statements   81
    1.4.8   Case Statement   85
    1.4.9   Loop Statements   89
1.5   Structural Modeling   91
    1.5.1   Module Instantiation   91
    1.5.2   Ports   92
    1.5.3   Design Examples   93
1.6   Problems   113

Chapter 2   Synthesis of Synchronous Sequential
               Machines 1 Using Verilog HDL   121

2.1   Synchronous Registers   122
    2.1.1   Parallel-In, Serial-Out Registers   122
    2.1.2   Serial-In, Parallel-Out Registers   131
    2.1.3   Serial-In, Serial-Out Registers   143
    2.1.4   Combinational Shifter   151

2.2	Synchronous Counters		173
	2.2.1	Modulo-8 Counter	174
	2.2.2	Modulo-10 Counter	186
	2.2.3	Johnson Counter	206
	2.2.4	Binary-to-Gray Code Converter	219
2.3	Moore Machines		233
	2.3.1	Design Using Behavioral Modeling	235
	2.3.2	Design Using Structural Modeling with D Flip-Flops, AND Gates, and an OR Gate	239
	2.3.3	Design Using Structural Modeling with D Flip-Flops, NOR Gates, and an OR Gate	244
	2.3.4	Design Using Structural Modeling with JK Flip-Flops, AND Gates, and an OR Gate	249
2.4	Mealy Machines		254
	2.4.1	Design Using Behavioral Modeling	255
	2.4.2	Design Using Structural Modeling with D Flip-Flops	259
	2.4.3	Design Using Structural Modeling with JK Flip-Flops	265
2.5	Moore–Mealy Equivalence		270
2.6	Output Glitches		317
	2.6.1	Glitch Elimination Using State Code Assignment	317
	2.6.2	Glitch Elimination Using Complemented Clock	322
	2.6.3	Glitch Elimination Using Delayed Clock	334
2.7	Problems		340

Chapter 3   Synthesis of Synchronous Sequential Machines 2 Using Verilog HDL 347

3.1	Multiplexers for δ Next-State Logic		348
	3.1.1	Linear-Select Multiplexers	349
	3.1.2	Nonlinear-Select Multiplexers	376
3.2	Decoders for λ Output Logic		392
3.3	Programmable Logic Devices		421
	3.3.1	Programmable Read-Only Memory	422
	3.3.2	Programmable Array Logic	426
	3.3.3	Programmable Logic Array	445
3.4	Iterative Networks		460
3.5	Error Detection in Synchronous Sequential Machines		473
	3.5.1	Overview of Error Detection and Correction	473
	3.5.2	Examples of Error Detection in Synchronous Sequential Machines	478
3.6	Problems		491

Chapter 4   Synthesis of Asynchronous Sequential
Machines Using Verilog HDL 497

    4.1   Introduction 497
        4.1.1   Built-In Primitive Gates 497
        4.1.2   Dataflow Modeling 498
        4.1.3   Behavioral Modeling 498
        4.1.4   Structural Modeling 501
    4.2   Synthesis Examples 501
    4.3   Problems 609

Chapter 5   Synthesis of Pulse-Mode Asynchronous
Sequential Machines Using Verilog HDL 617

    5.1   Introduction 617
        5.1.1   Built-In Primitives Gates 617
        5.1.2   Dataflow Modeling 618
        5.1.3   Behavioral Modeling 618
        5.1.4   Structural Modeling 620
    5.2   Synthesis Examples 620
    5.3   Problems 687

Appendix A   Event Queue 699

Appendix B   Verilog Project Procedure 715

Appendix C   Answers to Select Problems 717

    Chapter 1   Introduction to Verilog HDL 717
    Chapter 2   Synthesis of Synchronous Sequential Machines 1 Using
Verilog HDL 743
    Chapter 3   Synthesis of Synchronous Sequential Machines 2 Using
Verilog HDL 767
    Chapter 4   Synthesis of Asynchronous Sequential Machines Using
Verilog HDL 789
    Chapter 5   Synthesis of Pulse-Mode Asynchronous Sequential
Machines Using Verilog HDL 813

Index   843

PREFACE

The field of digital logic consists primarily of analysis and synthesis of combinational and sequential logic circuits, also referred to as finite-state machines. Finite-state machines are designed into every computer. They occur in the form of counters, shift registers, microprogram control sequencers, sequence detectors, and many other sequential structures. The principal characteristic of combinational logic is that the outputs are a function of the present inputs only, whereas, the outputs of sequential logic are a function of the input sequence; that is, the input history. Sequential logic, therefore, requires storage elements which indicate the present state of the machine relative to a unique sequence of inputs.

Sequential logic is partitioned into synchronous and asynchronous sequential machines. Synchronous sequential machines are controlled by a system clock which provides the triggering mechanism to cause state changes. Asynchronous sequential machines have no clocking mechanism — the machines change state upon the application of input signals. The input signals provide the means to enable the sequential machines to proceed through a prescribed sequence of states.

The purpose of this book is to present a thorough exposition of the analysis and synthesis of both synchronous and asynchronous sequential machines. The machines will be implemented using Verilog HDL (Hardware Description Language). Verilog HDL is an Institute of Electrical and Electronics Engineers (IEEE) standard: 1364-1995. The book concentrates on sequential logic design with emphasis on the detailed design of various Verilog HDL projects.

Emphasis is placed on structured and rigorous design principles that can be applied to practical applications. Each step of the analysis and synthesis procedures is clearly delineated. Each method that is presented is expounded in sufficient detail with accompanying examples. Many analysis and synthesis examples use mixed-logic symbols which incorporate both positive- and negative-input logic gates for NAND and NOR logic, while other examples utilize only positive-input logic gates. The use of mixed logic parallels the use of these symbols in the industry.

The book is intended to be tutorial, and as such, is comprehensive and self contained. All designs are carried through to completion — nothing is left unfinished or partially designed. Each chapter includes numerous problems of varying complexity to be designed by the reader using Verilog HDL design techniques. The Verilog HDL designs include the design module, the test bench module which tests the design for correct functionality, the outputs obtained from the test bench, and the waveforms obtained from the test bench.

It is assumed that the reader has an adequate knowledge of the topics listed in this paragraph and the following two paragraphs, which are prerequisites for any course in Verilog HDL: Number systems of different radices such as binary, octal, decimal, and hexadecimal, including conversion between radices. The number representations of sign magnitude, diminished-radix complement, and radix complement. Binary weighted and nonweighted codes, including conversion to and from binary-coded decimal (BCD), plus the Gray code.

Boolean algebra, which illustrates methods to minimize switching functions. These methods include algebraic minimization, Karnaugh maps, Karnaugh maps using map-entered variables, the Quine–McCluskey algorithm, and the Petrick algorithm.

Combinational logic and storage elements. This includes wired-AND logic gates, wired-OR logic gates, and three-state logic. Logic macro functions such as multiplexers, decoders, encoders, and comparators. Analysis and synthesis of combinational logic and sequential logic. Programmable logic devices. These include programmable read-only memory (PROM) devices, programmable array logic (PAL) devices, and programmable logic array (PLA) devices. The storage elements are *SR* latches, *D* flip-flops, *JK* flip-flops, and *T* flip-flops.

Chapter 1 introduces the Verilog Hardware Description Language, which will be used throughout the book to design the various types of sequential circuits. Verilog HDL is the state-of-the-art method for designing digital and computer systems and is ideally suited to describe both combinational, clocked sequential, and non-clocked logic sequential logic circuits. Verilog provides a clear relationship between the language syntax and the physical hardware. The Verilog simulator used in this book is easy to learn and use, yet powerful enough for any application. It is a logic simulator — called SILOS — developed by Silvaco Incorporated for use in the design and verification of digital systems.

The SILOS simulation environment is a method to quickly prototype and debug any logic function. It is an intuitive environment that displays every variable and port from a module to a logic gate. SILOS allows single-stepping through the Verilog source code, as well as drag-and-drop ability from the source code to a data analyzer for waveform generation and analysis. This chapter introduces the reader to the different modeling techniques, including built-in primitives for logic primitive gates and user-defined primitives for larger logic functions. The three main modeling methods of dataflow modeling, behavioral modeling, and structural modeling are introduced.

Chapter 2 designs synchronous sequential machines using Verilog HDL. The machines include different categories of synchronous registers, such as parallel-in serial-out registers; serial-in parallel-out registers; and serial-in serial-out registers. Different types of counters of various moduli are also designed in this chapter. These include: a modulo-8 counter, a modulo-10 counter, and a Johnson counter. Also included will be a binary-to-Gray code converter. Different versions of Moore and Mealy synchronous sequential machines will also be designed using Verilog together with different techniques to eliminate output glitches. Each step in the synthesis procedure employs several examples which help to clarify the corresponding step. Several examples are presented detailing the synthesis procedure in a step-by-step process.

Chapter 3 uses Verilog HDL to design alternative synchronous sequential machines. The devices include multiplexers for the δ next-state logic of both the linear-select and nonlinear-select category. Decoders are included for the λ output logic. Programmable logic devices are presented, which are used to synthesize synchronous sequential machines. These include: programmable read-only memories, programmable array logic devices, and programmable logic array devices. Sequential iterative machines are also used in the Verilog design process. A final section presents error detection in synchronous sequential machines using Verilog HDL.

Chapter 4 presents the synthesis of asynchronous sequential machines using Verilog HDL. The chapter includes numerous examples for a comparative study of the design methodologies. The designs will be accomplished by utilizing one or more of the following modeling methods for each design: built-in primitive gates, dataflow modeling, behavioral modeling, and structural modeling.

The examples illustrate the synthesis procedure for asynchronous sequential machines using a timing diagram and/or a verbal specification. In order to prevent possible race conditions and associated timing problems when two or more inputs change value simultaneously, it will be assumed that only one input variable will change state at a time. This is referred to as a *fundamental-mode model*.

Chapter 5 presents synthesis examples of pulse-mode asynchronous sequential machines using Verilog HDL. Moore and Mealy pulse-mode asynchronous sequential machines are designed using different Verilog HDL modeling constructs. The synthesis procedure is described using several different types of storage elements. The synthesis examples will utilize built-in primitives with *SR* latches and *D* flip-flops; *T* flip-flops only; dataflow modeling with *D* flip-flops; built-in primitives and *D* flip-flops; built-in primitives and *T* flip-flops.

The pulse width restrictions that are dominant in pulse-mode sequential machines can be eliminated by including *D* flip-flops in the feedback path from the *SR* latches to the δ next-state logic. Providing edge-triggered *D* flip-flops as a constituent part of the implementation negates the requirement of precisely controlled input pulse durations. This is by far the most reliable means of synthesizing pulse-mode machines. The *SR* latches — in conjunction with the *D* flip-flops — form a master-slave configuration.

Appendix A presents a brief discussion on event handling using the event queue. Operations that occur in a Verilog module are typically handled by an event queue.

Appendix B presents a procedure to implement a Verilog project.

Appendix C contains the solutions to select problems in each chapter.

The material presented in this book represents more than two decades of computer equipment design by the author. The book is not intended as a book on combinational logic, since it is assumed that the reader has an adequate background in the analysis and synthesis of combinational logic. The book is intended as a text for a course on sequential logic design using Verilog HDL. The book presents Verilog HDL with numerous design examples to help the reader thoroughly understand this popular hardware description language.

This book presents basic and advanced concepts in sequential machine analysis and synthesis and is designed for practicing electrical engineers, computer engineers, and computer scientists; for graduate students in electrical engineering, computer engineering, and computer science; and for senior-level undergraduate students.

A special thanks to David Dutton, CEO of Silvaco Incorporated, for allowing use of the SILOS Simulation Environment software for the examples in this book, and for all of my books that use Verilog HDL, and for his continued support. SILOS is an intuitive, easy to use, yet powerful Verilog HDL simulator for logic verification.

I would like to express my appreciation and thanks to the following people who gave generously of their time and expertise to review the manuscript and submit comments: Professor Daniel W. Lewis, Department of Computer Engineering, Santa Clara University, who supported me in all my endeavors; Geri Lamble; and Steve Midford. Thanks also to Nora Konopka and the staff at Taylor & Francis for their support.

Joseph Cavanagh

1.1   Built-In Primitives
1.2   User-Defined Primitives
1.3   Dataflow Modeling
1.4   Behavioral Modeling
1.5   Structural Modeling
1.6   Problems

1

Introduction to Verilog HDL

This chapter provides an introduction to the design methodologies and modeling constructs of the Verilog hardware description language (HDL). Modules, ports, and test benches will be presented. This chapter introduces Verilog in conjunction with combinational logic only.

A *module* is the basic unit of design in Verilog that describes the Verilog hardware and consists of the following types of modules: built-in logic primitives, user-defined logic primitives, dataflow modeling, behavioral modeling, and structural modeling. A module describes the functional operation of some logical entity and can be a stand-alone module or a collection of modules that are instantiated into a structural module. *Instantiation* means to use one or more lower-level modules in the construction of a higher-level structural module. A module can be a logic gate, an adder, a multiplexer, a counter, or some other logical function. Examples will be shown for each type of modeling.

Ports allow the modules to communicate with the external environment; that is, other modules and input/output signals. Ports, also referred to as terminals, can be declared as **input**, **output**, or **inout**. A port is a net by default; however, it can be declared explicitly as a net. A module contains an optional list of ports, as shown below for a full adder. Ports *a*, *b*, and *cin* are input ports; ports *sum* and *cout* are output ports.

module full_adder (a, b, cin, sum, cout);

Test benches will also be described. Test benches are used to apply input vectors to the design module in order to test the functional operation of the module in a simulation environment. The test bench for the full adder contains no ports as shown below because it does not communicate with the external environment.

module full_adder_tb;

When a Verilog module is finished, it must be tested to ensure that it operates according to the machine specifications. The functionality of the module can be tested by applying stimulus to the inputs and checking the outputs. The test bench will display the inputs and outputs in a radix (binary, octal, hexadecimal, or decimal) as well as the waveforms.

1.1 Built-In Primitives

Logic primitives such as **and**, **nand**, **or**, **nor**, **xor** (exclusive-OR), and **xnor** (exclusive-NOR) functions are classified as multiple-input gates. The **buf** and **not** functions have one input, but can have one or more outputs. These are all built-in primitives that can be instantiated into a module. The inputs of built-in primitives are declared as type **wire** or as type **reg** depending on whether they were generated by a structural or behavioral module.

Type **wire** represents a physical connection between hardware elements. The output of a logic gate is declared as **wire** and represents a net with a single driver. The connection can be a wire or a group of wires, both of which are called a net. Nets are 1-bit scalar values unless declared otherwise.

Type **reg** data types are registers that hold a value. The register value is retained in memory until it is changed by a subsequent assignment. A variable of type **reg** closely resembles a hardware register that is synthesized with D flip-flops, JK flip-flops, or SR latches.

This section presents a design methodology that is characterized by a low level of abstraction, where the logic hardware is described in terms of gates. Designing logic at this level is similar to designing logic by drawing gate symbols — there is a close correlation between the logic gate symbols and the Verilog built-in primitive gates.

The primitive gates are used to describe a net and have one or more scalar inputs, but only one scalar output. The output signal is listed first, followed by the inputs in any order. The outputs are declared as **wire;** the inputs can be declared as either **wire** or **reg**. The gates represent combinational logic functions and can be instantiated into a module, as follows, where the instance name (inst1) is optional:

gate_type inst1 (output, input_1, input_2, . . . , input_n);

Two or more instances of the same type of gate can be specified in the same construct, as shown below. Note that only the last instantiation has a semicolon terminating the line. All previous lines are terminated by a comma.

gate_type inst1 (output_1, input_11, input_12, . . . , input_1n),
inst2 (output_2, input_21, input_22, . . . , input_2n);

The best way to learn design methodologies using built-in primitives is by example. Therefore, combinational logic examples will be presented of varying complexety. When necessary, the theory for the examples will be presented prior to the Verilog design. All examples are carried through to completion at the gate level. Nothing is left unfinished or partially designed.

Example 1.1   The logic diagram of Figure 1.1 will be designed using built-in primitives for the logic gates which consist of NAND gates and one OR gate (inst3). These gates will generate the two outputs z_1 and z_2. The output of the gate labeled *inst2* (instantiation 2) will be at a high voltage level if either x_1, x_2, or x_3 is deasserted. Therefore, by DeMorgan's theorem, the output will be at a low voltage level if x_1, x_2, and x_3 are all asserted. Note that the gate labeled *inst3* is an OR gate that is drawn as an AND gate with active-low inputs and an active-low output. The output of each gate is assigned a net name, where a *net* is one or more interconnecting wires that connect the output of one logic element   to the input of one or more logic elements. The remaining gates in Figure 1.1 are drawn in the standard manner as NAND gates.

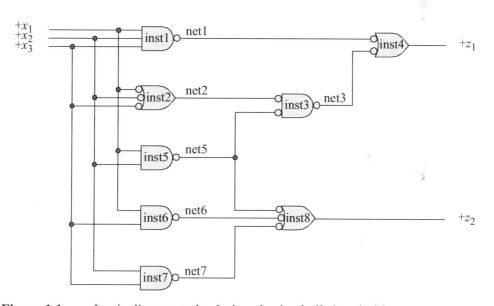

Figure 1.1   Logic diagram to be designed using built-in primitives.

The equations for z_1 and z_2 are shown in Equation 1.1 and Equation 1.2, respectively. If necessary, the laws of Boolean algebra should be reviewed in order to obtain the minimized expressions for the outputs.

$$z_1 = [x_1 x_2 x_3 + x_1 x_2 x_3 (x_1 x_2)]$$
$$= x_1 x_2 x_3 (1 + x_1 x_2)$$
$$= x_1 x_2 x_3 \qquad (1.1)$$

$$z_2 = x_1 x_2 + x_1 x_3 + x_2 x_3 \tag{1.2}$$

The Verilog design module is shown in Figure 1.2. The first line is usually reserved for a comment (//) and specifies the function of the module. Comments (//) can also be placed at the end of a line to indicate the function of a specific line of code. Line 2 is the beginning of the Verilog code and is indicated by the keyword **module** followed by the module name *log_eqn_sop15*. This is followed by the list of input and output ports placed within parentheses and terminated by a semicolon.

Verilog must know which ports are used for input and which ports are used for output; therefore, lines 4 and 5 list the input and output ports indicated by the keywords **input** and **output**, respectively. Line 7 begins the instantiation of the built-in primitives. The instantiation names and net names in the module correlate directly to the corresponding names in the logic diagram. Thus, line 7 in the module, which is

nand   inst1   (net1, x1, x2, x3);

represents NAND gate *inst1* with inputs x_1, x_2, and x_3 and output *net1* in the logic diagram. Line 14 in the module corresponds to OR function of instantiation *inst8* of the logic diagram whose inputs are *net5*, *net6*, and *net7* and whose active-high output is z_2. The end of the module is indicated by the keyword **endmodule** as shown in line 16. In this example, Figure 1.2 correctly describes the hardware that is represented by the logic diagram of Figure 1.1.

In order to verify that the module operates correctly, as specified in the logic diagram, the module must be tested. This is accomplished by means of a test bench. Test benches are used to apply input vectors to the module in order to test the functional operation of the module in a simulation environment. The functionality of the module can be tested by applying stimulus to the inputs and checking the outputs. The test bench can display the inputs and outputs in the following radices: binary (b), octal (o), hexadecimal (h), or decimal (d). Refer to the Verilog Project Procedure in Appendix B to review the procedure for generating the inputs, outputs, and waveforms of a Verilog design.

The test bench contains Verilog code to generate the input stimulus and code to display the output response to the stimulus. The test bench also provides code to instantiate the design module into the test bench. Figure 1.3 shows a test bench which applies input stimulus to test the validity of the Verilog design of Figure 1.2, which represents the logic diagram of Figure 1.1. Line 1 of the test bench is a comment indicating that the module is a test bench for the *log_eqn_sop15* module. Line 2 contains the keyword **module** followed by the module name, which includes *tb* indicating a test bench module. The name of the module and the name of the module under test are the same for ease of cross-referencing. The notations #0 and #10 specify the time at which values are assigned to the inputs. The keyword **endmodule** terminates the test bench module.

```
1    //logic diagram using built-in primitives
     module log_eqn_sop15 (x1, x2, x3, z1, z2);

     input x1, x2, x3;
5    output z1, z2;

     nand    inst1 (net1, x1, x2, x3);
     nand    inst2 (net2, x1, x2, x3);
     or      inst3 (net3, net2, net5);
10   nand    inst4 (z1, net1, net3);
     nand    inst5 (net5, x1, x2);
     nand    inst6 (net6, x1, x3);
     nand    inst7 (net7, x2, x3);
     nand    inst8 (z2, net5, net6, net7);
15

     endmodule
```

Figure 1.2   Design module for the logic diagram shown in Figure 1.1 using built-in primitives.

```
1    //test bench for log_eqn_sop_15
     module log_eqn_sop15_tb;

     reg x1, x2, x3;
5    wire z1, z2;

     //display variables
     initial
     $monitor ("x1=%b, x2=%b, x3=%b, z1=%b, z2=%b",
10             x1, x2, x3, z1, z2);

     //apply input vectors
     initial
     begin
15   #0   x1 = 1'b0;
          x2 = 1'b0;
          x3 = 1'b0;

     #10  x1 = 1'b0;
20        x2 = 1'b0;
          x3 = 1'b1;
22
                              //continued on next page
```

Figure 1.3   Test bench for the module of Figure 1.2.

```
23    #10  x1 = 1'b0;
            x2 = 1'b1;
25          x3 = 1'b0;

      #10  x1 = 1'b0;
            x2 = 1'b1;
            x3 = 1'b1;
30
      #10  x1 = 1'b1;
            x2 = 1'b0;
            x3 = 1'b0;

35    #10  x1 = 1'b1;
            x2 = 1'b0;
            x3 = 1'b1;

      #10  x1 = 1'b1;
40          x2 = 1'b1;
            x3 = 1'b0;

      #10  x1 = 1'b1;
            x2 = 1'b1;
45          x3 = 1'b1;

      #10  $stop;
      end

//instantiate the module into the test bench
log_eqn_sop15 inst1 (
52      .x1(x1),
        .x2(x2),
        .x3(x3),
55      .z1(z1),
        .z2(z2)
        );

59    endmodule
```

Figure 1.3   (Continued)

Values are assigned to the variables by the notation $1'b0$ for example, where the number 1 specifies the width of the variable (1 bit), b specifies the radix (binary), and 0 specifies the value (zero). The system task **$stop** causes simulation to stop.

Line 4 specifies that the inputs are **reg** type variables; that is, they contain their values until they are assigned new values. Outputs are assigned as type **wire** in test benches. Output nets are driven by the output ports of the module under test. Line 8 contains an **initial** statement, which executes only once.

Verilog provides a means to monitor a signal when its value changes. This is accomplished by the **$monitor** system task. The **$monitor** continuously monitors the values of the variables indicated in the parameter list that is enclosed in parentheses. It will display the value of the variables whenever a variable changes state. The string that is enclosed in quotes in the task is printed and specifies that the variables are to be shown in binary (%b). The **$monitor** is invoked only once. Line 13 is a second **initial** statement that allows the procedural code between the **begin** . . . **end** block statements to be executed only once. Every ten time units (#10) the input variables change state and are displayed by the system task **$monitor**.

Lines 51 through 57 instantiate the design module into the test bench module. The instantiation name is *inst1* followed by a left parenthesis. The port names of the design module are preceded by a period, which is followed by the corresponding port name in the test bench enclosed in parentheses; the port names in the module and the test bench do not necessarily have to be the same. A comma terminates each line of the port instantiation except the line containing the last port name — there is no termination character at the end of this line. This is followed by a right parenthesis followed by a semicolon. The keyword **endmodule** terminates the test bench module.

The logic shown in Figure 1.1 contains redundant gates to provide a review of Boolean algebra minimization, as shown in Equation 1.1 and Equation 1.2. The outputs obtained from the test bench are shown in Figure 1.4 and correspond to the equations for z_1 and z_2. There is a single active-high output for z_1 and four active-high outputs for z_2, including the case where the outputs overlap. The waveforms are shown in Figure 1.5. As mentioned previously, refer to Appendix B for the procedure to create a Verilog project and obtain the outputs and waveforms.

```
x1=0,  x2=0,  x3=0,  z1=0,  z2=0
x1=0,  x2=0,  x3=1,  z1=0,  z2=0
x1=0,  x2=1,  x3=0,  z1=0,  z2=0
x1=0,  x2=1,  x3=1,  z1=0,  z2=1
x1=1,  x2=0,  x3=0,  z1=0,  z2=0
x1=1,  x2=0,  x3=1,  z1=0,  z2=1
x1=1,  x2=1,  x3=0,  z1=0,  z2=1
x1=1,  x2=1,  x3=1,  z1=1,  z2=1
```

Figure 1.4   Outputs for the logic diagram of Figure 1.1 obtained from the test bench to Figure 1.3.

Example 1.2   The Karnaugh map of Figure 1.6 will be implemented using only NOR gates in a product-of-sums format. Equation 1.3 shows the product-of-sums expression obtained from the Karnaugh map. The minimal product-of-sums expression can be obtained by combining the 0s in Figure 1.6 to form sum terms in the same manner as the 1s were combined to form product terms. However, since 0s are being combined, each sum term must equal 0. Thus, the two 0s in row $x_1 x_2 = 00$ combine to yield the sum term $(x_1 + x_2 + x_4)$. In a similar manner, the remaining 0s are combined to yield the product-of-sums expression shown in Equation 1.3. When combining 0s to obtain sum terms, treat a variable with a value of 1 as false and a variable with a value of 0 as true. Thus, minterm locations 2 and 10 have variables $x_2 x_3 x_4 = 010$, providing a sum term of $(x_2 + x_3' + x_4)$. The logic diagram is shown in Figure 1.7, which includes the instantiation names and net names.

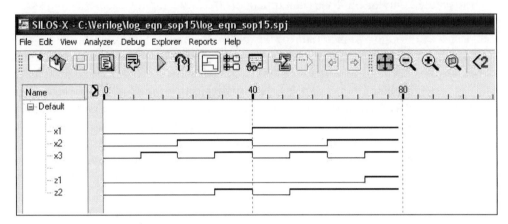

Figure 1.5   Waveforms for the logic diagram of Figure 1.1 obtained from the test bench of Figure 1.3.

$x_1 x_2$ \ $x_3 x_4$	0 0	0 1	1 1	1 0
0 0	0 (0)	1 (1)	1 (3)	0 (2)
0 1	1 (4)	1 (5)	0 (7)	1 (6)
1 1	1 (12)	1 (13)	0 (15)	1 (14)
1 0	1 (8)	1 (9)	1 (11)	0 (10)

z_1

Figure 1.6   Karnaugh map for Example 1.2.

$$z_1 = (x_1 + x_2 + x_4)(x_2 + x_3' + x_4)(x_2' + x_3' + x_4') \qquad (1.3)$$

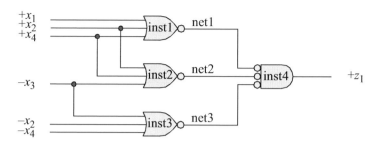

Figure 1.7   Logic diagram for Example 1.2.

The design module is shown in Figure 1.8 using NOR gate built-in primitives. The instantiation names and the net names shown in Figure 1.8 agree with the corresponding names in the logic diagram. A true value for a variable is indicated by the variable name, such as x_3; a false (complemented) value for a variable name is indicated by the symbol ($\sim$), such as $\sim x_3$. The test bench is shown in Figure 1.9. Since there are four inputs to the circuit, all 16 combinations of the four variables must be applied to the design module from the test bench in order to verify correct circuit operation. This is accomplished by assigning values to the four variables x_1, x_2, x_3, and x_4 in 16 separate lines of the test bench. The outputs obtained from the test bench are shown in Figure 1.10 and the waveforms are shown in Figure 1.11. The correct operation of the circuit can be verified by applying specific values to the inputs of the logic diagram and corroborate the resulting z_1 values with the outputs or the waveforms.

```
//logic diagram using built-in primitives
module log_eqn_pos8 (x1, x2, x3, x4, z1);


input x1, x2, x3, x4;
output z1;


//instantiate the nor built-in primitives
nor     inst1(net1, x1, x2, x4);
nor     inst2(net2, x2, x4, ~x3);
nor     inst3(net3, ~x3, ~x2, ~x4);
nor     inst4(z1, net1, net2, net3);


endmodule
```

Figure 1.8   Design module for the logic diagram of Figure 1.7 for Example 1.2.

```
//test bench for log_eqn_pos8
module log_eqn_pos8_tb;


reg x1, x2, x3, x4;   //inputs are reg for test bench
wire z1;              //outputs are wire for test bench

//display variables
//the brace ({) symbol specifies concatenation
initial
$monitor ("x1x2x3x4 = %b, z1 = %b",
         {x1, x2, x3, x4}, z1);

//apply input vectors
initial
begin
   #0    x1=1'b0; x2=1'b0; x3=1'b0; x4=1'b0;   //00
   #10   x1=1'b0; x2=1'b0; x3=1'b0; x4=1'b1;   //01
   #10   x1=1'b0; x2=1'b0; x3=1'b1; x4=1'b0;   //02
   #10   x1=1'b0; x2=1'b0; x3=1'b1; x4=1'b1;   //03
   #10   x1=1'b0; x2=1'b1; x3=1'b0; x4=1'b0;   //04
   #10   x1=1'b0; x2=1'b1; x3=1'b0; x4=1'b1;   //05
   #10   x1=1'b0; x2=1'b1; x3=1'b1; x4=1'b0;   //06
   #10   x1=1'b0; x2=1'b1; x3=1'b1; x4=1'b1;   //07
   #10   x1=1'b1; x2=1'b0; x3=1'b0; x4=1'b0;   //08
   #10   x1=1'b1; x2=1'b0; x3=1'b0; x4=1'b1;   //09
   #10   x1=1'b1; x2=1'b0; x3=1'b1; x4=1'b0;   //10
   #10   x1=1'b1; x2=1'b0; x3=1'b1; x4=1'b1;   //11
   #10   x1=1'b1; x2=1'b1; x3=1'b0; x4=1'b0;   //12
   #10   x1=1'b1; x2=1'b1; x3=1'b0; x4=1'b1;   //13
   #10   x1=1'b1; x2=1'b1; x3=1'b1; x4=1'b0;   //14
   #10   x1=1'b1; x2=1'b1; x3=1'b1; x4=1'b1;   //15

   #10   $stop;
end


//instantiate the module into the test bench
log_eqn_pos8 inst1 (
   .x1(x1),
   .x2(x2),
   .x3(x3),
   .x4(x4),
   .z1(z1)
   );

endmodule
```

Figure 1.9   Test bench for the module of Figure 1.8.

```
x1x2x3x4 = 0000,  z1 = 0
x1x2x3x4 = 0001,  z1 = 1
x1x2x3x4 = 0010,  z1 = 0
x1x2x3x4 = 0011,  z1 = 1
x1x2x3x4 = 0100,  z1 = 1
x1x2x3x4 = 0101,  z1 = 1
x1x2x3x4 = 0110,  z1 = 1
x1x2x3x4 = 0111,  z1 = 0
x1x2x3x4 = 1000,  z1 = 1
x1x2x3x4 = 1001,  z1 = 1
x1x2x3x4 = 1010,  z1 = 0
x1x2x3x4 = 1011,  z1 = 1
x1x2x3x4 = 1100,  z1 = 1
x1x2x3x4 = 1101,  z1 = 1
x1x2x3x4 = 1110,  z1 = 1
x1x2x3x4 = 1111,  z1 = 0
```

Figure 1.10   Outputs for the logic diagram of Figure 1.7 generated by the test bench of Figure 1.9.

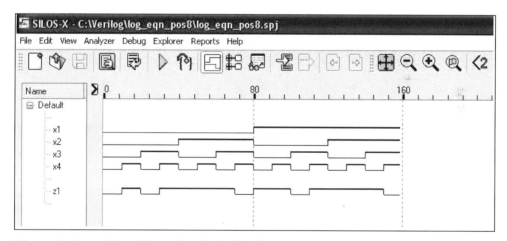

Figure 1.11   Waveforms for the logic diagram of Figure 1.7.

Example 1.3   A 4:1 multiplexer will be designed using built-in logic primitives. The 4:1 multiplexer of Figure 1.12 will be designed using built-in primitives of **and**, **or**, and **not**. The design is simpler and takes less code if a continuous assignment statement is used, but this section presents gate-level modeling only — continuous assignment statements are used in dataflow modeling.

The multiplexer has four data inputs, which are specified as a 4-bit vector $d[3:0]$, two select inputs, specified as a 2-bit vector $s[1:0]$, one scalar input *enable*, and one scalar output z_1. Also, the system function **$time** will be used in the test bench to return the current simulation time measured in nanoseconds (ns). The design module is shown in Figure 1.13, the test bench module in Figure 1.14 designating the appropriate inputs, the outputs in Figure 1.15, and the waveforms in Figure 1.16. The waveforms are shown as both hexadecimal values and as individual bits.

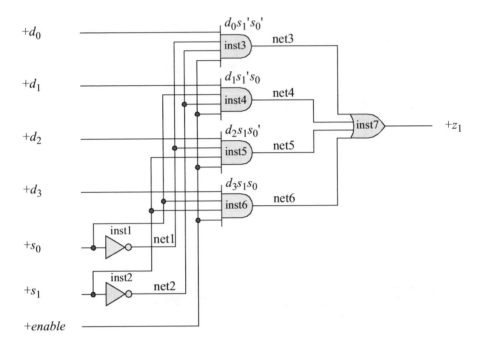

Figure 1.12   Logic diagram of a 4:1 multiplexer to be designed using built-in primitives.

```
//a 4:1 multiplexer using built-in primitives
module mux_4to1 (d, s, enbl, z1);

input [3:0] d;
input [1:0] s;
input enbl;
output z1;

not    inst1 (net1, s[0]),
       inst2 (net2, s[1]);

and    inst3 (net3, d[0], net1, net2, enbl),
       inst4 (net4, d[1], s[0], net2, enbl),
       inst5 (net5, d[2], net1, s[1], enbl),
       inst6 (net6, d[3], s[0], s[1], enbl);

or     inst7 (z1, net3, net4, net5, net6);
endmodule
```

Figure 1.13   Design module for a 4:1 multiplexer using built-in primitives.

```verilog
//test bench for 4:1 multiplexer
module mux_4to1_tb;

reg [3:0] d;
reg [1:0] s;
reg enbl;
wire z1;

initial
$monitor ($time,"ns, select:s=%b, inputs:d=%b, output:z1=%b",
         s, d, z1);
initial
begin
   #0    s[0]=1'b0;   s[1]=1'b0;
         d[0]=1'b0;   d[1]=1'b1;   d[2]=1'b0;   d[3]=1'b1;
         enbl=1'b1;   //d[0]=0; z1=0

   #10   s[0]=1'b0;   s[1]=1'b0;
         d[0]=1'b1;   d[1]=1'b1;   d[2]=1'b0;   d[3]=1'b1;
         enbl=1'b1;   //d[0]=1; z1=1

   #10   s[0]=1'b1;   s[1]=1'b0;
         d[0]=1'b1;   d[1]=1'b1;   d[2]=1'b0;   d[3]=1'b1;
         enbl=1'b1;   //d[1]=1; z1=1

   #10   s[0]=1'b0;   s[1]=1'b1;
         d[0]=1'b1;   d[1]=1'b1;   d[2]=1'b0;   d[3]=1'b1;
         enbl=1'b1;   //d[2]=0; z1=0

   #10   s[0]=1'b1;   s[1]=1'b0;
         d[0]=1'b1;   d[1]=1'b0;   d[2]=1'b0;   d[3]=1'b1;
         enbl=1'b1;   //d[1]=1; z1=0

   #10   s[0]=1'b1;   s[1]=1'b1;
         d[0]=1'b1;   d[1]=1'b1;   d[2]=1'b0;   d[3]=1'b1;
         enbl=1'b1;   //d[3]=1; z1=1

   #10   s[0]=1'b1;   s[1]=1'b1;
         d[0]=1'b1;   d[1]=1'b1;   d[2]=1'b0;   d[3]=1'b0;
         enbl=1'b1;   //d[3]=0; z1=0

   #10   s[0]=1'b1;   s[1]=1'b1;
         d[0]=1'b1;   d[1]=1'b1;   d[2]=1'b0;   d[3]=1'b0;
         enbl=1'b0;   //d[3]=0; z1=0

   #10   $stop;
end                                 //continued on next page
```

Figure 1.14   Test bench for the 4:1 multiplexer of Figure 1.12.

```
//instantiate the module into the test bench
mux_4to1 inst1 (
   .d(d),
   .s(s),
   .z1(z1),
   .enbl(enbl)
   );

endmodule
```

Figure 1.14   (Continued)

```
0  ns, select:s=00, inputs:d=1010, output:z1=0
10 ns, select:s=00, inputs:d=1011, output:z1=1
20 ns, select:s=01, inputs:d=1011, output:z1=1
30 ns, select:s=10, inputs:d=1011, output:z1=0
40 ns, select:s=01, inputs:d=1001, output:z1=0
50 ns, select:s=11, inputs:d=1011, output:z1=1
60 ns, select:s=11, inputs:d=0011, output:z1=0
```

Figure 1.15   Outputs for the 4:1 multiplexer of Figure 1.12.

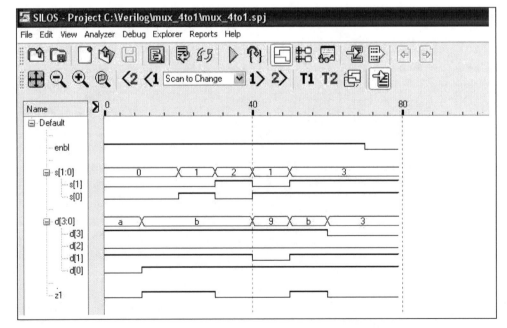

Figure 1.16   Waveforms for the 4:1 multiplexer of Figure 1.12.

Example 1.4 Recall that a decoder is a combinational logic macro device, which for every combination of inputs, a unique output is generated. Each output represents a minterm that corresponds to the binary representation of the input vector. A decoder with n inputs has a maximum of 2^n outputs, in which only one output signal is active — all other outputs are inactive.

A 3:8 decoder is shown in Figure 1.17, which will be designed using built-in primitives. The decoder will be designed using the following built-in primitives: **and** and **not**. The decoder has three data inputs, $x[2:0]$, where $x[0]$ is the low-order input. There is also an enable (*enbl*) input, which allows the appropriate output to be asserted. There are eight outputs, $z[7:0]$, where $z[0]$ is the low-order output. The design module is shown in Figure 1.18, the test bench module is shown in Figure 1.19, the outputs are shown in Figure 1.20, and waveforms are shown in Figure 1.21.

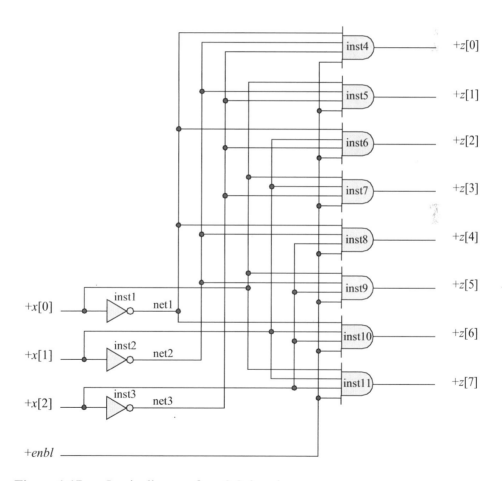

Figure 1.17 Logic diagram for a 3:8 decoder.

```
//3:8 decoder using built-in primitives
module decoder_3to8_bip5 (x, enbl, z);

input [2:0] x;
input enbl;
output [7:0] z;

//instantiate the inverters for the inputs
not    inst1 (net1, x[0]),
       inst2 (net2, x[1]),
       inst3 (net3, x[2]);

//instantiate the and gates for the decoder outputs
and    inst4 (z[0], net1, net2, net3, enbl),
       inst5 (z[1], net2, net3, x[0], enbl),
       inst6 (z[2], net1, x[1], net3, enbl),
       inst7 (z[3], net3, x[1], x[0], enbl),
       inst8 (z[4], x[2], net1, net2, enbl),
       inst9 (z[5], x[2], net2, x[0], enbl),
       inst10(z[6], x[2], x[1], net1, enbl),
       inst11(z[7], x[2], x[1], x[0], enbl);

endmodule
```

Figure 1.18   Design module for the 3:8 decoder using built-in primitives.

```
//test bench for decoder_3to8_bip5 module
module decoder_3to8_bip5_tb;

reg [2:0] x;        //inputs are reg for test bench
reg enbl;
wire [7:0] z;       //outputs are wire for test bench

//display inputs and outputs
initial
$monitor ("input = %b, enable = %b, output = %b",
          x [2:0], enbl, z [7:0]);

//apply input vectors
initial
begin
   #0    x [2:0] = 3'b000;    enbl = 1'b1;

                              //continued on next page
```

Figure 1.19   Test bench for the 3:8 decoder using built-in primitives.

```
      #10    x [2:0] = 3'b001;     enbl = 1'b1;

      #10    x [2:0] = 3'b010;     enbl = 1'b1;

      #10    x [2:0] = 3'b011;     enbl = 1'b1;

      #10    x [2:0] = 3'b100;     enbl = 1'b1;

      #10    x [2:0] = 3'b101;     enbl = 1'b1;

      #10    x [2:0] = 3'b110;     enbl = 1'b1;

      #10    x [2:0] = 3'b111;     enbl = 1'b1;

      #10    x [2:0] = 3'b111;     enbl = 1'b0;

      #10    $stop;
end

//instantiate the module into the test bench
decoder_3to8_bip5 inst1 (
    .x(x),
    .enbl(enbl),
    .z(z)
    );

endmodule
```

Figure 1.19   (Continued)

```
input = 000, enable = 1, output = 00000001
input = 001, enable = 1, output = 00000010
input = 010, enable = 1, output = 00000100
input = 011, enable = 1, output = 00001000
input = 100, enable = 1, output = 00010000
input = 101, enable = 1, output = 00100000
input = 110, enable = 1, output = 01000000
input = 111, enable = 1, output = 10000000
input = 111, enable = 0, output = 00000000
```

Figure 1.20   Outputs for the 3:8 decoder using built-in primitives.

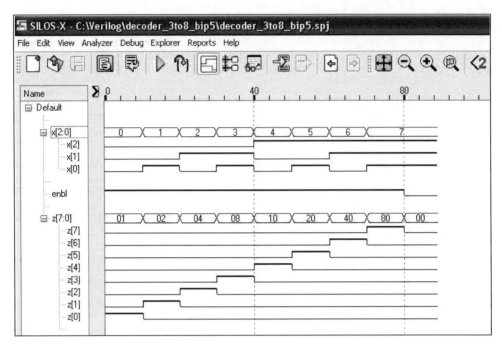

Figure 1.21   Waveforms for the 3:8 decoder using built-in primitives.

1.2 User-Defined Primitives

Verilog provides the capability to design primitives according to user specifications. These are called *user-defined primitives* (UDPs) and are usually at a higher-level logic function than built-in primitives. They are independent primitives and do not instantiate other primitives or modules. UDPs are instantiated into a module the same way as built-in primitives. A UDP is defined outside the module into which it is instantiated. There are two types of UDPs: combinational and sequential. Sequential primitives include level-sensitive and edge-sensitive circuits.

1.2.1 Defining a User-Defined Primitive

The syntax for a UDP is similar to that for declaring a module. The definition begins with the keyword **primitive** and ends with the keyword **endprimitive**. The UDP contains a name and a list of ports, which are declared as **input** or **output**. For a sequential UDP, the output port is declared as **reg**. UDPs can have one or more scalar inputs, but only one scalar output. The output port is listed first in the terminal list followed by the input ports, in the same way that the terminal list appears in built-in primitives. UDPs do not support **inout** ports.

The UDP table is an essential part of the internal structure and defines the functionality of the circuit. It is a lookup table similar in concept to a truth table. The table begins with the keyword **table** and ends with the keyword **endtable**. The contents of the table define the value of the output with respect to the inputs. The syntax for a UDP is shown below.

```
primitive udp_name (output, input_1, input_2, . . . , input_n);
    input input_1, input_2, . . . , input_n;
    output output;
    reg sequential_output;        //for sequential UDPs

    initial                       //for sequential UDPs

    table
        state table entries
    endtable
endprimitive
```

1.2.2 Combinational User-Defined Primitives

To illustrate the method for defining and using combinational UDPs, examples will be presented for designs of varying complexity. UDPs are not compiled separately. They are saved in the same project as the module with a *.v* extension; for example, *udp_and.v*.

Example 1.5 The Karnaugh map of Figure 1.22 will be designed as a sum-of-products expression using two UDPs: *udp_and2* and *udp_or3*. The UDPs will then be instantiated into the module *udp_sop*. The equation for the sum-of-products expression obtained from the Karnaugh map is shown in Equation 1.4 and the logic diagram is shown in Figure 1.23.

x_1x_2 \ x_3x_4	0 0	0 1	1 1	1 0
0 0	1	1	1	0
0 1	0	0	1	0
1 1	1	1	1	1
1 0	1	1	1	0

z_1

Figure 1.22 Karnaugh map to be implemented by UDPs.

$$z_1 = x_1 x_2 + x_3 x_4 + x_2' x_3' \qquad\qquad (1.4)$$

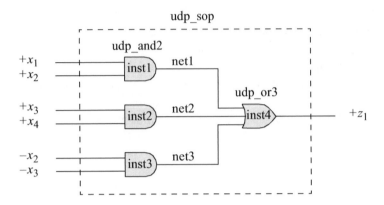

Figure 1.23   Logic diagram for sum-of-products UDP implementation.

UDPs will first be designed for a 2-input AND gate and a 3-input OR gate. The UDPs will then be instantiated into the sum-of-products design module *udp_sop*. The Verilog code for the *udp_and2* module is shown in Figure 1.24. The Verilog code for the *udp_or3* module is shown in Figure 1.25. The design module for *udp_sop*, the test bench module, the outputs, and the waveforms are shown in Figure 1.26, Figure 1.27, Figure 1.28, and Figure 1.29, respectively.

```
//UDP for a 2-input AND gate
primitive udp_and2 (z1, x1, x2);      //output is listed first

input x1, x2;
output z1;

//define state table
table
//inputs are in the same order as the input list
// x1 x2 :  z1;    comment is for readability
   0  0  :  0;
   0  1  :  0;
   1  0  :  0;
   1  1  :  1;
endtable

endprimitive
```

Figure 1.24   UDP for a 2-input AND gate.

```
//UDP for a 3-input OR gate
primitive udp_or3 (z1, x1, x2, x3);   //output is listed first

input x1, x2, x3;
output z1;

//define state table
table
//inputs are in the same order as the input list
// x1 x2 x3 :   z1;    comment is for readability
   0  0  0  :   0;
   0  0  1  :   1;
   0  1  0  :   1;
   0  1  1  :   1;
   1  0  0  :   1;
   1  0  1  :   1;
   1  1  0  :   1;
   1  1  1  :   1;
endtable

endprimitive
```

Figure 1.25   UDP for a 3-input OR gate.

```
//sum of products using UDPs for the AND gate and the OR gate
module udp_sop (x1, x2, x3, x4, z1);

input x1, x2, x3, x4;
output z1;

//define internal nets
wire net1, net2, net3;

//instantiate the udps
udp_and2 inst1 (net1, x1, x2);
udp_and2 inst2 (net2, x3, x4);
udp_and2 inst3 (net3, ~x2, ~x3);

udp_or3  inst4 (z1, net1, net2, net3);

endmodule
```

Figure 1.26   Design module for the sum-of-products logic diagram of Figure 1.23 using UDPs.

```verilog
//test bench for sum-of-products logic using UDPs
module udp_sop2_tb;

reg x1, x2, x3, x4;      //inputs are reg for test bench
wire z1;                 //outputs are wire for test bench

//display inputs and outputs
initial
$monitor ("x1 x2 x3 x4 = %b, z1 = %b",
            {x1, x2, x3, x4}, z1);

//apply input vectors
initial
begin
   #0     x1 = 1'b0; x2 = 1'b0; x3 = 1'b0; x4 = 1'b0;
   #10    x1 = 1'b0; x2 = 1'b0; x3 = 1'b0; x4 = 1'b1;
   #10    x1 = 1'b0; x2 = 1'b0; x3 = 1'b1; x4 = 1'b0;
   #10    x1 = 1'b0; x2 = 1'b0; x3 = 1'b1; x4 = 1'b1;

   #10    x1 = 1'b0; x2 = 1'b1; x3 = 1'b0; x4 = 1'b0;
   #10    x1 = 1'b0; x2 = 1'b1; x3 = 1'b0; x4 = 1'b1;
   #10    x1 = 1'b0; x2 = 1'b1; x3 = 1'b1; x4 = 1'b0;
   #10    x1 = 1'b0; x2 = 1'b1; x3 = 1'b1; x4 = 1'b1;

   #10    x1 = 1'b1; x2 = 1'b0; x3 = 1'b0; x4 = 1'b0;
   #10    x1 = 1'b1; x2 = 1'b0; x3 = 1'b0; x4 = 1'b1;
   #10    x1 = 1'b1; x2 = 1'b0; x3 = 1'b1; x4 = 1'b0;
   #10    x1 = 1'b1; x2 = 1'b0; x3 = 1'b1; x4 = 1'b1;

   #10    x1 = 1'b1; x2 = 1'b1; x3 = 1'b0; x4 = 1'b0;
   #10    x1 = 1'b1; x2 = 1'b1; x3 = 1'b0; x4 = 1'b1;
   #10    x1 = 1'b1; x2 = 1'b1; x3 = 1'b1; x4 = 1'b0;
   #10    x1 = 1'b1; x2 = 1'b1; x3 = 1'b1; x4 = 1'b1;

   #10    $stop;
end

//instantiate the module into the test bench
udp_sop2 inst1 (
   .x1(x1),
   .x2(x2),
   .x3(x3),
   .x4(x4),
   .z1(z1)
   );

endmodule
```

Figure 1.27   Test bench for the design module of Figure 1.26.

```
x1 x2 x3 x4 = 0000,  z1 = 1
x1 x2 x3 x4 = 0001,  z1 = 1
x1 x2 x3 x4 = 0010,  z1 = 0
x1 x2 x3 x4 = 0011,  z1 = 1

x1 x2 x3 x4 = 0100,  z1 = 0
x1 x2 x3 x4 = 0101,  z1 = 0
x1 x2 x3 x4 = 0110,  z1 = 0
x1 x2 x3 x4 = 0111,  z1 = 1

x1 x2 x3 x4 = 1000,  z1 = 1
x1 x2 x3 x4 = 1001,  z1 = 1
x1 x2 x3 x4 = 1010,  z1 = 0
x1 x2 x3 x4 = 1011,  z1 = 1

x1 x2 x3 x4 = 1100,  z1 = 1
x1 x2 x3 x4 = 1101,  z1 = 1
x1 x2 x3 x4 = 1110,  z1 = 1
x1 x2 x3 x4 = 1111,  z1 = 1
```

Figure 1.28   Outputs for the test bench of Figure 1.27 for the sum-of-products design module of Figure 1.26.

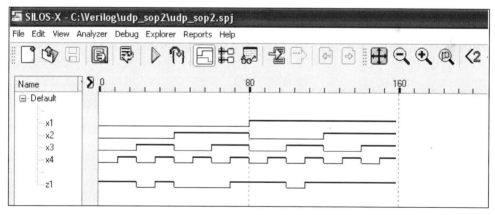

Figure 1.29   Waveforms for the test bench of Figure 1.28 for the sum-of-products design module of Figure 1.26.

Example 1.6   This example will design the sum-of-products equation shown in Equation 1.5 using user-defined-primitives. The logic diagram for Equation 1.5 is shown in Figure 1.30. The design will incorporate the following three NAND gate user-defined-primitives: *udp_nand2*, *udp_nand3*, and *udp_nand4*, as shown in Figure 1.31, Figure 1.32, and Figure 1.33, respectively.

$$z_1 = x_1'x_3 + x_1'x_2x_4 + x_1x_2'x_3' + x_2'x_3'x_4' \tag{1.5}$$

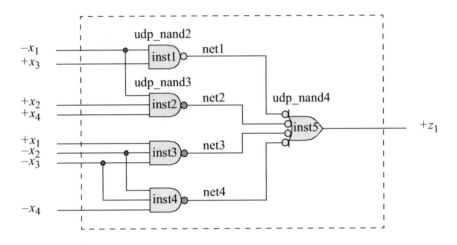

Figure 1.30   Logic circuit to represent Equation 1.5.

```
//UDP for a 2-input NAND gate
primitive udp_nand2 (z1, x1, x2);


input x1, x2;
output z1;


//define state table
table
//inputs are in the same order as the input list
// x1 x2 :  z1;   comment is for readability
   0  0  :  1;
   0  1  :  1;
   1  0  :  1;
   1  1  :  0;
endtable

endprimitive
```

Figure 1.31   UDP for a 2-input NAND gate.

```
//UDP for a 3-input NAND gate
primitive udp_nand3 (z1, x1, x2, x3);
input x1, x2, x3;
output z1;

//define state table
table
//inputs are in the same order as the input list
// x1 x2 x3 :   z1;   comment is for readability
   0  0  0  :   1;
   0  0  1  :   1;
   0  1  0  :   1;
   0  1  1  :   1;
   1  0  0  :   1;
   1  0  1  :   1;
   1  1  0  :   1;
   1  1  1  :   0;
endtable
endprimitive
```

Figure 1.32   UDP for a 3-input NAND gate.

```
//UDP for a 4-input NAND gate
primitive udp_nand4 (z1, x1, x2, x3, x4);

input x1, x2, x3, x4;
output z1;

//define state table
table
//inputs are in the same order as the input list
// x1 x2 x3 x4 :   z1;   comment is for readability
   0  0  0  0  :   1;
   0  0  0  1  :   1;
   0  0  1  0  :   1;
   0  0  1  1  :   1;
   0  1  0  0  :   1;
   0  1  0  1  :   1;
   0  1  1  0  :   1;
   0  1  1  1  :   1;
   1  0  0  0  :   1;
   1  0  0  1  :   1;
   1  0  1  0  :   1;
                              //continued on next page
```

Figure 1.33   UDP for a 4-input NAND gate.

```
    1   0   1   1   :   1;
    1   1   0   0   :   1;
    1   1   0   1   :   1;
    1   1   1   0   :   1;
    1   1   1   1   :   0;
endtable
endprimitive
```

Figure 1.33   (Continued)

The design module into which the UDPs will be instantiated is shown in Figure 1.34. The test bench, outputs, and waveforms are shown in Figure 1.35, Figure 1.36, and Figure 1.37, respectively.

```
//user-defined primitives to design Equation 1.5
module udp_sop3 (x1, x2, x3, x4, z1);

input x1, x2, x3, x4;
output z1;

//define internal nets
wire net1, net2, net3, net4;

//instantiate the udps
udp_nand2 (net1, ~x1, x3);
udp_nand3 (net2, ~x1, x2, x4);
udp_nand3 (net3, x1, ~x2, ~x3);
udp_nand3 (net4, ~x2, ~x3, ~x4);

udp_nand4 (z1, net1, net2, net3, net4);

endmodule
```

Figure 1.34   Module to design Equation 1.5 using user-defined primitives.

```
//test bench for udp_sop3
module udp_sop3_tb;

reg x1, x2, x3, x4;
wire z1;
                                //continued on next page
```

Figure 1.35   Test bench for the design module of Figure 1.34.

```verilog
//display inputs and outputs
initial
$monitor ("input = %b, z1 = %b", {x1, x2, x3, x4}, z1);


//apply input vectors
initial
begin
   #0      {x1, x2, x3, x4} = 4'b0000;
   #10     {x1, x2, x3, x4} = 4'b0001;
   #10     {x1, x2, x3, x4} = 4'b0010;
   #10     {x1, x2, x3, x4} = 4'b0011;


   #10     {x1, x2, x3, x4} = 4'b0100;
   #10     {x1, x2, x3, x4} = 4'b0101;
   #10     {x1, x2, x3, x4} = 4'b0110;
   #10     {x1, x2, x3, x4} = 4'b0111;


   #10     {x1, x2, x3, x4} = 4'b1000;
   #10     {x1, x2, x3, x4} = 4'b1001;
   #10     {x1, x2, x3, x4} = 4'b1010;
   #10     {x1, x2, x3, x4} = 4'b1011;


   #10     {x1, x2, x3, x4} = 4'b1100;
   #10     {x1, x2, x3, x4} = 4'b1101;
   #10     {x1, x2, x3, x4} = 4'b1110;
   #10     {x1, x2, x3, x4} = 4'b1111;

   #10     $stop;
end


//instantiate the module into the test bench
udp_sop3 inst1 (
   .x1(x1),
   .x2(x2),
   .x3(x3),
   .x4(x4),
   .z1(z1)
   );

endmodule
```

Figure 1.35   (Continued)

```
input = 0000, z1 = 1
input = 0001, z1 = 0
input = 0010, z1 = 1
input = 0011, z1 = 1
input = 0100, z1 = 0
input = 0101, z1 = 1
input = 0110, z1 = 1
input = 0111, z1 = 1
input = 1000, z1 = 1
input = 1001, z1 = 1
input = 1010, z1 = 0
input = 1011, z1 = 0
input = 1100, z1 = 0
input = 1101, z1 = 0
input = 1110, z1 = 0
input = 1111, z1 = 0
```

Figure 1.36   Outputs for the design module of Figure 1.34 for Equation 1.5.

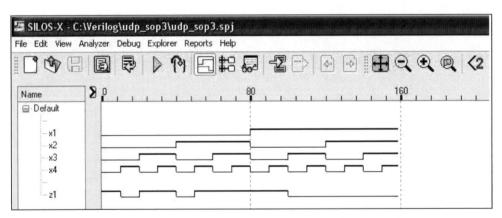

Figure 1.37   Waveforms for the test bench of Figure 1.36 for Equation 1.5.

Example 1.7   The Karnaugh map of Figure 1.38 will be implemented using a 4:1 multiplexer and additional logic. The equations for the multiplexer data inputs — d_0, d_1, d_2, and d_3 — are shown to the right of the Karnaugh map, where the multiplexer select inputs are $s_1 s_0 = x_1 x_2$, where s_0 (x_2) is the low-order select input. The circuit will be designed using user-defined primitives for the multiplexer and associated logic gates.

    Figure 1.39 depicts the logic diagram that is designed from the Karnaugh map. The *udp_and2* was previously designed. The user-defined primitive for a 4:1 multiplexer is shown in Figure 1.40. Note the entries in the table that contain the symbol (?), which indicates a "don't care" condition. Referring to the first line in the table, if

$s_1 s_0 = 00$, then it is irrelevant what the values are for inputs $d_1 d_2 d_3$, because only input d_0 is selected. The design module, test bench module, outputs, and waveforms are shown in Figure 1.41, Figure 1.42, Figure 1.43, and Figure 1.44, respectively.

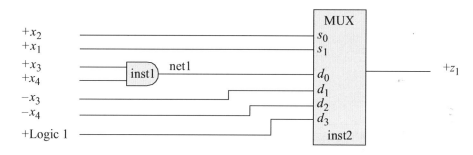

Figure 1.38   Karnaugh map for Example 1.7.

Figure 1.39   Logic diagram for the Karnaugh map of Figure 1.38.

```
//4:1 multiplexer as a UDP
primitive udp_mux4 (out, s1, s0, d0, d1, d2, d3);

input s1, s0, d0, d1, d2, d3;
output out;

table     //define state table
//inputs are in the same order as the input list
// s1 s0 d0 d1 d2 d3 :   out      comment is for readability
   0  0  1  ?  ?  ?  :   1;       //? is "don't care"
   0  0  0  ?  ?  ?  :   0;       //continued on next page
```

Figure 1.40   A UDP for a 4:1 multiplexer.

```
   0  1  ?  1  ?  ?  :  1;
   0  1  ?  0  ?  ?  :  0;

   1  0  ?  ?  1  ?  :  1;
   1  0  ?  ?  0  ?  :  0;

   1  1  ?  ?  ?  1  :  1;
   1  1  ?  ?  ?  0  :  0;

   ?  ?  0  0  0  0  :  0;
   ?  ?  1  1  1  1  :  1;
endtable
endprimitive
```

Figure 1.40   (Continued)

```
//logic circuit using a 4:1 multiplexer UDP
//together with other logic circuit UDPs

module mux4_kmap (x1, x2, x3, x4, z1);

input x1, x2, x3, x4;
output z1;

//instantiate the udps
udp_and2 inst1 (net1, x3, x4);

//the mux inputs are: s1, s0, d0,    d1,   d2,   d3
//they correspond to  x1, x2, net1, ~x3, ~x4, 1'b1

udp_mux4 inst2 (z1, x1, x2, net1, ~x3, ~x4, 1'b1);

endmodule
```

Figure 1.41   Module for the logic diagram of Figure 1.39.

```
//test bench for mux4_kmap
module mux4_kmap_tb;

reg x1, x2, x3, x4;        //inputs are reg for test bench
wire z1;                   //outputs are wire for test bench
                                    //continued on next page
```

Figure 1.42   Test bench for Figure 1.41 for the logic diagram of Figure 1.39.

```
//display inputs and outputs
initial
$monitor ("x1 x2 x3 x4 =%b, z1 = %b", {x1, x2, x3, x4}, z1);


//apply input vectors
initial
begin
    #0      x1 = 1'b0; x2 = 1'b0; x3 = 1'b0; x4 = 1'b0;
    #10     x1 = 1'b0; x2 = 1'b0; x3 = 1'b0; x4 = 1'b1;
    #10     x1 = 1'b0; x2 = 1'b0; x3 = 1'b1; x4 = 1'b0;
    #10     x1 = 1'b0; x2 = 1'b0; x3 = 1'b1; x4 = 1'b1;


    #10     x1 = 1'b0; x2 = 1'b1; x3 = 1'b0; x4 = 1'b0;
    #10     x1 = 1'b0; x2 = 1'b1; x3 = 1'b0; x4 = 1'b1;
    #10     x1 = 1'b0; x2 = 1'b1; x3 = 1'b1; x4 = 1'b0;
    #10     x1 = 1'b0; x2 = 1'b1; x3 = 1'b1; x4 = 1'b1;


    #10     x1 = 1'b1; x2 = 1'b0; x3 = 1'b0; x4 = 1'b0;
    #10     x1 = 1'b1; x2 = 1'b0; x3 = 1'b0; x4 = 1'b1;
    #10     x1 = 1'b1; x2 = 1'b0; x3 = 1'b1; x4 = 1'b0;
    #10     x1 = 1'b1; x2 = 1'b0; x3 = 1'b1; x4 = 1'b1;


    #10     x1 = 1'b1; x2 = 1'b1; x3 = 1'b0; x4 = 1'b0;
    #10     x1 = 1'b1; x2 = 1'b1; x3 = 1'b0; x4 = 1'b1;
    #10     x1 = 1'b1; x2 = 1'b1; x3 = 1'b1; x4 = 1'b0;
    #10     x1 = 1'b1; x2 = 1'b1; x3 = 1'b1; x4 = 1'b1;


    #10     $stop;
end

//instantiate the module into the test bench
mux4_kmap inst1 (
    .x1(x1),
    .x2(x2),
    .x3(x3),
    .x4(x4),
    .z1(z1)
    );

endmodule
```

Figure 1.42   (Continued)

```
s1  s0                s1 s0
x1  x2  x3  x4  =  0  0  00,  z1  =  0
x1  x2  x3  x4  =  0  0  01,  z1  =  0
x1  x2  x3  x4  =  0  0  10,  z1  =  0
x1  x2  x3  x4  =  0  0  11,  z1  =  1
x1  x2  x3  x4  =  0  1  00,  z1  =  1
x1  x2  x3  x4  =  0  1  01,  z1  =  1
x1  x2  x3  x4  =  0  1  10,  z1  =  0
x1  x2  x3  x4  =  0  1  11,  z1  =  0
x1  x2  x3  x4  =  1  0  00,  z1  =  1
x1  x2  x3  x4  =  1  0  01,  z1  =  0
x1  x2  x3  x4  =  1  0  10,  z1  =  1
x1  x2  x3  x4  =  1  0  11,  z1  =  0
x1  x2  x3  x4  =  1  1  00,  z1  =  1
x1  x2  x3  x4  =  1  1  01,  z1  =  1
x1  x2  x3  x4  =  1  1  10,  z1  =  1
x1  x2  x3  x4  =  1  1  11,  z1  =  1
```

Figure 1.43   Outputs for Figure 1.39 obtained from the test bench of Figure 1.42.

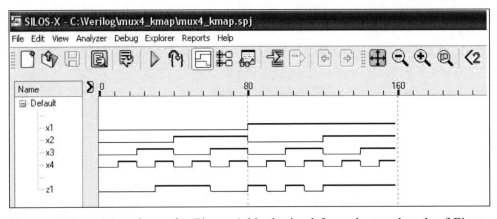

Figure 1.44   Waveforms for Figure 1.39 obtained from the test bench of Figure 1.42.

Example 1.8   This example will use user-defined primitives to design a full adder that is constructed from two half adders. A *half adder* is a combinational circuit that performs the addition of two operand bits and produces two outputs: a *sum* bit and a *carry-out* bit. The half adder does not accommodate a *carry-in* bit. A *full adder* is a combinational circuit that performs the addition of two operand bits plus a *carry-in* bit. The *carry-in* represents the *carry-out* of the previous lower-order stage. The full adder produces two outputs: a *sum* bit and a *carry-out* bit.

The logic diagram for a full adder that is designed from two half adders is shown in Figure 1.45. The full adder utilizes UDPs for all gates of the full adder. The design module is shown in Figure 1.46, the test bench module is shown in Figure 1.47, and the outputs are shown in Figure 1.48.

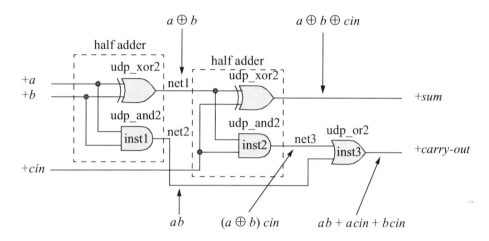

Figure 1.45   Full adder designed from two half adders.

```
//udp for a full adder designed from two half adder udps
module udp_full_adder3 (a, b, cin, sum, cout);

input a, b, cin;
output sum, cout;

//define internal nets
wire net1, net2, net3;
//udp for a full adder designed from two half adder udps
module udp_full_adder (a, b, cin, sum, cout);

input a, b, cin;
output sum, cout;

//define internal nets
wire net1, net2, net3;

//instantiate the udps for the full adder
//udps for the first half adder
udp_xor2 (net1, a, b);
udp_and2 (net2, a, b);

                              //continued on next page
```

Figure 1.46   Design module for the full adder of Figure 1.45.

```
//udps for the second half adder
udp_xor2 (sum, net1, cin);
udp_and2 (net3, net1, cin);

udp_or2 (cout, net3, net2);

endmodule
```

Figure 1.46   (Continued)

```
//test bench for the full adder
module udp_full_adder3_tb;

reg a, b, cin;
wire sum, cout;

//display inputs and outputs
initial
$monitor ("a b cin = %b, cout = %b, sum = %b",
          {a, b, cin}, cout, sum);

//apply input vectors
initial
begin
    #0      a = 1'b0;    b = 1'b0;    cin = 1'b0;
    #10     a = 1'b0;    b = 1'b0;    cin = 1'b1;
    #10     a = 1'b0;    b = 1'b1;    cin = 1'b0;
    #10     a = 1'b0;    b = 1'b1;    cin = 1'b1;

    #10     a = 1'b1;    b = 1'b0;    cin = 1'b0;
    #10     a = 1'b1;    b = 1'b0;    cin = 1'b1;
    #10     a = 1'b1;    b = 1'b1;    cin = 1'b0;
    #10     a = 1'b1;    b = 1'b1;    cin = 1'b1;
    #10     $stop;
end

//instantiate the module into the test bench
udp_full_adder3 inst1 (
    .a(a),
    .b(b),
    .cin(cin),
    .sum(sum),
    .cout(cout)
    );

endmodule
```

Figure 1.47   Test bench for the full adder of Figure 1.45.

```
a b cin = 000, cout = 0, sum = 0
a b cin = 001, cout = 0, sum = 1
a b cin = 010, cout = 0, sum = 1
a b cin = 011, cout = 1, sum = 0

a b cin = 100, cout = 0, sum = 1
a b cin = 101, cout = 1, sum = 0
a b cin = 110, cout = 1, sum = 0
a b cin = 111, cout = 1, sum = 1
```

Figure 1.48   Outputs for the full adder of Figure 1.45.

1.3 Dataflow Modeling

Loop statements are covered in detail in Section 1.4, but will be briefly described in
this section in order to minimize the code in test benches. The keyword **for** is used to
specify a loop. The **for** loop repeats the execution of a procedural statement or a block
of procedural statements a specified number of times — a procedural statement is a
synonym for instruction. The **for** loop is used when there is a specified beginning and
end to the loop. The format and function of a **for** loop is similar to the **for** loop used
in the C programming language. The parentheses following the keyword **for** contain
three expressions separated by semicolons, as shown below.

> **for** (register initialization; test condition; update register control variable)
> procedural statement or block of procedural statements

Gate-level modeling is an intuitive approach to digital design because it corresponds one-to-one with conventional digital logic design at the gate level. Dataflow
modeling, however, is at a higher level of abstraction than gate-level modeling.
Design automation tools are used to create gate-level logic from dataflow modeling by
a process called *logic synthesis*. Register transfer level (RTL) is a combination of
dataflow modeling and behavioral modeling — behavioral modeling is presented in
Section 1.4 — and characterizes the flow of data through logic circuits.

1.3.1 Continuous Assignment

The *continuous assignment* statement models dataflow behavior and is used to design
combinational logic without using gates and interconnecting nets. Continuous assignment statements provide a Boolean correspondence between the right-hand side
expression and the left-hand side target. The continuous assignment statement uses

the keyword **assign** and has the following syntax with optional drive strength and delay:

assign [drive_strength] [delay] left-hand side target = right-hand side expression

The continuous assignment statement assigns a value to a net (**wire**) that has been previously declared — it cannot be used to assign a value to a register. Therefore, the left-hand target must be a scalar or vector net or a concatenation of scalar and vector nets. The operands on the right-hand side can be registers, nets, or function calls. The registers and nets can be declared as either scalars or vectors.

The following are examples of continuous assignment statements for scalar nets:

$$\textbf{assign } z_1 = x_1 \ \& \ x_2 \ \& \ x_3; \qquad z_1 = x_1 \text{ AND } x_2 \text{ AND } x_3$$
$$\textbf{assign } z_1 = x_1 \ ^\wedge \ x_2; \qquad z_1 = x_1 \text{ XOR } x_2$$
$$\textbf{assign } z_1 = (x_1 \ \& \ x_2) \ | \ x_3; \qquad z_1 = (x_1 \text{ AND } x_2) \text{ OR } x_3$$

The **assign** statement continuously monitors the right-hand side expression. If a variable changes value, then the expression is evaluated and the result is assigned to the target after any specified delay. If no delay is specified, then the default delay is zero. The drive strength defaults to **strong0** and **strong1**. The continuous assignment statement can be considered to be a form of behavioral modeling, because the behavior of the circuit is specified, not the implementation.

Example 1.9   This example designs a 3-input AND gate using dataflow modeling which incorporates the continuous assignment statement. The AND function is also called the *conjunction* of two or more variables. The design module is shown in Figure 1.49, the test bench module is shown Figure 1.50, the outputs are shown in Figure 1.51, and the waveforms are shown in Figure 1.52.

```
//3-input AND gate using dataflow
module and3a_df (x1, x2, x3, z1);

input x1, x2, x3;
output z1;

//signals can be optionally declared as wire for dataflow
wire x1, x2, x3;
wire z1;

//use continuous assignment
assign z1 = x1 & x2 & x3;

endmodule
```

Figure 1.49   Module for a 3-input AND gate using continuous assignment.

```
//test bench for the 3-input AND gate
module and3a_df_tb;
reg x1, x2, x3;     //inputs are reg for test bench
wire z1;            //outputs are wire for test bench

//apply input vectors and display variables
initial
begin : apply_stimulus  //colon followed by a name
reg [3:0] invect;
   for (invect = 0; invect < 8; invect = invect + 1)
   begin
      {x1, x2, x3} = invect [3:0];
      #10 $display ("x1 x2 x3 = %b, z1 = %b", {x1, x2, x3}, z1);
   end
end

and3a_df inst1 (  //instantiate the module into the test bench
   .x1(x1),
   .x2(x2),
   .x3(x3),
   .z1(z1)
   );
endmodule
```

Figure 1.50   Test bench for the module of Figure 1.49.

```
x1 x2 x3 = 000, z1 = 0      x1 x2 x3 = 100, z1 = 0
x1 x2 x3 = 001, z1 = 0      x1 x2 x3 = 101, z1 = 0
x1 x2 x3 = 010, z1 = 0      x1 x2 x3 = 110, z1 = 0
x1 x2 x3 = 011, z1 = 0      x1 x2 x3 = 111, z1 = 1
```

Figure 1.51   Outputs for the 3-input AND gate using continuous assignment.

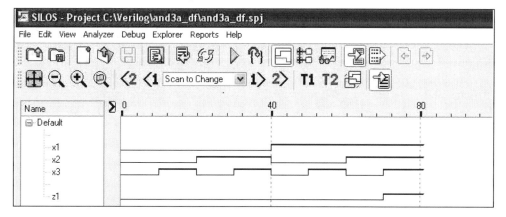

Figure 1.52   Waveforms for the 3-input AND gate using continuous assignment.

Example 1.10   A comparator will be designed that compares two 2-bit binary operands x_1x_2 and x_3x_4 and generates a high output for z_1 whenever $x_1x_2 \geq x_3x_4$. The comparator will be designed as a product of sums using NOR logic. A product of sums is an expression in which at least one term does not contain all the variables; that is, at least one term is a proper subset of the possible variables or their complements. For example, the equation shown below is a product of sums for the function z_1, because the second term does not contain the variable x_1.

$$z_1(x_1, x_2, x_3) = (x_1' + x_2 + x_3)(x_2' + x_3')(x_1 + x_2 + x_3)$$

The minimal product-of-sums expression can be obtained by combining 0s in a Karnaugh map to form sum terms in the same manner as 1s are combined to form product terms. However, since 0s are being combined, each sum term must equal 0. When combining 0s to obtain sum terms, treat a variable value of 1 as false and a variable value of 0 as true.

The Karnaugh map that represents the comparator is shown in Figure 1.53. The product-of-sums equation obtained from the Karnaugh map is shown in Equation 1.6. The logic diagram using NOR gates is shown in Figure 1.54. The design module is shown in Figure 1.55, the test bench module is shown in Figure 1.56, the outputs are shown in Figure 1.57, and the waveforms are shown in Figure 1.58.

Figure 1.53   Karnaugh map for the 2-bit comparator of Example 1.10.

$$z_1 = (x_1 + x_3')(x_1 + x_2 + x_4')(x_2 + x_3' + x_4') \tag{1.6}$$

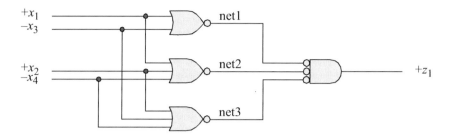

Figure 1.54 Logic diagram for the comparator of Example 1.10 that is implemented as a product of sums.

```
//dataflow for 2-bit comparator using nor logic
module comparator2a_nor (x1, x2, x3, x4, z1);

input x1, x2, x3, x4;    //define inputs and outputs
output z1;

//define inputs and output as wire
wire x1, x2, x3, x4;
wire z1;

//define internal nets
wire net1, net2, net3;

//define z1 using continuous assignment
assign    net1 = ~(x1 | ~x3),
          net2 = ~(x1 | x2 | ~x4),
          net3 = ~(x2 | ~x3 | ~x4);

assign    z1 = ~net1 & ~net2 & ~net3;

endmodule
```

Figure 1.55 Design module for the comparator of Example 1.10 using NOR logic.

```
//test bench for comparator2 using nor logic
module comparator2a_nor_tb;

reg x1, x2, x3, x4;
wire z1;
                                    //continued on next page
```

Figure 1.56 Test bench for the comparator of Example 1.10 using NOR logic.

```
//apply input vectors and display variables
initial
begin: apply_stimulus
   reg [4:0] invect;
   for (invect = 0; invect < 16; invect = invect + 1)
      begin
         {x1, x2, x3, x4} = invect [4:0];
         #10 $display ("x1 x2 x3 x4 = %b, z1 = %b",
                        {x1, x2, x3, x4}, z1);
      end
end

//instantiate the module into the test bench
comparator2a_nor inst1 (
   .x1(x1),
   .x2(x2),
   .x3(x3),
   .x4(x4),
   .z1(z1)
   );

endmodule
```

Figure 1.56   (Continued)

```
x1 x2 x3 x4 = 0000, z1 = 1
x1 x2 x3 x4 = 0001, z1 = 0
x1 x2 x3 x4 = 0010, z1 = 0
x1 x2 x3 x4 = 0011, z1 = 0

x1 x2 x3 x4 = 0100, z1 = 1
x1 x2 x3 x4 = 0101, z1 = 1
x1 x2 x3 x4 = 0110, z1 = 0
x1 x2 x3 x4 = 0111, z1 = 0

x1 x2 x3 x4 = 1000, z1 = 1
x1 x2 x3 x4 = 1001, z1 = 1
x1 x2 x3 x4 = 1010, z1 = 1
x1 x2 x3 x4 = 1011, z1 = 0

x1 x2 x3 x4 = 1100, z1 = 1
x1 x2 x3 x4 = 1101, z1 = 1
x1 x2 x3 x4 = 1110, z1 = 1
x1 x2 x3 x4 = 1111, z1 = 1
```

Figure 1.57   Outputs for the comparator of Example 1.10 using NOR logic.

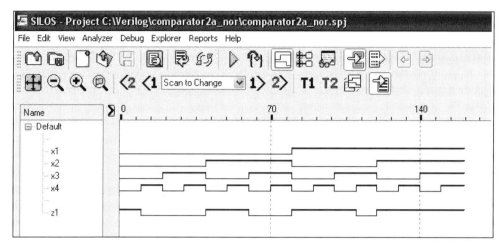

Figure 1.58   Waveforms for the comparator of Example 1.10 using NOR logic.

1.3.2 Reduction Operators

The reduction operators are: AND (&), NAND (~&), OR (|), NOR (~ |), exclusive-OR (^), and exclusive-NOR (^~ or ~^). Reduction operators are unary operators; that is, they operate on a single vector and produce a single-bit result. Reduction operators perform their respective operations on a bit-by-bit basis from right to left. If any bit in the operand is an unknown value (x) or a high impedance value (z), then the result of the operation is an x.

reduction AND   If any bit in the operand is 0, then the result is 0; otherwise, the result is 1. For example, let x_1 be the vector shown below.

1	1	1	0	1	0	1	1

The reduction AND (& x_1) operation is equivalent to the following operation:

$$1 \& 1 \& 1 \& 0 \& 1 \& 0 \& 1 \& 1$$

which returns a result of 0.

reduction NAND   If any bit in the operand is 0, then the result is 1; otherwise, the result is 0. For a vector x_1, the reduction NAND (~& x_1) is the inverse of the reduction AND operator.

reduction OR   If any bit in the operand is 1, then the result is 1; otherwise, the result is 0. For example, let x_1 be the vector shown below.

1	1	1	0	1	0	1	1

The reduction OR ($| x_1$) operation is equivalent to the following operation:

$$1 | 1 | 1 | 0 | 1 | 0 | 1 | 1$$

which returns a result of 1.

reduction NOR   If any bit in the operand is 1, then the result is 0; otherwise, the result is 1. For a vector x_1, the reduction NOR ($\sim | x_1$) is the inverse of the reduction OR operator.

reduction exclusive-OR   If there is an even number of 1s in the operand, then the result is 0; otherwise, the result is 1.   For example, let x_1 be the vector shown below.

1	1	1	0	1	0	1	1

The reduction exclusive-OR ($^\wedge x_1$) operation is equivalent to the following operation:

$$1 \wedge 1 \wedge 1 \wedge 0 \wedge 1 \wedge 0 \wedge 1 \wedge 1$$

which returns a result of 0. The reduction exclusive-OR operator can be used as an even parity generator.

reduction exclusive-NOR   If there is an odd number of 1s in the operand, then the result is 0; otherwise, the result is 1. For a vector x_1, the reduction exclusive-NOR ($^\wedge \sim x_1$) is the inverse of the reduction exclusive-OR operator. The reduction exclusive-NOR operator can be used as an odd parity generator.

Example 1.11 Figure 1.59 contains a module that illustrates the coding of the reduction operators. The test bench, outputs, and waveforms are shown in Figure 1.60, Figure 1.61, and Figure 1.62, respectively.

```
//module to illustrate the use of reduction operators
module reduction2 (a, red_and, red_nand, red_or, red_nor,
                   red_xor, red_xnor);
input [7:0] a;
output red_and, red_nand, red_or, red_nor, red_xor, red_xnor;
                                        //continued on next page
```

Figure 1.59   Design module to illustrate the utilization of the reduction operators.

```
wire [7:0] a;
wire red_and, red_nand, red_or, red_nor, red_xor, red_xnor;


assign    red_and  = &a,     //reduction AND
          red_nand = ~&a,    //reduction NAND
          red_or   = |a,     //reduction OR
          red_nor  = ~|a,    //reduction NOR
          red_xor  = ^a,     //reduction exclusive-OR
          red_xnor = ^~a;    //reduction exclusive-NOR

endmodule
```

Figure 1.59   (Continued)

```
//test bench for reduction2 module
module reduction2_tb;

reg [7:0] a;
wire red_and, red_nand, red_or, red_nor, red_xor, red_xnor;

initial
$monitor ("a=%b, red_and=%b, red_nand=%b,
             red_or=%b, red_nor=%b,
             red_xor=%b, red_xnor=%b",
             a, red_and, red_nand, red_or, red_nor,
             red_xor, red_xnor);

//apply input vectors
initial
begin
   #0    a = 8'b0011_0011;
   #10   a = 8'b1101_0011;
   #10   a = 8'b0000_0000;
   #10   a = 8'b0000_0001;
   #10   a = 8'b0001_0000;
   #10   a = 8'b0011_1100;
   #10   a = 8'b1111_0000;
   #10   a = 8'b0100_1111;
   #10   a = 8'b1101_1111;
   #10   a = 8'b1111_1111;
   #10   a = 8'b0111_1111;


   #10   $stop;
end                            //continued on next page
```

Figure 1.60   Test bench for the reduction operator module of Figure 1.59.

```
//instantiate the module into the test bench
reduction2 inst1 (
    .a(a),
    .red_and(red_and),
    .red_nand(red_nand),
    .red_or(red_or),
    .red_nor(red_nor),
    .red_xor(red_xor),
    .red_xnor(red_xnor)
    );
endmodule
```

Figure 1.60   (Continued)

```
a=00110011, red_and=0, red_nand=1, red_or=1,
            red_nor=0, red_xor=0, red_xnor=1

a=11010011, red_and=0, red_nand=1, red_or=1,
            red_nor=0, red_xor=1, red_xnor=0

a=00000000, red_and=0, red_nand=1, red_or=0,
            red_nor=1, red_xor=0, red_xnor=1

a=00000001, red_and=0, red_nand=1, red_or=1,
            red_nor=0, red_xor=1, red_xnor=0

a=00010000, red_and=0, red_nand=1, red_or=1,
            red_nor=0, red_xor=1, red_xnor=0

a=00111100, red_and=0, red_nand=1, red_or=1,
            red_nor=0, red_xor=0, red_xnor=1

a=11110000, red_and=0, red_nand=1, red_or=1,
            red_nor=0, red_xor=0, red_xnor=1

a=01001111, red_and=0, red_nand=1, red_or=1,
            red_nor=0, red_xor=1, red_xnor=0

a=11011111, red_and=0, red_nand=1, red_or=1,
            red_nor=0, red_xor=1, red_xnor=0

a=11111111, red_and=1, red_nand=0, red_or=1,
            red_nor=0, red_xor=0, red_xnor=1

a=01111111, red_and=0, red_nand=1, red_or=1,
            red_nor=0, red_xor=1, red_xnor=0

a=10101010, red_and=0, red_nand=1, red_or=1,
            red_nor=0, red_xor=0, red_xnor=1
```

Figure 1.61   Outputs for the reduction operator module of Figure 1.59.

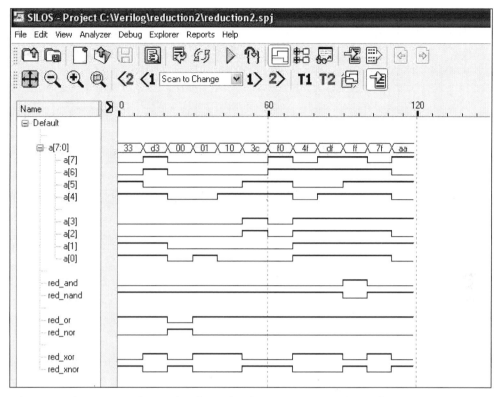

Figure 1.62   Waveforms for the reduction operator module of Figure 1.59.

1.3.3 Conditional Operator

The conditional operator (**? :**) has three operands, as shown in the syntax below. The *conditional_expression* is evaluated. If the result is true (1), then the *true_expression* is evaluated; if the result is false (0), then the *false_expression* is evaluated.

> conditional_expression **?** true_expression **:** false_expression;

The conditional operator can be used when one of two expressions is to be selected. For example, in the statement below, if x_1 is greater than or equal to x_2, then z_1 is assigned the value of x_3; if x_1 is less than x_2, then z_1 is assigned the value of x_4.

> z1 = (x1 >= x2) **?** x3 : x4;

Since the conditional operator selects one of two values, depending on the result of the conditional_expression evaluation, the operator can be used in place of the **if** . . . **else** construct. The **if** . . . **else** construct is presented in Section 1.47 entitled, Conditional Statements.

Conditional operators can be nested; that is, each true_expression and each false_expression can be a conditional operation, as shown below. This is useful for modeling a 4:1 multiplexer, as shown in Example 1.12.

conditional_expression **?** (cond_expr1 **?** true_expr1 **:** false_expr1)
                                  : (cond_expr2 **?** true_expr2 **:** false_expr2);

Example 1.12 A 4:1 multiplexer will be designed using the conditional operator. This design will declare the multiplexer inputs as scalars instead of vectors. The select inputs are: s_0 and s_1; the data inputs are: in_0, in_1, in_2, and in_3; the output is: *out*. The design module is shown in Figure 1.63. The **assign** statement is reproduced as shown below.

assign out = s1 **?** (s0 **?** in3 **:** in2) **:** (s0 **?** (in1 **:** in0);

The **assign** statement functions as shown below.

s1	s0	out
0	0	in0
0	1	in1
1	0	in2
1	1	in3

The test bench is shown in Figure 1.64. The outputs and waveforms are shown in Figure 1.65 and Figure 1.66, respectively.

```
//dataflow 4:1 mux using the conditional operator
module mux4to1_cond (s0, s1, in0, in1, in2, in3, out);


input s0, s1;
input in0, in1, in2, in3;
output out;


//use nested conditional operator
assign out = s1 ? (s0 ? in3 : in2) : (s0 ? in1 : in0);


endmodule
```

Figure 1.63   Design module for the conditional operator.

```
//mux4to1_cond test bench
module mux4to1_cond_tb;

reg in0, in1, in2, in3, s0, s1;   //inputs are reg
wire out;                         //outputs are wire

initial      //display signals
$monitor ("s1s0 = %b, in0in1in2in3 = %b, out = %b",
         {s1, s0}, {in0, in1, in2, in3}, out);

initial      //apply stimulus
begin
   #0    s1 = 1'b0;s0 = 1'b0;
         in0 = 1'b0;in1 = 1'b1;in2 = 1'b1;in3 = 1'b1;

   #10   s1 = 1'b0;s0 = 1'b1;
         in0 = 1'b0;in1 = 1'b1;in2 = 1'b1;in3 = 1'b0;

   #10   s1 = 1'b1;s0 = 1'b0;
         in0 = 1'b1;in1 = 1'b0;in2 = 1'b0;in3 = 1'b1;

   #10   s1 = 1'b1;s0 = 1'b1;
         in0 = 1'b0;in1 = 1'b1;in2 = 1'b0;in3 = 1'b1;

   #10   $stop;
end

mux4to1_cond inst1 ( //instantiate the module
   .s0(s0),
   .s1(s1),
   .in0(in0),
   .in1(in1),
   .in2(in2),
   .in3(in3),
   .out(out)
   );
endmodule
```

Figure 1.64   Test bench module for the conditional operator.

```
s1 s0 = 00, in0 in1 in2 in3 = 0111, out = 0
s1 s0 = 01, in0 in1 in2 in3 = 0110, out = 1
s1 s0 = 10, in0 in1 in2 in3 = 1001, out = 0
s1 s0 = 11, in0 in1 in2 in3 = 0101, out = 1
```

Figure 1.65   Outputs for the conditional operator.

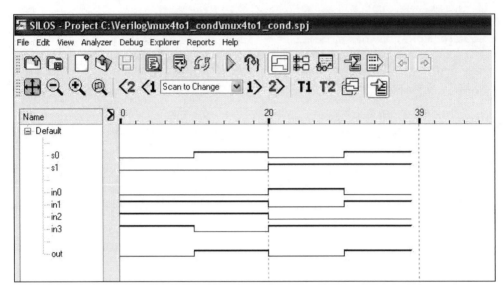

Figure 1.66   Waveforms for the conditional operator.

1.3.4 Relational Operators

Relational operators compare operands and return a Boolean result, either 1 (true) or 0 (false) indicating the relationship between the two operands. There are four relational operators: greater than (>), less than (<), greater than or equal (>=), and less than or equal (<=).

If the relationship is true, then the result is 1; if the relationship is false, then the result is 0. Net or register operands are treated as unsigned values; real or integer operands are treated as signed values. An **x** or **z** in any operand returns a result of **x**. When the operands are of unequal size, the smaller operand is zero-extended to the left.

Example 1.13   Figure 1.67 illustrates a design module showing examples of relational operators using dataflow modeling. The identifier *gt* means greater than, *gte* means greater than or equal, *lt* means less than, and *lte* means less than or equal. The test bench, which applies several different values to the two operands, is shown in Figure 1.68. The outputs and waveforms are shown in Figure 1.69 and Figure 1.70, respectively.

```
//relational operations
module relational_ops2 (a, b, gt, gte, lt, lte);

input [3:0] a, b;
output gt, gte, lt, lte;          //continued on next page
```

Figure 1.67   Design module to illustrate the relational operators.

```
//implement the relational operators using the assign statement
assign   gt = a > b,
         gte = a >= b,
         lt = a < b,
         lte = a <= b;
endmodule
```

Figure 1.67   (Continued)

```
//test bench for relational operations
module relational_ops2_tb;

reg [3:0] a, b;            //inputs are reg for test benches
wire gt, gte, lt, lte;   //outputs are wire for test benches

//display variables
initial
$monitor ("a=%b, b=%b, gt=%b, gte=%b, lt=%b, lte=%b",
            a, b, gt, gte, lt, lte);

//apply input vectors
initial
begin
   #0    a = 4'b0000;   b = 4'b0000;
   #10   a = 4'b1111;   b = 4'b1111;
   #10   a = 4'b0011;   b = 4'b0001;
   #10   a = 4'b1110;   b = 4'b1111;
   #10   a = 4'b1100;   b = 4'b1101;
   #10   a = 4'b1010;   b = 4'b1001;
   #10   a = 4'b1000;   b = 4'b1011;
   #10   a = 4'b1000;   b = 4'b0100;

   #10   $stop;
end

//instantiate the module into the test bench
relational_ops2 inst1 (
   .a(a),
   .b(b),
   .gt(gt),
   .gte(gte),
   .lt(lt),
   .lte(lte)
   );
endmodule
```

Figure 1.68   Test bench for the relational operators module of Figure 1.67.

```
a=0000, b=0000, gt=0, gte=1, lt=0, lte=1
a=1111, b=1111, gt=0, gte=1, lt=0, lte=1
a=0011, b=0001, gt=1, gte=1, lt=0, lte=0
a=1110, b=1111, gt=0, gte=0, lt=1, lte=1

a=1100, b=1101, gt=0, gte=0, lt=1, lte=1
a=1010, b=1001, gt=1, gte=1, lt=0, lte=0
a=1000, b=1011, gt=0, gte=0, lt=1, lte=1
a=1000, b=0100, gt=1, gte=1, lt=0, lte=0
```

Figure 1.69   Outputs for the relational operators test bench.

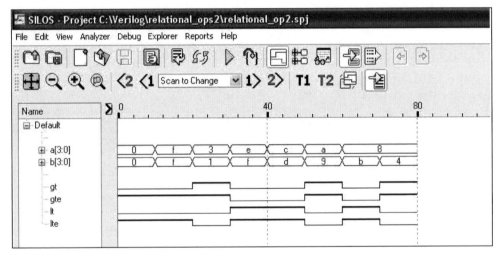

Figure 1.70   Waveforms for the relational operators test bench.

1.3.5 Logical Operators

There are three logical operators: the binary logical AND operator (&&), the binary logical OR operator (| |), and the unary logical negation operator (!). Logical operators evaluate to a logical 1 (true), a logical 0 (false), or an **x** (ambiguous). If a logical operation returns a nonzero value, then it is treated as a logical 1 (true); if a bit in an operand is **x** or **z**, then it is ambiguous and is normally treated as a false condition.

If a vector operand is nonzero, then it is treated as a logical 1 (true); if a vector operand is zero, then it is treated as a logical 0 (false). For example, let vector $a = 1000$ and vector $b = 1001$. Then a && b returns a value of 1, because both vector a and vector b are true. Similarly, $a | | b$ returns a value of 1. However, since vector a is nonzero (true), then $!a$ is zero (false).

Example 1.14 The design module of Figure 1.71 shows examples of the logical operators using dataflow modeling. Figure 1.72, Figure 1.73, and Figure 1.74 show the test bench, outputs, and waveforms, respectively. Refer to Figure 1.71 and assume that vector $a = 0110$ and vector $b = 1100$. The logical operation of a && b returns a value of 1, because both a and b are nonzero (true).

Now assume that $a = 0101$ and $b = 0000$. Thus, $z_1 = a$ && $b = 1$ && 0, which returns a value of 0, because 1 && $0 = 0$ — a is true and b is false. Output z_2, however, is equal to 1, because $z_2 = a \,||\, b = 1 \,||\, 0 = 1$. In a similar manner, $z_3 = !a = !1 = 0$, because a is true.

As a final example, assume that $a = 0000$ and $b = 0000$; that is, both variables are false. Therefore, $z_1 = a$ && $b = 0$ && 0, which returns a value of 0, because 0 && 0 $= 0$. Output $z_2 = a \,||\, b = 0 \,||\, 0 = 0$. In a similar manner, $z_3 = !a = !0 = 1$. If a bit in either operand is $\mathbf{x}$, then the result of a logical operation is $\mathbf{x}$. Also, $!\mathbf{x}$ is $\mathbf{x}$.

```
//examples of logical operators
module logical_ops (a, b, z1, z2, z3);

input [3:0] a, b;
output z1, z2, z3;

//perform the logical operations
assign    z1 = a && b,     //logical and
          z2 = a || b,     //logical or
          z3 = !a;         //logical negation

endmodule
```

Figure 1.71   Design module to illustrate the application of the logical operators.

```
//test bench for the logical operators
module logical_ops_tb;

reg [3:0] a, b;        //inputs are reg for test bench
wire z1, z2, z3;       //outputs are wire for test bench

//display variables
initial
$monitor ("a = %b, b = %b, z1 = %b, z2 = %b, z3 = %b",
          a, b, z1, z2, z3);

                              //continued on next page
```

Figure 1.72   Test bench for the logical operators.

```
//apply input vectors
initial
begin
   #0     a = 4'b0110;    b = 4'b1100;
   #10    a = 4'b0101;    b = 4'b0000;
   #10    a = 4'b0000;    b = 4'b0000;
   #10    a = 4'b1000;    b = 4'b1001;

   #10    a = 4'b1111;    b = 4'b1111;
   #10    a = 4'b0000;    b = 4'b0001;
   #10    a = 4'b0111;    b = 4'b0111;

   #10    $stop;
end


//instantiate the module into the test bench
logical_ops inst1 (
   .a(a),
   .b(b),
   .z1(z1),
   .z2(z2),
   .z3(z3)
   );

endmodule
```

Figure 1.72   (Continued)

```
z1 = &&,    z2 = ||,    z3 = !

a = 0110, b = 1100, z1 = 1, z2 = 1, z3 = 0

a = 0101, b = 0000, z1 = 0, z2 = 1, z3 = 0

a = 0000, b = 0000, z1 = 0, z2 = 0, z3 = 1

a = 1000, b = 1001, z1 = 1, z2 = 1, z3 = 0

a = 1111, b = 1111, z1 = 1, z2 = 1, z3 = 0

a = 0000, b = 0001, z1 = 0, z2 = 1, z3 = 1

a = 0111, b = 0111, z1 = 1, z2 = 1, z3 = 0
```

Figure 1.73   Outputs for the logical operators.

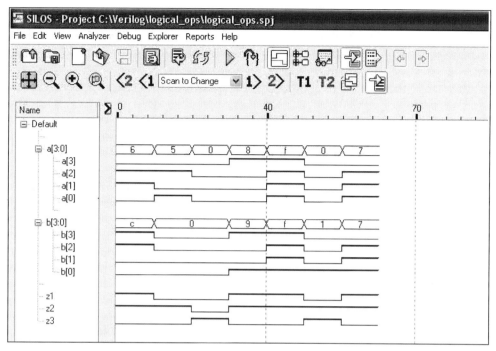

Figure 1.74   Waveforms for the logical operators.

1.3.6 Bitwise Operators

The bitwise operators are: AND (&), OR (|), negation (~), exclusive-OR (^), and exclusive-NOR (^~ or ~^). The bitwise operators perform logical operations on the operands on a bit-by-bit basis and produce a vector result. Except for negation, each bit in one operand is associated with the corresponding bit in the other operand. If one operand is shorter, then it is zero-extended to the left to match the length of the longer operand.

The *bitwise AND* operator and the *bitwise OR* operator perform their respective functional operations on two operands on a bit-by-bit basis. Examples of the bitwise AND operator and the bitwise OR operator are shown below.

$$
\begin{array}{ccccccccc}
   & 0 & 0 & 1 & 1 & 0 & 1 & 1 & 0 \\
\&) & 1 & 1 & 1 & 1 & 0 & 1 & 0 & 1 \\
\hline
   & 0 & 0 & 1 & 1 & 0 & 1 & 0 & 0
\end{array}
$$

$$
\begin{array}{ccccccccc}
   & 1 & 0 & 0 & 1 & 0 & 1 & 1 & 0 \\
|) & 0 & 1 & 0 & 1 & 0 & 1 & 0 & 1 \\
\hline
   & 1 & 1 & 0 & 1 & 0 & 1 & 1 & 1
\end{array}
$$

The *bitwise negation* operator performs the negation function on one operand on a bit-by-bit basis. Each bit in the operand is inverted. An example of the bitwise negation operator is shown below.

$$\sim) \quad \underline{0 \quad 1 \quad 0 \quad 1 \quad 0 \quad 1 \quad 0 \quad 1}$$
$$ \quad 1 \quad 0 \quad 1 \quad 0 \quad 1 \quad 0 \quad 1 \quad 0$$

The *bitwise exclusive-OR* operator and the *bitwise exclusive-NOR* operator perform their respective functional operations on two operands on a bit-by-bit basis. Examples of the bitwise exclusive-OR operator and the bitwise exclusive-NOR operator are shown below.

$$ \quad 0 \quad 0 \quad 1 \quad 1 \quad 0 \quad 1 \quad 1 \quad 0$$
$$^\wedge) \quad \underline{1 \quad 1 \quad 0 \quad 1 \quad 0 \quad 1 \quad 0 \quad 1}$$
$$ \quad 1 \quad 1 \quad 1 \quad 0 \quad 0 \quad 0 \quad 1 \quad 1$$

$$ \quad 1 \quad 0 \quad 1 \quad 1 \quad 0 \quad 1 \quad 0 \quad 0$$
$$^\wedge\sim) \quad \underline{1 \quad 1 \quad 0 \quad 1 \quad 0 \quad 1 \quad 0 \quad 1}$$
$$ \quad 1 \quad 0 \quad 0 \quad 1 \quad 1 \quad 1 \quad 1 \quad 0$$

Bitwise operators perform operations on operands on a bit-by-bit basis and produce a vector result. This is in contrast to logical operators, which perform operations on the operands in such a way that the truth or falsity of the result is determined by the truth or falsity of the operands. That is, the logical AND operator returns a value of 1 (true) only if both operands are nonzero (true); otherwise, it returns a value of 0 (false). If the result is ambiguous, it returns a value of **x**.

The logical OR operator returns a value of 1 (true) if either or both operands are true; otherwise, it returns a value of 0. The logical negation operator returns a value of 1 (true) if the operand has a value of zero and a value of 0 (false) if the operand is nonzero.

Example 1.15 Figure 1.75 shows a design module to illustrate the use of the five bitwise operators. The test bench is shown in Figure 1.76, which includes one case where the operands are of different lengths. The outputs and waveforms are shown in Figure 1.77 and Figure 1.78, respectively.

```
//dataflow example of the five bitwise operators
module bitwise2 (a, b, and_rslt, or_rslt, neg_rslt,
                xor_rslt, xnor_rslt);

                                    //continued on next page
```

Figure 1.75   Module to illustrate using the five bitwise operators.

```
//define inputs and outputs
input [7:0] a, b;
output [7:0] and_rslt, or_rslt, neg_rslt, xor_rslt, xnor_rslt;

wire [7:0] a, b;
wire [7:0] and_rslt, or_rslt, neg_rslt, xor_rslt, xnor_rslt;

//define outputs using continuous assignment
assign     and_rslt = a & b,          //bitwise AND
           or_rslt = a | b,           //bitwise OR
           neg_rslt = ~a,             //bitwise negation
           xor_rslt = a ^ b,          //bitwise exclusive-OR
           xnor_rslt = a ^~ b;        //bitwise exclusive-NOR

endmodule
```

Figure 1.75   (Continued)

```
//test bench for bitwise2 module
module bitwise2_tb;

reg [7:0] a, b;
wire [7:0] and_rslt, or_rslt, neg_rslt, xor_rslt, xnor_rslt;

initial
$monitor ("a=%b, b=%b, and_rslt=%b, or_rslt=%b, neg_rslt=%b,
           xor_rslt=%b, xnor_rslt=%b",
           a, b, and_rslt, or_rslt, neg_rslt,
           xor_rslt, xnor_rslt);

//apply input vectors
initial
begin
   #0     a = 8'b1100_0011;
          b = 8'b1001_1001;

   #10    a = 8'b1001_0011;
          b = 8'b1101_1001;

   #10    a = 8'b0000_1111;
          b = 8'b1101_1001;
                                   //continued on next page
```

Figure 1.76   Test bench for the five bitwise operators.

```
   #10    a = 8'b0100_1111;
          b = 8'b1101_1001;

   #10    a = 8'b1100_1111;
          b = 8'b1101_1001;

   #10    a = 8'b0000_0001;
          b = 8'b1000_0001;

   #10    a = 8'b0000_0000;
          b = 8'b0000_0000;

   #10    a = 8'b1111_1111;
          b = 8'b1111_1111;

   #10    a = 8'b1010_1010;
          b = 8'b1010_1010;

   #10    a = 8'b0101_0101;
          b = 8'b0101_0101;

   #10    a = 8'b0111_0101;
          b = 4'b0101;


   #10    $stop;
end


//instantiate the module into the test bench
bitwise2 inst1 (
   .a(a),
   .b(b),
   .and_rslt(and_rslt),
   .or_rslt(or_rslt),
   .neg_rslt(neg_rslt),
   .xor_rslt(xor_rslt),
   .xnor_rslt(xnor_rslt)
   );

endmodule
```

Figure 1.76   (Continued)

```
        a = 11000011,
        b = 10011001,

  and_rslt = 10000001,
  or_rslt  = 11011011,
  neg_rslt = 00111100,
  xor_rslt = 01011010,
  xnor_rslt= 10100101
------------------------------------------------------------
        a = 10010011,
        b = 11011001,

  and_rslt = 10010001,
  or_rslt  = 11011011,
  neg_rslt = 01101100,
  xor_rslt = 01001010,
  xnor_rslt= 10110101
------------------------------------------------------------
        a = 00001111,
        b = 11011001,

  and_rslt = 00001001,
  or_rslt  = 11011111,
  neg_rslt = 11110000,
  xor_rslt = 11010110,
  xnor_rslt= 00101001
------------------------------------------------------------
        a = 01001111,
        b = 11011001,

  and_rslt = 01001001,
  or_rslt  = 11011111,
  neg_rslt = 10110000,
  xor_rslt = 10010110,
  xnor_rslt= 01101001
------------------------------------------------------------
        a = 11001111,
        b = 11011001,

  and_rslt = 11001001,
  or_rslt  = 11011111,
  neg_rslt = 00110000,
  xor_rslt = 00010110,
  xnor_rslt= 11101001

                        //continued on next page
```

Figure 1.77   Outputs for the five bitwise operators.

```
         a = 00000001,
         b = 10000001,

 and_rslt = 00000001,
 or_rslt  = 10000001,
 neg_rslt = 11111110,
 xor_rslt = 10000000,
 xnor_rslt= 01111111
-------------------------------------------------------------
         a = 00000000,
         b = 00000000,

 and_rslt = 00000000,
 or_rslt  = 00000000,
 neg_rslt = 11111111,
 xor_rslt = 00000000,
 xnor_rslt= 11111111
-------------------------------------------------------------
         a = 11111111,
         b = 11111111,

 and_rslt = 11111111,
 or_rslt  = 11111111,
 neg_rslt = 00000000,
 xor_rslt = 00000000,
 xnor_rslt= 11111111
-------------------------------------------------------------
         a = 10101010,
         b = 10101010,

 and_rslt = 10101010,
 or_rslt  = 10101010,
 neg_rslt = 01010101,
 xor_rslt = 00000000,
 xnor_rslt= 11111111
-------------------------------------------------------------
         a = 01010101,
         b = 01010101,

 and_rslt = 01010101,
 or_rslt  = 01010101,
 neg_rslt = 10101010,
 xor_rslt = 00000000,
 xnor_rslt= 11111111
-------------------------------------------------------------

                        //continued on next page
```

Figure 1.77   (Continued)

```
         a = 01110101,
         b = 00000101,

 and_rslt = 00000101,
 or_rslt  = 01110101,
 neg_rslt = 10001010,
 xor_rslt = 01110000,
 xnor_rslt= 10001111
```

Figure 1.77   (Continued)

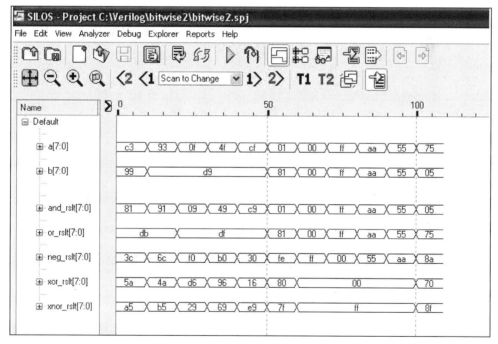

Figure 1.78   Waveforms for the five bitwise operators.

1.3.7 Shift Operators

The shift operators shift a single vector operand left or right a specified number of bit positions. These are logical shift operations, not algebraic; that is, as bits are shifted left or right, zeroes fill in the vacated bit positions. The bits shifted out of the operand are lost; they do not rotate to the high-order or low-order bit positions of the shifted operand. If the shift amount evaluates to **x** or **z**, then the result of the operation is **x**.

There are two shift operators, as shown below. The value in parentheses is the number of bits that the operand is shifted.

$$<< \text{(Left-shift amount)}$$
$$>> \text{(Right-shift amount)}$$

When an operand is shifted left, this is equivalent to a multiply-by-two operation for each bit position shifted. When an operand is shifted right, this is equivalent to a divide-by-two operation for each bit position shifted. The shift operators are useful to model the sequential add-shift multiplication algorithm and the sequential shift-subtract division algorithm.

Example 1.16 The design module illustrating examples of the shift-left and shift-right operators using dataflow modeling is shown in Figure 1.79. The shift-left operator shifts the bits two and four bit positions; the shift-right operator shifts the bits one and three bit positions. The test bench module is shown in Figure 1.80. The outputs and waveforms are shown in Figure 1.81 and Figure 1.82, respectively.

```
//dataflow module to illustrate the shift operators
module shift2 (a, b, a_rslt2, a_rslt4, b_rslt1, b_rslt3);

//define inputs and outputs
input [11:0] a, b;
output [11:0] a_rslt2, a_rslt4, b_rslt1, b_rslt3;

//define inputs and outputs as wire
wire a, b;
wire a_rslt2, a_rslt4, b_rslt1, b_rslt3;

//define outputs using continuous assignment
assign    a_rslt2 = a << 2,      //multiply by 4
          a_rslt4 = a << 4,      //multiply by 16

          b_rslt1 = b >> 1,      //divide by 2
          b_rslt3 = b >> 3;      //divide by 8
endmodule
```

Figure 1.79   Design module to illustrate the shift-left and shift-right operators.

```
//test bench for shift operators module
module shift2_tb;

reg [11:0] a, b;
wire [11:0] a_rslt2, a_rslt4, b_rslt1, b_rslt3;    //next page
```

Figure 1.80   Test bench for the shift-left and shift-right operators.

```
//display variables
initial
$monitor ("a = %b, b = %b, a_rslt2 = %b, a_rslt4 = %b,
            b_rslt1 = %b, b_rslt3 = %b",
        a, b, a_rslt2, a_rslt4, b_rslt1, b_rslt3);


//apply input vectors
initial
begin
   #0     a = 12'b0000_0000_0010;    //2
          b = 12'b0000_0000_1000;    //8


   #10    a = 12'b0000_0000_0110;    //6
          b = 12'b0000_0001_1000;    //24


   #10    a = 12'b0000_0000_1111;    //15
          b = 12'b0000_0011_1000;    //56


   #10    a = 12'b1111_1110_0000;    //-32
          b = 12'b0000_0000_0011;    //3


   #10    $stop;
end


//instantiate the module into the test bench
shift2 inst1 (
   .a(a),
   .b(b),
   .a_rslt2(a_rslt2),
   .a_rslt4(a_rslt4),
   .b_rslt1(b_rslt1),
   .b_rslt3(b_rslt3)
   );


endmodule
```

Figure 1.80   (Continued)

```
         a_rslt2 = a << 2,    //multiply by 4
         a_rslt4 = a << 4,    //multiply by 16

         b_rslt1 = b >> 1,    //divide by 2
         b_rslt3 = b >> 3;    //divide by 8
-------------------------------------------------------------
a = 000000000010,             //a = 2

a_rslt2 = 000000001000,       //a << 2 = 8; multiply by 4
a_rslt4 = 000000100000,       //a << 4 = 32; multiply by 16

b = 000000001000,             //b = 8

b_rslt1 = 000000000100,       //b >> 1 = 4; divide by 2
b_rslt3 = 000000000001        //b >> 3 = 1; divide by 8
-------------------------------------------------------------
a = 000000000110,             //a = 6

a_rslt2 = 000000011000,       //a << 2 = 24; multiply by 4
a_rslt4 = 000001100000,       //a << 4 = 96; multiply by 16

b = 000000011000,             //b = 24

b_rslt1 = 000000001100,       //b >> 1 = 12; divide by 2
b_rslt3 = 000000000011        //b >> 3 = 3; divide by 8
-------------------------------------------------------------
a = 000000001111,             //a = 15

a_rslt2 = 000000111100,       //a << 2 = 60; multiply by 4
a_rslt4 = 000011110000,       //a << 4 = 240; multiply by 16

b = 000000111000,             //b = 56

b_rslt1 = 000000011100,       //b >> 1 = 28; divide by 2
b_rslt3 = 000000000111        //b >> 3 = 7; divide by 8
-------------------------------------------------------------
a = 111111100000,             //a = -32

a_rslt2 = 111110000000,       //a << 2 = -128; multiply by 4
a_rslt4 = 111000000000,       //a << 4 = -512; multiply by 16

b = 000000000011,             //b = 3

b_rslt1 = 000000000001,       //b >> 1 = 1; divide by 2
b_rslt3 = 000000000000        //b >> 3 = 0; divide by 8
```

Figure 1.81   Outputs for the left-shift and right-shift operators.

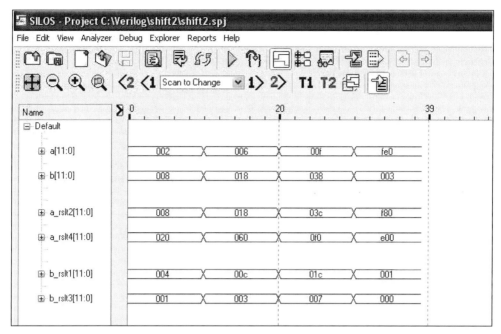

Figure 1.82 Waveforms for the left-shift and right-shift operators.

1.4 Behavioral Modeling

This section describes the *behavior* of a digital system and is not concerned with the direct implementation of logic gates, but more with the architecture of the system. This is an algorithmic approach to hardware implementation and represents a higher level of abstraction than previous modeling methods. Describing a module in *behavioral* modeling is an abstraction of the functional operation of the design. It does not describe the implementation of the design at the gate level.

In previous sections, built-in primitives, user-defined primitives (UDPs), and dataflow modeling were used to design hardware primarily at the gate level. A Verilog module may contain a mixture of built-in primitives, UDPs, dataflow constructs, and behavioral constructs. The constructs in behavioral modeling closely resemble those used in the C programming language.

A *procedure* is a series of operations taken to design a module. A Verilog module that is designed using behavioral modeling contains no internal structural details, it simply defines the behavior of the hardware in an abstract, algorithmic description. Verilog contains two structured procedure statements or behaviors: **initial** and **always**. A behavior may consist of a single statement or a block of statements delimited by the keywords **begin** . . . **end**. A module may contain multiple **initial** and **always** statements. These statements are the basic statements used in behavioral modeling and execute concurrently starting at time zero in which the order of execution is not important. All other behavioral statements are contained inside these structured pro-cedure statements.

1.4.1 Initial Statement

This section presents a recapitulation of the **initial** statement, which was originally presented in Section 1.1. All statements within an **initial** statement comprise an **initial** block. An **initial** statement executes only once beginning at time zero, then suspends execution. An **initial** statement provides a method to initialize and monitor variables before the variables are used in a module; it is also used to generate waveforms. For a given time unit, all statements within the **initial** block execute sequentially. Execu-tion or assignment is controlled by the # symbol, which is used to signify an optional time unit for timing control and delays. The syntax for an **initial** statement is shown below.

> **initial** [optional timing control] procedural statement or
> block of procedural statements

Each **initial** block executes concurrently at time zero and each block ends execu-tion independently. If there is only one procedural statement, then the statement does not require the keywords **begin** . . . **end**. However, if there are two or more procedural statements, then they are delimited by the keywords **begin** . . . **end**.

Example 1.17   A module showing the use of the **initial** statement is shown in Figure 1.83, where the variables x_1, x_2, x_3, x_4, and x_5 are initialized to specific values. Seven **initial** statements are used for both a single procedural statement and a block of pro-cedural statements. The outputs and waveforms are shown in Figure 1.84 and Figure 1.85, respectively.

```
//module showing use of the initial keyword
module initial_ex (x1, x2, x3, x4, x5);

output x1, x2, x3, x4, x5;

reg x1, x2, x3, x4, x5;


//display variables
initial
$monitor ($time, " x1x2x3x4x5 = %b", {x1, x2, x3, x4, x5});

                                    //continued on next page
```

Figure 1.83   Module to illustrate the use of the **initial** statement.

```verilog
//initialize variables to 0
//multiple statements require begin . . . end
initial
begin
   #0    x1 = 1'b0;
         x2 = 1'b0;
         x3 = 1'b0;
         x4 = 1'b0;
         x5 = 1'b0;
end


//set x1
//single statement requires no begin . . . end
initial
   #10   x1 = 1'b1;


//set x2 and x3
initial
begin
   #10   x2 = 1'b1;
   #10   x3 = 1'b1;
end


//set x4 and x5
initial
begin
   #10   x4 = 1'b1;
   #10   x5 = 1'b1;
end

//reset variables
initial
begin
   #20   x1 = 1'b0;
   #10   x2 = 1'b0;
   #10   x3 = 1'b0;
   #10   x4 = 1'b0;
   #10   x5 = 1'b0;
end

//determine length of simulation
initial
   #70   $finish;


endmodule
```

Figure 1.83   (Continued)

```
0   x1x2x3x4x5  =  00000
10  x1x2x3x4x5  =  11010
20  x1x2x3x4x5  =  01111
30  x1x2x3x4x5  =  00111
40  x1x2x3x4x5  =  00011
50  x1x2x3x4x5  =  00001
60  x1x2x3x4x5  =  00000
```

Figure 1.84   Outputs for the **initial** module of Figure 1.83.

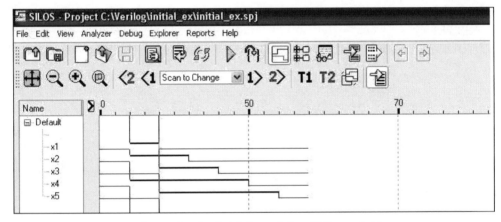

Figure 1.85   Waveforms for the **initial** module of Figure 1.83.

Figure 1.83 contains seven **initial** statements. The first **initial** statement invokes the system task **$monitor**, which causes the specified string (enclosed in quotation marks) to be printed whenever a variable changes in the argument list (enclosed in braces). The **$time** system function returns the simulation time as a decimal number.

The second **initial** statement initializes all variables to zero. The third **initial** statement sets x_1 at 10 time units. Since all **initial** statements begin execution at time zero, the fourth **initial** statement sets x_2 at 10 time units also, and sets x_3 at time 20 time units (#10 plus #10). This can be seen in the waveforms of Figure 1.85. Variable x_4 is set at 10 time units by the fifth **initial** statement, which also sets x_5 at 20 time units. The sixth **initial** statement resets all variables. The seventh **initial** statement invokes the system task **$finish**, which causes the simulator to exit the module and return control to the operating system.

1.4.2 Always Statement

The **always** statement executes the behavioral statements within the **always** block repeatedly in a looping manner and begins execution at time zero. Execution of the statements continues indefinitely until the simulation is terminated. The keywords **initial** and **always** specify a behavior and the statements within a behavior are classified as *behavioral* or *procedural*. The syntax for the **always** statement is shown below.

always [optional timing control] procedural statement or
block of procedural statements

An **always** statement is often used with an *event control list* — or *sensitivity list* — to execute a sequential block. When a change occurs to a variable in the sensitivity list, the statement or block of statements in the **always** block is executed. The keyword **or** is used to indicate multiple events. When one or more inputs change state, the statement in the **always** block is executed. The **begin** . . . **end** keywords are necessary only when there is more than one behavioral statement. Target variables used in an **always** statement are declared as type **reg**.

Example 1.18 A 5-input majority circuit will be designed that produces a high output on z_1 whenever the majority of inputs $x_1, x_2, x_3, x_4,$ and x_5 are at a logic 1, where x_5 is the low-order bit; otherwise, output z_1 will be at a logic 0. In order for there to be a majority, there must be an odd number of inputs. The circuit can be designed by plotting the five variables on a Karnaugh map and inserting 1s in minterm locations in which there are at least three 1s. Then the groups of 1s are combined to form a minimized sum-of-products expression. The 5-variable Karnaugh map is shown in Figure 1.86 and the resulting equation for the majority circuit is shown in Equation 1.7.

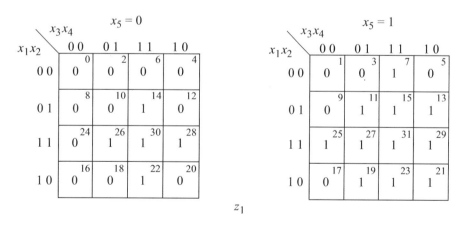

Figure 1.86 Karnaugh map for a 5-input majority circuit.

$$z_1 = x_3x_4x_5 + x_2x_3x_5 + x_2x_4x_5 + x_1x_3x_5 + x_1x_4x_5 + x_1x_2x_3 + x_1x_2x_4 +$$

$$x_2x_3x_4 + x_1x_3x_4 + x_1x_2x_5 \tag{1.7}$$

The behavioral design module using an **always** statement is shown in Figure 1.87. The entry of #5 to the immediate right of the equal sign specifies that the output is delayed by five time units to allow for the propagation delay — inertial delay — of the AND gate. The sensitivity list in the **always** statement lists the five inputs. Whenever one or more of the inputs changes value, the equation for z_1 is executed.

The test bench is shown in Figure 1.88. The system function **$time** obtains the current simulation time that is displayed in the outputs of Figure 1.89 every seven time units, as indicated by the #7 symbol immediately preceding the **$display** system task, which prints the inputs and output variables specified in the argument list. The waveforms are shown in Figure 1.90 and clearly show the propagation delay of five time units that occurs when an input changes; that is, output z_1 is asserted five time units after an input changes value if the input vector results in a majority of inputs.

```
//behavioral 5-input majority circuit
module maj5_bh (x1, x2, x3, x4, x5, z1);

input x1, x2, x3, x4, x5;
output z1;

wire x1, x2, x3, x4, x5;
reg z1;

always @ (x1 or x2 or x3 or x4 or x5)
   z1 = #5  (x3 & x4 & x5) | (x2 & x3 & x5) | (x2 & x4 & x5) |
            (x1 & x3 & x5) | (x1 & x4 & x5) | (x1 & x2 & x3) |
            (x1 & x2 & x4) | (x2 & x3 & x4) | (x1 & x3 & x4) |
            (x1 & x2 & x5);
endmodule
```

Figure 1.87   Design module for the 5-input majority circuit.

```
//test bench for the 5-input majority circuit
module maj5_bh_tb;

reg x1, x2, x3, x4, x5;
wire z1;

initial      //apply vectors and display variables
begin: apply_stimulus
   reg [5:0] invect;
   for (invect=0; invect<32; invect=invect+1)

                     //continued on next page
```

Figure 1.88   Test bench module for the 5-input majority circuit.

```
      begin
         {x1, x2, x3, x4, x5} = invect [5:0];
         #7 $display ($time, "input = %b, z1 = %b",
                     {x1, x2, x3, x4, x5}, z1);
      end
end


//instantiate the module into the test bench
maj5_bh inst1 (
   .x1(x1),
   .x2(x2),
   .x3(x3),
   .x4(x4),
   .x5(x5),
   .z1(z1)
   );

endmodule
```

Figure 1.88   (Continued)

```
7     input = 00000, z1 = 0    119    input = 10000, z1 = 0
14    input = 00001, z1 = 0    126    input = 10001, z1 = 0
21    input = 00010, z1 = 0    133    input = 10010, z1 = 0
28    input = 00011, z1 = 0    140    input = 10011, z1 = 1
35    input = 00100, z1 = 0    147    input = 10100, z1 = 0
42    input = 00101, z1 = 0    154    input = 10101, z1 = 1
49    input = 00110, z1 = 0    161    input = 10110, z1 = 1
56    input = 00111, z1 = 1    168    input = 10111, z1 = 1
63    input = 01000, z1 = 0    175    input = 11000, z1 = 0
70    input = 01001, z1 = 0    182    input = 11001, z1 = 1
77    input = 01010, z1 = 0    189    input = 11010, z1 = 1
84    input = 01011, z1 = 1    196    input = 11011, z1 = 1
91    input = 01100, z1 = 0    203    input = 11100, z1 = 1
98    input = 01101, z1 = 1    210    input = 11101, z1 = 1
105   input = 01110, z1 = 1    217    input = 11110, z1 = 1
112   input = 01111, z1 = 1    224    input = 11111, z1 = 1
```

Figure 1.89   Outputs for the 5-input majority circuit.

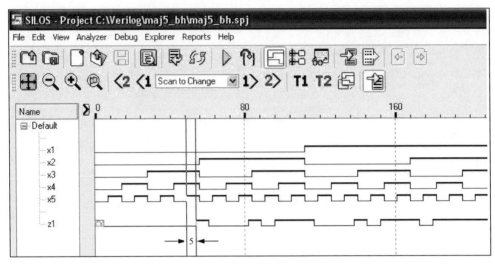

Figure 1.90   Waveforms for the 5-input majority circuit.

1.4.3 Intrastatement Delay

A procedural assignment may have an optional delay. A delay appearing to the right of an assignment operator is called an *intrastatement* delay. It is the delay by which the right-hand result is delayed before assigning it to the left-hand target. In the example below, the expression $(x_1 \& x_2)$ is evaluated, a delay of five time units is taken, then the result is assigned to z_1.

$$z_1 = \#5 \ (x_1 \ \& \ x_2);$$

The statement evaluates the logical function x_1 AND x_2, waits five time units, then assigns the result to z_1. If no delay is specified in a procedural assignment, then zero delay is the default delay and the assignment occurs immediately.

Example 1.19   This example will illustrate intrastatement delay for three operations: a statement consisting of two AND gates and an exclusive-OR gate; a statement using the conditional operator; and a statement for a 3-bit odd parity generator. There are three inputs: $x_1, x_2,$ and x_3. There are three outputs: $z_1, z_2,$ and z_3, which are defined as follows:

$$z_1 = \#2 \ (x_1 \ \& \ {\sim}x_2) \ {}^\wedge \ ({\sim}x_1 \ \& \ x_3);$$
$$z_2 = \#3 \ (x_1 \ {>}{=} \ x_2) \ ? \ x_2 : x_3;$$
$$z_3 = \#4 \ {\sim}(x_1 \ {}^\wedge \ x_2 \ {}^\wedge \ x_3);$$

The behavioral module is shown in Figure 1.91 in which intrastatement delays are assigned to the statements that generate $z_1, z_2,$ and z_3. The test bench module is shown in Figure 1.92 for all combinations of the inputs. Figure 1.93 shows the outputs for z_1, z_2, and z_3 based upon the definitions stated above. The waveforms are shown in Figure 1.94, which show the delays for each output. The values for the outputs are unknown until their respective delays have taken place. Since blocking assignments are used, the delays are cumulative; that is, z_1 receives its value two time units after the inputs change, z_2 receives its value at five time units, and z_3 receives its value at nine time units after the inputs change.

```
//behavioral model to demonstrate intrastatement delay
module intra_stmt_dly5 (x1, x2, x3, z1, z2, z3);

input x1, x2, x3;
output z1, z2, z3;

reg z1, z2, z3;

always @ (x1 or x2 or x3)
begin
   z1 = #2 (x1 & ~x2) ^ (~x1 & x3);
   z2 = #3 (x1 >= x2) ? x2 : x3;
   z3 = #4 ~(x1 ^ x2 ^ x3);
end
endmodule
```

Figure 1.91   Design module to illustrate the intrastatement delay.

```
//test bench for intrastatement delay
module intra_stmt_dly5_tb;

reg x1, x2, x3;
wire z1, z2, z3;

//apply input vectors and display variables
initial
begin: apply_stimulus
   reg [3:0] invect;
   for (invect=0; invect<8; invect=invect+1)
      begin
          {x1, x2, x3} = invect [3:0];
          #10  $display ("x1 x2 x3 = %b,
                 z1 = %b, z2 = %b, z3 = %b",
                 {x1, x2, x3}, z1, z2, z3);
      end
end                                 //continued on next page
```

Figure 1.92   Test bench module for the intrastatement delay module.

```
//instantiate the module into the test bench
intra_stmt_dly5 inst1 (
    .x1(x1),
    .x2(x2),
    .x3(x3),
    .z1(z1),
    .z2(z2),
    .z3(z3)
    );

endmodule
```

Figure 1.92   (Continued)

```
x1 x2 x3 = 000, z1 = 0, z2 = 0, z3 = 1
x1 x2 x3 = 001, z1 = 1, z2 = 0, z3 = 0
x1 x2 x3 = 010, z1 = 0, z2 = 0, z3 = 0
x1 x2 x3 = 011, z1 = 1, z2 = 1, z3 = 1

x1 x2 x3 = 100, z1 = 1, z2 = 0, z3 = 0
x1 x2 x3 = 101, z1 = 1, z2 = 0, z3 = 1
x1 x2 x3 = 110, z1 = 0, z2 = 1, z3 = 1
x1 x2 x3 = 111, z1 = 0, z2 = 1, z3 = 0
```

Figure 1.93   Outputs for the intrastatement delay module.

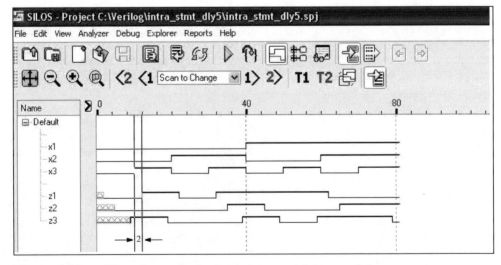

Figure 1.94   Waveforms for the intrastatement delay module.

1.4.4 Interstatement Delay

An *interstatement* delay is the delay by which a statement's execution is delayed; that is, it is the delay between statements. In the code segment shown in Equation 1.8, when the first statement has completed execution, a delay of five time units is taken before the second statement is executed. If no delays are specified in a procedural assignment, then there is zero delay in the assignment.

$$z_1 = (x_1 \mid x_2) \,\&\, x_3$$

$$\#5 \; z_2 = (x_1 \,\&\, x_2) \mid x_3 \tag{1.8}$$

The behavioral module of Figure 1.95 illustrates the use of an interstatement delay for Equation 1.8. The test bench is shown in Figure 1.96, the outputs are shown in Figure 1.97, and the waveforms are shown in Figure 1.98.

```
//behavioral module to illustrate interstatement delay
module inter_stmt_dly3 (x1, x2, x3, z1, z2);

input x1, x2, x3;
output z1, z2;

reg z1, z2;

always @ (x1 or x2 or x3)
begin
        z1 = (x1 | x2) & x3;
    #5 z2 = (x1 & x2) | x3;
end
endmodule
```

Figure 1.95   Design module to illustrate interstatement delay.

```
//test bench for interstatement delay
module inter_stmt_dly3_tb;

reg x1, x2, x3;
wire z1, z2;

//display variables
initial
$monitor ("x1 x2 x3 = %b, z1 = %b, z2 = %b",
           {x1, x2, x3}, z1, z2);   //continued on next page
```

Figure 1.96   Test bench for the interstatement delay design module.

```
//apply input vectors
initial
begin
   #0     x1 = 1'b0;   x2 = 1'b0;   x3 = 1'b0;
   #10    x1 = 1'b0;   x2 = 1'b0;   x3 = 1'b1;
   #10    x1 = 1'b0;   x2 = 1'b1;   x3 = 1'b0;
   #10    x1 = 1'b0;   x2 = 1'b1;   x3 = 1'b1;

   #10    x1 = 1'b1;   x2 = 1'b0;   x3 = 1'b0;
   #10    x1 = 1'b1;   x2 = 1'b0;   x3 = 1'b1;
   #10    x1 = 1'b1;   x2 = 1'b1;   x3 = 1'b0;
   #10    x1 = 1'b1;   x2 = 1'b1;   x3 = 1'b1;

   #10    $stop;
end

//instantiate the module into the test bench
inter_stmt_dly3 inst1 (
   .x1(x1),
   .x2(x2),
   .x3(x3),
   .z1(z1),
   .z2(z2)
   );
endmodule
```

Figure 1.96   (Continued)

```
z1 = (x1 | x2) & x3            z2 = (x1 & x2) | x3
The multiple x1 x2 x3 entries are the result of the interstate-
ment delay.  Observe the waveforms of Figure 1.98.
x1 x2 x3 = 000, z1 = 0, z2 = x
x1 x2 x3 = 000, z1 = 0, z2 = 0
x1 x2 x3 = 001, z1 = 0, z2 = 0
x1 x2 x3 = 001, z1 = 0, z2 = 1
x1 x2 x3 = 010, z1 = 0, z2 = 1
x1 x2 x3 = 010, z1 = 0, z2 = 0
x1 x2 x3 = 011, z1 = 1, z2 = 0
x1 x2 x3 = 011, z1 = 1, z2 = 1
x1 x2 x3 = 100, z1 = 0, z2 = 1
x1 x2 x3 = 100, z1 = 0, z2 = 0
x1 x2 x3 = 101, z1 = 1, z2 = 0
x1 x2 x3 = 101, z1 = 1, z2 = 1
x1 x2 x3 = 110, z1 = 0, z2 = 1
x1 x2 x3 = 111, z1 = 1, z2 = 1
```

Figure 1.97   Outputs for the interstatement delay module.

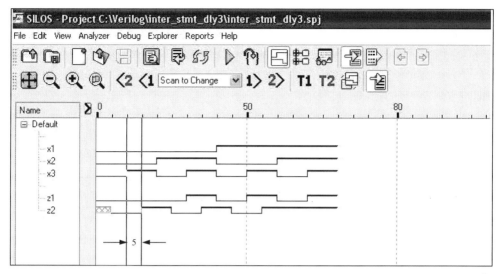

Figure 1.98   Waveforms for the interstatement delay module.

1.4.5 Blocking Assignments

A blocking procedural assignment completes execution before the next statement is executed. The assignment operator ($=$) is used for blocking assignments. The right-hand expression is evaluated, then the assignment is placed in an internal temporary register called the *event queue* and scheduled for assignment. If no time units are specified, then the scheduling takes place immediately. The event queue is covered in Appendix A.

In the code segment below, an interstatement delay of two time units is specified for the assignment to z_2. The evaluation of z_2 is delayed by the timing control; that is, the expression for z_2 will not be evaluated until the expression for z_1 has been executed, plus two time units. The execution of any of the following statements is blocked until the assignment occurs.

```
#2 z1 = x1 & x2;
#2 z2 = x1 & x3;
#2 z3 = x2 & x3;
```

Example 1.20   The module of Figure 1.99 shows delayed blocking assignments for the three statements shown above, each with an interstatement delay of two time units. The blocking statement for z_1 is assigned to be executed two time units later than the current simulation time t at $t + 2$. The right-hand side expression is evaluated at time $t + 2$ and assigned to z_1 at time $t + 2$. The statement for z_2 is evaluated at time $t + 4$, then assigned to z_2. The statement for z_3 is evaluated at time $t + 6$, then assigned to z_3.

The test bench is shown in Figure 1.100. The outputs and waveforms are shown in Figure 1.101 and Figure 1.102, respectively. The waveforms show the delay for each blocking statement. Observe the waveforms for output z_2. At 50 time units, both x_1 and x_3 are asserted. However, since the delays are cumulative, output z_2 is not asserted until 54 time units.

```
//example of blocking assignment
module blocking_7 (x1, x2, x3, z1, z2, z3);

input x1, x2, x3;
output z1, z2, z3;

reg z1, z2, z3;

always @ (x1 or x2 or x3)
begin
   #2 z1 = x1 & x2;
   #2 z2 = x1 & x3;
   #2 z3 = x2 & x3;
end

endmodule
```

Figure 1.99   Behavioral module to illustrate delayed blocking assignments.

```
//test bench for blocking assignment
module blocking_7_tb;

reg x1, x2, x3;
wire z1, z2, z3;

//apply input vectors and display variables
initial
begin: apply_stimulus
   reg [3:0] invect;
   for (invect = 0; invect < 8; invect = invect + 1)
      begin
         {x1, x2, x3} = invect [3:0];
         #10 $display ("x1 x2 x3 = %b,
                        z1 = %b, z2 = %b, z3 = %b",
                        {x1, x2, x3}, z1, z2, z3);
      end
end
                              //continued on next page
```

Figure 1.100   Test bench for the delayed blocking assignments of Figure 1.99.

```
//instantiate the module into the test bench
blocking_7 inst1 (
    .x1(x1),
    .x2(x2),
    .x3(x3),
    .z1(z1),
    .z2(z2),
    .z3(z3)
    );

endmodule
```

Figure 1.100 (Continued)

```
x1 x2 x3 = 000, z1 = 0, z2 = 0, z3 = 0
x1 x2 x3 = 001, z1 = 0, z2 = 0, z3 = 0
x1 x2 x3 = 010, z1 = 0, z2 = 0, z3 = 0
x1 x2 x3 = 011, z1 = 0, z2 = 0, z3 = 1
x1 x2 x3 = 100, z1 = 0, z2 = 0, z3 = 0
x1 x2 x3 = 101, z1 = 0, z2 = 1, z3 = 0
x1 x2 x3 = 110, z1 = 1, z2 = 0, z3 = 0
x1 x2 x3 = 111, z1 = 1, z2 = 1, z3 = 1
```

Figure 1.101 Outputs for the delayed blocking assignments of Figure 1.99.

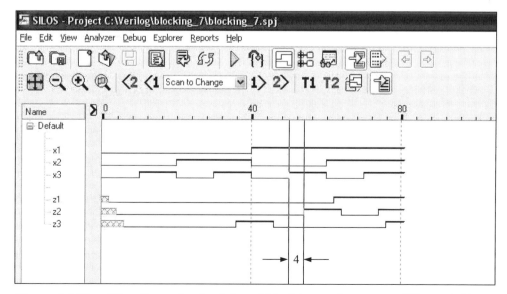

Figure 1.102 Waveforms for the delayed blocking assignments of Figure 1.99.

1.4.6 Nonblocking Assignments

The assignment symbol ($<=$) is used to represent a nonblocking procedural assignment. Nonblocking assignments allow the scheduling of assignments without blocking execution of the following statements in a sequential procedural block. A nonblocking assignment is used to synchronize assignment statements so that they appear to execute at the same time.

The Verilog simulator schedules a nonblocking assignment statement to execute, then proceeds to the next statement in the block without waiting for the previous nonblocking statement to complete execution. That is, the right-hand expression is evaluated and the value is stored in the event queue and is *scheduled* to be assigned to the left-hand target. The assignment is made at the end of the current time step if there are no intrastatement delays specified.

Nonblocking assignments are typically used to model several concurrent assignments that are caused by a common event such as the low-to-high transition of a clock pulse or a change to any variable in a sensitivity list (event control). The order of the assignments is irrelevant because the right-hand side evaluations are stored in the event queue before any assignments are made.

Example 1.21 A behavioral module will be used to design a full adder using nonblocking statements with intrastatement delays of 5 time units. A full adder has three scalar inputs: the augend a, the addend b, and the carry-in cin. There are two outputs: the sum designated as sum and the carry-out $cout$. The truth table for a full adder is shown in Table 1.1 for stage$_i$. The equations for sum_i and $cout_i$ are shown in Equation 1.9 and Equation 1.10, respectively.

Table 1.1 Truth Table for a Full Adder

a_i	b_i	cin_{i-1}	$cout_i$	sum_i
0	0	0	0	0
0	0	1	0	1
0	1	0	0	1
0	1	1	1	0
1	0	0	0	1
1	0	1	1	0
1	1	0	1	0
1	1	1	1	1

$$sum_i = a_i \oplus b_i \oplus cin_{i-1} \tag{1.9}$$

$$cout_i = a_i b_i + a_i cin_{i-1} + b_i cin_{i-1} \tag{1.10}$$

The behavioral module and test bench module are shown in Figure 1.103 and Figure 1.104, respectively. The outputs and waveforms are shown in Figure 1.105 and Figure 1.106, respectively. The waveforms show that when an input changes value, the outputs are delayed by the intrastatement delay of five time units, then the outputs are displayed simultaneously, because of the nonblocking assignment.

```verilog
//behavioral full adder using nonblocking assignments
module full_adder_nonblock (a, b, cin, sum, cout);

input a, b, cin;
output sum, cout;

//inputs are wire in behavioral (optional)
wire a, b, cin;

//reg used in always block
reg sum, cout;

//initialize sum and cout to avoid Xs until #10
initial
begin
   sum = 1'b0;
   cout = 1'b0;
end

always @ (a or b or cin)
begin
   sum  <= #5 (a ^ b ^ cin);       //nonblocking statement
   cout <= #5 ((a & b) | (a & cin) | (b & cin));
end
endmodule
```

Figure 1.103   Design module for a full adder using nonblocking assignments.

```verilog
//test bench for full adder using nonblocking statements
module full_adder_nonblock_tb;

reg a, b, cin;
wire sum, cout;

//apply stimulus and display variables
initial
begin: apply_stimulus
   reg [3:0] invect;
   for (invect = 0; invect < 8; invect = invect + 1)
                                //continued on next page
```

Figure 1.104   Test bench for the full adder of Figure 1.103.

```
      begin
         {a, b, cin} = invect [3:0];
         #10 $display ("a b cin = %b, cout = %b, sum = %b",
                    {a, b, cin}, cout, sum);
      end
end

//instantiate the module into the test bench
full_adder_nonblock inst1 (
   .a(a),
   .b(b),
   .cin(cin),
   .sum(sum),
   .cout(cout)
   );
endmodule
```

Figure 1.104   (Continued)

```
a b cin = 000, cout = 0, sum = 0
a b cin = 001, cout = 0, sum = 1
a b cin = 010, cout = 0, sum = 1
a b cin = 011, cout = 1, sum = 0
a b cin = 100, cout = 0, sum = 1
a b cin = 101, cout = 1, sum = 0
a b cin = 110, cout = 1, sum = 0
a b cin = 111, cout = 1, sum = 1
```

Figure 1.105   Outputs for the full adder of Figure 1.103.

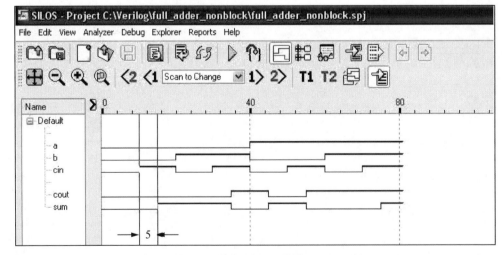

Figure 1.106   Waveforms for the full adder of Figure 1.103.

1.4.7 Conditional Statements

Conditional statements alter the flow within a behavior based upon certain conditions. The choice among alternative statements depends on the Boolean value of an expression. The alternative statements can be a single statement or a block of statements delimited by the keywords **begin** . . . **end**. The keywords **if** and **else** are used in conditional statements. There are three categories of the conditional statement as shown below. A true value is 1 or any nonzero value; a false value is 0, **x** (unknown), or **z** (high impedance). If the evaluation is false, then the next expression in the activity flow is evaluated.

No else statement
if (expression) statement1;   //if expression is true, then statement1 is executed.

One else statement   //choice of two statements. Only one is executed.
if (expression) statement1;   //if expression is true, then statement1 is executed.
else statement2;   //if expression is false, then statement2 is executed.

Nested if-else if statements   //choice of multiple statements. Only one is executed.
if (expression1) statement1;   //if expression1 is true, then statement1 is executed.
else if (expression2) statement2;   //if expression2 is true, then statement2 is executed.
else if (expression3) statement3;   //if expression3 is true, then statement3 is executed.
else default statement;

Examples of the three categories are shown below.

//no **else** statement
if $(x_1$ & $x_2) z_1 = 1;$

//one **else** statement
if (rst_n = = 0)
    ctr = 3'b000;
else ctr = next_count;

//nested **if-else if**
if (opcode = = 00)
    $z_1 = x_1 + x_2;$
else if (opcode = = 01)
    $z_1 = x_1 - x_2;$
else if (opcode = = 10)
    $z_1 = x_1 * x_2;$
else
    $z_1 = x_1 / x_2;$

Example 1.22   Figure 1.107 shows a behavioral design module using conditional statements that utilize one alternative **else** statement to illustrate an application of the four equations shown below. The equations use both the logical operators and the reduction operators.

$$z_1 = x_1 \& x_2$$
$$z_2 = x_2 \mid x_3$$
$$z_3 = x_3 \wedge x_4$$
$$z_4 = (x_1 \& x_4) \mid\mid (x_2 \& x_3)$$

Figure 1.108 shows the test bench that generates all 16 combinations of the four inputs $x_1, x_2, x_3,$ and x_4 and displays the four outputs $z_1, z_2, z_3,$ and z_4 for their respective equations. Figure 1.109 and Figure 1.110 display the corresponding outputs and waveforms, respectively.

```
//conditional statements using if ... else
module cond_stmt (x1, x2, x3, x4, z1, z2, z3, z4);

input x1, x2, x3, x4;
output z1, z2, z3, z4;

reg z1, z2, z3, z4;

always @ (x1 or x2)
begin
   if (x1 & x2)
      z1 = 1'b1;
   else
      z1 = 1'b0;
end

always @ (x2 or x3)
begin
   if (x2 | x3)
      z2 = 1'b1;
   else
      z2 = 1'b0;
end

always @ (x3 or x4)
begin
   if (x3 ^ x4)
      z3 = 1'b1;
   else
      z3 = 1'b0;
end                                    //continued on next page
```

Figure 1.107   Behavioral module using conditional statements.

```verilog
always @ (x1 or x2 or x3 or x4)
begin
   if ((x1 & x4) || (x2 & x3))
      z4 = 1'b1;
   else
      z4 = 1'b0;
end

endmodule
```

Figure 1.107   (Continued)

```verilog
//test bench for conditional statements module
module cond_stmt_tb;

reg x1, x2, x3, x4;
wire z1, z2, z3, z4;

//apply input vectors and display variables
initial
begin: apply_stimulus
   reg [4:0] invect;
   for (invect = 0; invect < 16; invect = invect + 1)
      begin
         {x1, x2, x3, x4} = invect [4:0];
         #10 $display ("{x1 x2 x3 x4} = %b,
                        z1 = %b, z2 = %b, z3 = %b, z4 = %b",
                        {x1, x2, x3, x4}, z1, z2, z3, z4);
   end
end

//instantiate the module into the test bench
cond_stmt inst1 (
   .x1(x1),
   .x2(x2),
   .x3(x3),
   .x4(x4),
   .z1(z1),
   .z2(z2),
   .z3(z3),
   .z4(z4)
   );

endmodule
```

Figure 1.108   Test bench for the module of Figure 1.107.

```
z1 = x1 & x2;
z2 = x2 | x3;
z3 = x3 ^ x4;
z4 = (x1 & x4) || (x2 & x3)

{x1 x2 x3 x4} = 0000,   z1 = 0,   z2 = 0,   z3 = 0,   z4 = 0
{x1 x2 x3 x4} = 0001,   z1 = 0,   z2 = 0,   z3 = 1,   z4 = 0
{x1 x2 x3 x4} = 0010,   z1 = 0,   z2 = 1,   z3 = 1,   z4 = 0
{x1 x2 x3 x4} = 0011,   z1 = 0,   z2 = 1,   z3 = 0,   z4 = 0

{x1 x2 x3 x4} = 0100,   z1 = 0,   z2 = 1,   z3 = 0,   z4 = 0
{x1 x2 x3 x4} = 0101,   z1 = 0,   z2 = 1,   z3 = 1,   z4 = 0
{x1 x2 x3 x4} = 0110,   z1 = 0,   z2 = 1,   z3 = 1,   z4 = 1
{x1 x2 x3 x4} = 0111,   z1 = 0,   z2 = 1,   z3 = 0,   z4 = 1

{x1 x2 x3 x4} = 1000,   z1 = 0,   z2 = 0,   z3 = 0,   z4 = 0
{x1 x2 x3 x4} = 1001,   z1 = 0,   z2 = 0,   z3 = 1,   z4 = 1
{x1 x2 x3 x4} = 1010,   z1 = 0,   z2 = 1,   z3 = 1,   z4 = 0
{x1 x2 x3 x4} = 1011,   z1 = 0,   z2 = 1,   z3 = 0,   z4 = 1

{x1 x2 x3 x4} = 1100,   z1 = 1,   z2 = 1,   z3 = 0,   z4 = 0
{x1 x2 x3 x4} = 1101,   z1 = 1,   z2 = 1,   z3 = 1,   z4 = 1
{x1 x2 x3 x4} = 1110,   z1 = 1,   z2 = 1,   z3 = 1,   z4 = 1
{x1 x2 x3 x4} = 1111,   z1 = 1,   z2 = 1,   z3 = 0,   z4 = 1
```

Figure 1.109   Outputs for the module of Figure 1.107.

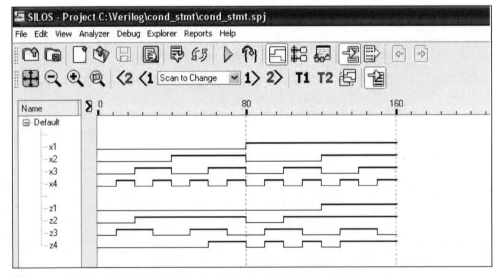

Figure 1.110   Waveforms for the module of Figure 1.107.

1.4.8 Case Statement

When there are many paths from which to chose, nested **if . . . else if** statements can be cumbersome. The **case** statement is an alternative to the **if . . . else if** construct and may simplify the readability of the Verilog code. The **case** statement is a multiple-way conditional branch and contains the keywords **case**, **endcase**, and **default**.

   It executes one of several different procedural statements depending on the comparison of an expression with a case item. The case expression may be an expression or a constant. The case items are evaluated in the order in which they are listed. The expression and the case item are compared bit-by-bit and must match exactly. The statement that is associated with a case item may be a single procedural statement or a block of statements delimited by the keywords **begin** . . . **end**. In the event that there is no match, the **default** statement is executed. The **endcase** keyword terminates the **case** statement. The **case** statement has the following syntax:

```
case (expression)
    case_item1 : procedural_statement1;
    case_item2 : procedural_statement2;
    case_item3 : procedural_statement3;
                 .
                 .
                 .
    case_itemn : procedural_statementn;
    default : default_statement;
endcase
```

Example 1.23 Figure 1.111 shows a behavioral module using the **case** statement to perform the following operations on two 3-bit operands, $a[2:0]$ and $b[2:0]$: AND, OR, XOR, NAND, NOR, XNOR, and NOT. The test bench is shown in Figure 1.112, where operand $a[2:0]$ is assigned the values 000, 010, 100, and 110; operand $b[2:0]$ is assigned the values 001, 011, 101, and 111. The outputs and waveforms are shown in Figure 1.113 and Figure 1.114, respectively.

```
//behavioral using the case statement for logical operations
module case_log_ops (a, b, opcode, rslt);

input [2:0] a, b, opcode;
output [2:0] rslt;

wire [2:0] a, b, opcode;    //inputs are wire (optional)
reg [2:0] rslt;             //outputs are reg

                            //continued on next page
```

Figure 1.111   Behavioral module using the **case** statement for logical operations.

```
parameter    and_op    = 3'b000,    //define operation codes
             or_op     = 3'b001,
             xor_op    = 3'b010,
             nand_op   = 3'b011,
             nor_op    = 3'b100,
             xnor_op   = 3'b101,
             not_op    = 3'b110;

//perform the logical operations
always @(a or b or opcode)
begin
   case (opcode)
      and_op:  rslt = a & b;
      or_op:   rslt = a | b;
      xor_op:  rslt = a ^ b;
      nand_op: rslt = ~(a & b);
      nor_op:  rslt = ~(a | b);
      xnor_op: rslt = ~(a ^ b);
      not_op:  rslt = ~a;
      default: rslt = 3'b000;
   endcase
end
endmodule
```

Figure 1.111   (Continued)

```
//test bench for logical operations using the case statement
module case_log_ops_tb;

reg [2:0] a, b, opcode;
wire [2:0] rslt;

//display variables
initial
$monitor ("a = %b, b = %b, op = %b, rslt = %b",
           a, b, opcode, rslt);

//apply input vectors
initial
begin
//and operation
   #0   a = 3'b000; b = 3'b001; opcode = 3'b000;
   #10  a = 3'b010; b = 3'b011; opcode = 3'b000;
   #10  a = 3'b100; b = 3'b101; opcode = 3'b000;
   #10  a = 3'b110; b = 3'b111; opcode = 3'b000;  //next page
```

Figure 1.112   Test bench for logical operations using the **case** statement.

```
//or operation
   #10   a = 3'b000; b = 3'b001; opcode = 3'b001;
   #10   a = 3'b010; b = 3'b011; opcode = 3'b001;
   #10   a = 3'b100; b = 3'b101; opcode = 3'b001;
   #10   a = 3'b110; b = 3'b111; opcode = 3'b001;

//xor operation
   #10   a = 3'b000; b = 3'b001; opcode = 3'b010;
   #10   a = 3'b010; b = 3'b011; opcode = 3'b010;
   #10   a = 3'b100; b = 3'b101; opcode = 3'b010;
   #10   a = 3'b110; b = 3'b111; opcode = 3'b010;

//nand operation
   #10   a = 3'b000; b = 3'b001; opcode = 3'b011;
   #10   a = 3'b010; b = 3'b011; opcode = 3'b011;
   #10   a = 3'b100; b = 3'b101; opcode = 3'b011;
   #10   a = 3'b110; b = 3'b111; opcode = 3'b011;

//nor operation
   #10   a = 3'b000; b = 3'b001; opcode = 3'b100;
   #10   a = 3'b010; b = 3'b011; opcode = 3'b100;
   #10   a = 3'b100; b = 3'b101; opcode = 3'b100;
   #10   a = 3'b110; b = 3'b111; opcode = 3'b100;

//xnor operation
   #10   a = 3'b000; b = 3'b001; opcode = 3'b101;
   #10   a = 3'b010; b = 3'b011; opcode = 3'b101;
   #10   a = 3'b100; b = 3'b101; opcode = 3'b101;
   #10   a = 3'b110; b = 3'b111; opcode = 3'b101;

//not operation
   #10   a = 3'b000; b = 3'b001; opcode = 3'b110;
   #10   a = 3'b010; b = 3'b011; opcode = 3'b110;
   #10   a = 3'b100; b = 3'b101; opcode = 3'b110;
   #10   a = 3'b110; b = 3'b111; opcode = 3'b110;

   #10   $stop;
end

//instantiate the module into the test bench
case_log_ops inst1 (
   .a(a),
   .b(b),
   .opcode(opcode),
   .rslt(rslt)
   );
endmodule
```

Figure 1.112   (Continued)

```
//and operation
a = 000, b = 001, op = 000, rslt = 000
a = 010, b = 011, op = 000, rslt = 010
a = 100, b = 101, op = 000, rslt = 100
a = 110, b = 111, op = 000, rslt = 110


//or operation
a = 000, b = 001, op = 001, rslt = 001
a = 010, b = 011, op = 001, rslt = 011
a = 100, b = 101, op = 001, rslt = 101
a = 110, b = 111, op = 001, rslt = 111


//xor operation
a = 000, b = 001, op = 010, rslt = 001
a = 010, b = 011, op = 010, rslt = 001
a = 100, b = 101, op = 010, rslt = 001
a = 110, b = 111, op = 010, rslt = 001


//nand operation
a = 000, b = 001, op = 011, rslt = 111
a = 010, b = 011, op = 011, rslt = 101
a = 100, b = 101, op = 011, rslt = 011
a = 110, b = 111, op = 011, rslt = 001


//nor operation
a = 000, b = 001, op = 100, rslt = 110
a = 010, b = 011, op = 100, rslt = 100
a = 100, b = 101, op = 100, rslt = 010
a = 110, b = 111, op = 100, rslt = 000


//xnor operation
a = 000, b = 001, op = 101, rslt = 110
a = 010, b = 011, op = 101, rslt = 110
a = 100, b = 101, op = 101, rslt = 110
a = 110, b = 111, op = 101, rslt = 110


//not operation
a = 000, b = 001, op = 110, rslt = 111
a = 010, b = 011, op = 110, rslt = 101
a = 100, b = 101, op = 110, rslt = 011
a = 110, b = 111, op = 110, rslt = 001
```

Figure 1.113   Outputs for logical operations using the **case** statement.

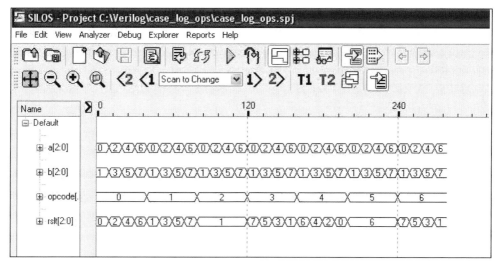

Figure 1.114   Waveforms for logical operations using the **case** statement.

1.4.9 Loop Statements

There are four types of loop statements in Verilog: **for**, **while**, **repeat**, and **forever**. Loop statements must be placed within an **initial** or an **always** block and may contain delay controls. The loop constructs allow for repeated execution of procedural statements within an **initial** or an **always** block.

For loop   The **for** loop contains three parts:

1. An *initial* condition to assign a value to a register control variable. This is executed once at the beginning of the loop to initialize a register variable that controls the loop.

2. A *test* condition to determine when the loop terminates. This is an expression that is executed before the procedural statements of the loop to determine if the loop should execute. The loop is repeated as long as the expression is true. If the expression is false, the loop terminates and the activity flow proceeds to the next statement in the module.

3. An *assignment* to modify the control variable, usually an increment or a decrement. This assignment is executed after each execution of the loop and before the next test to terminate the loop.

The **for** loop is generally used when there is a known beginning and an end to a loop. The **for** loop is similar in function to the **for** loop in the C programming language and has been used in the test bench of several previous examples.

While loop   The **while** loop executes a procedural statement or a block of procedural statements as long as a Boolean expression returns a value of true. When the procedural statements are executed, the Boolean expression is reevaluated. The loop is executed until the expression returns a value of false. If the evaluation of the expression is false, then the **while** loop is terminated and control is passed to the next statement in the module. If the expression is false before the loop is initially entered, then the **while** loop is never executed.

The Boolean expression may contain any of the following types: arithmetic, logical, relational, equality, bitwise, reduction, shift, concatenation, replication, or conditional. If the **while** loop contains multiple procedural statements, then they are contained within the **begin** . . . **end** keywords. The syntax for a **while** statement is as follows:

> **while** (expression)
> > procedural statement or block of procedural statements

Repeat loop   The **repeat** loop executes a procedural statement or a block of procedural statements a specified number of times. The **repeat** construct can contain a constant, an expression, a variable, or a signed value. The syntax for the **repeat** loop is as follows:

> **repeat** (loop count expression)
> > procedural statement or block of procedural statements

If the loop count is **x** (unknown value) or **z** (high impedance), then the loop count is treated as zero. The value of the loop count expression is evaluated once at the beginning of the loop.

Forever loop   The **forever** loop executes the procedural statement continuously until the system tasks **$finish** or **$stop** are encountered. It can also be terminated by the **disable** statement. The **disable** statement is a procedural statement; therefore, it must be used within an **initial** or an **always** block. It is used to prematurely terminate a block of procedural statements or a system task. When a **disable** statement is executed, control is transferred to the statement immediately following the procedural block or task. The **forever** loop is similar to a **while** loop in which the expression always evaluates to true (1). A timing control must be used with the **forever** loop; otherwise, the simulator would execute the procedural statement continuously without advancing the simulation time. The syntax of the **forever** loop is as follows:

> **forever**
> > procedural statement

The **forever** statement is typically used for clock generation as shown in Figure 1.115 together with the system task **$finish**. The variable *clk* will toggle every 10 time units for a period of 20 time units. The length of simulation is 100 time units.

```
//define clock
initial
begin
   clk = 1'b0;
   forever
      #10  clk = ~clk;
end

//define length of simulation
initial
   #100   $finish;
```

Figure 1.115   Clock generation using the **forever** statement.

1.5 Structural Modeling

Structural modeling consists of instantiation of one or more of the following design objects:

- Built-in primitives
- User-defined primitives (UDPs)
- Design modules

Instantiation means to use one or more lower-level modules — including logic primitives — that are interconnected in the construction of a higher-level structural module.   A module can be a logic gate, an adder, a multiplexer, a counter, or some other logical function. The objects that are instantiated are called *instances*. Structural modeling is described by the interconnection of these lower-level logic primitives or modules. The interconnections are made by wires that connect primitive terminals or module ports.

1.5.1 Module Instantiation

Design modules were instantiated into every test bench module in previous examples. The ports of the design module were instantiated by name and connected to the corresponding net names of the test bench. Each named instantiation was of the form

.design_module_port_name (test_bench_module_net_name)

Design module ports can be instantiated by name explicitly or by position. Instantiation by position is not recommended when a large number of ports are involved. Instantiation by name precludes the possibility of making errors in the instantiation process. Modules cannot be nested, but they can be instantiated into other modules.

Structural modeling is analogous to placing the instances on a logic diagram and then connecting them by wires. When instantiating built-in primitives, an instance name is optional; however, when instantiating a module, an instance name must be used. Instances that are instantiated into a structural module are connected by nets of type **wire**.

A structural module may contain behavioral statements (**always**), continuous assignment statements (**assign**), built-in primitives (**and**, **or**, **nand**, **nor**, etc.), UDPs (*mux4*, *half_adder*, *adder4*, etc.), design modules, or any combination of these objects. Design modules can be instantiated into a higher-level structural module in order to achieve a hierarchical design.

Each module in Verilog is either a top-level (higher-level) module or an instantiated module. There is only one top-level module and it is not instantiated anywhere else in the design project. Instantiated primitives or modules, however, can be instantiated many times into a top-level module and each instance of a module is unique and has a unique instance name.

1.5.2 Ports

Ports provide a means for the module to communicate with its external environment. Ports, also referred to as terminals, can be declared as **input**, **output**, or **inout**. A port is a net by default; however, it can be declared explicitly as a net. A module contains an optional list of ports, as shown below for a full adder.

module full_adder (a, b, cin, sum, cout);

Ports *a*, *b*, and *cin* are input ports; ports *sum* and *cout* are output ports. The test bench for the full adder contains no ports as shown below because it does not communicate with the external environment.

module full_adder_tb;

As mentioned previously, there are two methods of associating ports in the module being instantiated and the module doing the instantiation: instantiation by position and instantiation by name (the preferred method). The two methods cannot be mixed. Instantiation by position must have the ports in the module instantiation listed in the same order as in the module definition. Instantiation by name does not require the ports to be listed in the same order.

Input ports must always be of type net (**wire**) internally except for test benches; externally, input ports can be **reg** or **wire**. The input port names can be different, but the net (**wire**) names connecting the input ports must be the same. Output ports can be of type **reg** or **wire** internally; externally, output ports must always be connected to a **wire**.

When making intermodule port connections, it is permissible to connect ports of different widths. Port width matching occurs by right justification or truncation.

1.5.3 Design Examples

Examples will now be presented that illustrate the structural modeling technique. These examples include: converting from a 4-bit binary number to the excess-3 code, implementing a logic equation, the design of a majority circuit, a 3-bit comparator, and a nonlinear-select multiplexer. Each example will be completely designed in detail and will include appropriate theory where applicable.

Example 1.24   This example converts a 4-bit binary number to a 5-bit excess-3 code. The excess-3 code is obtained by adding three to the binary number and contains a fifth bit, the carry-out bit cy, which is set to a value of 1 for binary numbers equal to or greater than 13. Table 1.2 lists the binary numbers and the corresponding excess-3 numbers.

Table 1.2 Binary-to-Excess-3 Conversion

Binary				Excess 3				
x_1	x_2	x_3	x_4	cy	z_1	z_2	z_3	z_4
8	4	2	1	16	8	4	2	1
0	0	0	0	0	0	0	1	1
0	0	0	1	0	0	1	0	0
0	0	1	0	0	0	1	0	1
0	0	1	1	0	0	1	1	0
0	1	0	0	0	0	1	1	1
0	1	0	1	0	1	0	0	0
0	1	1	0	0	1	0	0	1
0	1	1	1	0	1	0	1	0
1	0	0	0	0	1	0	1	1
1	0	0	1	0	1	1	0	0
1	0	1	0	0	1	1	0	1
1	0	1	1	0	1	1	1	0
1	1	0	0	0	1	1	1	1
1	1	0	1	1	0	0	0	0
1	1	1	0	1	0	0	0	1
1	1	1	1	1	0	0	1	0

Figure 1.116 shows the Karnaugh maps used for the code conversion example. The coordinates of the Karnaugh maps correspond to the binary code; the map entries

in the minterm locations correspond to the excess-3 code for that particular bit. The equations for each of the five maps are shown in Equation 1.11.

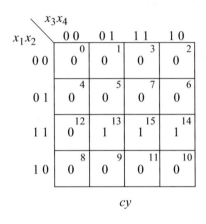

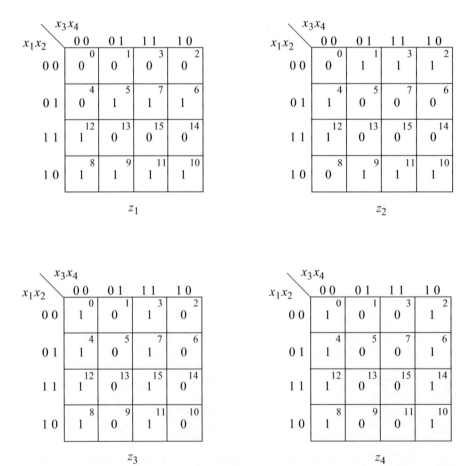

Figure 1.116   Karnaugh maps for the binary-to-excess-3 code conversion.

$$Cy = x_1 x_2 x_3 + x_1 x_2 x_4$$
$$= x_1 x_2 (x_3 + x_4)$$

$$z_1 = x_1' x_2 x_3 + x_1' x_2 x_4 + x_1 x_3' x_4' + x_1 x_2'$$

$$z_2 = x_2' x_3 + x_2' x_4 + x_2 x_3' x_4'$$

$$z_3 = x_3' x_4' + x_3 x_4$$
$$= (x_3 \oplus x_4)'$$

$$z_4 = x_4' \tag{1.11}$$

Figure 1.117 contains the structural design module for the binary-to-excess-3 code conversion. The module utilizes the continuous assignment statement of the dataflow construct to implement the AND and OR functions for the carry-out cy. It also uses built-in primitives for the implementation of the outputs z_1, z_2, z_3, and z_4. Figure 1.118 shows the test bench module. The outputs and waveforms are shown in Figure 1.119 and Figure 1.120, respectively.

```
//structural binary to excess-3 code conversion
module bin_excess3_struc (x1, x2, x3, x4, z1, z2, z3, z4, cy);

input x1, x2, x3, x4;
output cy, z1, z2, z3, z4;

wire net1, net2, net3, net4, net5, net6, net7;

//generate carry-out cy
assign cy = (x1 & x2 & x3) | (x1 & x2 & x4);

//generate output z1
and     inst1 (net1, x1, ~x2),
        inst2 (net2, x1, ~x3, ~x4),
        inst3 (net3, ~x1, x2, x4),
        inst4 (net4, ~x1, x2, x3);
or      inst5 (z1, net1, net2, net3, net4);

//generate output z2
and     inst6 (net5, ~x2, x3),
        inst7 (net6, ~x2, x4),
        inst8 (net7, x2, ~x3, ~x4);
or      inst9 (z2, net5, net6, net7);
                                    //continued on next page
```

Figure 1.117   Design module to convert from binary to excess-3.

```
//generate output z3
xnor   inst10 (z3, x3, x4);

//generate output z4
buf    inst11 (z4, ~x4);

endmodule
```

Figure 1.117   (Continued)

```
//test bench for binary to excess-3
module bin_excess3_struc_tb;

reg x1, x2, x3, x4;
wire z1, z2, z3, z4, cy;

//apply stimulus
initial
begin: apply_stimulus
   reg [4:0] invect;
   for (invect=0; invect<16; invect=invect+1)
      begin
         {x1, x2, x3, x4} = invect [4:0];
         #10   $display ("x1x2x3x4} = %b, cout = %b,
                        {z1z2z3z4} = %b",
                     {x1, x2, x3, x4}, cy, {z1, z2, z3, z4});
      end
end

//instantiate the module into the test bench
bin_excess3_struc inst1 (
   .x1(x1),
   .x2(x2),
   .x3(x3),
   .x4(x4),
   .cy(cy),
   .z1(z1),
   .z2(z2),
   .z3(z3),
   .z4(z4)
   );

endmodule
```

Figure 1.118   Test bench for the binary-to-excess-3 module.

```
x1x2x3x4} = 0000, cout = 0, {z1z2z3z4} = 0011
x1x2x3x4} = 0001, cout = 0, {z1z2z3z4} = 0100
x1x2x3x4} = 0010, cout = 0, {z1z2z3z4} = 0101
x1x2x3x4} = 0011, cout = 0, {z1z2z3z4} = 0110

x1x2x3x4} = 0100, cout = 0, {z1z2z3z4} = 0111
x1x2x3x4} = 0101, cout = 0, {z1z2z3z4} = 1000
x1x2x3x4} = 0110, cout = 0, {z1z2z3z4} = 1001
x1x2x3x4} = 0111, cout = 0, {z1z2z3z4} = 1010

x1x2x3x4} = 1000, cout = 0, {z1z2z3z4} = 1011
x1x2x3x4} = 1001, cout = 0, {z1z2z3z4} = 1100
x1x2x3x4} = 1010, cout = 0, {z1z2z3z4} = 1101
x1x2x3x4} = 1011, cout = 0, {z1z2z3z4} = 1110

x1x2x3x4} = 1100, cout = 0, {z1z2z3z4} = 1111
x1x2x3x4} = 1101, cout = 1, {z1z2z3z4} = 0000
x1x2x3x4} = 1110, cout = 1, {z1z2z3z4} = 0001
x1x2x3x4} = 1111, cout = 1, {z1z2z3z4} = 0010
```

Figure 1.119   Outputs for the binary-to-excess-3 module.

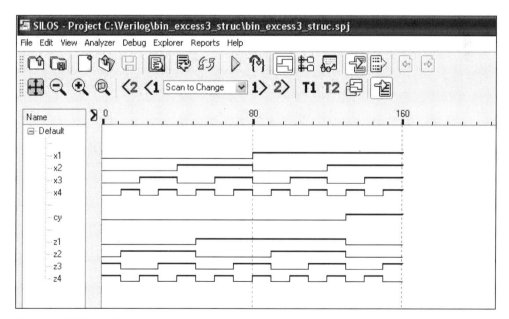

Figure 1.120   Waveforms for the binary-to-excess-3 module.

Example 1.25 A logic circuit will be designed using combinational logic gates to implement the two Karnaugh maps shown in Figure 1.121. The equations obtained from the maps are shown in Equation 1.12. The logic diagram is shown in Figure 1.122. Then the circuit will be designed using structural modeling.

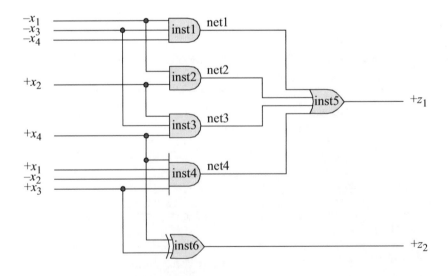

Figure 1.121 shows Karnaugh maps.

For z_1:

x_1x_2 \ x_3x_4	0 0	0 1	1 1	1 0
0 0	1 (0)	0 (1)	0 (3)	0 (2)
0 1	1 (4)	1 (5)	1 (7)	1 (6)
1 1	0 (12)	1 (13)	0 (15)	0 (14)
1 0	0 (8)	0 (9)	1 (11)	0 (10)

For z_2:

x_1x_2 \ x_3x_4	0 0	0 1	1 1	1 0
0 0	0 (0)	1 (1)	0 (3)	1 (2)
0 1	0 (4)	1 (5)	0 (7)	1 (6)
1 1	0 (12)	1 (13)	0 (15)	1 (14)
1 0	0 (8)	1 (9)	0 (11)	1 (10)

Figure 1.121   Karnaugh maps for Example 1.25.

$$z_1 = x_1'x_3'x_4' + x_1'x_2 + x_2x_3'x_4 + x_1x_2'x_3x_4$$

$$z_2 = x_3 \oplus x_4 \tag{1.12}$$

Figure 1.122   Logic diagram for Example 1.25.

Figure 1.123 and Figure 1.124 show the dataflow modules for the 3-input AND gate and the exclusive-OR gate, respectively. The other gates are designed in a similar manner. Figure 1.125 shows the structural design module for the logic diagram of Figure 1.122 using the AND, OR, and exclusive-OR gates that were designed using dataflow modeling and instantiated into the design module. Figure 1.126 shows the test bench. The outputs and waveforms are shown in Figure 1.127 and Figure 1.128, respectively.

```
//and3 dataflow
module and3_df (x1, x2, x3, z1);

//list inputs and output
input x1, x2, x3;
output z1;

//define signals as wire for dataflow
wire x1, x2, x3;
wire z1;

//continuous assign for dataflow
assign z1 = x1 & x2 & x3;

endmodule
```

Figure 1.123 Dataflow module for a 3-input AND gate.

```
//dataflow xor2_df
module xor2_df (x1, x2, z1);

//list inputs and outputs
input x1, x2;
output z1;

//define signals as wire for dataflow
wire x1, x2;
wire z1;

//continuous assignment for dataflow
assign z1 = x1 ^ x2;

endmodule
```

Figure 1.124 Dataflow module for an exclusive-OR gate.

```
//structural logic equation as a sum of products
module log_eqtn_sop (x1, x2, x3, x4, z1, z2);

input x1, x2, x3, x4;
output z1, z2;

//define internal nets
wire net1, net2, net3, net4;

//instantiate the logic gates for z1
and3_df inst1 (
    .x1(~x1),
    .x2(~x3),
    .x3(~x4),
    .z1(net1)
    );

and2_df inst2 (
    .x1(~x1),
    .x2(x2),
    .z1(net2)
    );

and3_df inst3 (
    .x1(x2),
    .x2(~x3),
    .x3(x4),
    .z1(net3)
    );

and4_df inst4 (
    .x1(x4),
    .x2(x1),
    .x3(~x2),
    .x4(x3),
    .z1(net4)
    );

or4_df inst5 (
    .x1(net1),
    .x2(net2),
    .x3(net3),
    .x4(net4),
    .z1(z1)
    );

                              //continued on next page
```

Figure 1.125   Design module for the logic diagram of Figure 1.122.

```
//instantiate the logic gates for z2
xor2_df inst6 (
   .x1(x4),
   .x2(x3),
   .z1(z2)
   );

endmodule
```

Figure 1.125   (Continued)

```
//test bench for logic equation as a sum of products
module log_eqtn_sop_tb;


reg x1, x2, x3, x4;
wire z1, z2;


//apply input vectors and display variables
initial
begin: apply_stimulus
   reg [4:0] invect;
   for (invect = 0; invect < 16; invect = invect + 1)
      begin
         {x1, x2, x3, x4} = invect [4:0];
         #10 $display ("x1 x2 x3 x4 = %b, z1 = %b, z2 = %b",
                        {x1, x2, x3, x4}, z1, z2);
      end
end


//instantiate the module into the test bench
log_eqtn_sop inst1 (
   .x1(x1),
   .x2(x2),
   .x3(x3),
   .x4(x4),
   .z1(z1),
   .z2(z2)
   );

endmodule
```

Figure 1.126   Test bench for the logic diagram of Figure 1.122.

```
x1 x2 x3 x4 = 0000, z1 = 1, z2 = 0
x1 x2 x3 x4 = 0001, z1 = 0, z2 = 1
x1 x2 x3 x4 = 0010, z1 = 0, z2 = 1
x1 x2 x3 x4 = 0011, z1 = 0, z2 = 0

x1 x2 x3 x4 = 0100, z1 = 1, z2 = 0
x1 x2 x3 x4 = 0101, z1 = 1, z2 = 1
x1 x2 x3 x4 = 0110, z1 = 1, z2 = 1
x1 x2 x3 x4 = 0111, z1 = 1, z2 = 0

x1 x2 x3 x4 = 1000, z1 = 0, z2 = 0
x1 x2 x3 x4 = 1001, z1 = 0, z2 = 1
x1 x2 x3 x4 = 1010, z1 = 0, z2 = 1
x1 x2 x3 x4 = 1011, z1 = 1, z2 = 0

x1 x2 x3 x4 = 1100, z1 = 0, z2 = 0
x1 x2 x3 x4 = 1101, z1 = 1, z2 = 1
x1 x2 x3 x4 = 1110, z1 = 0, z2 = 1
x1 x2 x3 x4 = 1111, z1 = 0, z2 = 0
```

Figure 1.127   Outputs for the logic diagram of Figure 1.122.

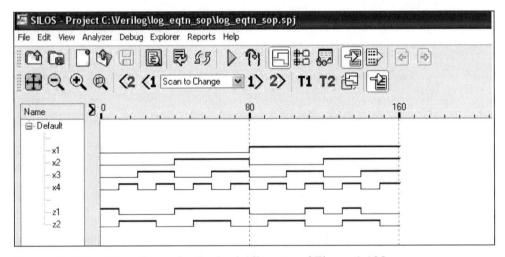

Figure 1.128   Waveforms for the logic diagram of Figure 1.122.

Example 1.26   This example illustrates the design of a 5-input majority circuit using dataflow modules that are instantiated into a structural module. The dataflow modules consist of nine 3-input AND gates and one 9-input OR gate. The output of a majority circuit is a logic 1 if the majority of the inputs is a logic 1; otherwise, the output is a logic 0. Therefore, a majority circuit must have an odd number of inputs in order to have a majority of the inputs at logic 1 level, as shown in Table 1.3 for a 5-input majority circuit. By analyzing Table 1.3 or by plotting it on a modified 5-variable Karnaugh map, as shown in Figure 1.129, Equation 1.13 can be realized which has the fewest number of terms. The equation can then be implemented with nine 3-input AND gates and one 9-input OR gate.

Table 1.3 Truth Table for a 5-Input Majority Circuit

Inputs					Output
x_1	x_2	x_3	x_4	x_5	z_1
0	0	1	1	1	1
0	1	0	1	1	1
0	1	1	0	1	1
0	1	1	1	0	1
0	1	1	1	1	1
1	0	0	1	1	1
1	0	1	0	1	1
1	0	1	1	0	1
1	0	1	1	1	1
1	1	0	0	1	1
1	1	0	1	0	1
1	1	0	1	1	1
1	1	1	0	0	1
1	1	1	0	1	1
1	1	1	1	0	1
1	1	1	1	1	1

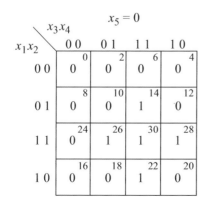
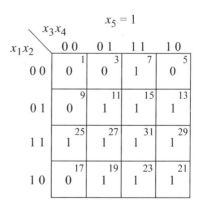

Figure 1.129   Karnaugh map for a 5-variable majority circuit.

$$z_1 = x_3 x_4 x_5 + x_2 x_3 x_5 + x_1 x_3 x_5 + x_2 x_4 x_5 + x_1 x_4 x_5$$

$$+ x_1 x_2 x_5 + x_1 x_2 x_4 + x_2 x_3 x_4 + x_1 x_3 x_4 \qquad (1.13)$$

The design module is shown in Figure 1.130, which instantiates nine 3-input dataflow AND gates, *and3_df*, and one 9-input dataflow OR gate, *or9_df*. The test bench module is shown in Figure 1.131. The outputs and waveforms are shown in Figure 1.132 and Figure 1.133, respectively.

```
//structural 5-input majority circuit
module majority5_struc (x1, x2, x3, x4, x5, z1);

input x1, x2, x3, x4, x5;
output z1;

//define internal nets
wire net1, net2, net3, net4, net5, net6, net7, net8, net9;

//instantiate the logic gates
and3_df inst1 (
    .x1(x3),
    .x2(x4),
    .x3(x5),
    .z1(net1)
    );

and3_df inst2 (
    .x1(x2),
    .x2(x3),
    .x3(x5),
    .z1(net2)
    );

and3_df inst3 (
    .x1(x1),
    .x2(x3),
    .x3(x5),
    .z1(net3)
    );

and3_df inst4 (
    .x1(x2),
    .x2(x4),
    .x3(x5),
    .z1(net4)
    );                          //continued on next page
```

Figure 1.130   Design module for a 5-input majority circuit.

```
and3_df inst5 (
   .x1(x1),
   .x2(x4),
   .x3(x5),
   .z1(net5)
   );

and3_df inst6 (
   .x1(x1),
   .x2(x2),
   .x3(x5),
   .z1(net6)
   );

and3_df inst7 (
   .x1(x1),
   .x2(x2),
   .x3(x4),
   .z1(net7)
   );

and3_df inst8 (
   .x1(x2),
   .x2(x3),
   .x3(x4),
   .z1(net8)
   );

and3_df inst9 (
   .x1(x1),
   .x2(x3),
   .x3(x4),
   .z1(net9)
   );

or9_df inst10 (
   .x1(net1),
   .x2(net2),
   .x3(net3),
   .x4(net4),
   .x5(net5),
   .x6(net6),
   .x7(net7),
   .x8(net8),
   .x9(net9),
   .z1(z1)
   );
endmodule
```

Figure 1.130   (Continued)

```
//test bench for 5-input majority circuit
module majority5_struc_tb;

reg x1, x2, x3, x4, x5;
wire z1;

//apply input vectors
initial
begin: apply_stimulus
   reg [5:0] invect;
   for (invect = 0; invect < 32; invect = invect + 1)
      begin
         {x1, x2, x3, x4, x5} = invect [5:0];
         #10 $display ("x1x2x3x4x5 = %b, z1 = %b",
                       {x1, x2, x3, x4, x5}, z1);
      end
end

//instantiate the module into the test bench
majority5_struc inst1 (
   .x1(x1),
   .x2(x2),
   .x3(x3),
   .x4(x4),
   .x5(x5),
   .z1(z1)
   );
endmodule
```

Figure 1.131   Test bench for the 5-input majority circuit.

```
x1x2x3x4x5 = 00000, z1 = 0        x1x2x3x4x5 = 10000, z1 = 0
x1x2x3x4x5 = 00001, z1 = 0        x1x2x3x4x5 = 10001, z1 = 0
x1x2x3x4x5 = 00010, z1 = 0        x1x2x3x4x5 = 10010, z1 = 0
x1x2x3x4x5 = 00011, z1 = 0        x1x2x3x4x5 = 10011, z1 = 1
x1x2x3x4x5 = 00100, z1 = 0        x1x2x3x4x5 = 10100, z1 = 0
x1x2x3x4x5 = 00101, z1 = 0        x1x2x3x4x5 = 10101, z1 = 1
x1x2x3x4x5 = 00110, z1 = 0        x1x2x3x4x5 = 10110, z1 = 1
x1x2x3x4x5 = 00111, z1 = 1        x1x2x3x4x5 = 10111, z1 = 1
x1x2x3x4x5 = 01000, z1 = 0        x1x2x3x4x5 = 11000, z1 = 0
x1x2x3x4x5 = 01001, z1 = 0        x1x2x3x4x5 = 11001, z1 = 1
x1x2x3x4x5 = 01010, z1 = 0        x1x2x3x4x5 = 11010, z1 = 1
x1x2x3x4x5 = 01011, z1 = 1        x1x2x3x4x5 = 11011, z1 = 1
x1x2x3x4x5 = 01100, z1 = 0        x1x2x3x4x5 = 11100, z1 = 1
x1x2x3x4x5 = 01101, z1 = 1        x1x2x3x4x5 = 11101, z1 = 1
x1x2x3x4x5 = 01110, z1 = 1        x1x2x3x4x5 = 11110, z1 = 1
x1x2x3x4x5 = 01111, z1 = 1        x1x2x3x4x5 = 11111, z1 = 1
```

Figure 1.132   Outputs for the 5-input majority circuit.

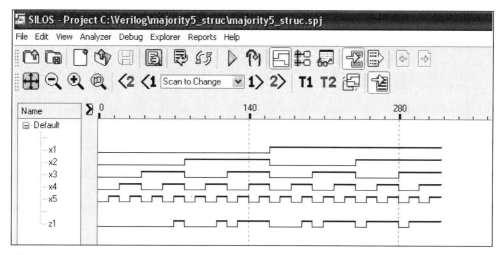

Figure 1.133   Waveforms for the 5-input majority circuit.

Example 1.27   This example converts a 4-bit Gray code to the corresponding 4-bit binary code. The Gray code is a nonweighted code that has the characteristic in which only one bit changes between adjacent code words. The Gray code belongs to a class of cyclic codes called *reflective codes*.

The general algorithm to convert from Gray code to binary code is shown in Equation 1.14, where n is the number of bits. The specific equations to convert a 4-bit Gray code segment to a 4-bit binary number are shown in Equation 1.15. The structural design will instantiate the exclusive-OR dataflow module, *xor2_df*, as shown in Figure 1.134. The structural design module, the test bench module, and the outputs are shown in Figure 1.135, Figure 1.136, and Figure 1.137.

$$b_{n-1} = g_{n-1}$$
$$b_i = b_{i+1} \oplus g_i \qquad\qquad (1.14)$$

$$b_3 = g_3$$
$$b_2 = b_3 \oplus g_2$$
$$b_1 = b_2 \oplus g_1$$
$$b_0 = b_1 \oplus g_0 \qquad\qquad (1.15)$$

```
//dataflow xor2_df
module xor2_df (x1, x2, z1);

//list inputs and outputs
input x1, x2;
output z1;

//define signals as wire for dataflow
wire x1, x2;
wire z1;

//continuous assignment for dataflow
assign z1 = x1 ^ x2;

endmodule
```

Figure 1.134   Dataflow module for a 2-input exclusive-OR circuit.

```
//structural gray-to-binary conversion
module gray_bin_struc (g3, g2, g1, g0, b3, b2, b1, b0);

input g3, g2, g1, g0;
output b3, b2, b1, b0;

assign b3 = g3;

xor2_df inst1 (         //instantiate the xor gates
    .x1(b3),
    .x2(g2),
    .z1(b2)
    );

xor2_df inst2 (
    .x1(b2),
    .x2(g1),
    .z1(b1)
    );

xor2_df inst3 (
    .x1(b1),
    .x2(g0),
    .z1(b0)
    );

endmodule
```

Figure 1.135   Design module for the Gray-to-binary conversion.

```
//test bench for gray-to-binary conversion
module gray_bin_struc_tb;

reg g3, g2, g1, g0;
wire b3, b2, b1, b0;


//apply input vectors
initial
begin: apply_stimulus
   reg[4:0] invect;
   for (invect = 0; invect < 16; invect = invect + 1)
      begin
         {g3, g2, g1, g0} = invect [4:0];
         #10 $display ("{g3g2g1g0} = %b, {b3b2b1b0} = %b",
                        {g3, g2, g1, g0}, {b3, b2, b1, b0});
      end
end

//instantiate the module into the test bench
gray_bin_struc inst1 (
   .g3(g3),
   .g2(g2),
   .g1(g1),
   .g0(g0),
   .b3(b3),
   .b2(b2),
   .b1(b1),
   .b0(b0)
   );

endmodule
```

Figure 1.136   Test bench for the Gray-to-binary conversion.

```
{g3g2g1g0} = 0000, {b3b2b1b0} = 0000
{g3g2g1g0} = 0001, {b3b2b1b0} = 0001
{g3g2g1g0} = 0010, {b3b2b1b0} = 0011
{g3g2g1g0} = 0011, {b3b2b1b0} = 0010

{g3g2g1g0} = 0100, {b3b2b1b0} = 0111
{g3g2g1g0} = 0101, {b3b2b1b0} = 0110
{g3g2g1g0} = 0110, {b3b2b1b0} = 0100
{g3g2g1g0} = 0111, {b3b2b1b0} = 0101
                              //continued on next page
```

Figure 1.137   Outputs for the Gray-to-binary conversion.

```
{g3g2g1g0} = 1000, {b3b2b1b0} = 1111
{g3g2g1g0} = 1001, {b3b2b1b0} = 1110
{g3g2g1g0} = 1010, {b3b2b1b0} = 1100
{g3g2g1g0} = 1011, {b3b2b1b0} = 1101

{g3g2g1g0} = 1100, {b3b2b1b0} = 1000
{g3g2g1g0} = 1101, {b3b2b1b0} = 1001
{g3g2g1g0} = 1110, {b3b2b1b0} = 1011
{g3g2g1g0} = 1111, {b3b2b1b0} = 1010
```

Figure 1.137   (Continued)

Example 1.28   As a final example for structural modeling, a *nonlinear-select multiplexer* will be used to implement the Karnaugh map shown in Figure 1.138, where y is a map-entered variable. A nonlinear-select multiplexer represents a smaller multiplexer than a linear-select multiplexer and has fewer data inputs. It can be effectively utilized to implement the same function with a corresponding reduction in machine cost.

If a multiplexer has unused data inputs — corresponding to unused states in the input map — then these unused inputs can be connected to logically adjacent multiplexer inputs. The resulting linked set of inputs can be addressed by a common select variable.

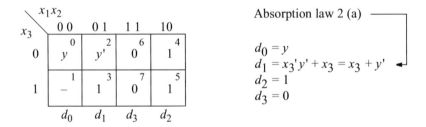

Figure 1.138   Karnaugh map to be implemented with a nonlinear-select multiplexer.

Figure 1.139 illustrates the nonlinear-select multiplexer that will be utilized in the design of the Karnaugh map of Figure 1.138. The select inputs are s_0 and s_1, where s_0 is the low-order select input that is selected by variable x_2. The data inputs are d_0, d_1, d_2, and d_3, where d_0 is the low-order data input. The outputs of the multiplexer are identical to the values in the corresponding minterm locations of the Karnaugh map. For example, in the Karnaugh map, if $x_1 x_2 = 00$, then minterm locations 0 and 1 contain the variable y, corresponding to input d_0. In the logic diagram of Figure 1.139, if $x_1 x_2 = 00$, then input d_0 is selected and output z_1 contains the value of y. All of the multiplexer outputs can be verified in a similar manner.

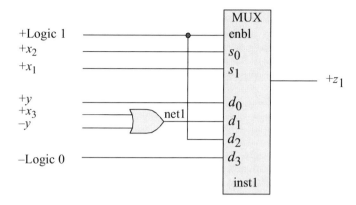

Figure 1.139   Logic diagram for Example 1.28 using a nonlinear-select multiplexer.

Figure 1.140 contains the structural module, which instantiates a dataflow 4:1 multiplexer *mux4_df* and utilizes the continuous assignment statement — both used in the design of the logic diagram of Figure 1.139. The test bench is shown in Figure 1.141. The outputs and waveforms are shown in Figure 1.142 and Figure 1.143, respectively.

```
//structural nonlinear-select multiplexer
module mux_nonlinear5 (x1, x2, x3, y, z1);

//define inputs and output
input x1, x2, x3, y;
output z1;

//define internal net
wire net1;

//use the continuous assign statement to design the or gate
assign net1 = (x3 | ~y);

//instantiate the 4:1 multiplexer
mux4_df inst1 (
    .s({x1, x2}),                   //({s1, s0})
    .d({1'b0, 1'b1, net1, y}),      //({d3, d2, d1, d0})
    .enbl(1'b1),
    .z1(z1)
    );
endmodule
```

Figure 1.140   Structural module for the nonlinear-select multiplexer.

```
//test bench for the nonlinear-select multiplexer circuit
module mux_nonlinear5_tb;

reg x1, x2, x3, y;

wire z1;

//apply input vectors and display variables
initial
begin: apply_stimulus
   reg [4:0] invect;
   for (invect = 0; invect < 16; invect = invect + 1)
      begin
         {x1, x2, x3, y} = invect [4:0];
         #10 $display ("x1 x2 x3 = %b, y = %b, z1 = %b",
                   {x1, x2, x3}, y, z1);
      end
end

//instantiate the module into the test bench
mux_nonlinear5 inst1 (
   .x1(x1),
   .x2(x2),
   .x3(x3),
   .y(y),
   .z1(z1)
   );

endmodule
```

Figure 1.141   Test bench for the nonlinear-select multiplexer.

$d_0 = x_1'x_2' = y$ $d_1 = x_1'x_2 = x_3 + y'$ $d_2 = x_1x_2' = 1$ $d_3 = x_1x_2 = 0$	$d_0 = x_1'x_2' = y$ $d_1 = x_1'x_2 = x_3 + y'$ $d_2 = x_1x_2' = 1$ $d_3 = x_1x_2 = 0$
x1 x2 x3 = 000, y = 0, z1 = 0	x1 x2 x3 = 100, y = 0, z1 = 1
x1 x2 x3 = 000, y = 1, z1 = 1	x1 x2 x3 = 100, y = 1, z1 = 1
x1 x2 x3 = 001, y = 0, z1 = 0	x1 x2 x3 = 101, y = 0, z1 = 1
x1 x2 x3 = 001, y = 1, z1 = 1	x1 x2 x3 = 101, y = 1, z1 = 1
x1 x2 x3 = 010, y = 0, z1 = 1	x1 x2 x3 = 110, y = 0, z1 = 0
x1 x2 x3 = 010, y = 1, z1 = 0	x1 x2 x3 = 110, y = 1, z1 = 0
x1 x2 x3 = 011, y = 0, z1 = 1	x1 x2 x3 = 111, y = 0, z1 = 0
x1 x2 x3 = 011, y = 1, z1 = 1	x1 x2 x3 = 111, y = 1, z1 = 0

Figure 1.142   Outputs for the nonlinear-select multiplexer.

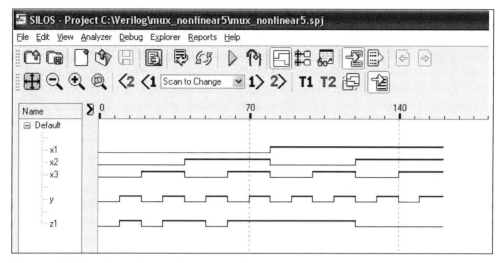

Figure 1.143   Waveforms for the nonlinear-select multiplexer.

1.6 Problems

1.1   Given the Karnaugh map shown below, obtain the logic diagram using NOR gates in a product-of-sums implementation. Then obtain the design module using built-in primitives, the test bench module, the outputs, and the waveforms.

	x_3x_4			
x_1x_2	0 0	0 1	1 1	1 0
0 0	1 [0]	0 [1]	1 [3]	1 [2]
0 1	0 [4]	0 [5]	0 [7]	0 [6]
1 1	1 [12]	1 [13]	1 [15]	0 [14]
1 0	1 [8]	0 [9]	1 [11]	0 [10]

$$z_1$$

1.2   Design a circuit using built-in primitive **nand** gates that satisfies the following specifications: $3 < N \le 8$ and $10 \le N < 15$. Obtain the Karnaugh map, the equation, and the logic diagram using NAND gates. Then obtain the design module using built-in primitives, the test bench module, the outputs, and the waveforms.

1.3   Given the equation shown below, obtain the Karnaugh map. Then obtain the sum-of-products equation from the Karnaugh map and generate the logic diagram using AND and OR gates, where output z_1 is asserted at a high logic level. Then obtain the design module using built-in primitives, the test bench module, the outputs, and the waveforms.

$$z_1(x_1, x_2, x_3, x_4) = \Sigma_m(1, 4, 7, 9, 11, 13) + \Sigma_d(5, 14, 15)$$

1.4   Repeat Problem 1.3, but generate the circuit as a product-of-sums design.

1.5   Design the logic for a 4-bit odd parity generator, then use built-in primitives to implement the design in Verilog. The output will be a logical 1 if there is an even number of 1s on the input; otherwise, the output will be a logical 0. Obtain the design module, the test bench module, the outputs, and the waveforms.

1.6   Design a circuit using dataflow modeling that satisfies the following specifications: $4 \leq N < 9$ and $10 < N < 14$. Derive the Karnaugh map and obtain the equation in a sum-of-products expression. Then design the logic diagram using NOR gates, where output z_1 is asserted at a high logic level. Generate the design module using the continuous assignment statement for NOR gates, the test bench module, the outputs, and the waveforms.

1.7   Given the Karnaugh map shown below, obtain the equation for output z_1 in a sum-of-products notation and the corresponding logic diagram using AND and OR gates. Then use dataflow modeling for the design module and generate a test bench. Obtain the outputs and the waveforms.

x_1x_2 \ x_3x_4	0 0	0 1	1 1	1 0
0 0	1 ^{0}	1 ^{1}	0 ^{3}	0 ^{2}
0 1	0 ^{4}	1 ^{5}	1 ^{7}	1 ^{6}
1 1	0 ^{12}	0 $^{13.}$	1 ^{15}	1 ^{14}
1 0	1 ^{8}	0 ^{9}	0 ^{11}	0 ^{10}

z_1

1.8   Given the Karnaugh map shown in Problem 1.7, obtain the equation for output z_1 in a product-of-sums notation and the corresponding logic diagram using NAND gates. Output z_1 is to be asserted at a high logic level. Then use dataflow modeling for the design module and generate a test bench. Obtain the outputs and the waveforms This problem is similar to Problem 1.7, but uses only NAND gates and generates the equation as a product-of-sums. Therefore, the outputs and waveforms should be identical to those of Problem 1.7.

1.9   Design a circuit using dataflow modeling to detect overflow in a fixed-point binary adder. The augend (a) and addend (b) are both four bits in the 2s complement number representation. Overflow occurs when the result of an arithmetic operation exceeds the word size of the machine. Overflow can be detected by the equation shown below, where $n-1$ is the high-order bit and s is the sum. Obtain the design module, the test bench module for eight variations of the two operands, the outputs, and the waveforms.

$$\text{Overflow} = (a_{n-1} \cdot b_{n-1} \cdot s_{n-1}') + (a_{n-1}' \cdot b_{n-1}' \cdot s_{n-1})$$

1.10   Use the three logical operators of AND (&&), OR ($||$), and negation ($!$) to implement the logical operations shown below. Obtain the dataflow design module, the test bench module for eight variations of the three 4-bit operands a, b, and c, the outputs, and the waveforms.

$z_1 = (a~\&\&~b)~\&\&~c$
$z_2 = (a~||~b)~\&\&~c$
$z_3 = (a~\&\&~c)~||~b$
$z_4 = !~(a~||~c)$

1.11   Use the three bitwise operators of AND (&), OR ($|$), and exclusive-OR ($\wedge$) to implement the logical operations shown below. Obtain the dataflow design module, the test bench module for eight variations of the three 4-bit operands a, b, and c, the outputs, and the waveforms

$z_1 = (a~\&~b)~|~c$
$z_2 = (a~\wedge~b)~\&~c$
$z_3 = (a~|~c)~\wedge~b$

1.12   Design a 4-bit odd parity generator using the exclusive-OR and exclusive-NOR operators. There are four data inputs and one output that is a logic 1 when the number of 1s in the input vector is even. Use dataflow modeling for the design module. Generate a test bench for all combinations of the inputs. Obtain the outputs and waveforms.

1.13   Design a 4-bit adder using dataflow modeling whose inputs are augend a and addend b with a carry-in cin. The outputs are sum and carry-out $cout$. Generate the design module, the test bench module containing eight variations of the augend and addend with specific values for carry-in. Obtain the outputs and the waveforms.

1.14   Use the bitwise AND and OR operators on the 8-bit operands a and b, then use the logical left shift and logical right shift operators on the results. Perform the operations shown below. Obtain the design module and the test bench module that applies eight sets of vectors to the two operands. Show the outputs and the waveforms.

$$z_1 = a \ \& \ b;$$
$$z_1\_sl = z_1 << 3; \qquad //\text{shift left } z_1 \text{ 3 bit positions}$$
$$z_1\_sr = z_1 >> 2; \qquad //\text{shift right } z_1 \text{ 2 bit positions}$$

$$z_2 = a \ | \ b;$$
$$z_2\_sl = z_2 << 4; \qquad //\text{shift left } z_2 \text{ 4 bit positions}$$
$$z_2\_sr = z_2 >> 3; \qquad //\text{shift right } z_2 \text{ 3 bit positions}$$

1.15   Use dataflow modeling to design a circuit that generates an output z_1 whenever a 4-bit unsigned binary number meets the following requirements, where $N > 0$: N is an odd number or N is evenly divisible by four. Obtain the design module using a sum-of-products expression, the test bench module for all sixteen combinations of the four bits, the outputs, and the waveforms.

1.16   Use behavioral modeling to design a circuit that counts the number of 1s in a 16-bit register x. A register is a logic macro device that stores data. The data is retained until new data is stored. Registers are implemented by means of storage elements. Registers are presented in this problem to illustrate one use of the **while** loop and the conditional statement **if**. Assume that the register contains the following contents: $f63f_{16}$. Display the individual counts — 1 through 12 — then the final count of the number of 1s. No test bench is required for this problem. The two counts are displayed in the design module by the **$display** system task.

1.17   Design a behavioral module that performs addition, shifting, and checks for overflow on two 8-bit operands a and b. Shift the sum left three bit positions and right two positions. Display the sum before and after the shift operations. Obtain the design module, the test bench module for eight variations of the augend and addend. Display the resulting outputs and the waveforms.

1.18   Implement the Karnaugh map shown below using a 4:1 multiplexer, where $x_1 x_2$ represent the select inputs $s_1 s_0$ and $x_3 x_4$ represent the data inputs $d_0 d_1 d_2 d_3$. The variable x_5 is a map-entered variable. Obtain the design module and the test bench module for all combinations of the three variables $x_3 x_4 x_5$ for the four combinations of the select inputs. Obtain the outputs and the waveforms.

$x_3 x_4$

$x_1 x_2$	0 0	0 1	1 1	1 0	
$s_0 = 0\,0$	x_5 _(0)_	0 _(1)_	0 _(3)_	x_5 _(2)_	$z_1 = x_4' x_5$
$s_1 = 0\,1$	1 _(4)_	1 _(5)_	0 _(7)_	0 _(6)_	$z_1 = x_3'$
$s_3 = 1\,1$	0 _(12)_	1 _(13)_	1 _(15)_	x_5' _(14)_	$z_1 = x_4 + x_3 x_5'$
$s_2 = 1\,0$	0 _(8)_	1 _(9)_	0 _(11)_	1 _(10)_	$z_1 = x_3 \wedge x_4$

1.19   Use behavioral modeling to design a full adder. A full adder has three scalar inputs a, b, and cin; there are two scalar outputs sum and $cout$. Obtain the design module, the test bench module for all combinations of the inputs, the outputs, and the waveforms. The equations for sum and $cout$ are shown below.

$$sum = a'b'cin + a'bcin' + ab'cin' + abcin$$

$$= a \oplus b \oplus cin$$

$$cout = a'bcin + ab'cin + abcin' + abcin$$

$$= ab + a\,cin + b\,cin$$

1.20   Use behavioral modeling to convert a 4-bit binary code word $binary[3:0]$ to the corresponding 4-bit Gray code word $gray[3:0]$. The general algorithm to convert an n-bit binary number to a Gray code number is shown below, where $n = 4$ for this problem. Obtain the design module, the test bench module for all 16 combinations of the four binary bits, the outputs, and the waveforms.

$$g_{n-1} = b_{n-1}$$
$$g_i = b_{i+1} \oplus b_i$$

1.21   Design a behavioral module using conditional statements to implement the equation shown below. The design module will use an intrastatement delay of five time units. Obtain the test bench, the outputs for all 16 combinations of the inputs, and the waveforms.

$$z_1 = x_1 x_2 + x_3 x_4$$

1.22   Design a 4:1 multiplexer using a combination of behavioral modeling and dataflow modeling. The multiplexer has four data inputs, which are specified as a 4-bit vector $d[3:0]$, two select inputs, specified as a 2-bit vector $s[1:0]$, one scalar enable input *enbl*, and one scalar output z_1. Obtain the design module and the test bench module containing eight combinations of the data inputs. Obtain the outputs and the waveforms.

1.23   Design a behavioral module that converts a 4-bit binary code to the excess-3 code. The excess-3 code is obtained by adding three to the binary code. Obtain the design module and the test bench module for all combinations of the four bits. Obtain the outputs and the waveforms.

1.24   Write a behavioral module to determine the decimal value of the following binary number: 0111_1110. No test bench is required.

1.25   Use behavioral modeling with the **case** statement to design a 6-function logic unit for the following six functions: add, subtract, multiply, AND, OR, and exclusive-OR. The operands are 4-bit vectors: *a[3:0]* and *b[3:0]*. Obtain the design module and the test bench module for four variations of the operands for each function. Obtain the outputs and waveforms.

1.26   Given the Karnaugh map shown below, obtain the equation for output z_1 in a sum-of-products form with the fewest number of terms. Then design the behavioral module and the test bench module for all combinations of the five variables $x_1, x_2, x_3, x_4,$ and x_5. Obtain the outputs and the waveforms.

x_1 \ $x_2 x_3$	0 0	0 1	1 1	1 0
0	$x_4 x_5' + x_4 x_5$ 0	0 1	1 3	1 2
1	$x_4 x_5 + x_4' + x_5'$ 4	1 5	0 7	0 6

z_1

1.27   Design a structural 4-bit, *[3:0]*, binary-to-excess-3 code converter by instantiating behavioral full adders into the design. The excess-3 code will contain five bits to include the carry out of the high-order bit position of *adder[3]*. For example, `binary = 1111`, `excess3 = 10010`. Obtain the design module and the test bench module for all 16 combinations of the binary inputs. Obtain the outputs and the waveforms.

1.28   Use structural modeling to design a 3-bit comparator for the following operands: *a[2:0]* and *b[2:0]*. Obtain the design module, the test bench module, the outputs, and the waveforms. Test the module with inputs that demonstrate the relative magnitude of the two operands for the categories shown below.

$$a[2:0] = b[2:0] \text{ and } a[2:0] > b[2:0]$$

1.29   Design a logic circuit that will generate a high logic level on output z_1 if a 4-bit binary number $x[3:0]$ has a value less than or equal to five or greater than nine. Obtain the structural design module and the test bench module for all 16 combinations of the inputs. Obtain the outputs and the waveforms.

1.30   Given the logic diagram shown below, obtain the Karnaugh map for output z_1. Then design the structural module that represents the logic diagram and the test bench module utilizing all eight combinations of the three inputs. Obtain the outputs and the waveforms.

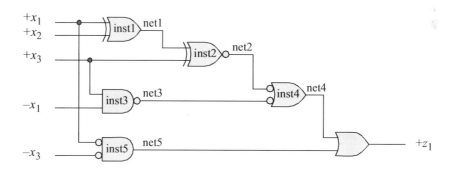

1.31   Given the logic diagram shown below, obtain the minimum product-of-sums equation, then design a structural module using NOR gates to implement the equation. Then design the test bench using all 16 combinations of the four input variables. Verify the results by displaying the outputs and the waveforms.

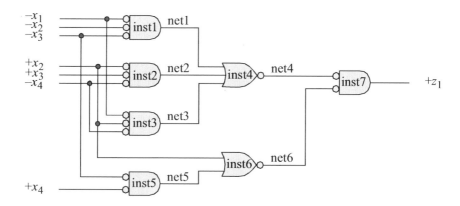

1.32   Given the Karnaugh map shown below, obtain the function z_1 in a minimum product-of-sums expression. Then implement the design as a structural module using NOR gates and design a test bench module that incorporates all 16 combinations of the four variables. Obtain the outputs and the waveforms.

x_3x_4

x_1x_2	0 0	0 1	1 1	1 0
0 0	0 (0)	0 (1)	1 (3)	0 (2)
0 1	0 (4)	1 (5)	0 (7)	0 (6)
1 1	0 (12)	1 (13)	1 (15)	0 (14)
1 0	1 (8)	1 (9)	1 (11)	1 (10)

z_1

2.1   *Synchronous Registers*
2.2   *Synchronous Counters*
2.3   *Moore Machines*
2.4   *Mealy Machines*
2.5   *Moore–Mealy Equivalence*
2.6   *Output Glitches*
2.7   *Problems*

2

Synthesis of Synchronous Sequential Machines 1 Using Verilog HDL

A *synchronous sequential machine* consists of storage elements, usually flip-flops, and δ next-state combinational logic that connects to the flip-flop data inputs. The machine may also contain combinational logic for the λ output function. In some cases, the output logic may require one or more storage elements, depending on the assertion and deassertion of the output signals. The number of flip-flops is determined by the number of states required by the machine. The combinational logic is derived directly from either the state diagram or from the state table.

This chapter implements synchronous sequential machine designs using Verilog HDL. The designs will be accomplished by utilizing built-in primitives, dataflow modeling, behavioral modeling, structural modeling, or a combination of these modeling techniques. Different types of synchronous registers will be designed. These include: parallel-in, serial-out registers; serial-in, parallel-out registers; and serial-in, serial-out registers. Also included will be high-speed combinational shifting techniques. These include: shift left logical, shift left algebraic, shift right logical, and shift right algebraic.

Different types of counters of various moduli are also designed in this chapter. These include: a modulo-8 counter, a modulo-10 counter, and a Johnson counter. Also included will be a binary-to-Gray code converter. Different versions of Moore and Mealy synchronous sequential machines will also be designed using Verilog together with different techniques to eliminate output glitches.

2.1 Synchronous Registers

The next state of a synchronous (clocked) register is usually a direct result of the input vector, whose binary variables connect to the flip-flop data inputs, either directly or indirectly through δ next-state logic. Most registers are used primarily for temporary storage of binary data and do not modify the data internally; that is, the state of the register is unchanged until the next active clock transition. An n-bit register requires n storage elements, either SR latches, D flip-flops, or JK flip-flops. There are 2^n different states in an n-bit register, where each n-tuple corresponds to a unique state of the register.

The simplest and most prevalent register is the *parallel-in, parallel-out* (PIPO) register used for temporary storage of binary data. There is a one-to-one correspondence between the input alphabet X, the state alphabet Y, and the output alphabet Z. The values of the present inputs $X_{i(t)}$ become the next state $Y_{k(t+1)}$ of the register at the next active clock transition. The synthesis procedure is not required for this type of register; therefore, the design will not be implemented in Verilog.

2.1.1 Parallel-In, Serial-Out Registers

A *parallel-in, serial-out* (PISO) register accepts binary input data in parallel and generates binary output data in serial form. The binary data can be shifted either left or right under control of a shift direction signal and a clock pulse, which is applied to all flip-flops simultaneously. The register shifts left or right one bit position at each active clock transition. Bits shifted out one end of the register are lost unless the register is cyclic, in which case, the bits are shifted (or rotated) into the other end.

If the PISO register is a right-shift register, then two conditions determine the value of the bits shifted into the vacated positions on the left. If the binary data represents an unsigned number, then 0s are shifted into the vacated positions. If the binary data represents a signed number — with the high-order bit specified as the sign of the number, where a 0 bit represents a positive number and a 1 bit represents a negative number —then the sign bit extends right one bit position for each active clock transition.

The state diagram for a parallel-in, serial-out shift right register for unsigned binary data is shown in Figure 2.1. Zeroes are shifted in to the vacated positions on the left. Upon completion of the load cycle, $y_i = x_i$. During the shift sequence, $y_i = y_{i-1}$ or 0, depending on the shift count. After four shift cycles, the state of the register is $y_1 y_2 y_3 y_4 = 0000$, and the process repeats with a new input vector X_i.

Examination of the state diagram reveals that each clock pulse shifts in 0s from the left and replaces the present state of a flip-flop with the present state of the flip-flop to its immediate left. Thus, the output of flip-flop y_i connects to the data input of flip-flop y_{i+1}. The Verilog design of the register will be implemented first using behavioral modeling then using structural modeling.

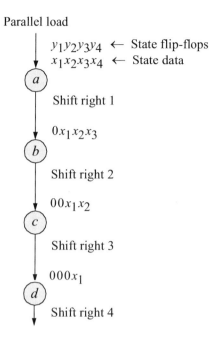

Parallel load

$y_1 y_2 y_3 y_4$ ← State flip-flops
$x_1 x_2 x_3 x_4$ ← State data

a

Shift right 1

$0 x_1 x_2 x_3$

b

Shift right 2

$0 0 x_1 x_2$

c

Shift right 3

$0 0 0 x_1$

d

Shift right 4

Figure 2.1   State diagram for a parallel-in, serial-out shift right register for unsigned binary data.

Example 2.1   This example designs a PISO register using behavioral modeling. Figure 2.2 illustrates the behavioral design module for the PISO shift register and Figure 2.3 shows the test bench with parallel binary input data of 1111. The outputs and waveforms are shown in Figure 2.4 and Figure 2.5, respectively.

```
//behavioral 4-bit shift right piso shift register
//for unsigned binary data
module shift_reg_piso4a (rst_n, clk, load, x, y, z);

input rst_n, clk, load;
input [1:4] x;
output [1:4] y;
output z;

reg [1:4] y;

assign z = y[4];

                              //continued on next page
```

Figure 2.2   Behavioral module for the PISO shift right register.

```verilog
always @ (negedge rst_n or posedge clk)
begin
   if (rst_n == 1'b0)
      y = 4'b0000;

   else
      y[1] <= ((load && x[1]) || (~load && 1'b0));
      y[2] <= ((load && x[2]) || (~load && y[1]));
      y[3] <= ((load && x[3]) || (~load && y[2]));
      y[4] <= ((load && x[4]) || (~load && y[3]));
end

endmodule
```

Figure 2.**2**   (Continued)

```verilog
//test bench for the 4-bit piso shift register
module shift_reg_piso4a_tb;

reg rst_n, clk, load;
reg [1:4] x;
wire [1:4] y;
wire z;

//define clock
initial
begin
   clk = 1'b0;
   forever
      #10 clk = ~clk;
end

//display variables
initial
$monitor ("x=%b, y=%b, z=%b", x, y, z);

//apply inputs
initial
begin
   #0     rst_n = 1'b0;  load = 1'b0;   x = 4'b0000;
   #3     rst_n = 1'b1;
   #2     x = 4'b1111;
   #3     load = 1'b1;
   #7     load = 1'b0;

   #100   $stop;
end                               //continued on next page
```

Figure 2.3   Test bench for the PISO shift right register.

```
//instantiate the module into the test bench
shift_reg_piso4a inst1 (
   .rst_n(rst_n),
   .clk(clk),
   .load(load),
   .x(x),
   .y(y),
   .z(z)
   );
endmodule
```

Figure 2.3 (Continued)

```
x=0000, y=0000, z=0
x=1111, y=0000, z=0
x=1111, y=1111, z=1
x=1111, y=0111, z=1
x=1111, y=0011, z=1
x=1111, y=0001, z=1
x=1111, y=0000, z=0
```

Figure 2.4 Outputs for the PISO shift right register.

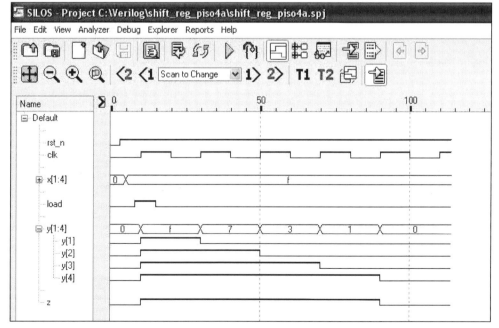

Figure 2.5 Waveforms for the PISO shift right register.

Example 2.2   This example designs a PISO register using structural modeling. The logic diagram for the PISO shift register using D flip-flops is shown in Figure 2.6 with implied reset inputs. Each stage (or cell) of the register is loaded with external data or receives data from the previous stage with the assertion of a clock signal. A *Load* signal is asserted to load the register with the binary input vector prior to the shift operation — this occurs at the first active *Clock* signal. Then the *Load* signal is deasserted and the shift operation begins.

The structural design module is shown in Figure 2.7 and the test bench module is shown in Figure 2.8. The same binary input vector that was used in the behavioral module is used in the structural module for comparison. The outputs and waveforms are shown in Figure 2.9 and Figure 2.10, respectively.

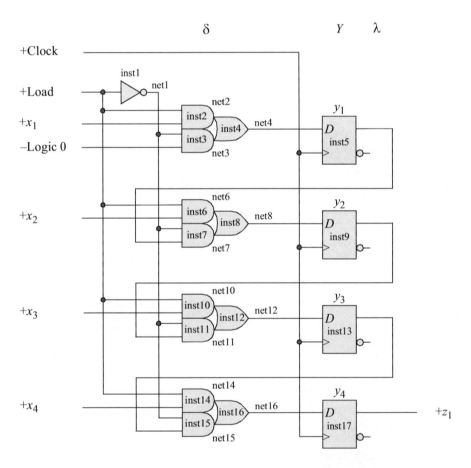

Figure 2.6   Implementation of a parallel-in, serial-out register using D flip-flops.

One application of a PISO register is to convert data from a parallel bus into serial data for use by a single-track device, such as a disk drive. The serialization process occurs during a write operation.

```
//structural 4-bit parallel-in, serial-out shift register
module piso4_struc (rst_n, clk, load, x, y, z1);


//define inputs and outputs
input rst_n, clk, load;
input [1:4] x;
output [1:4] y;
output z1;


//define internal nets
wire net1, net2, net3, net4, net6, net7, net8;
wire net10, net11, net12, net14, net15, net16;


//instantiate the load/shift logic
not inst1 (net1, load);


//instantiate the logic for flip-flop y[1]
and2_df inst2 (
   .x1(load),
   .x2(x[1]),
   .z1(net2)
   );

and2_df inst3 (
   .x1(net2),
   .x2(1'b0),
   .z1(net3)
   );

or2_df inst4 (
   .x1(net2),
   .x2(net3),
   .z1(net4)
   );

d_ff_bhinst5 (
   .rst_n(rst_n),
   .clk(clk),
   .d(net4),
   .q(y[1])
   );

                              //continued on next page
```

Figure 2.7   Structural design module for the PISO shift right register.

```
//instantiate the logic for flip-flop y[2]
and2_df inst6 (
   .x1(load),
   .x2(x[2]),
   .z1(net6)
   );

and2_df inst7 (
   .x1(net1),
   .x2(y[1]),
   .z1(net7)
   );

or2_df inst8 (
   .x1(net6),
   .x2(net7),
   .z1(net8)
   );

d_ff_bh inst9 (
   .rst_n(rst_n),
   .clk(clk),
   .d(net8),
   .q(y[2])
   );

//instantiate the logic for flip-flop y[3]
and2_df inst10 (
   .x1(load),
   .x2(x[3]),
   .z1(net10)
   );

and2_df inst11 (
   .x1(net1),
   .x2(y[2]),
   .z1(net11)
   );

                        //continued on next page
```

Figure 2.7   (Continued)

```
or2_df inst12 (
   .x1(net10),
   .x2(net11),
   .z1(net12)
   );

d_ff_bh inst13 (
   .rst_n(rst_n),
   .clk(clk),
   .d(net12),
   .q(y[3])
   );

//instantiate the logic for flip-flop y[4]
and2_df inst14 (
   .x1(load),
   .x2(x[4]),
   .z1(net14)
   );

and2_df inst15 (
   .x1(net1),
   .x2(y[3]),
   .z1(net15)
   );

or2_df inst16 (
   .x1(net14),
   .x2(net15),
   .z1(net16)
   );

d_ff_bh inst17 (
   .rst_n(rst_n),
   .clk(clk),
   .d(net16),
   .q(y[4])
   );

assign z1 = y[4];

endmodule
```

Figure 2.7   (Continued)

```
//test bench for parallel-in, serial-out shift register
module piso4_struc_tb;

//inputs are reg for test benches, outputs are wire
reg rst_n, clk, load;
reg [1:4] x;

wire [1:4] y;
wire z1;

//display variables
initial
$monitor ("y = %b, z = %b", y, z1);

//generate clock
initial
begin
   clk = 1'b0;
   forever
      #10clk = ~clk;
end

//apply inputs
initial
begin
   #0      rst_n = 1'b0;  load = 1'b0;   x = 4'b0000;
   #3      rst_n = 1'b1;

   #2      x = 4'b1111;

   #3      load = 1'b1;
   #7      load = 1'b0;

   #100    $stop;
end

//instantiate the module into the test bench
piso4_struc inst1 (
   .rst_n(rst_n),
   .clk(clk),
   .load(load),
   .x(x),
   .y(y),
   .z1(z1)
   );

endmodule
```

Figure 2.8   Test bench module for the PISO shift right register.

```
y = 0000,  z = 0
y = 1111,  z = 1
y = 0111,  z = 1
y = 0011,  z = 1
y = 0001,  z = 1
y = 0000,  z = 0
```

Figure 2.9   Outputs for the PISO shift right register.

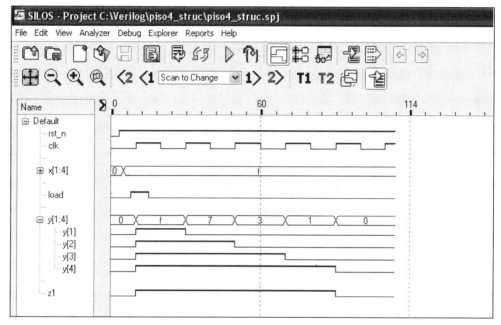

Figure 2.10   Waveforms for the PISO shift right register.

2.1.2 Serial-In, Parallel-Out Registers

The *serial-in, parallel-out* (SIPO) register is another typical synchronous iterative net-work containing p identical cells. Data enters the register from the left and shifts se-rially to the right through all p stages, one bit position per clock pulse. After p shifts, the register is fully loaded and the bits are transferred in parallel to the destination. A typical application is to change serial data read from a disk drive to parallel data to be sent to a processor.

An example of a 4-bit SIPO register is shown in the state diagram of Figure 2.11, in which four bits of serial data, x_1, x_2, x_3, and x_4 are shifted into a register from the left, where x_4 is the first bit entered. The initial state of the register is either unknown or reset to $y_1 y_2 y_3 y_4 = 0000$. During the shift sequence, $y_1 = x_i$ and $y_i = y_{i-1}$. After four shift cycles, the state of the register is $y_1 y_2 y_3 y_4 = x_1 x_2 x_3 x_4$ and the 4-bit word is transferred in parallel to the destination.

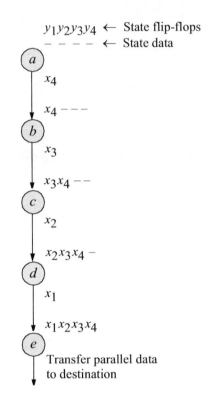

Figure 2.11   State diagram for a serial-in, parallel-out register.

Example 2.3   This example designs a SIPO register using behavioral modeling. Figure 2.12 shows a behavioral design module that implements the serial-in, parallel-out register of Figure 2.11. The test bench is shown in Figure 2.13 and provides an input sequence to illustrate more than four serial bits for the input data. The outputs and waveforms are shown in Figure 2.14 and Figure 2.15, respectively.

```
//behavioral 4-bit serial-in, parallel-out shift register
module shift_reg_sipo4_bh (rst_n, clk, x, y);

input rst_n, clk, x;
output [1:4] y;

reg [1:4] y;                            //continued on next page
```

Figure 2.12   Behavioral design module for the 4-bit SIPO register.

```verilog
always @ (rst_n)
begin
   if (rst_n == 0)
      y = 4'b0000;
end


always @ (posedge clk)
begin
   y[1] <= x;
   y[2] <= y[1];
   y[3] <= y[2];
   y[4] <= y[3];
end

endmodule
```

Figure 2.12   (Continued)

```verilog
//test bench for 4-bit serial-in parallel-out shift register
module shift_reg_sipo4_bh_tb;

reg rst_n, clk, x;
wire [1:4] y;

//define clock
initial
begin
   clk = 1'b0;
   forever
      #10 clk = ~clk;
end

//display variables
initial
$monitor ("ser_in = %b, shift_reg = %b", x, y);

//apply inputs
initial
begin
   #0      rst_n = 1'b0;
           x = 1'b0;

   #5      rst_n = 1'b1;
           x = 1'b1;
                                    //continued on next page
```

Figure 2.13   Test bench module for the 4-bit SIPO register.

```
   #10    x = 1'b1;
   #10    x = 1'b0;
   #10    x = 1'b1;

   #10    x = 1'b0;
   #10    x = 1'b1;
   #10    x = 1'b0;
   #10    x = 1'b1;

   #90    $stop;
end

//instantiate the module into the test bench
shift_reg_sipo4_bh inst1 (
   .rst_n(rst_n),
   .clk(clk),
   .x(x),
   .y(y)
   );

endmodule
```

Figure 2.13   (Continued)

```
ser_in = 0, shift_reg = 0000
ser_in = 1, shift_reg = 0000
ser_in = 1, shift_reg = 1000
ser_in = 0, shift_reg = 1000

ser_in = 0, shift_reg = 0100
ser_in = 1, shift_reg = 0100
ser_in = 0, shift_reg = 0100
ser_in = 0, shift_reg = 0010

ser_in = 1, shift_reg = 0010
ser_in = 0, shift_reg = 0010
ser_in = 0, shift_reg = 0001
ser_in = 1, shift_reg = 0001

ser_in = 1, shift_reg = 1000
ser_in = 1, shift_reg = 1100
ser_in = 1, shift_reg = 1110
ser_in = 1, shift_reg = 1111
```

Figure 2.14   Outputs for the 4-bit SIPO register.

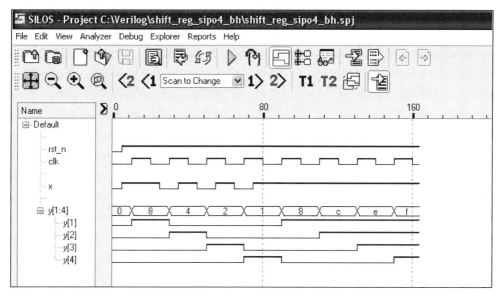

Figure 2.15   Waveforms for the 4-bit SIPO register.

Like the PISO register, the synthesis of a SIPO register is intuitively obvious and can be designed from the state diagram without any intermediate steps. The data input of each flip-flop is connected directly to the output of the preceding flip-flop with the exception of flip-flop y_1, which receives the external serial binary data. A typical application of a serial-in, parallel-out shift register is to deserialize binary data from a single-track peripheral subsystem as illustrated in Figure 2.16. The resulting word of parallel bits is placed on the system data bus.

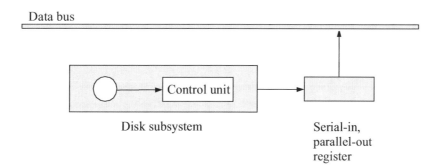

Figure 2.16   A serial-in, parallel-out register to deserialize data from a disk subsystem.

Figure 2.17 shows the implementation of a 4-bit serial-in, parallel-out shift register using JK flip-flops, where y_4 is the low-order flip-flop. D flip-flops or SR latches are equally acceptable storage elements. Each stage of the machine is required to perform only one function: Store the state of the preceding storage element. Data bits at the serial input are changed at the positive clock transition to allow bit x_i to be stable at the JK inputs of flip-flop y_1 before the active negative clock transition.

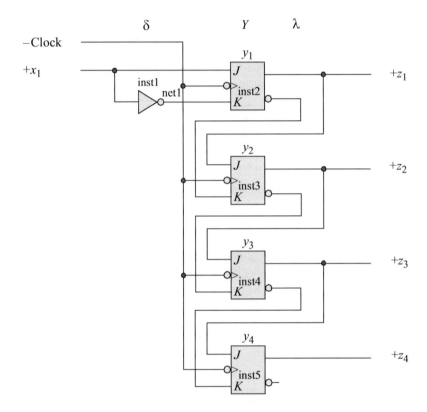

Figure 2.17   Implementation of a 4-bit serial-in, parallel-out shift register using JK flip-flops. The flip-flops have implied active-low *Reset* inputs.

Example 2.4   This example designs a SIPO register using structural modeling with JK flip-flops. The structural design module of the 4-bit serial-in, parallel-out shift register is shown in Figure 2.18 using a **not** gate in the implementation. The **not** gate is an inverting built-in primitive with one scalar input and one or more scalar outputs. The output terminal is listed first when it is instantiated into the module; the input is listed last. A negative-edge triggered JK flip-flop, *jkff-neg-clk*, is also used and is instantiated four times to implement the 4-bit serial-in, parallel-out register. The test bench module is shown in Figure 2.19 and provides an input sequence to illustrate serial bits for the input data. The outputs and waveforms are shown in Figure 2.20 and Figure 2.21, respectively.

```verilog
//structural 4-bit serial-in, parallel-out shift register
module shift_reg_sipo4_struc (rst_n, clk, x1, y);

input rst_n, clk, x1;      //define inputs and outputs
output [1:4] y;

wire net1;                 //define internal nets

//instantiate the logic for flip-flop y[1]
not inst1 (net1, x1);

jkff_neg_clk inst2 (
   .rst_n(rst_n),
   .clk(clk),
   .j(x1),
   .k(net1),
   .q(y[1])
   );

//instantiate the logic for flip-flop y[2]
jkff_neg_clk inst3 (
   .rst_n(rst_n),
   .clk(clk),
   .j(y[1]),
   .k(~y[1]),
   .q(y[2])
   );

//instantiate the logic for flip-flop y[3]
jkff_neg_clk inst4 (
   .rst_n(rst_n),
   .clk(clk),
   .j(y[2]),
   .k(~y[2]),
   .q(y[3])
   );

//instantiate the logic for flip-flop y[4]
jkff_neg_clk inst5 (
   .rst_n(rst_n),
   .clk(clk),
   .j(y[3]),
   .k(~y[3]),
   .q(y[4])
   );
endmodule
```

Figure 2.18   Structural design module for the 4-bit serial-in, parallel-out shift register using *JK* flip-flops

```verilog
//test bench for 4-bit serial_in, parallel-out
//shift register using JK flip-flops
module shift_reg_sipo4_struc_tb;

reg rst_n, clk, x1;
wire [1:4] y;

//define clock
initial
begin
   clk = 1'b0;
   forever
      #10 clk = ~clk;
end

//display variables
initial
$monitor ("ser_in = %b, shift_reg = %b", x1, y);

//apply inputs
initial
begin
   #0      rst_n = 1'b0;
           x1 = 1'b0;

   #5      rst_n = 1'b1;
           x1 = 1'b1;

   #10     x1 = 1'b1;
   #10     x1 = 1'b0;
   #10     x1 = 1'b0;
   #10     x1 = 1'b1;

   #50     $stop;
end

//instantiate the module into the test bench
shift_reg_sipo4_struc inst1 (
   .rst_n(rst_n),
   .clk(clk),
   .x1(x1),
   .y(y)
   );

endmodule
```

Figure 2.19   Test bench module for the 4-bit serial-in, parallel-out shift register using *JK* flip-flops.

```
ser_in = 0, shift_reg = 0000
ser_in = 1, shift_reg = 0000
ser_in = 1, shift_reg = 1000
ser_in = 0, shift_reg = 1000

ser_in = 0, shift_reg = 0100
ser_in = 1, shift_reg = 0100
ser_in = 1, shift_reg = 1010
ser_in = 1, shift_reg = 1101
```

Figure 2.20   Outputs for the 4-bit serial-in, parallel-out shift register using *JK* flip-flops.

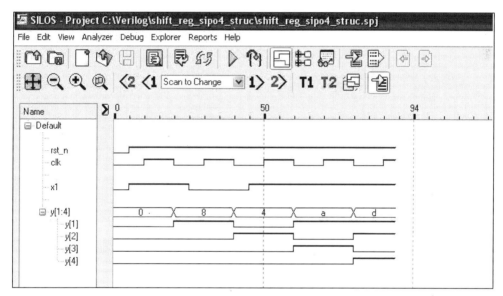

Figure 2.21   Waveforms for the 4-bit serial-in, parallel-out shift register using *JK* flip-flops.

Example 2.5   This example designs a SIPO register using structural modeling with *D* flip-flops. Another useful application of a SIPO register is to generate a sequence of nonoverlapping pulses for system timing. This provides a simple, yet effective state machine, where each pulse represents a different state. A small amount of additional logic is required as shown in Figure 2.22 (a). The flip-flops have an implied active-low *Reset* input. The machine outputs are presented in Figure 2.22 (b). The machine is initially reset to $y_1 y_2 y_3 y_4 = 0000$. Whenever $y_1 y_2 y_3 = 000$, a 1 bit will be shifted into flip-flop y_1 at the next positive clock transition. If either $y_1, y_2,$ or $y_3 = 1$, then a 0 bit will be shifted into flip-flop y_1, and $y_i = y_{i-1}$ at the next positive clock transition. Thus, the required four nonoverlapping pulses are generated.

The structural design module is shown in Figure 2.23 and the test bench module is shown in Figure 2.24. The outputs and waveforms are shown in Figure 2.25 and Figure 2.26, respectively.

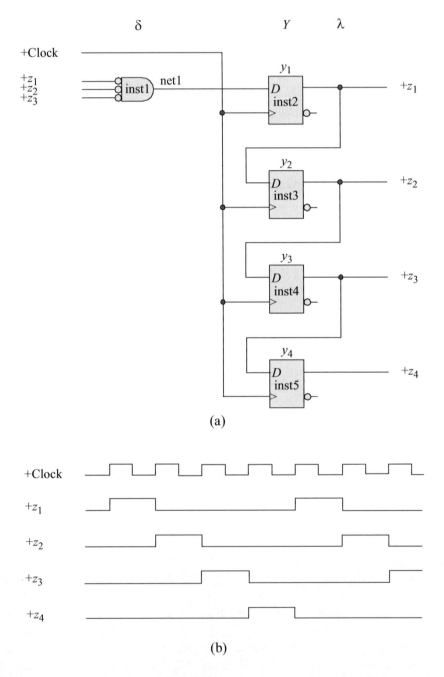

(a)

(b)

Figure 2.22   A serial-in, parallel-out register configured to generate a sequence of nonoverlapping pulses: (a) logic diagram and (b) timing diagram.

```verilog
//structural 4-bit serial-in, parallel-out shift register
//using D flip-flops to generate nonoverlapping pulses

module sipo4_struc (rst_n, clk, y);

input rst_n, clk;       //define inputs and outputs
output [1:4] y;

wire net1;              //define internal nets

//instantiate the logic for flip-flop y[1]
nor3_df inst1 (
   .x1(y[1]),
   .x2(y[2]),
   .x3(y[3]),
   .z1(net1)
   );

d_ff inst2 (
   .rst_n(rst_n),
   .clk(clk),
   .d(net1),
   .q(y[1])
   );

//instantiate the logic for flip-flop y[2]
d_ff inst3 (
   .rst_n(rst_n),
   .clk(clk),
   .d(y[1]),
   .q(y[2])

   );
//instantiate the logic for flip-flop y[3]
d_ff inst4 (
   .rst_n(rst_n),
   .clk(clk),
   .d(y[2]),
   .q(y[3])
   );

//instantiate the logic for flip-flop y[4]
d_ff inst5 (
   .rst_n(rst_n),
   .clk(clk),
   .d(y[3]),
   .q(y[4])
   );
endmodule
```

Figure 2.23   Structural design module to generate a sequence of four nonoverlapping pulses.

```verilog
//test bench for serial-in, parallel-out
//shift register for nonoverlapping pulses

module sipo4_struc_tb;

//inputs are reg for test benches
reg rst_n, clk;

//outputs are wire for test benches
wire [1:4] y;


//display outputs
initial
$monitor ("out = %b", y);


//generate reset
initial
begin
    #0      rst_n = 1'b0;
    #2      rst_n = 1'b1;
end


//generate clock
initial
begin
    clk = 1'b0;
    forever
        #10    clk = ~clk;
end


//determine length of simulation
initial
    #110    $stop;

//instantiate the module into the test bench
sipo4_struc inst1 (
    .rst_n(rst_n),
    .clk(clk),
    .y(y)
    );


endmodule
```

Figure 2.24   Test bench module to generate a sequence of four nonoverlapping pulses.

```
out = 0000
out = 1000
out = 0100
out = 0010
out = 0001
out = 1000
```

Figure 2.25   Outputs to generate a sequence of four nonoverlapping pulses.

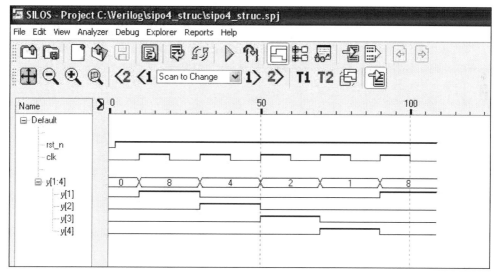

Figure 2.26   Waveforms to generate a sequence of four nonoverlapping pulses.

2.1.3 Serial-In, Serial-Out Registers

The synthesis of a *serial-in, serial-out* (SISO) register is identical to that of a SIPO register, with the exception that only one output is required. The low-order flip-flop provides the single output for the register as shown in the logic diagram of Figure 2.27 for a 4-bit SISO register using *JK* flip-flops with implied reset inputs.

One application of a SISO register is in the design of a queue in which parallel bytes are shifted into a matrix of SISO registers, where each bit of a byte is shifted into a particular column of the matrix. In this application, the SISO registers perform the function of a first-in, first-out (FIFO) queue, which acts as a buffer between a single-track input/output (I/O) device and the system I/O data bus. Information is read from the device into a SIPO register, then into the FIFO, and then transferred to the destination by means of a parallel data bus.

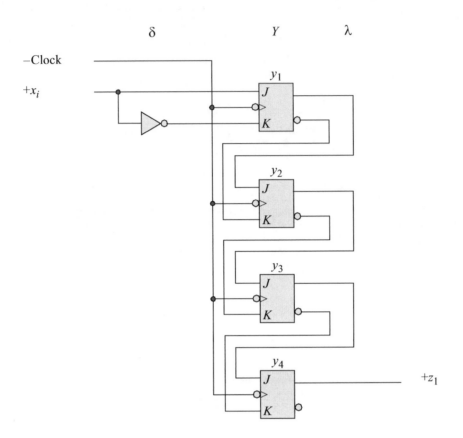

Figure 2.27   Implementation of a 4-bit serial-in, serial-out register using *JK* flip-flops.

Example 2.6   This example designs a SISO register using behavioral modeling. The behavioral design module for the serial-in, serial-out shift register is shown in Figure 2.28, where the statement shown below indicates that *y[1:4]* is assigned the concatenated contents of the current input x_i and the contents of *y[1:3]*. Thus, *y[4]* is shifted out of the register.

$$y <= \{x, y[1:3]\};$$

The test bench is shown in Figure 2.29, in which an input sequence of serial data bits is applied to flip-flop *y[1]*. The outputs and waveforms are shown in Figure 2.30 and Figure 2.31, respectively.

This mode of data transfer between a single-track I/O device and a destination allows the I/O device to be logically removed from the system data bus temporarily without losing any data, because the data is stored in the FIFO queue. In this situation, data continues to be read from the device and is transferred to the FIFO, where the bytes are retained until the device control unit again gains control of the bus. The FIFO prevents data from being lost while the control unit is arbitrating for bus control.

The same implementation of a SISO register matrix can be used as an instruction queue in a CPU instruction pipeline. The CPU prefetches instructions from memory during unused memory cycles and stores the instructions in the FIFO queue. Thus, an instruction stream can be placed in the instruction queue to wait for decoding and execution by the processor. Instruction queueing provides an effective method to increase system throughput.

```verilog
//behavioral 4-bit serial-in, serial-out shift register
module shift_reg_siso4_bh (rst_n, clk, x, y, z1);

input rst_n, clk, x;
output [1:4] y;
output z1;

reg [1:4] y;        //variables are reg in always

assign z1 = y[4];

always @ (negedge rst_n or posedge clk)
begin
   if (rst_n == 1'b0)
      y = 4'b000;
   else
      y = {x, y[1:3]};
end
endmodule
```

Figure 2.28   Behavioral design module for the SISO register of Figure 2.27.

```verilog
//test bench for serial-in, serial-out shift register
module shift_reg_siso4_bh_tb;

reg rst_n, clk, x;
wire [1:4] y;
wire z1;

//define clock
initial
begin
   clk = 1'b0;
   forever
      #10 clk = ~clk;
end                           //continued on next page
```

Figure 2.29   Test bench module for the SISO register.

```
initial      //display variables
$monitor ("ser_in = %b, siso_reg = %b, z = %b", x, y, z1);

initial      //apply inputs
begin
   #0     rst_n = 1'b0;   x = 1'b0;
   #5     rst_n = 1'b1;
   #3     x = 1'b1;
   #17    x = 1'b1;
   #20    x = 1'b0;
   #20    x = 1'b1;
   #20    x = 1'b0;
   #20    x = 1'b1;
   #20    x = 1'b0;
   #20    x = 1'b1;
   #40    $stop;
end

//instantiate the module into the test bench
shift_reg_siso4_bh inst1 (
   .rst_n(rst_n),
   .clk(clk),
   .x(x),
   .y(y),
   .z1(z1)
   );
endmodule
```

Figure 2.29   (Continued)

```
ser_in = 0, siso_reg = 0000, z = 0
ser_in = 1, siso_reg = 0000, z = 0
ser_in = 1, siso_reg = 1000, z = 0
ser_in = 1, siso_reg = 1100, z = 0
ser_in = 0, siso_reg = 1100, z = 0
ser_in = 0, siso_reg = 0110, z = 0
ser_in = 1, siso_reg = 0110, z = 0
ser_in = 1, siso_reg = 1011, z = 1
ser_in = 0, siso_reg = 1011, z = 1
ser_in = 0, siso_reg = 0101, z = 1
ser_in = 1, siso_reg = 0101, z = 1
ser_in = 1, siso_reg = 1010, z = 0
ser_in = 0, siso_reg = 1010, z = 0
ser_in = 0, siso_reg = 0101, z = 1
ser_in = 1, siso_reg = 0101, z = 1
ser_in = 1, siso_reg = 1010, z = 0
ser_in = 1, siso_reg = 1101, z = 1
```

Figure 2.30   Outputs for the SISO register.

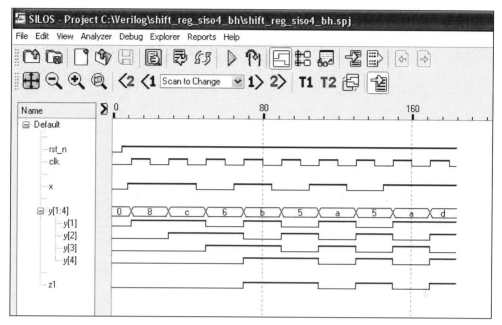

Figure 2.31   Waveforms for the SISO register.

Example 2.7   This example designs a SISO register using structural modeling with *JK* flip-flops.   The same serial-in, serial-out register of Figure 2.27 will now be designed using structural modeling.   The structural design module is shown in Figure 2.32 and the test bench is shown in Figure 2.33.   The outputs and waveforms are shown in Figure 2.34 and Figure 2.35, respectively.

```
//structural 4-bit serial-in, serial-out
//shift register using JK flip-flops
module shift_reg_siso4_jk (rst_n, clk, x, y, z1);

//define inputs and outputs
input rst_n, clk, x;

output [1:4] y;
output z1;

//define internal nets
wire net1;

assign z1 = y[4];
                              //continued on next page
```

Figure 2.32   Structural design module for the SISO register of Figure 2.27.

```verilog
//instantiate the logic for flip-flop y[1]
not inst1 (net1, x);


jkff_neg_clk inst2 (
   .rst_n(rst_n),
   .clk(clk),
   .j(x),
   .k(net1),
   .q(y[1])
   );


//instantiate the logic for flip-flop y[2]
jkff_neg_clk inst3 (
   .rst_n(rst_n),
   .clk(clk),
   .j(y[1]),
   .k(~y[1]),
   .q(y[2])
   );


//instantiate the logic for flip-flop y[3]
jkff_neg_clk inst4 (
   .rst_n(rst_n),
   .clk(clk),
   .j(y[2]),
   .k(~y[2]),
   .q(y[3])
   );


//instantiate the logic for flip-flop y[4]
jkff_neg_clk inst5 (
   .rst_n(rst_n),
   .clk(clk),
   .j(y[3]),
   .k(~y[3]),
   .q(y[4])
   );

endmodule
```

Figure 2.32   (Continued)

```
//test bench for serial-in, serial-out shift register
module shift_reg_siso4_jk_tb;

reg rst_n, clk, x;
wire [1:4] y;
wire z1;

//define clock
initial
begin
   clk = 1'b0;
   forever
      #10 clk = ~clk;
end

//display variables
initial
$monitor ("ser_in = %b, siso_reg = %b, z = %b", x, y, z1);

//apply inputs
initial
begin
   #0     rst_n = 1'b0;   x = 1'b0;
   #5     rst_n = 1'b1;

   #3     x = 1'b1;
   #17    x = 1'b1;
   #20    x = 1'b0;
   #20    x = 1'b1;
   #20    x = 1'b0;
   #20    x = 1'b1;
   #20    x = 1'b0;
   #20    x = 1'b1;

   #50    $stop;
end

//instantiate the module into the test bench
shift_reg_siso4_jk inst1 (
   .rst_n(rst_n),
   .clk(clk),
   .x(x),
   .y(y),
   .z1(z1)
   );

endmodule
```

Figure 2.33   Test bench module for the SISO register of Figure 2.27.

```
ser_in = 0, siso_reg = 0000, z = 0
ser_in = 1, siso_reg = 0000, z = 0
ser_in = 1, siso_reg = 1000, z = 0
ser_in = 1, siso_reg = 1100, z = 0

ser_in = 0, siso_reg = 1100, z = 0
ser_in = 0, siso_reg = 0110, z = 0
ser_in = 1, siso_reg = 0110, z = 0
ser_in = 1, siso_reg = 1011, z = 1

ser_in = 0, siso_reg = 1011, z = 1
ser_in = 0, siso_reg = 0101, z = 1
ser_in = 1, siso_reg = 0101, z = 1
ser_in = 1, siso_reg = 1010, z = 0

ser_in = 0, siso_reg = 1010, z = 0
ser_in = 0, siso_reg = 0101, z = 1
ser_in = 1, siso_reg = 0101, z = 1
ser_in = 1, siso_reg = 1010, z = 0

ser_in = 1, siso_reg = 1101, z = 1
```

Figure 2.34   Outputs for the SISO register of Figure 2.27.

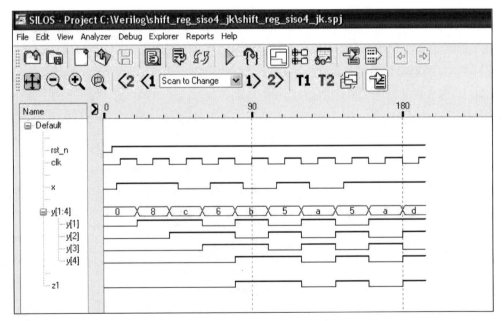

Figure 2.35   Waveforms for the SISO register of Figure 2.27.

2.1.4 Combinational Shifter

A combinational shifter will now be presented. Although not sequential in structure, it is used extensively in high-speed processors, specifically for machines with long word sizes such as, 32- or 64-bit operands. The shifter accomplishes all shift operations, whether left or right, algebraic or logical, by shifting left only. This results in considerable hardware savings, especially for large operands.

There are four basic shift operations: *shift left logical* (SLL), *shift left algebraic* (SLA) for unsigned and signed operands, respectively; *shift right logical* (SRL) and *shift right algebraic* (SRA) for unsigned and signed operands, respectively. The four shift operations are stated below.

Shift left logical (SLL)   The logical shift operations are much simpler to implement than the arithmetic shift operations. For SLL, the high-order bit of the unsigned operand is shifted out of the left end of the shifter for each shift cycle. Zeroes are entered from the right and fill the vacated low-order bit positions.

Shift left algebraic (SLA)   SLA operates on signed operands in 2s complement representation. The numeric part of the operand is shifted left the number of bit positions specified in the shift count field. The sign remains unchanged and does not participate in the shift operation. All remaining bits participate in the left shift operation. The bits are shifted out of the high-order numeric bit position and 0s are shifted in to the vacated register positions on the right. An overflow occurs if a bit shifted out of the high-order numeric position is different than the sign bit.

Shift right logical (SRL)   Any right shift operation can be implemented by shifting left an amount that is the 2s complement of the right shift count. For example, if the right shift count is 011_2 (3_{10}), then the equivalent left shift count is $100 + 1 = 101$, which is the 2s complement of the right shift count. The equivalent left shift operation is implemented in two levels of hardware, as will be explained subsequently.

Shift right algebraic (SRA)   The numeric part of the signed operand is shifted right the number of bits specified by the shift count. The sign of the operand remains unchanged. All numeric bits participate in the right shift. The sign bit propagates right to fill in the vacated high-order numeric bit positions. When the operation is executed by shifting left, it is identical to SRL with the exception that the high-order bits in the second level are set to the value of the sign bit, as will be explained subsequently.

Before presenting the individual structural modules for each of the four shift operations, a behavioral module will be implemented that performs all four shift operations. This method utilizes the **case** statement. Recall that the **case** statement is a multi-way conditional branch that executes one of several different procedural statements depending on the comparison of an expression with a **case** item.

Figure 2.36 shows the behavioral design module for 8-bit operands and Figure 2.37 shows the test bench module, which provides several different operands to be shifted and also provides different shift amounts. The outputs and waveforms are shown in Figure 2.38 and Figure 2.39, respectively.

```
//behavioral logical and algebraic shifter
module comb_shifter (a, shft_code, shft_amt, shft_rslt);

input [7:0] a;
input [1:0] shft_code;
input [3:0] shft_amt;
output [7:0] shft_rslt;

wire [7:0] a;
wire [3:0] shft_amt;

//variables used in always are declared as registers
reg [7:0] reg_a;
reg [7:0] shft_rslt;
reg [15:0] sra_reg;

//define shift codes
parameter    sll = 2'b00,
             sla = 2'b01,
             srl = 2'b10,
             sra = 2'b11;

//perform the shift operations
always @ (a or shft_code)
begin
   case (shft_code)
      sll:
         begin
            reg_a = a << shft_amt;
            shft_rslt = reg_a;
         end

      sla:
         begin
            reg_a = a;
            reg_a = reg_a << shft_amt;
            reg_a[7] = a[7];
            shft_rslt = reg_a;
         end

      srl:
         begin
            reg_a = a >> shft_amt;
            shft_rslt = reg_a;
         end                          //continued on next page
```

Figure 2.36   Behavioral module to implement the four shift operations of SLL, SLA, SRL, and SRA.

```
        sra:
            begin
                sra_reg[15:8] = {8{a[7]}};
                sra_reg[7:0] = a;
                sra_reg = sra_reg >> shft_amt;
                shft_rslt = sra_reg[7:0];
            end
    endcase
end
endmodule
```

Figure 2.36   (Continued)

```
//test bench for logical and algebraic shifter
module comb_shifter_tb;

reg [7:0] a;
reg [1:0] shft_code;
reg [3:0] shft_amt;
wire [7:0] shft_rslt;

initial      //display variables
$monitor ("a=%b, shft_code=%b, shft_amt=%b, shft_rslt=%b",
            a, shft_code, shft_amt, shft_rslt);

initial      //apply input vectors
begin
          //shift left logical
   #0     a = 8'b0000_1111;
          shft_code = 2'b00; shft_amt = 4'b0010;

          //shift left algebraic
   #10    a = 8'b1000_1111;
          shft_code = 2'b01; shft_amt = 4'b0010;

          //shift right logical
   #10    a = 8'b0000_1111;
          shft_code = 2'b10; shft_amt = 4'b0010;

          //shift right algebraic
   #10    a = 8'b1000_1111;
          shft_code = 2'b11; shft_amt = 4'b0010;
-------------------------------------------------------------
                                   //continued on next page
```

Figure 2.37   Test bench module for the four shift operations of SLL, SLA, SRL, and SRA.

```
               //shift left logical
    #10        a = 8'b1111_1111;
               shft_code = 2'b00;shft_amt = 4'b0100;

               //shift left algebraic
    #10        a = 8'b1111_1111;
               shft_code = 2'b01;shft_amt = 4'b0100;

               //shift right logical
    #10        a = 8'b1111_1111;
               shft_code = 2'b10;shft_amt = 4'b0100;

               //shift right algebraic
    #10        a = 8'b1111_1111;
               shft_code = 2'b11;shft_amt = 4'b0100;
-------------------------------------------------------------
               //shift left logical
    #10        a = 8'b1100_0011;
               shft_code = 2'b00;shft_amt = 4'b0101;

               //shift left algebraic
    #10        a = 8'b1100_0011;
               shft_code = 2'b01;shft_amt = 4'b0101;

               //shift right logical
    #10        a = 8'b1100_0011;
               shft_code = 2'b10;shft_amt = 4'b0101;

               //shift right algebraic
    #10        a = 8'b0111_1111;
               shft_code = 2'b11;shft_amt = 4'b0101;
-------------------------------------------------------------

    #10        $stop;

end

//instantiate the module into the test bench
comb_shifter inst1 (
    .a(a),
    .shft_code(shft_code),
    .shft_amt(shft_amt),
    .shft_rslt(shft_rslt)
    );

endmodule
```

Figure 2.37   (Continued)

```
Shift operation codes
sll = 00, sla = 01, srl = 10, sra = 11

a=00001111, shft_code=00, shft_amt=0010, shft_rslt=00111100
a=10001111, shft_code=01, shft_amt=0010, shft_rslt=10111100
a=00001111, shft_code=10, shft_amt=0010, shft_rslt=00000011
a=10001111, shft_code=11, shft_amt=0010, shft_rslt=11100011
-------------------------------------------------------------
Shift operation codes
sll = 00, sla = 01, srl = 10, sra = 11

a=11111111, shft_code=00, shft_amt=0100, shft_rslt=11110000
a=11111111, shft_code=01, shft_amt=0100, shft_rslt=11110000
a=11111111, shft_code=10, shft_amt=0100, shft_rslt=00001111
a=11111111, shft_code=11, shft_amt=0100, shft_rslt=11111111
-------------------------------------------------------------
Shift operation codes
sll = 00, sla = 01, srl = 10, sra = 11

a=11000011, shft_code=00, shft_amt=0101, shft_rslt=01100000
a=11000011, shft_code=01, shft_amt=0101, shft_rslt=11100000
a=11000011, shft_code=10, shft_amt=0101, shft_rslt=00000110
a=01111111, shft_code=11, shft_amt=0101, shft_rslt=00000011
-------------------------------------------------------------
```

Figure 2.38   Outputs for the four shift operations of SLL, SLA, SRL, and SRA.

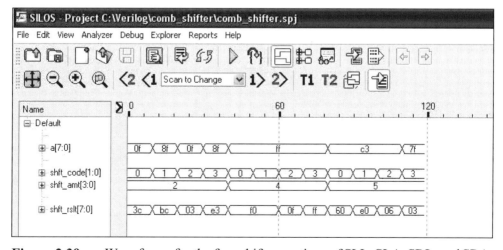

Figure 2.39   Waveforms for the four shift operations of SLL, SLA, SRL, and SRA.

Shift left logical (SLL)   Figure 2.40 shows an 8-bit register with a left shift count of 3 (011) for a shift left logical operation. The operand to be shifted is loaded into the shift left register, then shifted left the requisite number of bits as specified by the shift count.

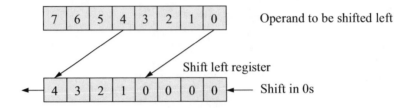

Figure 2.40   Shift left logical 3 (011) bit positions.

The design of a combinational shift operation is more easily accomplished with the utilization of multiplexers. Therefore, a 4:1 multiplexer will be designed using behavioral modeling, then instantiated into a structural module the requisite number of times to accommodate the operand size. The structural design module utilizes only one byte in the shifter; however, the concept can be easily extended for larger operands. A block diagram of the multiplexer is shown in Figure 2.41 using the ANSI/IEEE Std. 91-1984 format.

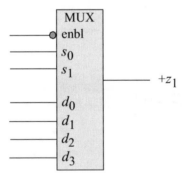

Figure 2.41   Block diagram for a 4:1 multiplexer.

A block diagram of the required 4:1 multiplexers is shown in Figure 2.42. If the shift amount is *shft_amt[00]*, then no shifting occurs — the input operand, *a[7:0]*, is passed through the 0 input of the multiplexers of the shifting element unchanged. If the shift amount is *shft_amt[01]*, then all bit positions of operand *a[7:0]* are shifted left one bit position by assigning a logic 0 to input 1 of the *inst0* multiplexer, bit *a[0]* to input 1 of the *inst1* multiplexer, and the remaining bits assigned to the appropriate data inputs of the remaining multiplexers. Bit *a[7]* is shifted off the left end of the shifting element, representing a logical left shift operation.

If the shift amount is *shft_amt[11]*, then operand *a[7:0]* is shifted left three bit positions with zeroes filling the vacated low-order bit positions. In this case, input 3 of multiplexers 7 through 0 are assigned the values *a[4] a[3] a[2] a[1] a[0] 0 0 0*.

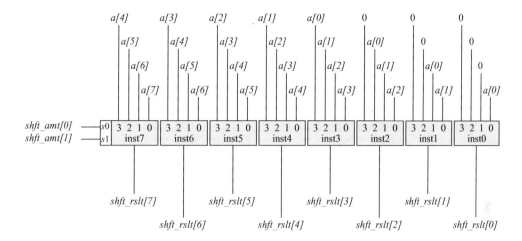

Figure 2.42   Block diagram for the logical organization for a high-speed shifter.

The behavioral module for a 4:1 multiplexer using the **case** statement is shown in Figure 2.43. Note that the data inputs for the multiplexer are labelled *[3:0] data*. Therefore, when the multiplexer is instantiated into the structural module representing Figure 2.42, the data inputs must be listed in the same sequence.

```
//behavioral 4:1 multiplexer using a case statement
module mux_4_1_case (sel, data, out);

input [1:0] sel;
input [3:0] data;
output out;
reg out;

always @ (sel or data)
begin
   case (sel)
      (0) : out = data[0];
      (1) : out = data[1];
      (2) : out = data[2];
      (3) : out = data[3];
   endcase
end
endmodule
```

Figure 2.43   Four-to-one multiplexer to be used in the combinational shifter.

The structural module for the shift left logical operation is shown in Figure 2.44 using the 4:1 multiplexers. The test bench module is shown in Figure 2.45 and applies several input vectors to be shifted left. The outputs and waveforms are shown in Figure 2.46 and Figure 2.47, respectively.

```
//structural combinational shift left logical
//shifter using multiplexers
module shifter_usg_mux_sll (a, shft_amt, shft_rslt);

input [7:0] a;
input [1:0] shft_amt;
output [7:0] shft_rslt;

//instantiate the multiplexers
mux_4_1_case inst0 (
   .sel(shft_amt),
   .data({{3{1'b0}}, a[0]}),
   .out(shft_rslt[0])
   );

mux_4_1_case inst1 (
   .sel(shft_amt),
   .data({{2{1'b0}}, a[0], a[1]}),
   .out(shft_rslt[1])
   );

mux_4_1_case inst2 (
   .sel(shft_amt),
   .data({1'b0, a[0], a[1], a[2]}),
   .out(shft_rslt[2])
   );

mux_4_1_case inst3 (
   .sel(shft_amt),
   .data({a[0], a[1], a[2], a[3]}),
   .out(shft_rslt[3])
   );

mux_4_1_case inst4 (
   .sel(shft_amt),
   .data({a[1], a[2], a[3], a[4]}),
   .out(shft_rslt[4])
   );

                           //continued on next page
```

Figure 2.44   Structural module for the shift left logical operation.

```
mux_4_1_case inst5 (
    .sel(shft_amt),
    .data({a[2], a[3], a[4], a[5]}),
    .out(shft_rslt[5])
    );

mux_4_1_case inst6 (
    .sel(shft_amt),
    .data({a[3], a[4], a[5], a[6]}),
    .out(shft_rslt[6])
    );

mux_4_1_case inst7 (
    .sel(shft_amt),
    .data({a[4], a[5], a[6], a[7]}),
    .out(shft_rslt[7])
    );

endmodule
```

Figure 2.44   (Continued)

```
//test bench for shifter using multiplexers
module shifter_usg_mux_sll_tb;

reg[7:0] a;
reg [1:0] shft_amt;
wire [7:0] shft_rslt;

//display variables
initial
$monitor ("a=%b, shft_amt=%b, shft_rslt=%b",
            a, shft_amt, shft_rslt);

//apply input vectors
initial
begin
    #0    a = 8'b0000_0000;
          shft_amt = 2'b00;

    #10   a = 8'b0000_1111;
          shft_amt = 2'b01;
                                    //continued on next page
```

Figure 2.45   Test bench module for the shift left logical operation.

```
    #10    a = 8'b0000_1111;
           shft_amt = 2'b10;

    #10    a = 8'b0000_1111;
           shft_amt = 2'b11;

    #10    a = 8'b1111_0000;
           shft_amt = 2'b00;

    #10    a = 8'b1111_0000;
           shft_amt = 2'b01;

    #10    a = 8'b1111_0000;
           shft_amt = 2'b10;

    #10    a = 8'b1111_0000;
           shft_amt = 2'b11;

    #10    $stop;

end

//instantiate the module into the test bench
shifter_usg_mux_sll inst1 (
    .a(a),
    .shft_amt(shft_amt),
    .shft_rslt(shft_rslt)
    );

endmodule
```

Figure 2.45   (Continued)

```
a=00000000, shft_amt=00, shft_rslt=00000000  //shift left 0
a=00001111, shft_amt=01, shft_rslt=00011110  //shift left 1
a=00001111, shft_amt=10, shft_rslt=00111100  //shift left 2
a=00001111, shft_amt=11, shft_rslt=01111000  //shift left 3

a=11110000, shft_amt=00, shft_rslt=11110000  //shift left 0
a=11110000, shft_amt=01, shft_rslt=11100000  //shift left 1
a=11110000, shft_amt=10, shft_rslt=11000000  //shift left 2
a=11110000, shft_amt=11, shft_rslt=10000000  //shift left 3
```

Figure 2.46   Outputs for the shift left logical operation.

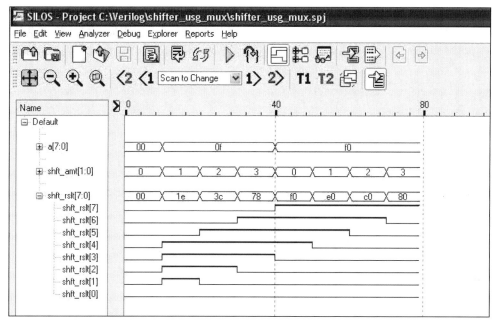

Figure 2.47   Waveforms for the shift left logical operation.

Shift left algebraic (SLA)   Recall that for a shift left algebraic operation, the numeric part of the operand is shifted left the number of bit positions specified by the shift amount. The logic diagram is similar to Figure 2.40 except that the sign remains unchanged and does not participate in the shift operation. The bits are shifted out of the high-order numeric bit position and 0s are shifted into the vacated register positions on the right. If a bit shifted out is different than the sign bit, then an overflow has occurred.

The structural design module for a shift left algebraic operation using multiplexers is shown in Figure 2.48. The test bench module is shown in Figure 2.49 and provides several operands to be shifted left. The outputs and waveforms are shown in Figure 2.50 and Figure 2.51, respectively.

```
//structural shifter using multiplexers for
//shift left algebraic
module shifter_usg_mux_sla (a, shft_amt, shft_rslt);

input [7:0] a;
input [1:0] shft_amt;

output [7:0] shft_rslt;          //continued on next page
```

Figure 2.48   Structural design module for a shift left algebraic operation.

```
//instantiate the multiplexers
mux_4_1_case inst0 (
   .sel(shft_amt),
   .data({{3{1'b0}}, a[0]}),
   .out(shft_rslt[0])
   );

mux_4_1_case inst1 (
   .sel(shft_amt),
   .data({{2{1'b0}}, a[0], a[1]}),
   .out(shft_rslt[1])
   );

mux_4_1_case inst2 (
   .sel(shft_amt),
   .data({1'b0, a[0], a[1], a[2]}),
   .out(shft_rslt[2])
   );

mux_4_1_case inst3 (
   .sel(shft_amt),
   .data({a[0], a[1], a[2], a[3]}),
   .out(shft_rslt[3])
   );

mux_4_1_case inst4 (
   .sel(shft_amt),
   .data({a[1], a[2], a[3], a[4]}),
   .out(shft_rslt[4])
   );

mux_4_1_case inst5 (
   .sel(shft_amt),
   .data({a[2], a[3], a[4], a[5]}),
   .out(shft_rslt[5])
   );

mux_4_1_case inst6 (
   .sel(shft_amt),
   .data({a[3], a[4], a[5], a[6]}),
   .out(shft_rslt[6])
   );

mux_4_1_case inst7 (
   .sel(shft_amt),
   .data({4{a[7]}}),
   .out(shft_rslt[7])
   );

endmodule
```

Figure 2.48   (Continued)

```
//test bench for structural shifter using multiplexers
//for shift left algebraic
module shifter_usg_mux_sla_tb;

reg[7:0] a;
reg [1:0] shft_amt;
wire [7:0] shft_rslt;

initial      //display variables
$monitor ("a=%b, shft_amt=%b, shft_rslt=%b",
          a, shft_amt, shft_rslt);

//apply input vectors
initial
begin
   #0    a = 8'b0000_0000;
         shft_amt = 2'b00;

   #10   a = 8'b0000_1111;
         shft_amt = 2'b01;

   #10   a = 8'b0000_1111;
         shft_amt = 2'b10;

   #10   a = 8'b0000_1111;
         shft_amt = 2'b11;

   #10   a = 8'b0100_1111;
         shft_amt = 2'b01;
//----------------------------
   #10   a = 8'b1111_0000;
         shft_amt = 2'b00;

   #10   a = 8'b1111_0000;
         shft_amt = 2'b01;

   #10   a = 8'b1111_0000;
         shft_amt = 2'b10;

   #10   a = 8'b1111_0000;
         shft_amt = 2'b11;

   #10   a = 8'b1000_0000;
         shft_amt = 2'b11;

   #10   $stop;
end                          //continued on next page
```

Figure 2.49   Test bench module for the shift left algebraic operation.

```
//instantiate the module into the test bench
shifter_usg_mux_sla inst1 (
   .a(a),
   .shft_amt(shft_amt),
   .shft_rslt(shft_rslt)
   );
endmodule
```

Figure 2.49   (Continued)

```
a=00000000, shft_amt=00, shft_rslt=00000000
a=00001111, shft_amt=01, shft_rslt=00011110
a=00001111, shft_amt=10, shft_rslt=00111100
a=00001111, shft_amt=11, shft_rslt=01111000
a=01001111, shft_amt=01, shft_rslt=00011110

a=11110000, shft_amt=00, shft_rslt=11110000
a=11110000, shft_amt=01, shft_rslt=11100000
a=11110000, shft_amt=10, shft_rslt=11000000
a=11110000, shft_amt=11, shft_rslt=10000000
a=10000000, shft_amt=11, shft_rslt=10000000
```

Figure 2.50   Outputs for the shift left algebraic operation.

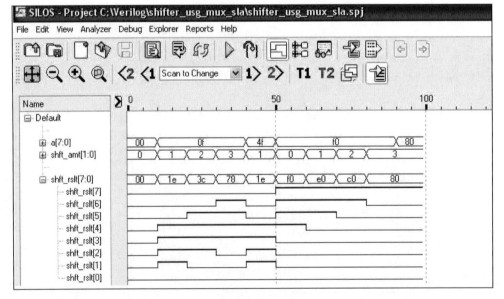

Figure 2.51   Waveforms for the shift left algebraic operation.

Shift right logical (SRL)   Recall that any right shift operation can be implemented by shifting left an amount that is the 2s complement of the right shift count. For example, if the right shift count is 011_2 (3_{10}), then the equivalent left shift count is $100 + 1 = 101$, which is the 2s complement of the right shift count. The shift right logical operation is implemented in two levels of hardware, as shown in Figure 2.52, which shifts right an 8-bit operand three bit positions.

In level A, the operand is offset to the right by a number of bit positions equal to the operand length, minus one bit position. The "minus one bit" represents a left shift of one bit position when 2s complementing the right shift count; that is, it is the "+1" in the 2s complementation process. This built-in left shift of one bit position reduces the amount hardware by one cell. The remaining high-order bit positions are set to zero, because the operation is a logical right shift of an unsigned number.

In level B, the operand is shifted left by an amount equal to the equivalent left shift count minus 1; that is, the 1s complement of the right shift count. The resultant operand in level B is identical to the shifted operand that would have been obtained by a right shift operation without utilizing two levels.

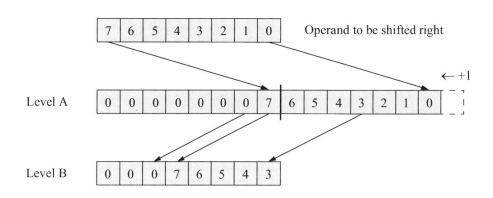

Figure 2.52   Shift right logical 3 (011) bit positions.

The structural design module is shown in Figure 2.53 using 4:1 multiplexers for 8-bit operands. The test bench module is shown in Figure 2.54 and provides several operands to be shifted right logically. The outputs and waveforms are shown in Figure 2.55 and Figure 2.56, respectively

```
//structural shifter using multiplexers
//for shift right logical
module shifter_usg_mux_srl (a, shft_amt, shft_rslt);

input [7:0] a;
input [1:0] shft_amt;
output [7:0] shft_rslt;              //continued on next page
```

Figure 2.53   Structural design module for a shift right logical operation.

```
//instantiate the multiplexers
mux_4_1_case inst0 (
   .sel(shft_amt),
   .data({a[3], a[2], a[1], a[0]}),
   .out(shft_rslt[0])
   );

mux_4_1_case inst1 (
   .sel(shft_amt),
   .data({a[4], a[3], a[2], a[1]}),
   .out(shft_rslt[1])
   );

mux_4_1_case inst2 (
   .sel(shft_amt),
   .data({a[5], a[4], a[3], a[2]}),
   .out(shft_rslt[2])
   );

mux_4_1_case inst3 (
   .sel(shft_amt),
   .data({a[6], a[5], a[4], a[3]}),
   .out(shft_rslt[3])
   );

mux_4_1_case inst4 (
   .sel(shft_amt),
   .data({a[7], a[6], a[5], a[4]}),
   .out(shft_rslt[4])
   );

mux_4_1_case inst5 (
   .sel(shft_amt),
   .data({1'b0, a[7], a[6], a[5]}),
   .out(shft_rslt[5])
   );

mux_4_1_case inst6 (
   .sel(shft_amt),
   .data({1'b0, 1'b0, a[7], a[6]}),
   .out(shft_rslt[6])
   );

mux_4_1_case inst7 (
   .sel(shft_amt),
   .data({{3{1'b0}}, a[7]}),
   .out(shft_rslt[7])
   );

endmodule
```

Figure 2.53   (Continued)

```
//test bench for shifter using multiplexers
//for shift right logical
module shifter_usg_mux_srl_tb;

reg[7:0] a;
reg [1:0] shft_amt;
wire [7:0] shft_rslt;

initial      //display variables
$monitor ("a=%b, shft_amt=%b, shft_rslt=%b",
          a, shft_amt, shft_rslt);

initial      //apply input vectors
begin
   #0    a = 8'b0000_0000;
         shft_amt = 2'b00;

   #10   a = 8'b0000_1111;
         shft_amt = 2'b01;

   #10   a = 8'b0000_1111;
         shft_amt = 2'b10;

   #10   a = 8'b0000_1111;
         shft_amt = 2'b11;

   #10   a = 8'b1111_0000;
         shft_amt = 2'b00;

   #10   a = 8'b1111_0000;
         shft_amt = 2'b01;

   #10   a = 8'b1111_0000;
         shft_amt = 2'b10;

   #10   a = 8'b1111_0000;
         shft_amt = 2'b11;
   #10   $stop;
end

//instantiate the module into the test bench
shifter_usg_mux_srl inst1 (
   .a(a),
   .shft_amt(shft_amt),
   .shft_rslt(shft_rslt)
   );
endmodule
```

Figure 2.54   Test bench module for the shift right logical operation.

```
a=00000000, shft_amt=00, shft_rslt=00000000
a=00001111, shft_amt=01, shft_rslt=00000111
a=00001111, shft_amt=10, shft_rslt=00000011
a=00001111, shft_amt=11, shft_rslt=00000001

a=11110000, shft_amt=00, shft_rslt=11110000
a=11110000, shft_amt=01, shft_rslt=01111000
a=11110000, shft_amt=10, shft_rslt=00111100
a=11110000, shft_amt=11, shft_rslt=00011110
```

Figure 2.55   Outputs for the shift right logical operation.

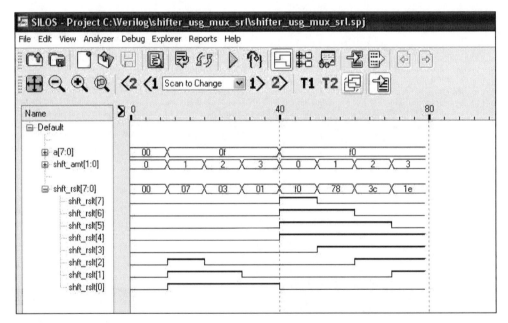

Figure 2.56   Waveforms for the shift right logical operation.

Shift right algebraic (SRA)   Recall that the numeric part of the signed operand is shifted right the number of bits specified by the shift count for a shift right algebraic operation. The sign of the operand remains unchanged. All numeric bits participate in the right shift. The sign bit propagates right to fill in the vacated high-order numeric bit positions. When the operation is executed by shifting left, the high-order bits in level B are set to the value of the sign bit, as shown in the logical configuration of Figure 2.57, which shifts right algebraic an 8-bit operand five bit positions.

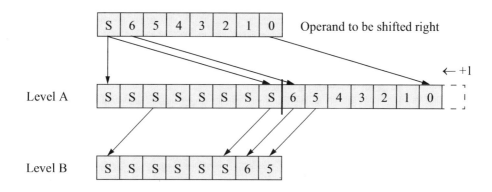

Figure 2.57   Shift right algebraic 5 (101) bit positions.

Figure 2.57 illustrates a shift right algebraic operation with a right shift count of 5 (101). The equivalent left shift count is 010 + 1, or simply 010 after the "+1" left shift has been implemented. The operand in level B is identical to the operand that would have been obtained by a shift right algebraic operation without utilizing two levels.

Since the sign bit must be inserted into the high-order positions of level A, an 8:1 multiplexer is used in the design, which is shown in Figure 2.58 as a behavioral module using the **case** statement. Figure 2.59 shows the structural design module for a shift right algebraic operation. The test bench module is shown in Figure 2.60 and provides several operands to be shifted right algebraically. The outputs and waveforms are shown in Figure 2.61 and Figure 2.62, respectively.

```
//behavioral 8:1 multiplexer using the case statement
module mux_8to1_case2 (sel, data, out);

input [2:0] sel;
input [7:0] data;
output out;

reg out;

always @ (sel or data)
begin
   case (sel)
         (0) : out = data[0];
         (1) : out = data[1];
         (2) : out = data[2];
         (3) : out = data[3];            //continued on next page
```

Figure 2.58   Behavioral module for an 8:1 multiplexer using the **case** statement.

```
      (4):  out = data[4];
      (5):  out = data[5];
      (6):  out = data[6];
      (7):  out = data[7];
      default: out = 1'b0;
   endcase
end
endmodule
```

Figure 2.58   (Continued)

```
//structural shifter using multiplexers
//for shift right algebraic
module shifter_usg_mux_sra (a, shift_amt, shift_rslt);

input [7:0] a;
input [2:0] shift_amt;
output [7:0] shift_rslt;

//instantiate the multiplexers
mux_8to1_case2 inst0 (
   .sel(shift_amt),
   .data({a[7], a[6], a[5], a[4], a[3], a[2], a[1], a[0]}),
   .out(shift_rslt[0])
   );

mux_8to1_case2 inst1 (
   .sel(shift_amt),
   .data({a[7], a[7], a[6], a[5], a[4], a[3], a[2], a[1]}),
   .out(shift_rslt[1])
   );

mux_8to1_case2 inst2 (
   .sel(shift_amt),
   .data({{3{a[7]}}, a[6], a[5], a[4], a[3], a[2]}),
   .out(shift_rslt[2])
   );

mux_8to1_case2 inst3 (
   .sel(shift_amt),
   .data({{4{a[7]}}, a[6], a[5], a[4], a[3]}),
   .out(shift_rslt[3])
   );                              //continue on next page
```

Figure 2.59   Structural design module for a shift right algebraic operation.

```
mux_8to1_case2 inst4 (
   .sel(shift_amt),
   .data({{5{a[7]}}, a[6], a[5], a[4]}),
   .out(shift_rslt[4])
   );

mux_8to1_case2 inst5 (
   .sel(shift_amt),
   .data({{6{a[7]}}, a[6], a[5]}),
   .out(shift_rslt[5])
   );

mux_8to1_case2 inst6 (
   .sel(shift_amt),
   .data({{7{a[7]}}, a[6]}),
   .out(shift_rslt[6])
   );

mux_8to1_case2 inst7 (
   .sel(shift_amt),
   .data({8{a[7]}}),
   .out(shift_rslt[7])
   );

endmodule
```

Figure 2.59   (Continued)

```
//test bench for shifter using multiplexers
//for shift right algebraic
module shifter_usg_mux_sra_tb;

reg [7:0] a;
reg [2:0] shift_amt;

wire [7:0] shift_rslt;

//display variables
initial
$monitor ("a=%b, shift_amt=%b, shift_rslt=%b",
          a, shift_amt, shift_rslt);

                            //continued on next page
```

Figure 2.60   Test bench module for the shift right algebraic operation.

```
//apply input vectors
initial
begin
   #0     a = 8'b0000_0000;     shift_amt = 3'b000;
   #10    a = 8'b0000_1111;     shift_amt = 3'b001;

   #10    a = 8'b0000_1111;     shift_amt = 3'b010;
   #10    a = 8'b0000_1111;     shift_amt = 3'b011;

   #10    a = 8'b1111_0000;     shift_amt = 3'b100;
   #10    a = 8'b1111_0000;     shift_amt = 3'b101;

   #10    a = 8'b1111_0000;     shift_amt = 3'b110;
   #10    a = 8'b1111_0000;     shift_amt = 3'b111;

   #10    a = 8'b1000_0000;     shift_amt = 3'b111;
   #10    a = 8'b0111_1111;     shift_amt = 3'b111;

   #10    $stop;
end

//instantiate the module into the test bench
shifter_usg_mux_sra inst1 (
   .a(a),
   .shift_amt(shift_amt),
   .shift_rslt(shift_rslt)
   );
endmodule
```

Figure 2.60   (Continued)

```
a=00000000, shift_amt=000, shift_rslt=00000000
a=00001111, shift_amt=001, shift_rslt=00000111

a=00001111, shift_amt=010, shift_rslt=00000011
a=00001111, shift_amt=011, shift_rslt=00000001

a=11110000, shift_amt=100, shift_rslt=11111111
a=11110000, shift_amt=101, shift_rslt=11111111

a=11110000, shift_amt=110, shift_rslt=11111111
a=11110000, shift_amt=111, shift_rslt=11111111

a=10000000, shift_amt=111, shift_rslt=11111111
a=01111111, shift_amt=111, shift_rslt=00000000
```

Figure 2.61   Outputs for the shift right algebraic operation.

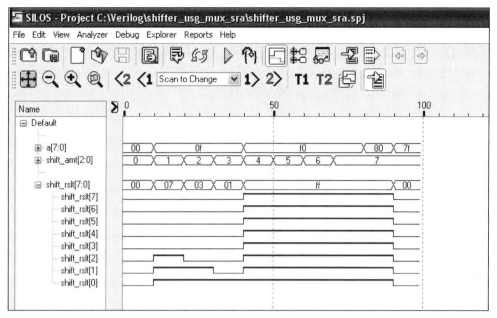

Figure 2.62   Waveforms for the shift right algebraic operation.

2.2 Synchronous Counters

Counters are designed in this section using Verilog. The designs illustrated in this section will be a modulo-8 counter, a modulo-10 counter, and a Johnson counter. Also presented will be a binary-to-Gray code converter using *JK* flip-flops to illustrate the versatility of counters.

Counters are usually clocked synchronous devices used in the design of digital systems and have a finite number of states. The λ output logic is usually a function of the present state only; that is, $\lambda(Y_{j(t)})$. The state of the counter is interpreted as an integer with respect to a modulus. A number A modulo n is defined as the remainder after dividing A by n. Some counters accommodate a set of binary input variables which provides an initial state for the counter. There are also asynchronous counters, which are inherently slow, because of the ripple effect caused by the output of stage y_i functioning as the clock input for stage y_{i+1}.

This section will discuss only synchronous counters. Counters are associated with a set of transformations on a set of states and follow a prescribed sequence of states under control of a clock input signal. When the active clock transition occurs at the input, the state of the machine changes to some predetermined value as defined by the machine specifications. The counting sequence is usually an increment or decrement by one, or an arbitrary prescribed sequence, or a state in which only one flip-flop changes state, as in a Gray code counter.

2.2.1 Modulo-8 Counter

A modulo-8 counter counts in the following sequence: 000 001 010 . . . 110 111 000. The state diagram for a modulo-8 counter is shown in Figure 2.63. The counter is initially reset to state a ($y_1 y_2 y_3 = 000$), then increments by one at each clock transition until state h ($y_1 y_2 y_3 = 111$) is reached. At the next clock transition, the counter sequences to state a ($y_1 y_2 y_3 = 000$). The counter will be designed using behavioral modeling, structural modeling using D flip-flops, and structural modeling using JK flip-flops.

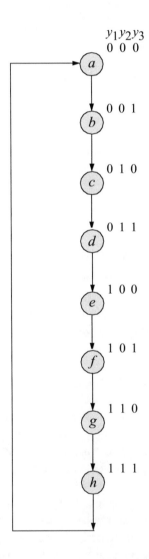

Figure 2.63   State diagram for a modulo-8 counter.

Example 2.8   This example designs a modulo-8 counter using behavioral modeling. The behavioral design module is shown in Figure 2.64. The operator symbol for modulus is the percent symbol (%). As previously stated, a number A modulo n is defined as the remainder after dividing A by n. Therefore, the statement $y = (y + 1)$ % 8; in Figure 2.64 specifies that the counter is incremented by one, then the count y modulus 8 is obtained. The test bench module is shown in Figure 2.65. The outputs and waveforms are shown in Figure 2.66 and Figure 2.67, respectively.

```
//behavioral modulo-8 counter
module ctr_mod8_bh (rst_n, clk, y);

input rst_n, clk;       //define inputs and outputs
output [2:0] y;

wire rst_n, clk;        //or do not declare inputs as wire,
                        //because inputs are wire by default
reg [2:0] y;            //variables in always declared as reg

//define counting sequence
always @ (posedge clk or negedge rst_n)
begin
   if (rst_n == 0)
      y = 3'b000;
   else
      y = (y + 1) % 8;
end
endmodule
```

Figure 2.64   Behavioral design module for the modulo-8 counter.

```
//test bench for modulo-8 behavioral counter
module ctr_mod8_bh_tb;

reg rst_n, clk;    //inputs are reg in test bench
wire [2:0] y;      //outputs are wire in test bench

initial            //display outputs
$monitor ("count = %b", y);

//define reset
initial
begin
   #0     rst_n = 1'b0;
   #5     rst_n = 1'b1;
end                          //continued on next page
```

Figure 2.65   Test bench module for the modulo-8 counter.

```
initial              //define clock
begin
   clk = 1'b0;
   forever
      #10 clk = ~clk;
end

initial              //define length of simulation
begin
   #160 $finish;
end

//instantiate the module into the test bench
ctr_mod8_bh inst1 (
   .rst_n(rst_n),
   .clk(clk),
   .y(y)
   );
endmodule
```

Figure 2.65   (Continued)

```
count = 000    count = 011    count = 110
count = 001    count = 100    count = 111
count = 010    count = 101    count = 000
```

Figure 2.66   Outputs for the modulo-8 counter.

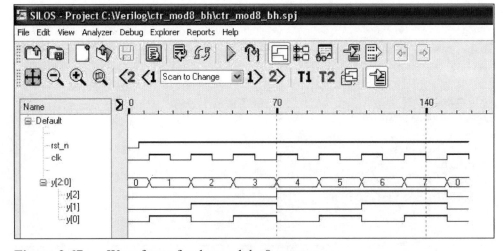

Figure 2.67   Waveforms for the modulo-8 counter.

Example 2.9   This example designs a modulo-8 counter using structural modeling with D flip-flops. The input maps for the D flip-flops are obtained from the state diagram of Figure 2.63 and are shown in Figure 2.68. The logic diagram is shown in Figure 2.69 with implied reset inputs.

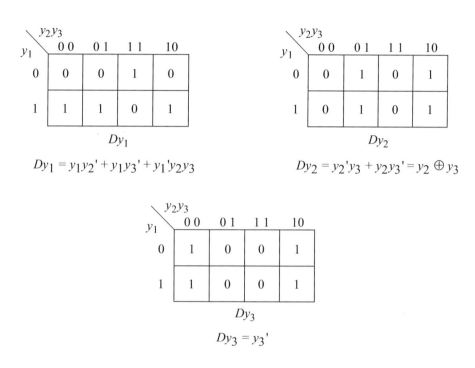

$$Dy_1 = y_1 y_2' + y_1 y_3' + y_1' y_2 y_3$$

$$Dy_2 = y_2' y_3 + y_2 y_3' = y_2 \oplus y_3$$

$$Dy_3 = y_3'$$

Figure 2.68   Input maps for the modulo-8 counter using D flip-flops

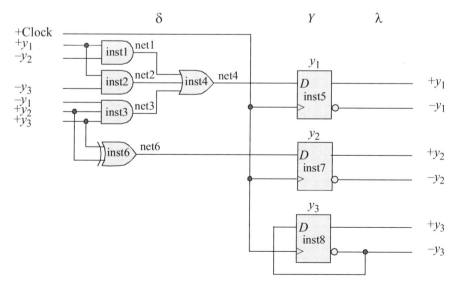

Figure 2.69   Logic diagram for the modulo-8 counter using D flip-flops.

The structural design module for the modulo-8 counter is shown in Figure 2.70. The module instantiates 2- and 3-input AND gates, a 3-input OR gate, an exclusive-OR function, and *D* flip-flops. The test bench module is shown in Figure 2.71. The outputs and waveforms are shown in Figure 2.72 and Figure 2.73, respectively.

```
//structural modulo-8 counter using D flip-flops
module ctr_mod8_d_st (rst_n, clk, y);

//define inputs and outputs
input rst_n, clk;
output [1:3] y;

//define wires
wire rst_n, clk;
wire [1:3] y;

//define internal nets
wire net1, net2, net3, net4, net6;

//instantiate the logic for flip-flop y[1]
and2_df inst1 (
    .x1(y[1]),
    .x2(~y[2]),
    .z1(net1)
    );

and2_df inst2 (
    .x1(y[1]),
    .x2(~y[3]),
    .z1(net2)
    );

and3_df inst3 (
    .x1(~y[1]),
    .x2(y[2]),
    .x3(y[3]),
    .z1(net3)
    );

or3_df inst4 (
    .x1(net1),
    .x2(net2),
    .x3(net3),
    .z1(net4)
    );
                            //continued on next page
```

Figure 2.70   Structural design module for the modulo-8 counter using *D* flip-flops.

```
d_ff_bh inst5 (
   .rst_n(rst_n),
   .clk(clk),
   .d(net4),
   .q(y[1])
   );

//instantiate the logic for flip-flop y[2]
xor2_df inst6 (
   .x1(y[3]),
   .x2(y[2]),
   .z1(net6)
   );

d_ff_bh inst7 (
   .rst_n(rst_n),
   .clk(clk),
   .d(net6),
   .q(y[2])
   );

//instantiate the logic for flip-flop y[3]
d_ff_bh inst8 (
   .rst_n(rst_n),
   .clk(clk),
   .d(~y[3]),
   .q(y[3])
   );

endmodule
```

Figure 2.70 (Continued)

```
//test bench for the structural modulo-8 counter
module ctr_mod8_d_st_tb;

reg rst_n, clk;    //inputs are reg in test bench
wire [1:3] y;      //outputs are wire in test bench

//display outputs
initial
$monitor ("count = %b", y);

                              //continued on next page
```

Figure 2.71 Test bench module for the modulo-8 counter using *D* flip-flops.

```
//define reset
initial
begin
    #0      rst_n = 1'b0;
    #3      rst_n = 1'b1;
end

//define clock
initial
begin
    clk = 1'b0;
    forever
        #10 clk = ~clk;
end

//define length of simulation
initial
begin
    #140 $finish;
end

//instantiate the module into the test bench
ctr_mod8_d_st inst1 (
    .rst_n(rst_n),
    .clk(clk),
    .y(y)
    );

endmodule
```

Figure 2.71   (Continued)

```
count = 000
count = 001
count = 010
count = 011

count = 100
count = 101
count = 110
count = 111

count = 000
```

Figure 2.72   Outputs for the modulo-8 counter using D flip-flops.

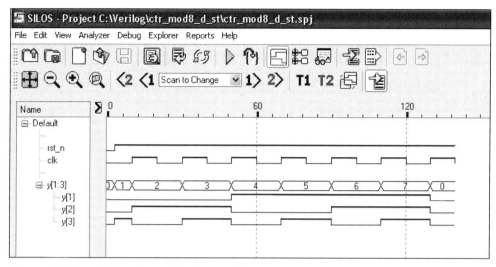

Figure 2.73   Waveforms for the modulo-8 counter using *D* flip-flops.

Example 2.10   This example designs a modulo-8 counter using structural modeling with *JK* flip-flops. The excitation table for a *JK* flip-flop is shown in Table 2.1. The input maps for the *JK* flip-flops are obtained from the state diagram of Figure 2.63 using the *JK* excitation table and are shown in Figure 2.74. As can be seen from the equations, only one AND (y_2y_3) is required for the δ next-state logic.

The logic diagram is shown in Figure 2.75 with implied reset inputs. The flip-flops are clocked on the negative clock transition. Notice that the low-order flip-flop y_3 is implemented in toggle mode ($JK = 11$). Also, when $y_3 = 1$, y_2 toggles, and when $y_2y_3 = 11$, y_1 toggles. Circuit action takes place only on the negative edge of the clock. Once the negative transition has occurred, no further change of the input values will cause a change in circuit activity until the following negative clock transition.

The structural design module is shown in Figure 2.76 and the test bench module is shown in Figure 2.77. The outputs and waveforms are shown in Figure 2.78 and Figure 2.79, respectively. The resulting waveforms are essentially the same as for the structural design using *D* flip-flops.

Table 2.1 Excitation Table for a *JK* Flip-Flop

Present state $Y_{j(t)}$	Next state $Y_{k(t+1)}$	Flip-flop inputs $J\,K$
0	0	0 –
0	1	1 –
1	0	– 1
1	1	– 0

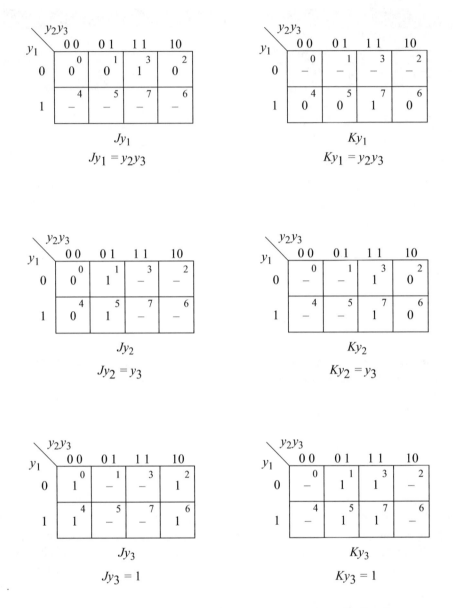

Figure 2.74   Input maps for the modulo-8 counter using *JK* flip-flops.

As the counter progresses through the counting sequence, the machine specifications may require an indication when a particular state has been reached. This is easily implemented by using an AND gate to detect, for example, state $f(y_1 y_2 y_3 = 101)$. The $-Clock$ signal is included as an AND gate input to prevent an erroneous output caused by a momentary transition through state $y_1 y_2 y_3 = 101$ when two or more flip-flops change state for a state transition sequence that does not end in state f.

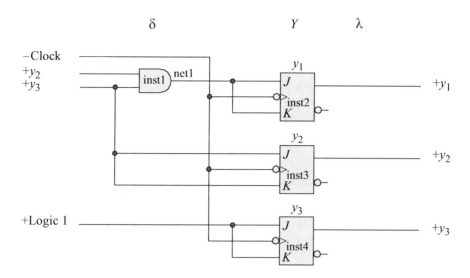

Figure 2.75   Logic diagram for the modulo-8 counter using *JK* flip-flops.

```
//structural for a modulo-8 counter using jk flip-flops
module ctr_mod8_jk_struc (rst_n, clk, y);

//define inputs and outputs
input rst_n, clk;
output [1:3]y;

wire net1;      //define internal nets

//instantiate the logic for flip-flop y[1]
and2_df inst1 (
   .x1(y[2]),
   .x2(y[3]),
   .z1(net1)
   );

jkff_neg_clk inst2 (
   .rst_n(rst_n),
   .clk(clk),
   .j(net1),
   .k(net1),
   .q(y[1])
   );
                            //continued on next page
```

Figure 2.76   Structural design module for the modulo-8 counter using *JK* flip-flops.

```
//instantiate the logic for flip-flop y[2]
jkff_neg_clk inst3 (
   .rst_n(rst_n),
   .clk(clk),
   .j(y[3]),
   .k(y[3]),
   .q(y[2])
   );

//instantiate the logic for flip-flop y[3]
jkff_neg_clk inst4 (
   .rst_n(rst_n),
   .clk(clk),
   .j(1'b1),
   .k(1'b1),
   .q(y[3])
   );

endmodule
```

Figure 2.76   (Continued)

```
//test bench for modul0-8 structural counter
//using JK flip-flops
module ctr_mod8_jk_struc_tb;

reg rst_n, clk;
wire [1:3] y;

initial      //display outputs
$monitor("count = %b", y);

initial      //define reset
begin
   #0 rst_n = 1'b0;
   #5 rst_n = 1'b1;
end

initial      //define clock
begin
   clk = 1'b0;
   forever
      #10clk = ~clk;
end                              //continued on next page
```

Figure 2.77   Test bench module for the modulo-8 counter using *JK* flip-flops.

```
//define length of simulation
initial
begin
    #175 $finish;
end

//instantiate the module into the test bench
ctr_mod8_jk_struc inst1 (
    .rst_n(rst_n),
    .clk(clk),
    .y(y)
    );

endmodule
```

Figure 2.77 (Continued)

```
count = 000        count = 100
count = 001        count = 101
count = 010        count = 110
count = 011        count = 111
                   count = 000
```

Figure 2.78 Outputs for the modulo-8 counter using *JK* flip-flops.

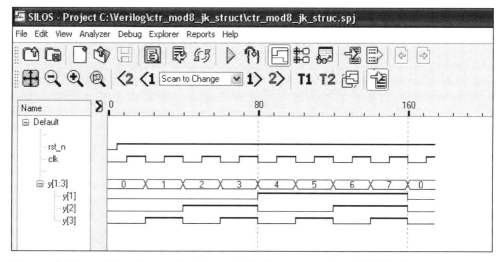

Figure 2.79 Waveforms for the modulo-8 counter using *JK* flip-flops.

2.2.2 Modulo-10 Counter

A modulo-10, or binary-coded decimal (BCD) decade counter, generates ten states in the following sequence: 0000, 0001, 0010, 0011, 0100, 0101, 0110, 0111, 1000, 1001, 0000, ... Thus, each decade requires four flip-flops. There are six unused states, 1010 through 1111, that represent invalid numbers for BCD. These unused states can be regarded as "don't care" states for the purpose of minimizing the δ next-state logic, unless the counter is self-starting, in which case, all unused states contain entries which cause the counter to proceed to a predetermined state at the next active clock transition.

A modulo-10 counter will be designed using the following four Verilog modeling constructs: behavioral modeling using the modulus operator (%), behavioral modeling using the **case** statement, structural modeling with a self-starting state of $y_1y_2y_3y_4 = 0000$, and structural modeling that with no self-starting state.

Example 2.11   This example designs a modulo-10 counter using behavioral modeling with the modulus operator. The behavioral design module using the modulus operator is shown in Figure 2.80 and the test bench module is shown in Figure 2.81. A state diagram is not required for the behavioral module. The outputs and waveforms are shown in Figure 2.82 and Figure 2.83, respectively.

```verilog
//behavioral modulo-10 counter using the modulus operator
module ctr_mod10_bh (rst_n, clk, y);

//define inputs and outputs
input rst_n, clk;
output [1:4] y;

//variable y is used in always statement
reg [1:4] y;

//define counting sequence
always @ (posedge clk or negedge rst_n)
begin
   if (rst_n == 0)
      y = 4'b0000;
   else
      y = (y + 1) % 10;
end

endmodule
```

Figure 2.80   Behavioral design module for the modulo-10 counter using the modulus operator.

```
//test bench for the modulo-10 counter

module ctr_mod10_bh_tb;

//inputs are reg for test bench
reg rst_n, clk;


//outputs are wire for test bench
wire [1:4] y;


//display outputs
initial
$monitor ("count = %b", y);


//define reset
initial
begin
   #0      rst_n = 1'b0;
   #5      rst_n = 1'b1;
end


//define clock
initial
begin
   clk = 1'b0;
   forever
      #10    clk = ~clk;
end


//define length of simulation
initial
   #200  $stop;


//instantiate the module into the test bench
ctr_mod10_bh inst1 (
   .rst_n(rst_n),
   .clk(clk),
   .y(y)
   );

endmodule
```

Figure 2.81   Test bench module for the modulo-10 counter using the modulus operator.

```
count = 0000     count = 0110
count = 0001     count = 0111
count = 0010     count = 1000
count = 0011     count = 1001
count = 0100     count = 0000
count = 0101
```

Figure 2.82   Outputs for the modulo-10 counter using the modulus operator.

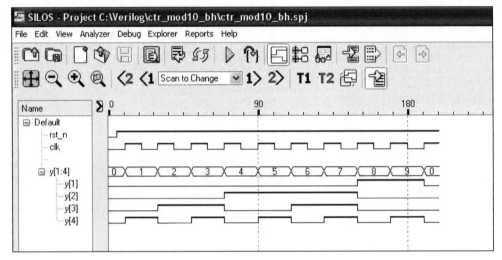

Figure 2.83   Waveforms for the modulo-10 counter using the modulus operator.

Example 2.12   This example designs a modulo-10 counter using behavioral modeling with the **case** statement. The behavioral design module is shown in Figure 2.84. The test bench module is shown in Figure 2.85. The outputs and waveforms are shown in Figure 2.86 and Figure 2.87, respectively.

```
//behavioral counter modulo-10 using case statement
module ctr_mod10_bh_case (rst_n, clk, count);

input rst_n, clk;
output [1:4] count;

//variables used in always are declared as reg
reg [1:4] count, next_count;       //continued on next page
```

Figure 2.84   Behavioral design module for the modulo-10 counter using the **case** statement.

```verilog
always @ (negedge rst_n or negedge clk)
begin
   if (~rst_n)                  //if reset = 0
      count = 4'b0000;
   else
      count = next_count;
end

//define the counting sequence
always @ (count)
begin
   case (count)
      4'b0000 : next_count = 4'b0001;
      4'b0001 : next_count = 4'b0010;
      4'b0010 : next_count = 4'b0011;
      4'b0011 : next_count = 4'b0100;
      4'b0100 : next_count = 4'b0101;
      4'b0101 : next_count = 4'b0110;
      4'b0110 : next_count = 4'b0111;
      4'b0111 : next_count = 4'b1000;
      4'b1000 : next_count = 4'b1001;
      4'b1001 : next_count = 4'b0000;
      default : next_count = 4'b0000;
endcase
end
endmodule
```

Figure 2.84   (Continued)

```verilog
//test bench for modulo-10 counter
module ctr_mod10_bh_case_tb;

reg rst_n, clk;       //define inputs and outputs
wire [1:4] count;

initial                //display outputs
$monitor ("count = %b", count);

initial                //define reset
begin
   #0 rst_n = 1'b0;
   #5 rst_n = 1'b1;
end
                        //continued on next page
```

Figure 2.85   Test bench module for the modulo-10 counter using the **case** statement.

```
initial                //define clock
begin
   clk = 1'b0;
   forever
      #10 clk = ~clk;
end

initial                //define length of simulation
begin
   #210 $finish;
end

//instantiate the module into the test bench
ctr_mod10_bh_case inst1 (
   .rst_n(rst_n),
   .clk(clk),
   .count(count)
   );
endmodule
```

Figure 2.85   (Continued)

```
count = 0000      count = 0100      count = 1000
count = 0001      count = 0101      count = 1001
count = 0010      count = 0110      count = 0000
count = 0011      count = 0111
```

Figure 2.86   Outputs for the modulo-10 counter using the **case** statement.

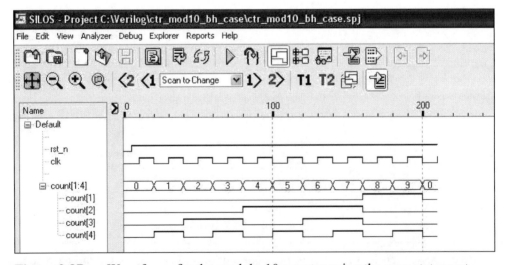

Figure 2.87   Waveforms for the modulo-10 counter using the **case** statement.

Example 2.13 This example designs a modulo-10 counter using structural modeling with *JK* flip-flops and a self-starting state. The state diagram for the modulo-10 counter with a self-starting state of $y_1 y_2 y_3 y_4 = 0000$ is shown in Figure 2.88. The state diagram also shows the unused states of 1010, 1011, 1100, 1101, 1110, and 1111. If the counter enters an unused state due to noise or any other transient condition, the next clock pulse will return the counter to a predetermined valid state, in this case $y_1 y_2 y_3 y_4 = 0000$. All unused states, *k* through *p*, will sequence to $y_1 y_2 y_3 y_4 = 0000$ as the self-starting state. The unused states correspond to minterms 10 through 15.

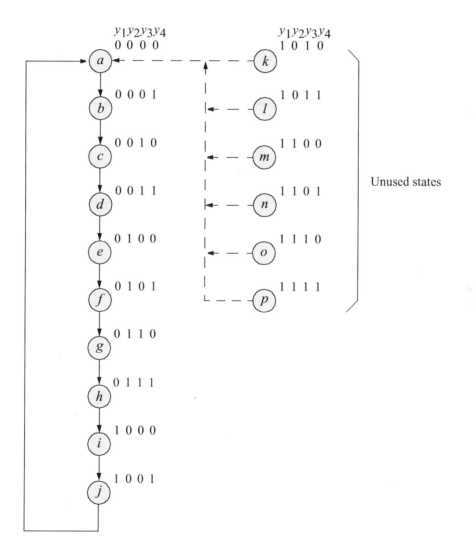

Figure 2.88 State diagram for the modulo-10 counter with a self-starting state of $y_1 y_2 y_3 y_4 = 0000$.

Using the excitation table for a JK flip-flop, reproduced in Table 2.2, the next-state table is obtained as shown in Table 2.3. Note that states $y_1y_2y_3y_4 = 1010$ through $y_1y_2y_3y_4 = 1111$ proceed to state $y_1y_2y_3y_4 = 0000$ as the next state, since this is a self-starting counter and state $y_1y_2y_3y_4 = 0000$ was chosen as the self-starting state.

Table 2.2 Excitation Table for a JK Flip-Flop

Present state $Y_{j(t)}$	Next state $Y_{k(t+1)}$	Flip-flop inputs $J\,K$
0	0	0 –
0	1	1 –
1	0	– 1
1	1	– 0

Table 2.3 Next-State Table for the Modulo-10 Counter of Figure 2.88 Using JK Flip-Flops

Present state $y_1y_2y_3y_4$	Next state $y_1y_2y_3y_4$	$Jy_1\,Ky_1$	$Jy_2\,Ky_2$	$Jy_3\,Ky_3$	$Jy_4\,Ky_4$
0 0 0 0	0 0 0 1	0 –	0 –	0 –	1 –
0 0 0 1	0 0 1 0	0 –	0 –	1 –	– 1
0 0 1 0	0 0 1 1	0 –	0 –	– 0	1 –
0 0 1 1	0 1 0 0	0 –	1 –	– 1	– 1
0 1 0 0	0 1 0 1	0 –	– 0	0 –	1 –
0 1 0 1	0 1 1 0	0 –	– 0	1 –	– 1
0 1 1 0	0 1 1 1	0 –	– 0	– 0	1 –
0 1 1 1	1 0 0 0	1 –	– 1	– 1	– 1
1 0 0 0	1 0 0 1	– 0	0 –	0 –	1 –
1 0 0 1	0 0 0 0	– 1	0 –	0 –	– 1
1 0 1 0	0 0 0 0	– 1	0 –	– 1	0 –
1 0 1 1	0 0 0 0	– 1	0 –	– 1	– 1
1 1 0 0	0 0 0 0	– 1	– 1	0 –	0 –
1 1 0 1	0 0 0 0	– 1	– 1	0 –	– 1
1 1 1 0	0 0 0 0	– 1	– 1	– 1	0 –
1 1 1 1	0 0 0 0	– 1	– 1	– 1	– 1

The input maps of Figure 2.89 are obtained directly from the next-state table. Referring to the state diagram, the entries for Jy_1 and Ky_1 for location $y_1y_2y_3y_4 = 0111$ are obtained as follows: Flip-flop y_1 changes from 0 to 1 as the machine progresses from state h to state i. From the excitation table, a transition from 0 to 1 results in $Jy_1 Ky_1 = 1-$, as indicated in the input maps for Jy_1 and Ky_1. Four sets of input maps are necessary: two maps, Jy_i and Ky_i, for each flip-flop y_i.

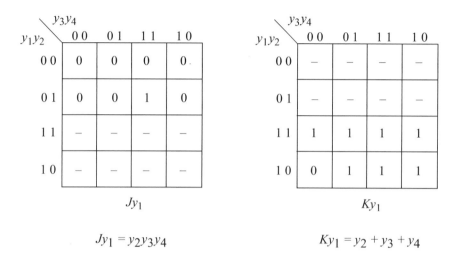

$$Jy_1 = y_2 y_3 y_4 \qquad\qquad Ky_1 = y_2 + y_3 + y_4$$

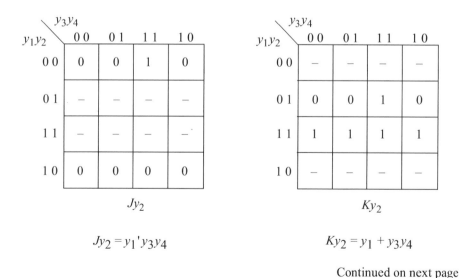

$$Jy_2 = y_1' y_3 y_4 \qquad\qquad Ky_2 = y_1 + y_3 y_4$$

Continued on next page

Figure 2.89 Input maps for the modulo-10 counter using JK flip-flops with a self-starting state of $y_1 y_2 y_3 y_4 = 0000$.

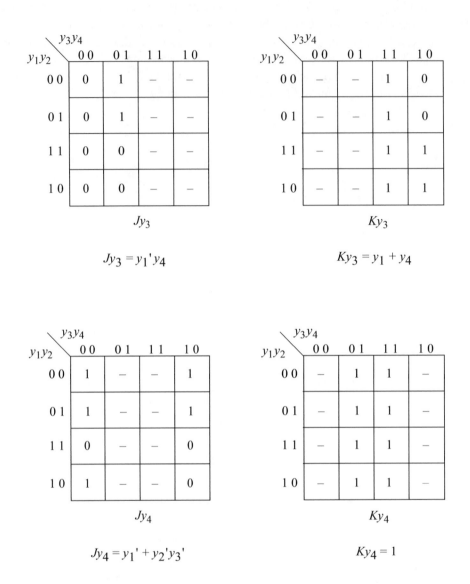

Figure 2.89   (Continued)

The logic diagram of Figure 2.90 is derived from the *JK* input equations of Figure 2.89. Although not shown in the logic diagram, it is assumed that the counter flip-flops have a set and reset function. The flip-flops were designed using behavioral modeling. The counter is reset initially to $y_1 y_2 y_3 y_4 = 0000$. State changes occur only on the negative clock transition. By examining the logic diagram, the counter can be shown to increment through the modulo-10 counting sequence, then return to 0000 at the next negative clock transition. Since the counter is self-starting, the next state should be 0000 from any invalid state 1010 through 1111.

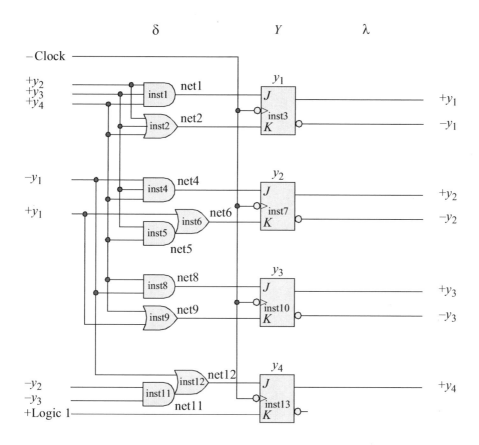

Figure 2.90   Logic diagram for the modulo-10 counter using *JK* flip-flops, where y_4 is the low-order stage. The counter has a self-starting state of $y_1y_2y_3y_4 = 0000$.

The structural design module for the modulo-10 counter with a self-starting state of $y_1y_2y_3y_4 = 0000$, is shown in Figure 2.91. The test bench module is shown in Figure 2.92. The outputs and waveforms are shown in Figure 2.93 and Figure 2.94, respectively.

```
//structural for modulo-10 counter using JK flip-flops
module ctr_mod10_struc (rst_n, clk, y);

//define inputs and outputs
input rst_n, clk;
output [1:4] y;                        //continued on next page
```

Figure 2.91   Structural design module for the modulo-10 counter using *JK* flip-flops with a self-starting state of $y_1y_2y_3y_4$ 0000.

```
//define internal nets
wire net1, net2, net4, net5, net6, net8, net9, net11, net12;

//instantiate the logic for flip-flop y[1]
and3_df inst1 (
    .x1(y[2]),
    .x2(y[3]),
    .x3(y[4]),
    .z1(net1)
    );

or3_df inst2 (
    .x1(y[2]),
    .x2(y[3]),
    .x3(y[4]),
    .z1(net2)
    );

jkff inst3 (           //set_n input is not instantiated
    .rst_n(rst_n),
    .clk(clk),
    .j(net1),
    .k(net2),
    .q(y[1])
    );

//instantiate the logic for flip-flop y[2]
and3_df inst4 (
    .x1(~y[1]),
    .x2(y[3]),
    .x3(y[4]),
    .z1(net4)
    );

and2_df inst5 (
    .x1(y[3]),
    .x2(y[4]),
    .z1(net5)
    );

or2_df inst6 (
    .x1(y[1]),
    .x2(net5),
    .z1(net6)
    );

                                    //continued on next page
```

Figure 2.91   (Continued)

```
jkff inst7 (
   .rst_n(rst_n),
   .clk(clk),
   .j(net4),
   .k(net6),
   .q(y[2])
   );

//instantiate the logic for flip-flop y[3]
and2_df inst8 (
   .x1(y[4]),
   .x2(~y[1]),
   .z1(net8)
   );

or2_df inst9 (
   .x1(y[4]),
   .x2(y[1]),
   .z1(net9)
   );

jkff inst10 (
   .rst_n(rst_n),
   .clk(clk),
   .j(net8),
   .k(net9),
   .q(y[3])
   );

//instantiate the logic for flip-flop y[4]
and2_df inst11 (
   .x1(~y[2]),
   .x2(~y[3]),
   .z1(net11)
   );

or2_df inst12 (
   .x1(~y[1]),
   .x2(net11),
   .z1(net12)
   );

                              //continued on next page
```

Figure 2.91   (Continued)

```
jkff inst13 (
   .rst_n(rst_n),
   .clk(clk),
   .j(net12),
   .k(1'b1),
   .q(y[4])
   );
endmodule
```

Figure 2.91   (Continued)

```
//test bench for the modulo-10 counter using JK flip-flops
module ctr_mod10_struc_tb;

reg rst_n, clk;        //inputs are reg for test bench
wire [1:4] y;          //outputs are wire for test bench

initial                //display count
$monitor ("Count = %b,", y);

initial                //generate reset
begin
   #0     rst_n = 1'b0;
   #5     rst_n = 1'b1;
end

initial                //generate clock
begin
   clk = 1'b0;
   forever
      #10    clk = ~clk;
end

initial                //determine length of simulation
   #200   $stop;

//instantiate the module into the test bench
ctr_mod10_struc inst1 (
   .rst_n(rst_n),
   .clk(clk),
   .y(y)
   );
endmodule
```

Figure 2.92   Test bench module for the modulo-10 counter using *JK* flip-flops with a self-starting state of $y_1 y_2 y_3 y_4$ 0000.

```
Count = 0000,
Count = 0001,
Count = 0010,
Count = 0011,
Count = 0100,
Count = 0101,
Count = 0110,
Count = 0111,
Count = 1000,
Count = 1001,
Count = 0000,
```

Figure 2.93   Outputs for the modulo-10 counter using *JK* flip-flops with a self-starting state of $y_1y_2y_3y_4$ 0000.

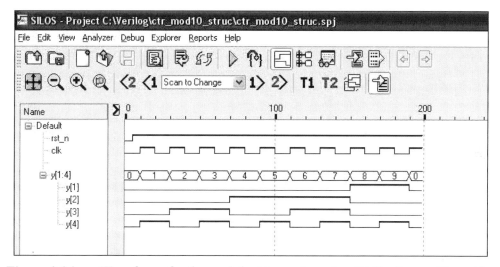

Figure 2.94   Waveforms for the modulo-10 counter using *JK* flip-flops with a self-starting state of $y_1y_2y_3y_4$ 0000.

Example 2.14   This example designs a modulo-10 counter using structural modeling with *JK* flip-flops and no self-starting state. The counter logic can be minimized considerably by not allowing the self-starting attribute; that is, the unused states — corresponding to minterm locations 10 through 15, which represent invalid BCD digits — are treated as "don't care" states. The 1s in the input maps can now combine with more minterm locations to provide a minimal number of logic gates for the δ next-state function. The input maps with the additional "don't care" states are shown in Figure 2.95.

The logic diagram with no self-starting state is shown in Figure 2.96 and is implemented from the JK input equations of Figure 2.95. Although not shown in the logic diagram, it is assumed that the JK flip-flops have a set and reset function. The following logic elements are utilized in the logic design: a 3-input dataflow AND gate (*and3_df*), a 2-input dataflow AND gate (*and2_df*), and a negative-edge triggered JK flip-flop (*jkff_neg_clk*).

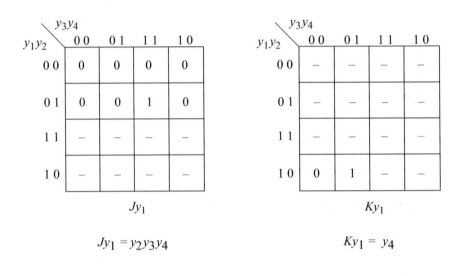

$$Jy_1 = y_2 y_3 y_4 \qquad\qquad Ky_1 = y_4$$

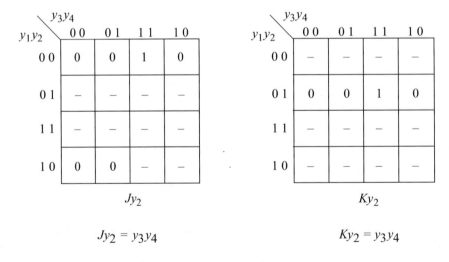

$$Jy_2 = y_3 y_4 \qquad\qquad Ky_2 = y_3 y_4$$

Continued on next page

Figure 2.95   Input maps for the modulo-10 counter of Figure 2.88 with no self-starting state.

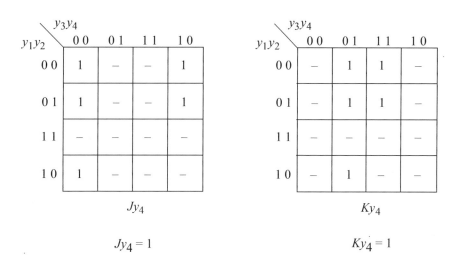

Figure 2.95   (Continued)

The λ output logic is not mandatory for a single-stage 4-bit modulo-10 counter. For a multi-stage counter, however, a signal must be made available from decade$_i$ to decade$_{i+1}$ to indicate when decade$_i$ has reached a terminal count of 1001. The logic diagram for a 3-digit, modulo-10 counter has a range of 000 to 999. Each stage of the counter contains the internal logic as shown in Figure 2.96.

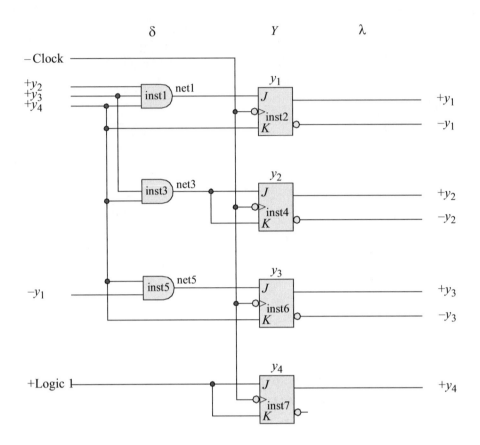

Figure 2.96   Logic diagram for the modulo-10 counter of Figure 2.88 using *JK* flip-flops, where flip-flop y_4 is the low-order stage. The counter has no self-starting state.

The structural design module is shown in Figure 2.97 and the test bench module is shown in Figure 2.98. The outputs and waveforms are shown in Figure 2.99 and Figure 2.100, respectively.

```
//structural modulo-10 counter using JK flip-flops
//no self-starting state
module ctr_mod10_jk_struc2 (rst_n, clk, y);

//define inputs and outputs
input   rst_n, clk;
output [1:4] y;
                                //continued on next page
```

Figure 2.97   Structural design module for the modulo-10 counter of Figure 2.96.

```
wire net1, net3, net5;      //define internal nets

//instantiate the logic for flip-flop y[1]
and3_df inst1 (
   .x1(y[2]),
   .x2(y[3]),
   .x3(y[4]),
   .z1(net1)
   );

jkff_neg_clk inst2 (    //set_n input is not instantiated
   .rst_n(rst_n),
   .clk(clk),
   .j(net1),
   .k(y[4]),
   .q(y[1])
   );

//instantiate the logic for flip-flop y[2]
and2_df inst3 (
   .x1(y[3]),
   .x2(y[4]),
   .z1(net3)
   );

jkff_neg_clk inst4 (
   .rst_n(rst_n),
   .clk(clk),
   .j(net3),
   .k(net3),
   .q(y[2])
   );

//instantiate the logic for flip-flop y[3]
and2_df inst5 (
   .x1(y[4]),
   .x2(~y[1]),
   .z1(net5)
   );

jkff_neg_clk inst6 (
   .rst_n(rst_n),
   .clk(clk),
   .j(net5),
   .k(y[4]),
   .q(y[3])
   );                          //continued on next page
```

Figure 2.97   (Continued)

```
//instantiate the logic for flip-flop y[4]
jkff_neg_clk inst7 (
   .rst_n(rst_n),
   .clk(clk),
   .j(1'b1),
   .k(1'b1),
   .q(y[4])
   );

endmodule
```

Figure 2.97   (Continued)

```
//test bench for modulo-10 counter using JK flip-flops
module ctr_mod10_jk_struc2_tb;

reg rst_n, clk;        //define inputs and outputs
wire [1:4] y;

//display outputs
initial
$monitor ("count = %b", y);

//define reset
initial
begin
   #0 rst_n = 1'b0;
   #5 rst_n = 1'b1;
end

//define clock
initial
begin
   clk = 1'b0;
   forever
      #10 clk = ~clk;
end

//define length of simulation
initial
begin
   #210 $finish;
end
                              //continued on next page
```

Figure 2.98   Test bench module for the modulo-10 counter of Figure 2.96.

```
//instantiate the module into the test bench
ctr_mod10_jk_struc2 inst1 (
   .rst_n(rst_n),
   .clk(clk),
   .y(y)
   );

endmodule
```

Figure 2.98   (Continued)

```
count = 0000
count = 0001
count = 0010
count = 0011
count = 0100
count = 0101
count = 0110
count = 0111
count = 1000
count = 1001
count = 0000
```

Figure 2.99   Outputs for the modulo-10 counter of Figure 2.96.

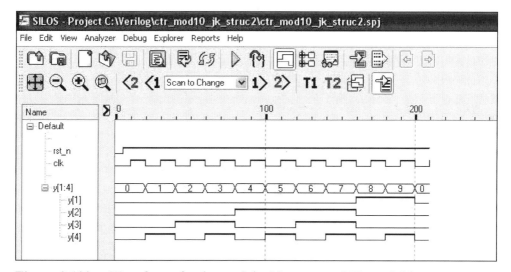

Figure 2.100   Waveforms for the modulo-10 counter of Figure 2.96.

2.2.3 Johnson Counter

Counters have a counting sequence that increases or decreases in a binary manner from a beginning value to some predefined end value. Still other counters can be designed for a unique application in which the counting sequence is neither entirely up nor entirely down. These have a nonsequential counting sequence that is prescribed by external requirements.

One such counter is the Johnson counter shown in Figure 2.101, in which the counting sequence is $y_1 y_2 y_3 = 000, 100, 110, 111, 011, 001, 000, \ldots$. The counter is reset initially to $y_1 y_2 y_3 = 000$. For six of the eight possible states for three variables, the state transitions are completely defined. The remaining two states are unspecified and can be regarded as "don't care" states in order to minimize the δ next-state logic.

The Johnson counter has the characteristic in which any two contiguous state codes (or code words) differ by only one variable. It is similar, in this respect, to a Gray code counter, which is used in Karnaugh maps.

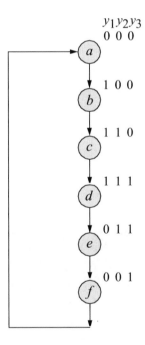

Figure 2.101   State diagram for a Johnson counter with a nonsequential counting sequence. There are two unused states: $y_1 y_2 y_3 = 010$ and 101.

The Johnson counter is also referred to as a Möbius counter, because the output of the last stage is inverted and fed back to the first stage. August F. Möbius was a German mathematician who discovered a one-sided surface that is constructed from a rectangle by holding one end fixed, rotating the opposite end through 180 degrees, and applying it to the first end.

Example 2.15 This example designs a Johnson counter using behavioral modeling with conditional statements. The behavioral design module of a 3-stage Johnson counter using the **if-else if** *conditional statements* is shown in Figure 2.102. These keywords are used as conditional statements to alter the flow of activity through a behavioral module. They permit a choice of alternative paths based upon a Boolean value obtained from a condition that is evaluated. A true value from a condition is 1 or any nonzero value; a false value is 0, **x**, or **z**. If the evaluation is false, then the next expression in the activity flow is evaluated. The test bench is shown in Figure 2.103. The outputs and waveforms are shown in Figure 2.104 and Figure 2.105, respectively.

An example of conditional statements of the type used in Figure 2.102 is shown below, which provides a choice of multiple statements. The alternative statements can be a single statement or a block of statements delimited by the keywords **begin** . . . **end**.

if (expression1) statement1;   //if expression1 is true, then statement1 is executed.
else if (expression2) statement2; //if expression2 is true, then statement2 is executed.
else if (expression3) statement3; //if expression3 is true, then statement3 is executed.

```
//behavioral for 3-bit Johnson counter using
//the if, else if conditional statements
module ctr_johnson3_bh (rst_n, clk, y);

input rst_n, clk;
output [1:3] y;

reg [1:3] y;        //outputs are reg
reg [1:3] d;        //internal next state

//reset counter and set y
always @ (posedge clk or negedge rst_n)
begin
   if (rst_n == 0)
      y <= 3'b000;
   else
      y <= d;
end

//determine counting sequence
always @ (y)
begin
      if (y == 3'b000)
            d <= 3'b100;

      else if (y == 3'b100)
            d <= 3'b110;            //continued on next page
```

Figure 2.102   Behavioral design module for the 3-stage Johnson counter.

```
      else if (y == 3'b110)
            d <= 3'b111;

      else if (y == 3'b111)
            d <= 3'b011;

      else if (y == 3'b011)
            d <= 3'b001;

      else if (y == 3'b001)
            d <= 3'b000;
end

endmodule
```

Figure 2.102   (Continued)

```
//test bench for 3-bit Johnson counter
module ctr_johnson3_bh_tb;

reg rst_n, clk;    //inputs are reg for test bench
wire [1:3] y;      //outputs are wire for test bench

initial            //display count
$monitor ("Count = %b", y);

//define reset
initial
begin
    #0 rst_n = 1'b0;
    #5 rst_n = 1'b1;
end

initial            //define clock
begin
    clk = 1'b0;
    forever
        #10 clk = ~clk;
end

initial            //define length of simulation
begin
    #140 $finish;
end                            //continued on next page
```

Figure 2.103   Test bench module for the 3-stage Johnson counter.

```
//instantiate the module into the test bench
ctr_johnson3_bh inst1 (
    .rst_n(rst_n),
    .clk(clk),
    .y(y)
    );

endmodule
```

Figure 2.103   (Continued)

```
Count = 000
Count = 100
Count = 110
Count = 111
Count = 011
Count = 001
Count = 000
Count = 100
```

Figure 2.104   Outputs for the 3-stage Johnson counter.

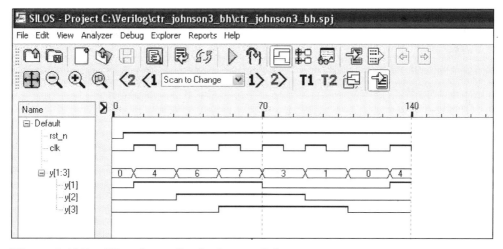

Figure 2.105   Waveforms for the 3-stage Johnson counter.

Example 2.16   This example designs a Johnson counter using behavioral modeling with the **case** statement. The 3-stage Johnson counter will use the multiple-way conditional branch characteristics of the **case** statement. The behavioral design module is shown in Figure 2.106 and the test bench module is shown in Figure 2.107. The next count of the counter is a function of the current state of the counter and is determined by the **case** statement. The outputs and waveforms are shown in Figure 2.108 and Figure 2.109, respectively.

```verilog
//behavioral for 3-bit Johnson counter
//using the case statement

module ctr_johnson3_case (rst_n, clk, y);

//define inputs and output
input rst_n, clk;
output [1:3] y;

//variables used in an always block
//are declared as registers
reg [1:3] y, next_count;

//set next count
always @ (posedge clk or negedge rst_n)
begin
   if(~rst_n)
      y <= 4'b0000;
   else
      y <= next_count;
end

//determine next count
always @ (y)
begin
   case (y)
      3'b000 : next_count = 3'b100;
      3'b100 : next_count = 3'b110;
      3'b110 : next_count = 3'b111;
      3'b111 : next_count = 3'b011;
      3'b011 : next_count = 3'b001;
      3'b001 : next_count = 3'b000;

      default : next_count = 3'b000;
   endcase

end

endmodule
```

Figure 2.106   Behavioral design module for the 3-stage Johnson counter using the **case** statement.

```verilog
//test bench for 3-bit Johnson counter

module ctr_johnson3_case_tb;

//inputs are reg for test bench
reg rst_n, clk;

//outputs are wire for test bench
wire [1:3] y;


//display count
initial
$monitor ("count = %b", y);


//define reset
initial
begin
   #0      rst_n = 1'b0;
   #5      rst_n = 1'b1;
end


//define clock
initial
begin
   clk = 1'b0;
   forever
      #10 clk = ~clk;
end


//define length of simulation
initial
   #140 $finish;


//instantiate the module into the test bench
ctr_johnson3_case inst1 (
   .rst_n(rst_n),
   .clk(clk),
   .y(y)
   );


endmodule
```

Figure 2.107   Test bench module for the 3-stage Johnson counter using the **case** statement.

```
count = 000        count = 011
count = 100        count = 001
count = 110        count = 000
count = 111        count = 100
```

Figure 2.108   Outputs for the 3-stage Johnson counter using the **case** statement.

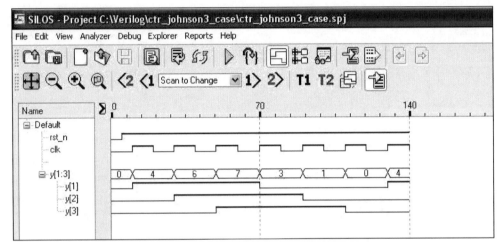

Figure 2.109   Waveforms for the 3-stage Johnson counter using the **case** statement.

Example 2.17   This example designs a Johnson counter using structural modeling with D flip-flops. The input maps are obtained from the state diagram of Figure 2.101 using D flip-flops, as shown in Figure 2.110. The maps can be derived directly from the state diagram without the necessity of generating a next-state table.

For example, from state a ($y_1 y_2 y_3 = 000$), the machine sequences to state b ($y_1 y_2 y_3 = 100$) where the next state for flip-flop y_1 is 1. Thus, a 1 is entered in minterm location $y_1 y_2 y_3 = 000$ for flip-flop y_1. Likewise, from state c the machine proceeds to state d where the next state for y_1 is 1; therefore, a 1 is entered in minterm location $y_1 y_2 y_3 = 110$ for flip-flop y_1. In a similar manner, the remaining entries are obtained for the input map for y_1, as well as for the input maps for y_2 and y_3.

The logic diagram using D flip-flops is shown in Figure 2.111. Although not shown in the logic diagram, it is assumed that the D flip-flops have a reset input. The structural design module is shown in Figure 2.112 and the test bench module is shown in Figure 2.113. The test bench uses the **$time** system function to return the current simulation time. The outputs and waveforms are shown in Figure 2.114 and Figure 2.115, respectively.

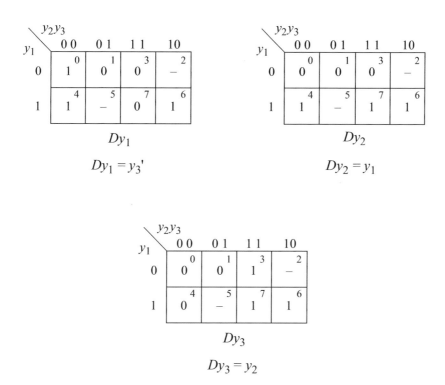

$$Dy_1 = y_3'$$

$$Dy_2 = y_1$$

$$Dy_3 = y_2$$

Figure 2.110 Input maps for the 3-stage Johnson counter using D flip-flops.

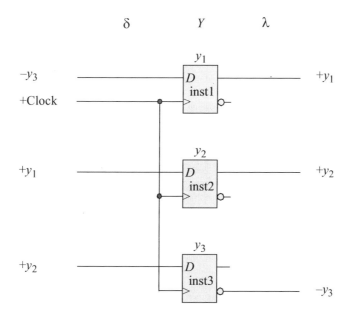

Figure 2.111 Logic diagram for the 3-stage Johnson counter using D flip-flops.

```
//structural johnson 3-bit counter
module ctr_johnson3_struc (rst_n, clk, y);

//define inputs and outputs
input rst_n, clk;
output [1:3] y;

//instantiate the logic for flip-flop y[1]
d_ff_bh inst1 (
    .rst_n(rst_n),
    .clk(clk),
    .d(~y[3]),
    .q(y[1])
    );

//instantiate the logic for flip-flop y[2]
d_ff_bh inst2 (
    .rst_n(rst_n),
    .clk(clk),
    .d(y[1]),
    .q(y[2])
    );

//instantiate the logic for flip-flop y[3]
d_ff_bh inst3 (
    .rst_n(rst_n),
    .clk(clk),
    .d(y[2]),
    .q(y[3])
    );
endmodule
```

Figure 2.112   Structural design module for the 3-stage Johnson counter using *D* flip-flops.

```
//test bench for the 3-bit johnson counter
module ctr_johnson3_struc_tb;

reg clk, rst_n;
wire [1:3] y;

//display outputs at simulation time
initial
$monitor ($time, "Count = %b", y);   //continued on next page
```

Figure 2.113   Test bench for the 3-stage Johnson counter using *D* flip-flops.

```
//define reset
initial
begin
   #0    rst_n = 1'b0;
   #10   rst_n = 1'b1;
end


//define clk
initial
begin
   clk = 1'b0;
   forever
      #10 clk = ~clk;
end


//define length of simulation
initial
begin
   #120 $finish;
end


//instantiate the module into the test bench
ctr_johnson3_struc inst1 (
   .clk(clk),
   .rst_n(rst_n),
   .y(y)
   );


endmodule
```

Figure 2.113   (Continued)

```
0       Count = 000
10      Count = 100
30      Count = 110
50      Count = 111

70      Count = 011
90      Count = 001
110     Count = 000
130     Count = 100
```

Figure 2.114   Outputs for the 3-stage Johnson counter using *D* flip-flops.

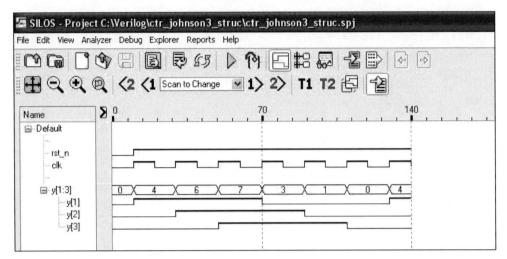

Figure 2.115   Waveforms for the 3-stage Johnson counter using *D* flip-flops.

Example 2.18 This example designs a Johnson counter using structural modeling with *JK* flip-flops. The counter is a 3-stage Johnson counter. The flip-flops in this design contain active-low set and reset inputs, and a negative-edge clock input, as shown below.

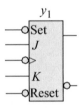

The structural design module is shown in Figure 2.116 and the test bench module is shown in Figure 2.117. The outputs and waveforms are shown in Figure 2.118 and Figure 2.119, respectively.

```
//structural johnson 3-stage counter using JK flip-flops
module ctr_johnson3_jk (clk, set_n, rst_n, y);

//define inputs and outputs
input clk, set_n, rst_n;
output [1:3] y;
                                    //continued on next page
```

Figure 2.116   Structural design module for the 3-stage Johnson counter using *JK* flip-flops.

```
//instantiate the logic for flip-flop y[1]
jkff_neg_clk inst1 (
   .set_n(1'b1),
   .rst_n(rst_n),
   .clk(clk),
   .j(~y[3]),
   .k(y[3]),
   .q(y[1])
   );

//instantiate the logic for flip-flop y[2]
jkff_neg_clk inst2 (
   .set_n(1'b1),
   .rst_n(rst_n),
   .clk(clk),
   .j(y[1]),
   .k(~y[1]),
   .q(y[2])
   );

//instantiate the logic for flip-flop y[3]
jkff_neg_clk inst3 (
   .set_n(1'b1),
   .rst_n(rst_n),
   .clk(clk),
   .j(y[2]),
   .k(~y[2]),
   .q(y[3])
   );
endmodule
```

Figure 2.116   (Continued)

```
//test bench for johnson 3-stage counter using JK flip-flops
module ctr_johnson3_jk_tb;

//define inputs and outputs
reg clk, set_n, rst_n;      //inputs are reg for test bench
wire [1:3] y;               //outputs are wire for test bench

//display outputs
initial
$monitor ("Count = %b", y);
                                    //continued on next page
```

Figure 2.117   Test bench module for the 3-stage Johnson counter using *JK* flip-flops.

```
//define reset
initial
begin
    #0  rst_n = 1'b0;
    #5  rst_n = 1'b1;
end

//define clock
initial
begin
    clk = 1'b0;
    forever
        #10   clk = ~clk;
end

//define length of simulation
initial
begin
    #150   $finish;
end

//instantiate the module into the test bench
ctr_johnson3_jk inst1 (
    .clk(clk),
    .set_n(set_n),
    .rst_n(rst_n),
    .y(y)
    );

endmodule
```

Figure 2.117   (Continued)

```
Count = 000
Count = 100
Count = 110
Count = 111

Count = 011
Count = 001
Count = 000
Count = 100
```

Figure 2.118   Outputs for the 3-stage Johnson counter using *JK* flip-flops.

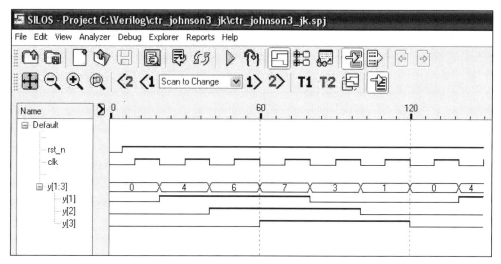

Figure 2.119   Waveforms for the 3-stage Johnson counter using *JK* flip-flops.

2.2.4 Binary-to-Gray Code Converter

One final application of counters is presented for the design of code converters. Although a code converter is more easily implemented by means of a read-only memory (ROM), it is presented in this section to illustrate the versatility of counters. It may seem inappropriate to classify a code converter as a counter, but it falls into the general category of machines that are loaded in parallel and then sequence to a new state upon the application of an active clock transition. Unlike registers such as parallel-in, parallel-out registers, where the next state is a function of the present inputs, a code converter of this type has a next state that is a function of the present state only.

The next-state table of Table 2.4 shows the relationship between the binary 8421 code and the Gray code. The Gray code belongs to a class of cyclic codes called reflective codes. Notice in the first four rows, that y_4 reflects across the reflecting axis; that is, y_4 in rows 2 and 3 is the mirror image of y_4 in rows 0 and 1. In the same manner, y_3 and y_4 reflect across the reflecting axis drawn under row 3. Thus, rows 4 through 7 reflect the state of rows 0 through 3 for y_3 and y_4. The same is true for y_2, y_3, and y_4 relative to rows 8 through 15 and rows 0 through 7.

The Gray code is an unweighted code and a Gray code counter has significant applications in sequential machine testing. By applying the outputs of an n-bit Gray code counter to the inputs of a synchronous sequential machine under test, the machine's behavior can be more easily monitored, since only one input changes during each test cycle.

A state diagram is not relevant in this application, since the machine will not sequence through a series of states. Rather, a binary input vector X_i is loaded into the machine, and after a clock pulse is applied, the corresponding Gray code word becomes the next state.

Table 2.4 Next-State Table for Converting from the Binary 8421 Code to the Gray Code

Row	Present state (Binary Code $b_1 b_2 b_3 b_4$)				Next state (Gray Code $g_1 g_2 g_3 g_4$)				
	y_1	y_2	y_3	y_4	y_1	y_2	y_3	y_4	
0	0	0	0	0	0	0	0	0	
1	0	0	0	1	0	0	0	1	
2	0	0	1	0	0	0	1	1	$\leftarrow y_4$ is reflected
3	0	0	1	1	0	0	1	0	
4	0	1	0	0	0	1	1	0	$\leftarrow$ y3 and y_4
5	0	1	0	1	0	1	1	1	are reflected
6	0	1	1	0	0	1	0	1	
7	0	1	1	1	0	1	0	0	
8	1	0	0	0	1	1	0	0	$\leftarrow y_2, y_3,$ and y_4
9	1	0	0	1	1	1	0	1	are reflected
10	1	0	1	0	1	1	1	1	
11	1	0	1	1	1	1	1	0	
12	1	1	0	0	1	0	1	0	
13	1	1	0	1	1	0	1	1	
14	1	1	1	0	1	0	0	1	
15	1	1	1	1	1	0	0	0	

Example 2.19   This example designs a binary-to-Gray code converter using structural modeling with JK flip-flops. For convenience, the excitation table for a JK flip-flop is reproduced in Table 2.5. The input maps for the code converter are shown in Figure 2.120, using JK flip-flops.

Consider the present state in row 3 ($y_1 y_2 y_3 y_4 = 0011$) of Table 2.4. Flip-flop y_3 proceeds from a present state of 1 to a next state of 1. From Table 2.5, a 1-to-1 transition specifies the JK input values to be $JK = -0$. These values are entered in the map for Jy_3 and Ky_3. Now examine row 12 ($y_1 y_2 y_3 y_4 = 1100$) of Table 2.4. Flip-flop y_3 moves from a present state of 0 to a next state of 1. This transition yields JK values of $JK = 1-$, which are entered in minterm location 1100 for Jy_3 and Ky_3, respectively.

Table 2.5 Excitation Table for a JK Flip-Flop

Present state $Y_{j(t)}$	Next state $Y_{k(t+1)}$	Flip-flop inputs JK
0	0	0 –
0	1	1 –
1	0	– 1
1	1	– 0

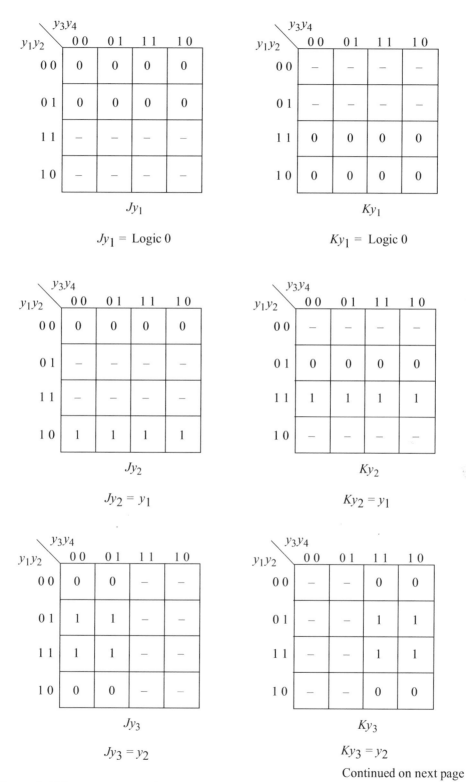

Figure 2.120   Input maps for the binary-to-Gray code converter.

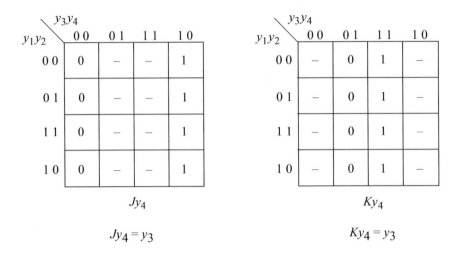

Figure 2.120   (Continued)

The ith Gray code bit g_i can be obtained from the corresponding binary code word by the following algorithm:

$$g_{n-1} = b_{n-1}$$
$$g_i = b_i \oplus b_{i+1}$$

for $0 \le i \le n - 2$, where the symbol $\oplus$ denotes modulo-2 addition. For example, using the algorithm, the 4-bit binary code word $b_3\, b_2\, b_1\, b_0 = 1010$ translates to the 4-bit Gray code word $g_3\, g_2\, g_1\, g_0 = 1111$ as follows:

$$g_3 = b_3 \qquad\qquad = 1$$
$$g_2 = b_2 \oplus b_3 = 0 \oplus 1 = 1$$
$$g_1 = b_1 \oplus b_2 = 1 \oplus 0 = 1$$
$$g_0 = b_0 \oplus b_1 = 0 \oplus 1 = 1$$

The logic diagram is shown in Figure 2.121. The machine is initially reset. The binary code word $b_1\, b_2\, b_3\, b_4$ is loaded into the code converter by generating a positive pulse on the +Load signal. The next negative clock transition performs the requisite binary-to-Gray code conversion. The machine is reset before each binary vector is applied.

From Table 2.4, it is observed that the high-order bit of the Gray code word is the same as the high-order bit of the corresponding binary code word. Therefore, all the logic associated with flip-flop y_1 can be eliminated, and $g_1 = b_1$. If all four flip-flops are an integral part of a macro logic function, then flip-flop y_1 is retained and acts as a 1-bit parallel-in, parallel-out register.

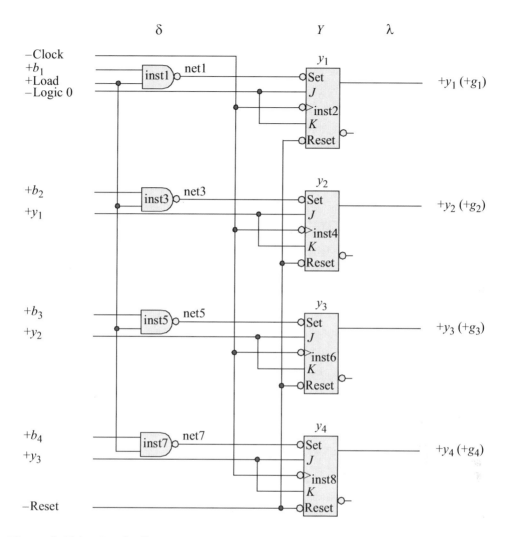

Figure 2.121 Logic diagram using *JK* flip-flops for the binary-to-Gray code converter. Flip-flop y_4 is the low-order stage.

The structural design module is shown in Figure 2.122, which instantiates the NAND gates as dataflow modules (*nand2-df*) and the negative-edge *JK* flip-flops as *jkff-neg-clk* modules. The test bench module is shown in Figure 2.123, which applies five binary vectors to the design module. The outputs and waveforms are shown in Figure 2.124 and Figure 2.125, respectively.

The reverse algorithm to convert from the Gray code to the binary 8421 code is defined as follows:

$$b_{n-1} = g_{n-1}$$
$$b_i = b_{i+1} \oplus g_i$$

```
//structural binary-to-gray code converter
//using negative edge-triggered JK flip-flops
module bin_to_gray_struc2_jk (rst_n, clk, load, b, y);

//define inputs and outputs
input rst_n, clk, load;
input [1:4] b;
output [1:4] y;

//define internal nets
wire net1, net3, net5, net7;

//instantiate the logic for flip-flop y[1]
nand2_df inst1 (
   .x1(b[1]),
   .x2(load),
   .z1(net1)
   );

jkff_neg_clk inst2 (
   .rst_n(rst_n),
   .clk(clk),
   .j(1'b0),
   .k(1'b0),
   .set_n(net1),
   .q(y[1])
   );

//instantiate the logic for flip-flop y[2]
nand2_df inst3 (
   .x1(b[2]),
   .x2(load),
   .z1(net3)
   );

jkff_neg_clk inst4 (
   .rst_n(rst_n),
   .clk(clk),
   .j(y[1]),
   .k(y[1]),
   .set_n(net3),
   .q(y[2])
   );

                              //continued on next page
```

Figure 2.122   Structural design module to convert from the 8421 binary code to the corresponding Gray code.

```
//instantiate the logic for flip-flop y[3]
nand2_df inst5 (
    .x1(b[3]),
    .x2(load),
    .z1(net5)
    );

jkff_neg_clk inst6 (
    .rst_n(rst_n),
    .clk(clk),
    .j(y[2]),
    .k(y[2]),
    .set_n(net5),
    .q(y[3])
    );

//instantiate the logic for flip-flop y[4]
nand2_df inst7 (
    .x1(b[4]),
    .x2(load),
    .z1(net7)
    );

jkff_neg_clk inst8 (
    .rst_n(rst_n),
    .clk(clk),
    .j(y[3]),
    .k(y[3]),
    .set_n(net7),
    .q(y[4])
    );

endmodule
```

Figure 2.122 (Continued)

```
//test bench for the binary-to-gray code converter
module bin_to_gray_struc2_jk_tb;

reg rst_n, clk, load;     //inputs are reg for test bench
reg[1:4] b;
wire [1:4] y;             //outputs are wire for test bench
                                    //continued on next page
```

Figure 2.123 Test bench module for converting from the 8421 binary code to the corresponding Gray code.

```
//display variables
initial
$monitor ("Binary = %b, Gray = %b", b, y);

//define clock
initial
begin
   clk = 1'b1;
   forever
      #10 clk = ~clk;
end

//apply input vectors
initial
begin
//--------------------------------
   #0    rst_n = 1'b0;   //0 time
   #3    rst_n = 1'b1;   //3 time
         b = 4'b1010;
   #2    load = 1'b1;    //5 time
   #10   load = 1'b0;    //15 time
//--------------------------------
   #5    rst_n = 1'b0;   //20 time
   #3    rst_n = 1'b1;   //23 time
         b = 4'b1000;
   #2    load = 1'b1;    //25 time
   #10   load = 1'b0;    //35 time
//--------------------------------
   #5    rst_n = 1'b0;
   #3    rst_n = 1'b1;
         b = 4'b1001;
   #2    load = 1'b1;
   #10   load = 1'b0;
//--------------------------------
   #5    rst_n = 1'b0;
   #3    rst_n = 1'b1;
         b = 4'b0100;
   #2    load = 1'b1;
   #10   load = 1'b0;
//--------------------------------
   #5    rst_n = 1'b0;
   #3    rst_n = 1'b1;
         b = 4'b0010;
   #2    load = 1'b1;
   #10   load = 1'b0;
   #10   $stop;
end                              //continue on next page
```

Figure 2.123   (Continued)

```
//instantiate the module into the test bench
bin_to_gray_struc2_jk inst1 (
    .rst_n(rst_n),
    .clk(clk),
    .load(load),
    .b(b),
    .y(y)
    );
endmodule
```

Figure 2.123   (Continued)

```
Binary = xxxx, Gray = 0000      Binary = 1001, Gray = 0000
Binary = 1010, Gray = 0000      Binary = 1001, Gray = 1001
Binary = 1010, Gray = 1010      Binary = 1001, Gray = 1101
Binary = 1010, Gray = 1111      Binary = 1001, Gray = 0000
Binary = 1010, Gray = 0000
                                Binary = 0100, Gray = 0000
Binary = 1000, Gray = 0000      Binary = 0100, Gray = 0100
Binary = 1000, Gray = 1000      Binary = 0100, Gray = 0110
Binary = 1000, Gray = 1100      Binary = 0100, Gray = 0000
Binary = 1000, Gray = 0000
                                Binary = 0010, Gray = 0000
                                Binary = 0010, Gray = 0010
                                Binary = 0010, Gray = 0011
```

Figure 2.124   Outputs for the binary-to-Gray code converter.

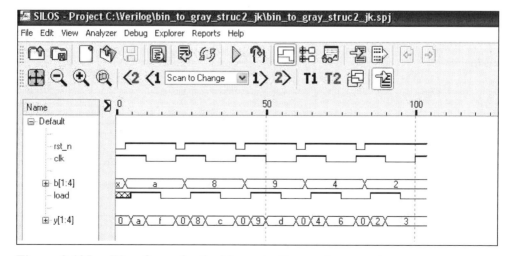

Figure 2.125   Waveforms for the binary-to-Gray code converter.

A much simpler way to convert from binary code to Gray code is using built-in primitives or behavioral modeling, both of which will be shown in the next two examples.

Example 2.20   This example designs a binary-to-Gray code converter using built-in primitives. The table to convert from binary to Gray is reproduced in Table 2.6. The logic diagram for the binary-to-Gray code converter is shown in Figure 2.126, as derived from Table 2.6 and the algorithm previously presented.

Table 2.6 Binary-to-Gray Code Conversion

Binary				Gray			
b_3	b_2	b_1	b_0	g_3	g_2	g_1	g_0
0	0	0	0	0	0	0	0
0	0	0	1	0	0	0	1
0	0	1	0	0	0	1	1
0	0	1	1	0	0	1	0
0	1	0	0	0	1	1	0
0	1	0	1	0	1	1	1
0	1	1	0	0	1	0	1
0	1	1	1	0	1	0	0
1	0	0	0	1	1	0	0
1	0	0	1	1	1	0	1
1	0	1	0	1	1	1	1
1	0	1	1	1	1	1	0
1	1	0	0	1	0	1	0
1	1	0	1	1	0	1	1
1	1	1	0	1	0	0	1
1	1	1	1	1	0	0	0

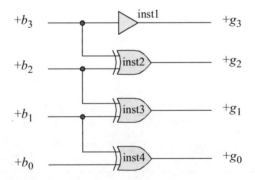

Figure 2.126   Logic diagram for a binary-to-Gray code converter.

The built-in primitive design module is shown in Figure 2.127 and the test bench module is shown in Figure 2.128. The outputs and waveforms are shown in Figure 2.129 and Figure 2.130, respectively.

```verilog
//built-in primitives binary-to-gray converter
module bin_to_gray_bip (bin, gray);

input [3:0] bin;
output [3:0] gray;

buf     inst1 (gray[3], bin[3]);

xor     inst2 (gray[2], bin[3], bin[2]),
        inst3 (gray[1], bin[2], bin[1]),
        inst4 (gray[0], bin[1], bin[0]);
endmodule
```

Figure 2.127 Design module for binary-to-Gray conversion using built-in primitives.

```verilog
//test bench for binary-to-gray converter using
//built-in primitives
module bin_to_gray_bip_tb;

reg [3:0] bin;
wire [3:0] gray;

//apply input vectors and display variables
initial
begin: apply_stimulus
   reg [4:0] invect;
   for (invect = 0; invect < 16; invect = invect + 1)
      begin
         bin = invect [4:0];
         #10 $display ("binary = %b, gray = %b", bin, gray);
      end
end

//instantiate the module into the test bench
bin_to_gray_bip inst1 (
   .bin(bin),
   .gray(gray)
   );
endmodule
```

Figure 2.128 Test bench module for binary-to-Gray conversion using built-in primitives.

```
binary = 0000, gray = 0000
binary = 0001, gray = 0001
binary = 0010, gray = 0011
binary = 0011, gray = 0010

binary = 0100, gray = 0110
binary = 0101, gray = 0111
binary = 0110, gray = 0101
binary = 0111, gray = 0100

binary = 1000, gray = 1100
binary = 1001, gray = 1101
binary = 1010, gray = 1111
binary = 1011, gray = 1110

binary = 1100, gray = 1010
binary = 1101, gray = 1011
binary = 1110, gray = 1001
binary = 1111, gray = 1000
```

Figure 2.129   Outputs for the binary-to-Gray conversion.

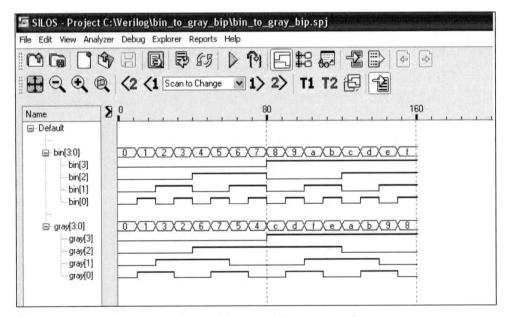

Figure 2.130   Waveforms for the binary-to-Gray conversion.

Example 2.21 This example designs a binary-to-Gray code converter using behavioral modeling. The design module for the behavioral design is shown in Figure 2.131 and the test bench module is shown in Figure 2.132. The outputs and waveforms are shown in Figure 2.133 and Figure 2.134, respectively.

```
//behavioral binary-to-gray converter
module bin_to_gray_bh (binary, gray);

//define inputs and outputs
input [3:0] binary;
output [3:0] gray;

//variables used in always are declared as type reg
reg [3:0] gray;

//define gray code
always @ (binary)
begin
   gray[3] <= binary[3];
   gray[2] <= binary[3] ^ binary[2];
   gray[1] <= binary[2] ^ binary[1];
   gray[0] <= binary[1] ^ binary[0];
end
endmodule
```

Figure 2.131 Behavioral design module for binary-to-Gray code conversion.

```
//test bench for behavioral binary-to-gray converter
module bin_to_gray_bh_tb;

reg [3:0] binary;      //inputs are reg for test bench
wire [3:0] gray;       //outputs are wire for test bench

//apply input vectors and display outputs
initial
begin: apply_stimulus
   reg [4:0] invect;
   for (invect = 0; invect < 16; invect = invect + 1)
      begin
         binary = invect [4:0];
         #10 $display ("binary = %b, gray = %b", binary, gray);
      end
end                                    //continued on next page
```

Figure 2.132 Test bench module for binary-to-Gray code conversion.

```
//instantiate the module into the test bench
bin_to_gray_bh inst1 (
   .binary(binary),
   .gray(gray)
   );

endmodule
```

Figure 2.132   (Continued)

```
binary = 0000,  gray = 0000      binary = 1000,  gray = 1100
binary = 0001,  gray = 0001      binary = 1001,  gray = 1101
binary = 0010,  gray = 0011      binary = 1010,  gray = 1111
binary = 0011,  gray = 0010      binary = 1011,  gray = 1110

binary = 0100,  gray = 0110      binary = 1100,  gray = 1010
binary = 0101,  gray = 0111      binary = 1101,  gray = 1011
binary = 0110,  gray = 0101      binary = 1110,  gray = 1001
binary = 0111,  gray = 0100      binary = 1111,  gray = 1000
```

Figure 2.133   Outputs for binary-to-Gray code conversion.

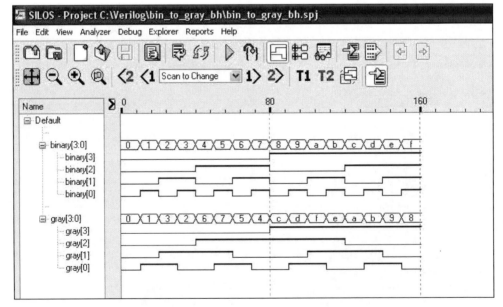

Figure 2.134   Waveforms for binary-to-Gray code conversion.

2.3 Moore Machines

This section presents procedures for synthesizing Moore machines using behavioral modeling, D flip-flops, and JK flip-flops. The primary focus will be on the synthesis of *deterministic synchronous sequential machines*, in which the next state is uniquely determined by the present state $Y_{j(t)}$ and the present inputs $X_{i(t)}$. Moore machines are synchronous sequential machines in which the output function λ produces an output vector Z_r which is determined by the present state only, and is not a function of the present inputs.

The state diagram, shown in Figure 2.135, will be used for all design methodologies. To minimize the amount of hardware, the state diagram utilizes adjacent state codes; that is, when a state has two possible next states, then the two next states should be adjacent — differ by only one variable. This will provide a maximum number of 1s in adjacent squares of the input maps.

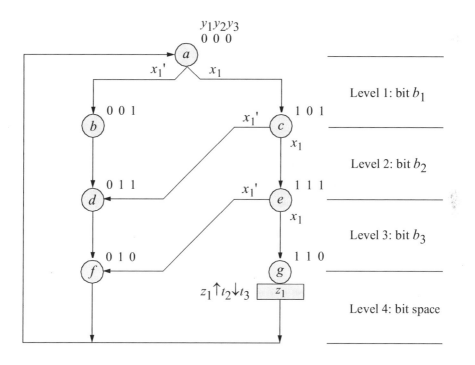

Figure 2.135   State diagram for a Moore machine, which generates an output z_1 whenever a 3-bit word $x_1 = 111$. The state codes are adjacent and there is one unused state, $y_1 y_2 y_3 = 100$.

The state diagram graphically describes the machine's behavior. Seven states are required, providing four state levels — one level for each bit in the 3-bit words and one level for the bit space between words, at which time output z_1 will be displayed if the word contained a sequence of $x_1 = 111$.

For a Moore machine, the outputs can be asserted for segments of the clock period rather than for the entire clock period only. This is illustrated below where the clock transitions (t_i) define the clock cycles for an active positive clock transition, and hence, the state times. This applies to both positive and negative active clocks. Two clock cycles are shown, one for the present state $Y_{j(t)}$ and one for the next state $Y_{k(t+1)}$.

$$t_1 = \text{beginning of the present state } Y_{j(t)}$$
$$t_2 = \text{middle of the present state } Y_{j(t)}$$
$$t_3 = \text{end of the present state } Y_{j(t)}$$
$$t_4 = \text{middle of the next state } Y_{k(t+1)}$$

Output z_1 can be active for segments of the clock period, as shown below.

$$z_1 \uparrow t_1 \downarrow t_2$$
$$z_1 \uparrow t_2 \downarrow t_3$$
$$z_1 \uparrow t_1 \downarrow t_3$$
$$z_1 \uparrow t_2 \downarrow t_4$$

The leading edge of the clock pulse, which defines the beginning of the present state, is labeled t_1. The leading edge may be a positive or negative clock transition and is used for clocking positive- or negative-edge-triggered devices, respectively. All assertion/deassertion times are referenced to the present state $Y_{j(t)}$. Time t_2 occurs at the middle of the present state; time t_3 occurs at the end of the present state; and time t_4 occurs at the midpoint of the next state $Y_{k(t+1)}$.

The assertion of an output is indicated by an up-arrow ($\uparrow$); deassertion is indicated by a down-arrow ($\downarrow$). The output assertion/deassertion times for a Mealy machine cannot be uniquely specified as for a Moore machine, because the outputs are contingent not only upon a specific state but also upon the input variables, whose assertion times may not be known.

The *output symbol* is represented by a rectangle and is placed immediately following the state symbol — designated by a circle — for a Moore machine, or placed immediately after an input variable that causes the output to become active for a Mealy machine. Thus, the output for a Moore machine is a function of the present state only, whereas the output for a Mealy machine is a function of both the present state and the present input.

Asserting the output signals at various times and for different durations provides more flexibility in the λ output logic. Waveforms that are asserted during the following times $z_1 \uparrow t_2 \downarrow t_3$ and $z_1 \uparrow t_2 \downarrow t_4$ are especially useful in avoiding glitches. Glitches are discussed in detail in a later chapter.

2.3.1 Design Using Behavioral Modeling

Behavioral modeling describes the *behavior* of a digital system and is not concerned with the direct implementation of logic gates but more on the architecture of the system. There is no requirement for input maps or output maps. This is an algorithmic approach to hardware implementation and represents a higher level of abstraction than gate-level design.

Using the state diagram of Figure 2.135, the behavioral design module of Figure 2.136 is obtained. The test bench module is shown in Figure 2.137. The system task **\$random** is used in the test bench to randomly select a value for input x_1 from the values 0 and 1. The outputs and waveforms are shown in Figure 2.138 and Figure 2.139, respectively.

```
//behavioral moore synchronous sequential machine
//to detect a sequence of 111 on a serial data line
module moore_ssm3 (clk, rst_n, x1, y, z1);

input clk, rst_n, x1;

output [2:0] y;                  //y is an array of 3 bits
output z1;

wire clk, rst_n, x1;

reg [2:0] y, next_state;         //outputs are reg in always
reg z1;


//assign state codes
parameter    state_a = 3'b000,   //parameter defines a constant
             state_b = 3'b001,   //state names must have at
             state_c = 3'b101,   //least two characters
             state_d = 3'b011,
             state_e = 3'b111,
             state_f = 3'b010,
             state_g = 3'b110;


//set next state
always @ (posedge clk or negedge rst_n)
begin
   if (~rst_n)                   //if (~rst_n) is true (1),
      y <= #3 state_a;           //then y <= state_a #3 later
   else
      y <= #3 next_state;
end                              //continued on next page
```

Figure 2.136   Behavioral design module for the Moore machine of Figure 2.135.

```verilog
always @ (y or clk)      //determine output
begin
   if (y == state_g)
       begin
          if (~clk)
             z1 = 1'b1;
          else
             z1 = 1'b0;
       end

   else
       z1 = 1'b0;
end

always @ (y or x1)//determine next state
begin
   case (y)                             //case is a multiple-way
       state_a:                         //conditional branch.
          if (x1)                       //if y = state_a, then
             next_state = state_c;      //execute if ... else
          else
             next_state = state_b;

       state_b: next_state = state_d;

       state_c:
          if (x1)
             next_state = state_e;
          else
             next_state = state_d;

       state_d: next_state = state_f;

       state_e:
          if (x1)
             next_state = state_g;
          else
             next_state = state_f;

       state_f: next_state = state_a;

       state_g: next_state = state_a;

       default: next_state = state_a;
   endcase
end
endmodule
```

Figure 2.136   (Continued)

```
//test bench for moore_ssm3
module moore_ssm3_tb;

reg clk, x1, rst_n;
wire [2:0] y;
wire z1;

//display variables
initial
$monitor ("x1 = %b, state = %b, z1 = %b", x1, y, z1);

//define clock
initial
begin
   clk = 1'b0;
   forever                  //forever continually executes
      #10 clk = ~clk;    //the procedural statement
end

//define input vectors
initial
begin
   #0 x1 = 1'b0;
      rst_n = 1'b0;
   #5 rst_n = 1'b1;

   @ (posedge clk)              //if x1=0 in state_a,
                                //go to state_b (001)
   x1 = $random;@(posedge clk)  //if x1=0/1 in state_b,
                                //go to state_d (011)
   x1 = $random;@(posedge clk)  //if x1=0/1 in state_d,
                                //go to state_f (010)
   x1 = $random;@(posedge clk)  //if x1=0/1 in state_f,
                                //go to state_a (000)
   x1 = 1'b1;@(posedge clk)     //if x1=1 in state_a,
                                //go to state_c (101)
   x1 = 1'b1;@(posedge clk)     //if x1=1 in state_c,
                                //go to state_e (111)
   x1 = 1'b1;@(posedge clk)     //if x1=1 in state_e,
                                //go to state_g (110);z1=1
   x1 = $random;@(posedge clk)  //if x1=0/1 in state_g,
                                //go to state_a (000)
   x1 = 1'b1;@(posedge clk)     //if x1=1 in state_a,
                                //go to state_c (101)
   x1 = 1'b0;@(posedge clk)     //if x1=0 in state_c,
                                //go to state_d (011)
                                     //continued on next page
```

Figure 2.137   Test bench module for the Moore machine of Figure 2.135.

```
    x1 = $random;@(posedge clk)    //if x1=0/1 in state_d,
                                   //go to state_f (010)
    x1 = $random;@(posedge clk)    //if x1=0/1 in state_f,
                                   //go to state_a (000)
    x1 = 1'b1;@(posedge clk)       //if x1=1 in state_a,
                                   //go to state_c (101)
    x1 = 1'b1;@(posedge clk)       //if x1=1 in state_c,
                                   //go to state_e (111)
    x1 = 1'b0;@(posedge clk)       //if x1=0 in state_e,
                                   //go to state_f (010)
    x1 = $random;@(posedge clk)    //if x1=0/1 in state_f,
                                   //go to state_a (000)

    #100   $stop;
end

//instantiate the module into the test bench
moore_ssm3 inst1 (
    .clk(clk),
    .rst_n(rst_n),
    .x1(x1),
    .y(y),
    .z1(z1)
    );

endmodule
```

Figure 2.137   (Continued)

```
x1 = 0, state = xxx, z1 = 0    x1 = 1, state = 101, z1 = 0
x1 = 0, state = 000, z1 = 0    x1 = 1, state = 011, z1 = 0
x1 = 0, state = 001, z1 = 0    x1 = 1, state = 010, z1 = 0
x1 = 1, state = 001, z1 = 0    x1 = 1, state = 000, z1 = 0
x1 = 1, state = 011, z1 = 0    x1 = 1, state = 101, z1 = 0
x1 = 1, state = 010, z1 = 0    x1 = 0, state = 101, z1 = 0
x1 = 1, state = 000, z1 = 0    x1 = 0, state = 111, z1 = 0
x1 = 1, state = 101, z1 = 0    x1 = 1, state = 111, z1 = 0

x1 = 1, state = 111, z1 = 0    x1 = 1, state = 010, z1 = 0
x1 = 1, state = 110, z1 = 0    x1 = 1, state = 000, z1 = 0
x1 = 1, state = 110, z1 = 1    x1 = 1, state = 101, z1 = 0
x1 = 1, state = 110, z1 = 0    x1 = 1, state = 111, z1 = 0
x1 = 1, state = 000, z1 = 0    x1 = 1, state = 110, z1 = 0
x1 = 0, state = 000, z1 = 0    x1 = 1, state = 110, z1 = 1
x1 = 0, state = 101, z1 = 0    x1 = 1, state = 110, z1 = 0
                               x1 = 1, state = 000, z1 = 0
```

Figure 2.138   Outputs for the Moore machine of Figure 2.135.

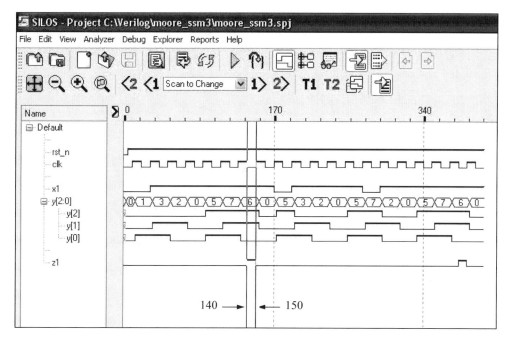

Figure 2.139   Waveforms for the Moore machine of Figure 2.135.

2.3.2 Design Using Structural Modeling with *D* Flip-Flops, AND Gates, and an OR Gate

The input maps, shown in Figure 2.140, are derived from the state diagram of Figure 2.135. Input x_1 is used as a map-entered-variable. Refer to the input map for flip-flop y_1. Since the purpose of an input map is to obtain the flip-flop input equations by combining 1s in the minterm locations, the variable x_1 is entered as the value in minterm location $y_1 y_2 y_3 = 000$. That is, y_1 has a next value of 1 if and only if x_1 has a value of 1. The same reasoning applies to all other minterm entries, except minterm location 4, which represents the unused state and is treated as a "don't care" state.

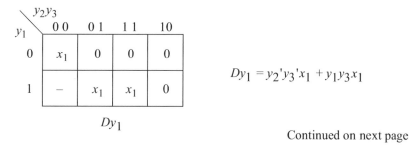

$$Dy_1 = y_2' y_3' x_1 + y_1 y_3 x_1$$

Continued on next page

Figure 2.140   Input maps for the Moore machine of Figure 2.135.

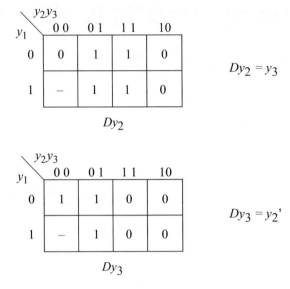

$Dy_2 = y_3$

$Dy_3 = y_2'$

Figure 2.140   (Continued)

The logic diagram is shown in Figure 2.141 using AND gates, OR gates, and D flip-flops with an implied reset input. The structural design module is shown in Figure 2.142. The test bench module is shown in Figure 2.143. The outputs and waveforms are shown in Figure 2.144 and Figure 2.145, respectively.

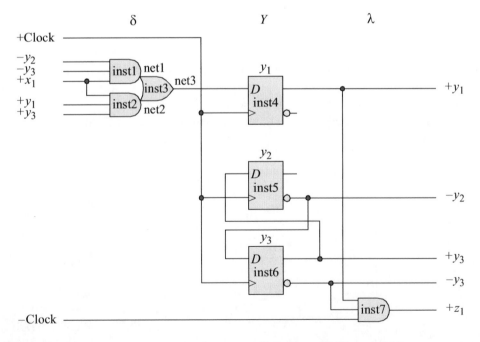

Figure 2.141   Logic diagram for the Moore machine of Figure 2.135.

```verilog
//structural moore machine using D flip-flops
module moore_ssm24a (rst_n, clk, x1, y, z1);

//define inputs and outputs
input rst_n, clk, x1;
output [1:3] y;
output z1;

//define internal nets
wire net1, net2, net3;

//instantiate the logic for flip-flop y[1]
and3_df inst1 (
   .x1(~y[2]),
   .x2(~y[3]),
   .x3(x1),
   .z1(net1)
   );

and3_df inst2 (
   .x1(x1),
   .x2(y[1]),
   .x3(y[3]),
   .z1(net2)
   );

or2_df inst3 (
   .x1(net1),
   .x2(net2),
   .z1(net3)
   );

d_ff_bh inst4 (
   .rst_n(rst_n),
   .clk(clk),
   .d(net3),
   .q(y[1])
   );

//instantiate the logic for flip-flop y[2]
d_ff_bh inst5 (
   .rst_n(rst_n),
   .clk(clk),
   .d(y[3]),
   .q(y[2])
   );
                           //continued on next page
```

Figure 2.142   Structural design module for the Moore machine of Figure 2.141.

```
//instantiate the logic for flip-flop y[3]
d_ff_bh inst6 (
   .rst_n(rst_n),
   .clk(clk),
   .d(~y[2]),
   .q(y[3])
   );

//instantiate the logic for output z1
and3_df inst7 (
   .x1(y[1]),
   .x2(~y[3]),
   .x3(~clk),
   .z1(z1)
   );

endmodule
```

Figure 2.142   (Continued)

```
//test bench for moore_ssm24a
module moore_ssm24a_tb;

reg rst_n, clk, x1;
wire [1:3] y;
wire z1;

initial                    //define clock
begin
   clk = 1'b0;
   forever
      #10 clk = ~clk;
end

initial                    //display variables
$monitor ("x1 = %b, state = %b, z1 = %b", x1, y, z1);

//define input sequence
initial
begin
   #0     rst_n = 1'b0;  //reset to state_a (000)
          clk = 1'b0;
          x1 = 1'b0;
   #5     rst_n = 1'b1;  //deassert reset
                                   //continued on next page
```

Figure 2.143   Test bench module for the Moore machine of Figure 2.142.

```
//-------------------------------------------------------
   #60    x1 = 1'b1;
   #40    x1 = 1'b0;

   #40    x1 = 1'b1;
   #100   x1 = 1'b0;
//-------------------------------------------------------
   #60    $stop;
end

//instantiate the module into the test bench
moore_ssm24a inst1 (
   .rst_n(rst_n),
   .clk(clk),
   .x1(x1),
   .y(y),
   .z1(z1)
   );

endmodule
```

Figure 2.143   (Continued)

```
x1 = 0,  state = 000,  z1 = 0
x1 = 0,  state = 001,  z1 = 0
x1 = 0,  state = 011,  z1 = 0
x1 = 0,  state = 010,  z1 = 0
x1 = 0,  state = 000,  z1 = 0
x1 = 1,  state = 000,  z1 = 0
x1 = 1,  state = 101,  z1 = 0
x1 = 1,  state = 111,  z1 = 0
x1 = 0,  state = 111,  z1 = 0
x1 = 0,  state = 010,  z1 = 0
x1 = 0,  state = 000,  z1 = 0
x1 = 1,  state = 000,  z1 = 0
x1 = 1,  state = 101,  z1 = 0
x1 = 1,  state = 111,  z1 = 0
x1 = 1,  state = 110,  z1 = 0
x1 = 1,  state = 110,  z1 = 1
x1 = 1,  state = 000,  z1 = 0
x1 = 1,  state = 101,  z1 = 0
x1 = 0,  state = 101,  z1 = 0
x1 = 0,  state = 011,  z1 = 0
x1 = 0,  state = 010,  z1 = 0
x1 = 0,  state = 000,  z1 = 0
```

Figure 2.144   Outputs for the Moore machine of Figure 2.142.

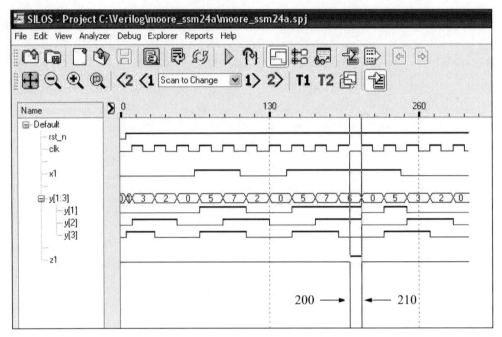

Figure 2.145   Waveforms for the Moore machine of Figure 2.142.

2.3.3 Design Using Structural Modeling with *D* Flip-Flops, NOR Gates, and an OR Gate

The characteristic table for a *D* flip-flop is reproduced in Table 2.7. The state diagram of Figure 2.135 will now be designed using structural modeling with *D* flip-flops, NOR gates, and an OR gate, as shown in the logic diagram of Figure 2.146. Although not shown, it is assumed that the *D* flip-flops have a reset input. The output will assume the state of the *D* input at the next positive clock transition. After the occurrence of the clock's positive edge, any change to the *D* input will not affect the output until the next active clock transition.

Table 2.7 *D* Flip-Flop Characteristic Table

Data input D	Present state $Y_{j(t)}$	Next state $Y_{k(t+1)}$
0	0	0
0	1	0
1	0	1
1	1	1

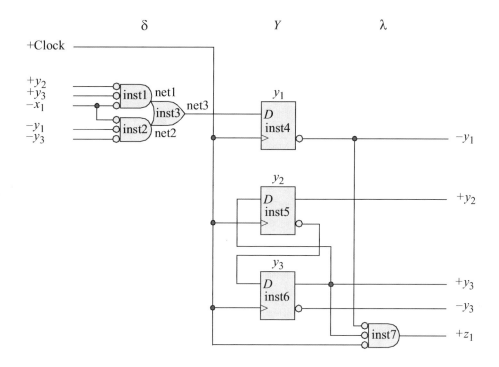

Figure 2.146 Logic diagram for the Moore machine of Figure 2.135 using NOR gates and an OR gate for the input logic.

The structural design module is shown in Figure 2.147 and the test bench module is shown in Figure 2.148. The outputs and waveforms are shown in Figure 2.149 and Figure 2.150, respectively.

```verilog
//structural moore machine using D flip-flops and NOR gates
module moore_ssm24_nor (rst_n, clk, x1, y, z1);

//define inputs and outputs
input rst_n, clk, x1;

output [1:3] y;
output z1;

//define internal nets
wire net1, net2, net3;

                                    //continued on next page
```

Figure 2.147 Structural design module for the Moore machine of Figure 2.146.

```
//instantiate the logic for flip-flop y[1]
nor3_df inst1 (
   .x1(y[2]),
   .x2(y[3]),
   .x3(~x1),
   .z1(net1)
   );

nor3_df inst2 (
   .x1(~x1),
   .x2(~y[1]),
   .x3(~y[3]),
   .z1(net2)
   );

or2_df inst3 (
   .x1(net1),
   .x2(net2),
   .z1(net3)
   );

d_ff_bh inst4 (
   .rst_n(rst_n),
   .clk(clk),
   .d(net3),
   .q(y[1])
   );

//instantiate the logic for flip-flop y[2]
d_ff_bh inst5 (
   .rst_n(rst_n),
   .clk(clk),
   .d(y[3]),
   .q(y[2])
   );

//instantiate the logic for flip-flop y[3]
d_ff_bh inst6 (
   .rst_n(rst_n),
   .clk(clk),
   .d(~y[2]),
   .q(y[3])
   );

                                    //continued on next page
```

Figure 2.147   (Continued)

```
//instantiate the logic for output z1
nor3_df inst7 (
   .x1(~y[1]),
   .x2(y[3]),
   .x3(clk),
   .z1(z1)
   );

endmodule
```

Figure 2.147   (Continued)

```
//test bench for moore_ssm24a
module moore_ssm24_nor_tb;

reg rst_n, clk, x1;
wire [1:3] y;
wire z1;

//define clock
initial
begin
   clk = 1'b0;
   forever
      #10 clk = ~clk;
end

//display variables
initial
$monitor ("x1 = %b, state = %b, z1 = %b", x1, y, z1);

//define input sequence
initial
begin
   #0     rst_n = 1'b0;      //reset to state_a (000)
          clk = 1'b0;
          x1 = 1'b0;
   #5     rst_n = 1'b1;      //deassert reset

   #60    x1 = 1'b1;
   #40    x1 = 1'b0;

   #40    x1 = 1'b1;
   #100   x1 = 1'b0;
   #60    $stop;
end                                    //continued on next page
```

Figure 2.148   Test bench module for the Moore machine of Figure 2.146.

```
//instantiate the module into the test bench
moore_ssm24_nor inst1 (
    .rst_n(rst_n),
    .clk(clk),
    .x1(x1),
    .y(y),
    .z1(z1)
    );
endmodule
```

Figure 2.148   (Continued)

```
x1 = 0, state = 000, z1 = 0      x1 = 1, state = 000, z1 = 0
x1 = 0, state = 001, z1 = 0      x1 = 1, state = 101, z1 = 0
x1 = 0, state = 011, z1 = 0      x1 = 1, state = 111, z1 = 0
x1 = 0, state = 010, z1 = 0      x1 = 1, state = 110, z1 = 0
x1 = 0, state = 000, z1 = 0      x1 = 1, state = 110, z1 = 1
x1 = 1, state = 000, z1 = 0      x1 = 1, state = 000, z1 = 0
x1 = 1, state = 101, z1 = 0      x1 = 1, state = 101, z1 = 0
x1 = 1, state = 111, z1 = 0      x1 = 0, state = 101, z1 = 0
x1 = 0, state = 111, z1 = 0      x1 = 0, state = 011, z1 = 0
x1 = 0, state = 010, z1 = 0      x1 = 0, state = 010, z1 = 0
x1 = 0, state = 000, z1 = 0      x1 = 0, state = 000, z1 = 0
```

Figure 2.149   Outputs for the Moore machine of Figure 2.146.

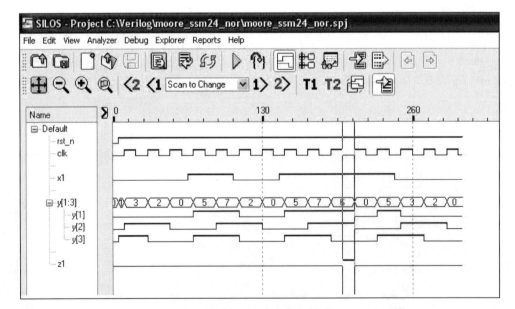

Figure 2.150   Waveforms for the Moore machine of Figure 2.146.

2.3.4 Design Using Structural Modeling with *JK* Flip-Flops, AND Gates, and an OR Gate

The Moore machine represented by the state diagram of Figure 2.135 will now be designed using *JK* flip-flops and is functionally equivalent to the design using *D* flip-flops. The excitation table for a *JK* flip-flop is reproduced in Table 2.8. The input maps are derived from the state diagram and are shown in Figure 2.151. The output map will contain a "don't care" condition in minterm location $y_1 y_2 y_3 = 100$ and can be combined with the 1 in minterm location $y_1 y_2 y_3 = 110$ to yield the output equation of $z_1 = y_1 y_3{}'$.

Table 2.8 Excitation Table for a *JK* Flip-Flop

Present state $Y_{j(t)}$	Next state $Y_{k(t+1)}$	Flip-flop inputs JK
0	0	0 –
0	1	1 –
1	0	– 1
1	1	– 0

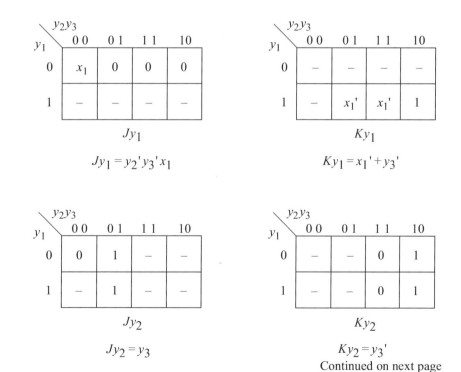

Figure 2.151 Input maps for the Moore machine of Figure 2.135 using *JK* flip-flops.

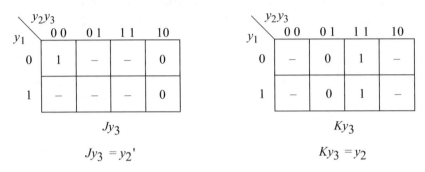

$$Jy_3$$

$$Jy_3 = y_2'$$

$$Ky_3$$

$$Ky_3 = y_2$$

Figure 2.151   (Continued)

The δ input logic incorporates less hardware than the equivalent design using D flip-flops. This is due to the inherent added logic functions supplied by the JK data inputs. As stated previously, the unused state can be utilized in the minimization process only because the assertion for z_1 occurs after the machine has stabilized.

The logic diagram is shown in Figure 2.152. The structural design module is shown in Figure 2.153 and the test bench module is shown in Figure 2.154. The JK flip-flops have an implied reset. The outputs and waveforms are shown in Figure 2.155 and Figure 2.156, respectively. Since the flip-flops are negative-edge triggered devices, the output will be asserted when the clock pulse is at a positive voltage level.

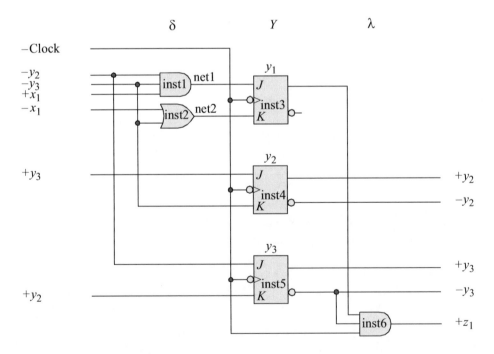

Figure 2.152   Logic diagram for the Moore machine of Figure 2.135 using JK flip-flops.

```
//Structural moore machine using JK flip-flops

module moore_ssm24_jk (rst_n, set_n, clk, x1, y, z1);

//define inputs and outputs
input rst_n, set_n, clk, x1;
output [1:3] y;
output z1;

//define internal nets
wire net1, net2;

//instantiate the logic for flip-flop y[1]
and3_df inst1 (
    .x1(~y[2]),
    .x2(~y[3]),
    .x3(x1),
    .z1(net1)
    );

or2_df inst2 (
    .x1(~x1),
    .x2(~y[3]),
    .z1(net2)
    );

jkff_neg_clk inst3 (
    .rst_n(rst_n),
    .set_n(set_n),
    .clk(clk),
    .j(net1),
    .k(net2),
    .q(y[1])
    );

//instantiate the logic for flip-flop y[2]
jkff_neg_clk inst4 (
    .rst_n(rst_n),
    .set_n(set_n),
    .clk(clk),
    .j(y[3]),
    .k(~y[3]),
    .q(y[2])
    );

                            //continued on next page
```

Figure 2.153   Structural design module for the Moore machine of Figure 2.152 using *JK* flip-flops.

```
//instantiate the logic for flip-flop y[3]
jkff_neg_clk inst5 (
   .rst_n(rst_n),
   .set_n(set_n),
   .clk(clk),
   .j(~y[2]),
   .k(y[2]),
   .q(y[3])
   );

//instantiate the logic for output z1
and3_df inst6 (
   .x1(y[1]),
   .x2(~y[3]),
   .x3(clk),
   .z1(z1)
   );

endmodule
```

Figure 2.153   (Continued)

```
//test bench for moore_ssm24_jk

module moore_ssm24_jk_tb;

reg rst_n, set_n, clk, x1;

wire [1:3] y;
wire z1;

//define clock
initial
begin
   clk = 1'b0;
   forever
      #10 clk = ~clk;
end

//display variables
initial
$monitor ("x1 = %b, state = %b, z1 = %b", x1, y, z1);

                                    //continued on next page
```

Figure 2.154   Test bench module for the Moore machine of Figure 2.152.

```
//define input sequence
initial
begin
   #0     rst_n = 1'b0;  //reset to state_a (000)
          set_n = 1'b1;
          clk = 1'b0;
          x1 = 1'b0;
   #5     rst_n = 1'b1;  //deassert reset
//----------------------------------------------------
   #80    x1 = 1'b1;
   #40    x1 = 1'b0;

   #40    x1 = 1'b1;
   #100   x1 = 1'b0;
//----------------------------------------------------
   #80    $stop;
end

//instantiate the module into the test bench
moore_ssm24_jk inst1 (
   .rst_n(rst_n),
   .set_n(set_n),
   .clk(clk),
   .x1(x1),
   .y(y),
   .z1(z1)
   );

endmodule
```

Figure 2.154   (Continued)

```
x1 = 0, state = 000, z1 = 0    x1 = 1, state = 000, z1 = 0
x1 = 0, state = 001, z1 = 0    x1 = 1, state = 101, z1 = 0
x1 = 0, state = 011, z1 = 0    x1 = 1, state = 111, z1 = 0
x1 = 0, state = 010, z1 = 0    x1 = 1, state = 110, z1 = 0
x1 = 0, state = 000, z1 = 0    x1 = 1, state = 110, z1 = 1
x1 = 1, state = 000, z1 = 0    x1 = 1, state = 000, z1 = 0
x1 = 1, state = 101, z1 = 0    x1 = 1, state = 101, z1 = 0
x1 = 1, state = 111, z1 = 0    x1 = 0, state = 101, z1 = 0
x1 = 0, state = 111, z1 = 0    x1 = 0, state = 011, z1 = 0
x1 = 0, state = 010, z1 = 0    x1 = 0, state = 010, z1 = 0
x1 = 0, state = 000, z1 = 0    x1 = 0, state = 000, z1 = 0
```

Figure 2.155   Outputs for the Moore machine of Figure 2.152.

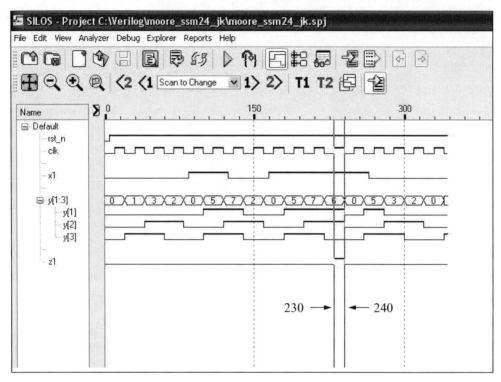

Figure 2.156   Waveforms for the Moore machine of Figure 2.152.

2.4 Mealy Machines

The next-state function δ for Mealy machines maps the Cartesian product of X and Y into Y, and thus, is determined by both the present inputs and the present state. The output function λ maps the Cartesian product of X and Y into Z, such that the output vector is a function of both the present inputs and the present state. This is the underlying difference between Moore and Mealy machines — *the outputs of a Moore machine are directly related to the present state only, whereas, the outputs of a Mealy machine are a function of both the present state and the present inputs.*

This section will provide three examples of Mealy machines designed using behavioral modeling, structural modeling using D flip-flops, and structural modeling using JK flip-flops. The state diagram for all designs is shown in Figure 2.157 using adjacent state codes.

The Mealy machine accepts serial data on an input line x_1 which consists of 3-bit words. The words are contiguous with no space between adjacent words. The machine is controlled by a periodic clock, where one clock period is equal to one bit cell. The format for the 3-bit words is shown below, where $b_i = 0$ or 1.

$$x_1 = \quad \cdots \quad \Big| \, b_1 b_2 b_3 \, \Big| \, b_1 b_2 b_3 \, \Big| \, b_1 b_2 b_3 \, \Big| \quad \cdots$$

Whenever a word contains the bit pattern $b_1 b_2 b_3 = 111$, the machine will assert output z_1 during the b_3 bit cell according to the following assertion/deassertion statement: $z_1 \uparrow t_2 \downarrow t_3$. Thus, z_1 is active for the last half of bit cell b_3. An example of a valid word in a series of words is as follows:

$$x_1 = \quad \cdots \quad \Big| \, 001 \, \Big| \, 101 \, \Big| \, 011 \, \Big| \, 101 \, \Big| \, 111 \, \Big| \, 010 \, \Big| \quad \cdots$$

$$z_1 \uparrow t_2 \downarrow t_3$$

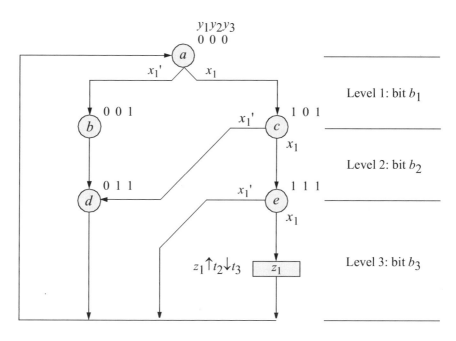

Figure 2.157   State diagram for the Mealy machine, which generates an output z_1 whenever a 3-bit word $x_1 = 111$. Unused states are: $y_1 y_2 y_3 = 010$, 100, and 110.

2.4.1 Design Using Behavioral Modeling

The behavioral design module is shown in Figure 2.158 using the **case** statement. The test bench module is shown in Figure 2.159 and takes the machine through three sequences, or paths, of the state diagram. The outputs are shown in Figure 2.160 and indicate that output z_1 is asserted in state code $y_1 y_2 y_3 = 111$. The waveforms are shown in Figure 2.161 showing the three paths. Note that output z_1 is asserted during the last half of the clock cycle in state (e) $y_1 y_2 y_3 = 111$.

```
//behavioral mealy synchronous sequential machine

module mealy_ssm12 (rst_n, clk, x1, y, z1);

//define inputs and outputs
input rst_n, clk, x1;

output [1:3] y;
output z1;

//variables are reg in always
reg [1:3] y, next_state;
reg z1;

//assign state codes
//parameter defines a constant
parameter    state_a = 3'b000,
             state_b = 3'b001,
             state_c = 3'b101,
             state_d = 3'b011,
             state_e = 3'b111;

//set next state
always @ (posedge clk)
begin
   if (~rst_n)            //if (~rst_n) is true,
      y <= state_a;       //... go to state_a
   else
      y <= next_state;
end

//determine output
always @ (y or clk)
begin
   if (y == state_e)      //== specifies logical
         begin            //... equality or compare
            if (~clk)
               z1 = 1'b1;  //assert z1 at t2,
            else           //deassert at t3
               z1 = 1'b0;
         end
   else
      z1 = 1'b0;
end

                          //continued on next page
```

Figure 7.158   Behavioral design module for a Mealy machine, which asserts output z_1 whenever a sequence of $x_1 = 111$ occurs on input x_1.

```
//determine next state
always @ (y)
begin
   case (y)                            //case is a multi-way
      state_a:                         //... conditional branch
         if (x1)                       //if y = state_a, then
            next_state = state_c;      //... do if ... else
         else
            next_state = state_b;

      state_b: next_state = state_d;

      state_c:
         if (x1)
            next_state = state_e;
         else
            next_state = state_d;

      state_d: next_state = state_a;

      state_e: next_state = state_a;

      default: next_state = state_a;
   endcase
end
endmodule
```

Figure 2.158 (Continued)

```
//test bench for mealy synchronous sequential machine
module mealy_ssm12_tb;

reg rst_n, clk, x1;          //inputs are reg for test bench
wire [1:3] y;                //outputs are wire for test bench
wire z1;

initial                      //display variables
$monitor ("x1 = %b, state = %b, z1 = %b", x1, y, z1);

initial                      //define clock
begin
   clk = 1'b0;
   forever
      #10 clk = ~clk;
end                          //continued on next page
```

Figure 2.159 Test bench module for the Mealy machine of Figure 2.158.

```
//define input sequence
initial
begin
    #0      rst_n = 1'b0;               //reset to state_a (000)
            x1 = 1'b0;
    #15     rst_n = 1'b1;               //deassert reset

    x1 = 1'b0;@ (posedge clk)       //go to state_b (001)
    x1 = $random;@ (posedge clk)    //go to state_d (011)
    x1 = $random;@ (posedge clk)    //go to state_a (000)

//----------------------------------------------------------
    x1 = 1'b1;@ (posedge clk)       //go to state_c (101)
    x1 = 1'b0;@ (posedge clk)       //go to state_d (011)
    x1 = $random;@ (posedge clk)    //go to state_a (000)

//----------------------------------------------------------
    x1 = 1'b1;@ (posedge clk)       //go to state_c (101)
    x1 = 1'b1;@ (posedge clk)       //go to state_e (111)
    x1 = 1'b0;@ (posedge clk)       //assert z1
                                    //go to state_a (000)

//----------------------------------------------------------
    #10     $stop;
end

//----------------------------------------------------------
//instantiate the module into the test bench
mealy_ssm12 inst1 (
    .rst_n(rst_n),
    .clk(clk),
    .x1(x1),
    .y(y),
    .z1(z1)
    );
```

Figure 2.159   (Continued)

```
x1 = 0, state = xxx, z1 = 0    x1 = 1, state = 011, z1 = 0
x1 = 0, state = 000, z1 = 0    x1 = 1, state = 000, z1 = 0
x1 = 0, state = 001, z1 = 0    x1 = 1, state = 101, z1 = 0
x1 = 1, state = 011, z1 = 0    x1 = 0, state = 111, z1 = 0
x1 = 1, state = 000, z1 = 0    x1 = 0, state = 111, z1 = 1
x1 = 0, state = 101, z1 = 0    x1 = 0, state = 000, z1 = 0
```

Figure 2.160   Outputs for the Mealy machine of Figure 2.158.

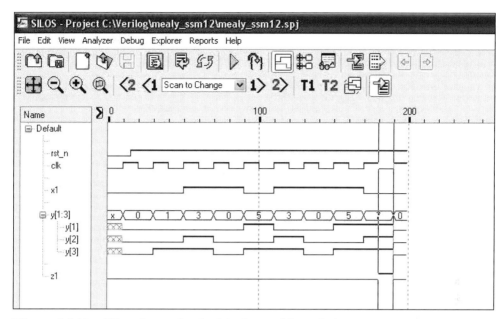

Figure 2.161   Waveforms for the Mealy machine of Figure 2.158.

2.4.2 Design Using Structural Modeling with *D* Flip-Flops

The Mealy machine for the state diagram of Figure 2.157 will now be designed using structural modeling with D flip-flops. The input maps are derived from the state diagram and are shown in Figure 2.162 using input x_1 as a map-entered variable. The logic diagram, designed from the input equations, is shown in Figure 2.163 containing the instantiations for the input gates and the net names. The D flip-flops have an implied reset input.

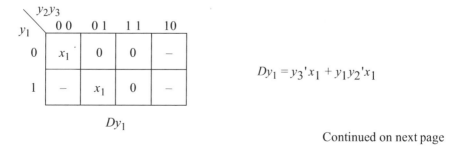

$$Dy_1 = y_3'x_1 + y_1y_2'x_1$$

Continued on next page

Figure 2.162   Input maps for the Mealy machine of Figure 2.157.

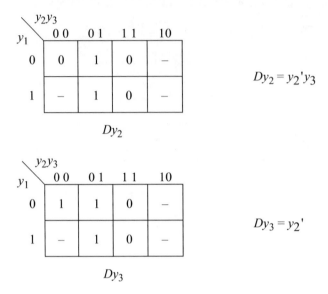

$$Dy_2 = y_2'y_3$$

$$Dy_3 = y_2'$$

Figure 2.162   (Continued)

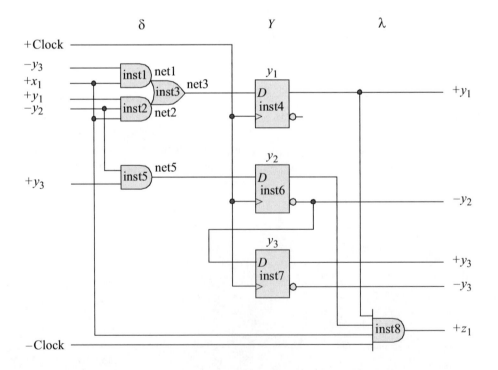

Figure 2.163   Logic diagram for the Mealy machine of Figure 2.157.

The structural design module is shown in Figure 2.164. The test bench module is shown in Figure 2.165 and takes the machine through three paths of the state diagram. The outputs and waveforms are shown in Figure 2.166 and Figure 2.167, respectively. The waveforms clearly show the three paths through the state diagram and output z_1 being asserted during the last half of the clock cycle.

```
//structural mealy to detect 111
module mealy_ssm13 (rst_n, clk, x1, y, z1);

input rst_n, clk, x1;
output [1:3] y;

output z1;

//define internal nets
wire net1, net2, net3, net5;

//design for flip-flop y[1]
and2_df inst1 (
   .x1(~y[3]),
   .x2(x1),
   .z1(net1)
   );

and3_df inst2 (
   .x1(y[1]),
   .x2(~y[2]),
   .x3(x1),
   .z1(net2)
   );

or2_df inst3 (
   .x1(net1),
   .x2(net2),
   .z1(net3)
   );

d_ff_bh inst4 (
   .rst_n(rst_n),
   .clk(clk),
   .d(net3),
   .q(y[1])
   );

                              //continued on next page
```

Figure 2.164 Structural design module for the Mealy machine of Figure 2.157.

```
//design for flip-flop y[2]
and2_df inst5 (
    .x1(~y[2]),
    .x2(y[3]),
    .z1(net5)
    );

d_ff_bh inst6 (
    .rst_n(rst_n),
    .clk(clk),
    .d(net5),
    .q(y[2])
    );

//design for flip-flop y[3]
d_ff_bh inst7 (
    .rst_n(rst_n),
    .clk(clk),
    .d(~y[2]),
    .q(y[3])
    );

//design for output z1
and4_df inst8 (
    .x1(y[1]),
    .x2(y[2]),
    .x3(x1),
    .x4(~clk),
    .z1(z1)
    );
endmodule
```

Figure 2.164   (Continued)

```
//test bench for mealy ssm to detect 111
module mealy_ssm13_tb;

reg rst_n, clk, x1;        //inputs are reg for test bench
wire [1:3] y;              //outputs are wire for test bench
wire z1;

//display variables
initial
$monitor ("x1 = %b, state = %b, z1 = %b", x1, y, z1);
                                        //continued on next page
```

Figure 2.165   Test bench for the Mealy machine of Figure 2.164.

```
//define clock
initial
begin
   clk = 1'b0;
   forever
      #10 clk = ~clk;
end


//define input sequence
initial
begin
   #0    x1 = 1'b0;
         rst_n = 1'b0;          //reset to state_a (000)
   #5    rst_n = 1'b1;          //deassert reset

   x1 = 1'b0;@ (posedge clk)  //go to state_b (001)
   x1 = 1'b0;@ (posedge clk)  //go to state_d (011)
   x1 = 1'b1;@ (posedge clk)  //go to state_a (000)
//--------------------------------------------------------
   x1 = 1'b1;@ (posedge clk)  //go to state_c (101)
   x1 = 1'b0;@ (posedge clk)  //go to state_d (011)
   x1 = 1'b0;@ (posedge clk)  //go to state_a (000)
//--------------------------------------------------------
   x1 = 1'b1;@ (posedge clk)  //go to state_c (101)
   x1 = 1'b1;@ (posedge clk)  //go to state_e (111)
   x1 = 1'b1;@ (posedge clk)  //assert z1
                              //go to state_a (000)
//--------------------------------------------------------
   #60   $stop;
end
//--------------------------------------------------------

//instantiate the module into the test bench
mealy_ssm13 inst1 (
   .rst_n(rst_n),
   .clk(clk),
   .x1(x1),
   .y(y),
   .z1(z1)
   );

endmodule
```

Figure 2.165   (Continued)

```
x1 = 0, state = 000, z1 = 0
x1 = 0, state = 001, z1 = 0
x1 = 0, state = 011, z1 = 0

x1 = 1, state = 000, z1 = 0
x1 = 1, state = 101, z1 = 0
x1 = 0, state = 111, z1 = 0

x1 = 0, state = 000, z1 = 0
x1 = 1, state = 001, z1 = 0
x1 = 1, state = 011, z1 = 0

x1 = 1, state = 000, z1 = 0
x1 = 1, state = 101, z1 = 0
x1 = 1, state = 111, z1 = 0

x1 = 1, state = 111, z1 = 1
x1 = 1, state = 000, z1 = 0
```

Figure 2.166   Outputs for the Mealy machine of Figure 2.164.

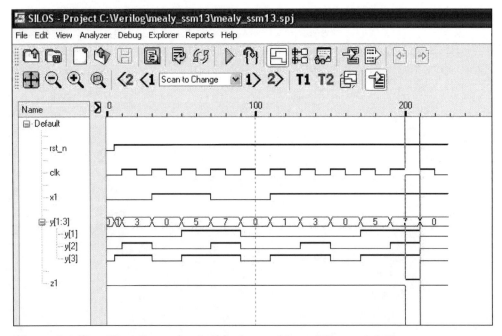

Figure 2.167   Waveforms for the Mealy machine of Figure 2.164.

2.4.3 Design Using Structural Modeling with *JK* Flip-Flops

The Mealy machine for the state diagram of Figure 2.157 will now be designed using *JK* flip-flops. Step 1 through step 3 of the synthesis procedure using *D* flip-flops remain unchanged for *JK* flip-flops. The input maps will be generated from the state diagram using input x_1 as a map-entered variable and are shown in Figure 2.168.

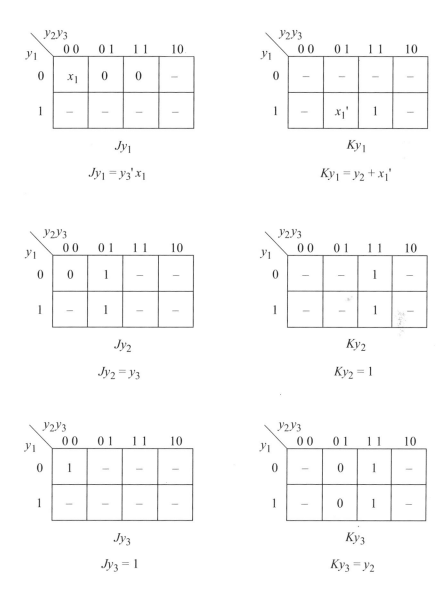

Figure 2.168 Input maps for the Mealy machine of Figure 2.157 using *JK* flip-flops and adjacent state codes for state pairs (b, c) and (d, e). Input x_1 is a map-entered variable.

The logic diagram is shown in Figure 2.169 indicating the instantiation of the δ input logic gates, the net names, and the *JK* flip-flops which are instantiated as negative-edge triggered flip-flops. Although not shown, it is assumed that the flip-flops have a set and a reset input. Since the clock is a negative-edge input, therefore, output z_1 will be asserted during the positive voltage level of the clock.

The structural design module is shown in Figure 2.170 and the test bench module is shown in Figure 2.171. The outputs and waveforms are shown in Figure 2.172 and Figure 2.173, respectively.

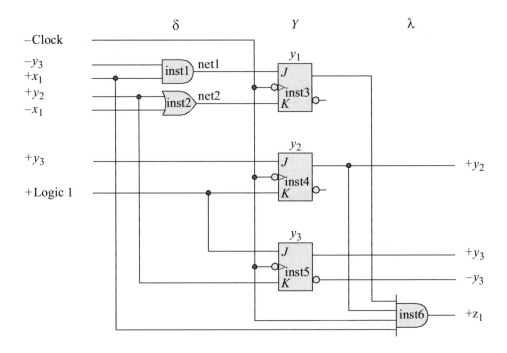

Figure 2.169   Logic diagram for the Mealy machine of Figure 2.157 using *JK* flip-flops. Output z_1 is asserted when a sequence of 111 has been detected in a 3-bit word on a serial data line x_1. Output z_1 is asserted at time t_2 and deasserted at time t_3.

```
//structural mealy using jk flip-flops to detect 111
module mealy_ssm13_jk (set_n, rst_n, clk, x1, y, z1);

input set_n, rst_n, clk, x1;
output [1:3] y;
output z1;

//define internal nets
wire net1, net2;                        //continued on next page
```

Figure 2.170   Structural design module for the Mealy machine of Figure 2.169.

```
and2_df inst1 (            //design for flip-flop y[1]
   .x1(~y[3]),
   .x2(x1),
   .z1(net1)
   );

or2_df inst2 (
   .x1(y[2]),
   .x2(~x1),
   .z1(net2)
   );

jkff_neg_clk inst3 (
   .set_n(set_n),
   .rst_n(rst_n),
   .clk(clk),
   .j(net1),
   .k(net2),
   .q(y[1])
   );
//------------------------------------------------------------
jkff_neg_clk inst4 (    //design for flip-flop y[2]
   .set_n(set_n),
   .rst_n(rst_n),
   .clk(clk),
   .j(y[3]),
   .k(1'b1),
   .q(y[2])
   );
//------------------------------------------------------------
jkff_neg_clk inst5 (    //design for flip-flop y[3]
   .set_n(set_n),
   .rst_n(rst_n),
   .clk(clk),
   .j(1'b1),
   .k(y[2]),
   .q(y[3])
   );
//------------------------------------------------------------
and4_df inst6 (            //design for output z1
   .x1(y[1]),
   .x2(y[2]),
   .x3(clk),
   .x4(x1),
   .z1(z1)
   );
endmodule
```

Figure 2.170   (Continued)

```
//test bench for mealy ssm using jk flip-flops to detect 111

module mealy_ssm13_jk_tb;

reg set_n, rst_n, clk, x1;  //inputs are reg for test bench
wire [1:3] y;               //outputs are wire for test bench
wire z1;

//display variables
initial
$monitor ("x1 = %b, state = %b, z1 = %b", x1, y, z1);

//define clock
initial
begin
   clk = 1'b0;
   forever
      #10 clk = ~clk;
end

//define input sequence
initial
begin
   #0      x1 = 1'b0;
           set_n = 1'b1;
           rst_n = 1'b0;           //reset to state_a (000)
   #5      rst_n = 1'b1;           //deassert reset

   x1 = 1'b0;@ (negedge clk)   //go to state_b (001)
   x1 = 1'b0;@ (negedge clk)   //go to state_d (011)
   x1 = 1'b1;@ (negedge clk)   //go to state_a (000)
//---------------------------------------------------------
   x1 = 1'b1;@ (negedge clk)   //go to state_c (101)
   x1 = 1'b0;@ (negedge clk)   //go to state_d (011)
   x1 = 1'b0;@ (negedge clk)   //go to state_a (000)
//---------------------------------------------------------
   x1 = 1'b1;@ (negedge clk)   //go to state_c (101)
   x1 = 1'b1;@ (negedge clk)   //go to state_e (111)
   x1 = 1'b1;@ (negedge clk)   //assert z1
                               //go to state_a (000)
//---------------------------------------------------------

   #10     $stop;
end
//---------------------------------------------------------

                               //continued on next page
```

Figure 2.171   Test bench module for the Mealy machine of Figure 2.170.

```
//instantiate the module into the test bench
mealy_ssm13_jk inst1 (
   .set_n(set_n),
   .rst_n(rst_n),
   .clk(clk),
   .x1(x1),
   .y(y),
   .z1(z1)
   );

endmodule
```

Figure 2.171   (Continued)

```
x1 = 0, state = 000, z1 = 0    x1 = 0, state = 011, z1 = 0
x1 = 0, state = 001, z1 = 0    x1 = 1, state = 000, z1 = 0
x1 = 1, state = 011, z1 = 0    x1 = 1, state = 101, z1 = 0
x1 = 1, state = 000, z1 = 0    x1 = 1, state = 111, z1 = 0
x1 = 0, state = 101, z1 = 0    x1 = 1, state = 111, z1 = 1
                               x1 = 1, state = 000, z1 = 0
```

Figure 2.172   Outputs for the Mealy machine of Figure 2.170.

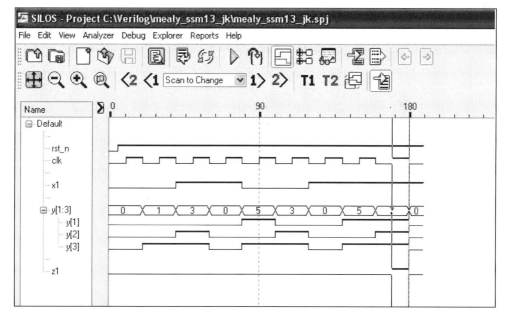

Figure 2.173   Waveforms for the Mealy machine of Figure 2.170.

2.5 Moore–Mealy Equivalence

This section presents Verilog design examples of equivalent Moore and Mealy synchronous sequential machines. The examples include behavioral modeling, structural modeling, and built-in primitives.

Example 2.22 This example designs a Mealy machine to detect a 1101 sequence using behavioral modeling. The state diagram for a Mealy machine which detects an input sequence of 1101 on a serial data line x_1 is shown in Figure 2.174. Whenever $x_1 = 1101$ *anywhere* in the bit stream, output z_1 is asserted at time t_2 and deasserted at time t_3. An example of a sequence of bits is shown below.

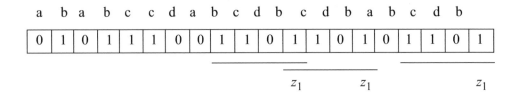

The behavioral design module is shown in Figure 2.175 and the test bench module is shown in Figure 2.176. The outputs and waveforms are shown in Figure 2.177 and Figure 2.178, respectively.

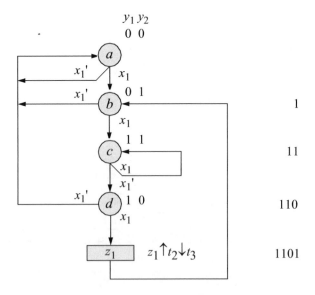

Figure 2.174   State diagram for a Mealy machine to detect an input sequence of x_1 = 1101 anywhere in the bit stream.

```
//behavioral module for a mealy ssm to detect 1101

module mealy_ssm14_bh (rst_n, clk, x1, y, z1);

input clk, rst_n, x1;
output [1:2] y;
output z1;

reg [1:2] y, next_state;
wire z1;

//assign state codes
parameter   state_a = 2'b00,
            state_b = 2'b01,
            state_c = 2'b11,
            state_d = 2'b10;

//set next state
always @ (posedge clk)
begin
   if (~rst_n)
      y <= state_a;
   else
      y <= next_state;
end

assign z1 = ((y[1]) && (~y[2]) && (x1)  && (~clk));

//determine next state
always @ (y or x1)
begin
   case (y)
      state_a:
         if (x1)
            next_state = state_b;
         else
            next_state = state_a;

      state_b:
         if (x1)
            next_state = state_c;
         else
            next_state = state_a;

                              //continued on next page
```

Figure 2.175   Behavioral design module for a Mealy machine to detect a sequence of 1101 on a serial input x_1 anywhere in the bit stream.

```
      state_c:
         if (x1)
            next_state = state_c;
         else
            next_state = state_d;

      state_d:
         if (x1)
            next_state = state_b;
         else
            next_state = state_a;

      default: next_state = state_a;
   endcase

end

endmodule
```

Figure 2.175   (Continued)

```
//test bench for mealy machine to detect 1101
module mealy_ssm14_bh_tb;

reg rst_n, clk, x1;
wire [1:2] y;
wire z1;

//display variables
initial
$monitor ("x1 = %b, state = %b, z1 = %b", x1, y, z1);

//define clock
initial
begin
   clk = 1'b0;
   forever
      #10 clk = ~clk;
end

                              //continued on next page
```

Figure 2.176   Test bench module for the Mealy machine to detect a sequence of 1101 on a serial input x_1 anywhere in the bit stream.

```
//define input sequence
initial
begin
   #0      rst_n = 1'b0;              //reset to state_a
           x1 = 1'b0;
   #5      rst_n = 1'b1;              //deassert reset

   x1 = 1'b0;@ (posedge clk)  //go to state_a
   x1 = 1'b1;@ (posedge clk)  //go to state_b
   x1 = 1'b0;@ (posedge clk)  //go to state_a
   x1 = 1'b1;@ (posedge clk)  //go to state_b
   x1 = 1'b1;@ (posedge clk)  //go to state_c
   x1 = 1'b1;@ (posedge clk)  //go to state_c
   x1 = 1'b0;@ (posedge clk)  //go to state_d
   x1 = 1'b0;@ (posedge clk)  //go to state_a
   x1 = 1'b1;@ (posedge clk)  //go to state_b
   x1 = 1'b1;@ (posedge clk)  //go to state_c
   x1 = 1'b0;@ (posedge clk)  //go to state_d
   x1 = 1'b1;@ (posedge clk)  //go to state_b, assert z1 at t2

   x1 = 1'b1;@ (posedge clk)  //go to state_c
   x1 = 1'b0;@ (posedge clk)  //go to state_d
   x1 = 1'b1;@ (posedge clk)  //go to state_b, assert z1 at t2

   x1 = 1'b0;@ (posedge clk)  //go to state_a
   x1 = 1'b1;@ (posedge clk)  //go to state_b
   x1 = 1'b1;@ (posedge clk)  //go to state_c
   x1 = 1'b0;@ (posedge clk)  //go to state_d
   x1 = 1'b1;@ (posedge clk)  //go to state_b, assert z1 at t2

   x1 = 1'b0;@ (posedge clk)  //go to state_a

   #10     $stop;

end

//instantiate the module into the test bench
mealy_ssm14_bh inst1 (
   .rst_n(rst_n),
   .clk(clk),
   .x1(x1),
   .y(y),
   .z1(z1)
   );

endmodule
```

Figure 2.176   (Continued)

```
x1 = 0, state = xx, z1 = 0   x1 = 0, state = 01, z1 = 0
x1 = 1, state = 00, z1 = 0   x1 = 1, state = 00, z1 = 0
x1 = 0, state = 01, z1 = 0   x1 = 1, state = 01, z1 = 0
x1 = 1, state = 00, z1 = 0   x1 = 0, state = 11, z1 = 0
x1 = 1, state = 01, z1 = 0   x1 = 1, state = 10, z1 = 0
x1 = 1, state = 11, z1 = 0   x1 = 1, state = 10, z1 = 1
x1 = 0, state = 11, z1 = 0
x1 = 0, state = 10, z1 = 0   x1 = 0, state = 01, z1 = 0
x1 = 1, state = 00, z1 = 0   x1 = 0, state = 00, z1 = 0
x1 = 1, state = 01, z1 = 0
x1 = 0, state = 11, z1 = 0
x1 = 1, state = 10, z1 = 0
x1 = 1, state = 10, z1 = 1

x1 = 1, state = 01, z1 = 0
x1 = 0, state = 11, z1 = 0
x1 = 1, state = 10, z1 = 0
x1 = 1, state = 10, z1 = 1
```

Figure 2.177   Outputs for the Mealy machine to detect a sequence of 1101 on a serial input x_1 anywhere in the bit stream.

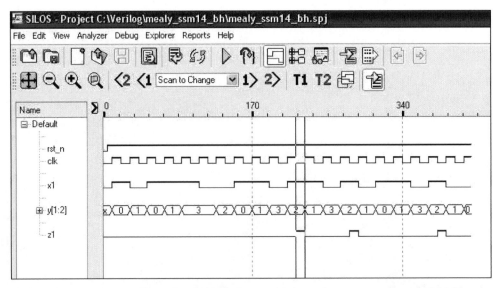

Figure 2.178   Waveforms for the Mealy machine to detect a sequence of 1101 on a serial input x_1 anywhere in the bit stream.

Example 2.23   This example designs an equivalent Moore machine to detect a 1101 sequence using behavioral modeling. The characteristics of state d in the state diagram of Figure 2.174 are preserved by two new states, d and e, in the equivalent Moore state diagram of Figure 2.179. State d no longer produces an output in the Moore model. The output is generated in state e if $x_1 = 1$ in state d. The addition of the extra state (e) permits conformation to the definition of a Moore machine.

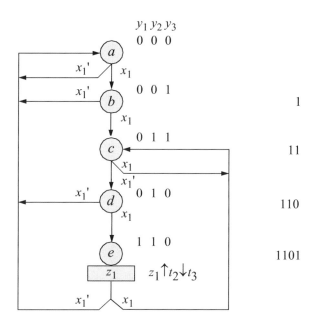

Figure 2.179   Equivalent Moore state diagram for the Mealy machine of Figure 2.174. Output z_1 is asserted whenever the input sequence is $x_1 = 1101$.

The behavioral design module is shown in Figure 2.180 using the **always** statement. The **always** statement executes the behavioral statements within the **always** block repeatedly in a looping manner and begins execution at time zero. The **always** construct never exits the corresponding block. Execution of the statements continues indefinitely until the simulation is terminated.

The keyword **always** specifies a behavior and the statements within a behavior are classified as *behavioral* or *procedural*. There can be more than one **always** statement in a behavioral module and, together with the **initial** statement, is one of the basic constructs for representing concurrency.

The test bench module is shown in Figure 2.181 and takes the machine through a series of bits on input x_1 The series of bits includes nonoverlapping and overlapping 1101 bits. The outputs and waveforms are shown in Figure 2.182 and Figure 2.183, respectively.

```
//behavioral for moore ssm to detect 1101

module moore_ssm27_bh (rst_n, clk, x1, y, z1);

//define inputs and outputs
input rst_n, clk, x1;
output [1:3] y;
output z1;

//variables are reg in always
reg [1:3] y, next_state;
reg z1;

//assign state codes
//parameter defines a constant
parameter   state_a = 3'b000,
            state_b = 3'b001,
            state_c = 3'b011,
            state_d = 3'b010,
            state_e = 3'b110;

//set next state
always @ (posedge clk)
begin
    if (~rst_n)              //if (~rst_n) is true
        y <= state_a;        //go to state_a
    else
        y <= next_state;
end

//determine output
always @ (y or clk)
begin
    if (y == state_e)        //== specifies logical
        begin                //equality or compare
            if (~clk)
                z1 = 1'b1;    //assert z1 at t2
            else
                z1 = 1'b0;
        end
    else
        z1 = 1'b0;
end

                            //continued on next page
```

Figure 2.180   Behavioral design module for the Moore state diagram of Figure 2.179 that asserts output z_1 whenever the input sequence is $x_1 = 1101$.

```
//determine next state
always @ (y or x1)
begin
   case (y)                  //case is a multiple-way
      state_a:               //... conditional branch
         if (~x1)
            next_state = state_a;
         else
            next_state = state_b;


      state_b:
         if (~x1)
            next_state = state_a;
         else
            next_state = state_c;


      state_c:
         if (x1)
            next_state = state_c;
         else
            next_state = state_d;


      state_d:
         if (~x1)
            next_state = state_a;
         else
            next_state = state_e;


      state_e:
         if (~x1)
            next_state = state_a;
         else
            next_state = state_c;


      default: next_state = state_a;

   endcase

end

endmodule
```

Figure 2.180   (Continued)

```
//test for moore ssm to detect 1101
module moore_ssm27_bh_tb;

reg rst_n, clk, x1;        //inputs are reg for test bench
wire [1:3] y;              //outputs are wire for test bench
wire z1;

//display variables
initial
$monitor ("x1 = %b, state = %b, z1 = %b", x1, y, z1);

//define clock
initial
begin
   clk = 1'b0;
   forever
      #10 clk = ~clk;
end

//define input sequence
initial
begin
   #0     rst_n = 1'b0;          //reset to state_a
          x1 = 1'b0;
   #5     rst_n = 1'b1;          //deassert reset
//--------------------------------------------------------------
   x1 = 1'b0;@ (posedge clk)  //go to state_a
   x1 = 1'b1;@ (posedge clk)  //go to state_b
   x1 = 1'b0;@ (posedge clk)  //go to state_a
//--------------------------------------------------------------
   x1 = 1'b1;@ (posedge clk)  //go to state_b
   x1 = 1'b1;@ (posedge clk)  //go to state_c
   x1 = 1'b1;@ (posedge clk)  //go to state_c
   x1 = 1'b0;@ (posedge clk)  //go to state_d
   x1 = 1'b0;@ (posedge clk)  //go to state_a
//--------------------------------------------------------------
   x1 = 1'b1;@ (posedge clk)  //go to state_b
   x1 = 1'b1;@ (posedge clk)  //go to state_c
   x1 = 1'b0;@ (posedge clk)  //go to state_d
   x1 = 1'b1;@ (posedge clk)  //go to state_e, assert z1 at t2
   x1 = 1'b0;@ (posedge clk)  //go to state_a
//--------------------------------------------------------------

                              //continued on next page
```

Figure 2.181   Test bench module for the Moore state diagram of Figure 2.179 that asserts output z_1 whenever the input sequence is $x_1 = 1101$.

```
//----------------------------------------------------------
   x1 = 1'b1;@ (posedge clk)   //go to state_b
   x1 = 1'b1;@ (posedge clk)   //go to state_c
   x1 = 1'b0;@ (posedge clk)   //go to state_d
   x1 = 1'b1;@ (posedge clk)   //go to state_e, assert z1 at t2
   x1 = 1'b1;@ (posedge clk)   //go to state_c
   x1 = 1'b0;@ (posedge clk)   //go to state_d
   x1 = 1'b1;@ (posedge clk)   //go to state_e, assert z1 at t2
   x1 = 1'b0;@ (posedge clk)   //go to state_a
//----------------------------------------------------------
   #10    $stop;
end

//instantiate the module into the test bench
moore_ssm27_bh inst1 (
   .rst_n(rst_n),
   .clk(clk),
   .x1(x1),
   .y(y),
   .z1(z1)
   );

endmodule
```

Figure 2.181   (Continued)

```
x1 = 0, state = xxx, z1 = 0    x1 = 1, state = 000, z1 = 0
x1 = 1, state = 000, z1 = 0    x1 = 1, state = 001, z1 = 0
x1 = 0, state = 001, z1 = 0    x1 = 0, state = 011, z1 = 0
x1 = 1, state = 000, z1 = 0    x1 = 1, state = 010, z1 = 0
x1 = 1, state = 001, z1 = 0    x1 = 1, state = 110, z1 = 0
x1 = 1, state = 011, z1 = 0    x1 = 1, state = 110, z1 = 1
x1 = 0, state = 011, z1 = 0
x1 = 0, state = 010, z1 = 0    x1 = 0, state = 011, z1 = 0
x1 = 1, state = 000, z1 = 0    x1 = 1, state = 010, z1 = 0
x1 = 1, state = 001, z1 = 0    x1 = 0, state = 110, z1 = 0
x1 = 0, state = 011, z1 = 0    x1 = 0, state = 110, z1 = 1
x1 = 1, state = 010, z1 = 0    x1 = 0, state = 000, z1 = 0
x1 = 0, state = 110, z1 = 0
x1 = 0, state = 110, z1 = 1
```

Figure 2.182   Outputs for the Moore state diagram of Figure 2.179 that asserts output z_1 whenever the input sequence is $x_1 = 1101$.

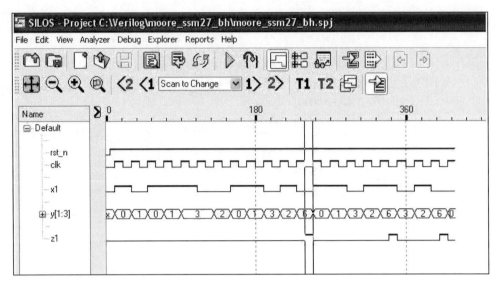

Figure 2.183   Waveforms for the Moore state diagram of Figure 2.179 that asserts output z_1 whenever the input sequence is $x_1 = 1101$.

Example 2.24 This example designs a Mealy machine to detect a 101 and a 110 sequence using behavioral modeling. The same technique that was used for a single sequence with a single output also applies to multiple input sequences that generate multiple outputs, as shown for the Mealy machine in the state diagram of Figure 2.184. This machine detects two different word configurations on a serial input line x_1, as shown below. The information on input x_1 consists of 3-bit words. There is no space between adjacent words. Outputs z_1 and z_2 are asserted during the third bit time of a valid word.

$$x_1 = \cdots \left| b_1 b_2 b_3 \right| b_1 b_2 b_3 \left| b_1 b_2 b_3 \right| \cdots$$

If $x_1 = 101$, then assert z_1
If $x_1 = 110$, then assert z_2 z_1/z_2

     The behavioral design module is shown in Figure 2.185 using the **case** statement, the **always** statement, and the **assign** statement. The keyword **assign** is classified as a continuous assignment statement used to design combinational logic. The test bench module is shown in Figure 2.186 and applies different bit sequences which assert both output z_1 and output z_2. The outputs and waveforms are shown in Figure 2.187 and Figure 2.188, respectively.

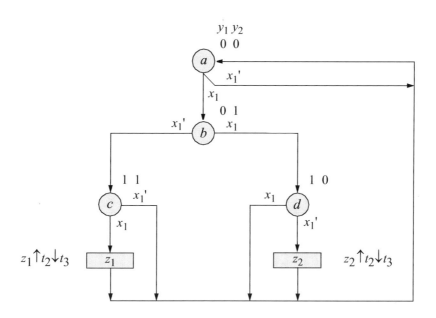

Figure 2.184   State diagram for a Mealy machine. Output z_1 is asserted whenever $x_1 = 101$; output z_2 is asserted whenever $x_1 = 110$.

```
//behavioral module for a mealy ssm to detect 101 or 110

module mealy_ssm15 (rst_n, clk, x1, y, z1, z2);

input rst_n, clk, x1;
output [1:2] y;
output z1, z2;

//variables are reg in always
reg [1:2] y, next_state;

//assign state codes
//parameter defines a constant
parameter    state_a = 2'b00,
             state_b = 2'b01,
             state_c = 2'b11,
             state_d = 2'b10;
                              //continued on next page
```

Figure 2.185   Behavioral design module for a Mealy machine that asserts output z_1 whenever $x_1 = 101$ and asserts output z_2 whenever $x_1 = 110$.

```
//set next state
always @ (posedge clk)
begin
    if (~rst_n)
        y <= state_a;
    else
        y <= next_state;
end


//determine outputs
assign z1 = ((y[1]) && (y[2]) && (x1) && (~clk));
assign z2 = ((y[1]) && (~y[2]) && (~x1) && (~clk));


//determine next state
always @ (y or x1)
begin
    case (y)
        state_a:
            if (~x1)
                next_state = state_a;
            else
                next_state = state_b;


        state_b:
            if (~x1)
                next_state = state_c;
            else
                next_state = state_d;


        state_c: next_state = state_a;


        state_d: next_state = state_a;


        default: next_state = state_a;

    endcase

end

endmodule
```

Figure 2.185   (Continued)

```
//test bench for mealy machine to detect 101 or 110
module mealy_ssm15_tb;

reg rst_n, clk, x1;      //inputs are reg for test bench
wire [1:2] y;            //outputs are wire for test bench
wire z1, z2;

initial                  //display variables
$monitor ("x1 = %b, state = %b, z1 = %b, z2 = %b",
           x1, y, z1, z2);

//define clock
initial
begin
   clk = 1'b0;
   forever
      #10 clk = ~clk;
end


//define input sequence
initial
begin
   #0    rst_n = 1'b0;         //reset to state_a
         x1 = 1'b0;
   #5    rst_n = 1'b1;         //deassert reset

   x1 = 1'b0;@ (posedge clk)  //go to state_a
   x1 = 1'b1;@ (posedge clk)  //go to state_b
   x1 = 1'b0;@ (posedge clk)  //go to state_c
   x1 = 1'b0;@ (posedge clk)  //go to state_a

   x1 = 1'b1;@ (posedge clk)  //go to state_b
   x1 = 1'b0;@ (posedge clk)  //go to state_c
   x1 = 1'b1;@ (posedge clk)  //go to state_a; assert z1 at t2

   x1 = 1'b1;@ (posedge clk)  //go to state_b
   x1 = 1'b1;@ (posedge clk)  //go to state_d
   x1 = 1'b0;@ (posedge clk)  //go to state_a; assert z2 at t2

   x1 = 1'b1;@ (posedge clk)  //go to state_b
   x1 = 1'b1;@ (posedge clk)  //go to state_d
   x1 = 1'b1;@ (posedge clk)  //go to state_a

   #10    $stop;
end
                              //continued on next page
```

Figure 2.186   Test bench module for the Mealy machine that asserts output z_1 whenever $x_1 = 101$ and asserts output z_2 whenever $x_1 = 110$.

```
//instantiate the module into the test bench
mealy_ssm15 inst1 (
    .rst_n(rst_n),
    .clk(clk),
    .x1(x1),
    .y(y),
    .z1(z1),
    .z2(z2)
    );

endmodule
```

Figure 2.186   (Continued)

```
x1 = 0, state = xx, z1 = 0, z2 = x
x1 = 1, state = 00, z1 = 0, z2 = 0
x1 = 0, state = 01, z1 = 0, z2 = 0
x1 = 0, state = 11, z1 = 0, z2 = 0
x1 = 1, state = 00, z1 = 0, z2 = 0
x1 = 0, state = 01, z1 = 0, z2 = 0
x1 = 1, state = 11, z1 = 0, z2 = 0

x1 = 1, state = 11, z1 = 1, z2 = 0

x1 = 1, state = 00, z1 = 0, z2 = 0
x1 = 1, state = 01, z1 = 0, z2 = 0
x1 = 0, state = 10, z1 = 0, z2 = 0

x1 = 0, state = 10, z1 = 0, z2 = 1

x1 = 1, state = 00, z1 = 0, z2 = 0
x1 = 1, state = 01, z1 = 0, z2 = 0
x1 = 1, state = 10, z1 = 0, z2 = 0
x1 = 1, state = 00, z1 = 0, z2 = 0
```

Figure 2.187   Outputs for the Mealy machine of Figure 2.184 that asserts output z_1 whenever $x_1 = 101$ and asserts output z_2 whenever $x_1 = 110$.

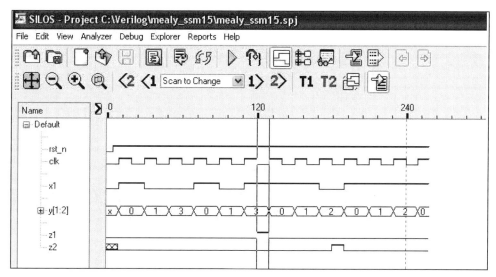

Figure 2.188   Waveforms for the Mealy machine of Figure 2.184 that asserts output z_1 whenever $x_1 = 101$ and asserts output z_2 whenever $x_1 = 110$.

Example 2.25   This example designs a Mealy machine to detect a 101 and a 110 sequence using structural modeling with D flip-flops. The state diagram is shown in Figure 2.189. This machine detects two different word configurations on a serial input line x_1. The information on input x_1 consists of 3-bit words. There is no space between adjacent words. Outputs z_1 and z_2 are asserted during the third bit time of a valid word, as shown below.

$$x_1 = \cdots \left| b_1 b_2 b_3 \right| b_1 b_2 b_3 \left| b_1 b_2 b_3 \right| \cdots$$

If $x_1 = 101$, then assert z_1

If $x_1 = 110$, then assert z_2 z_1/z_2

The input maps for the D flip-flops, the input equations, and the output equations are shown in Figure 2.190. The output equations are obtained directly from the state diagram. The logic diagram containing the δ input logic, the state flip-flops, and the λ output logic is shown in Figure 2.191.

The structural design module is shown in Figure 2.192, and shows the instantiations and net names on the various components. The test bench module is shown in Figure 2.193, which takes the machine through three different paths in the state diagram, two of which assert z_1 and z_2. The outputs and waveforms are shown in Figure 2.194 an Figure 2.195, respectively.

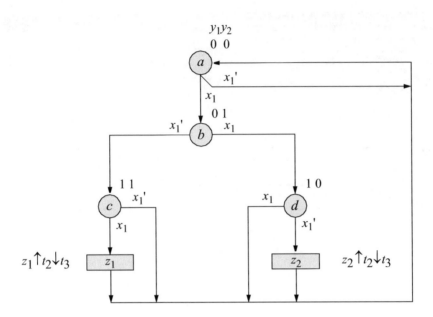

Figure 2.189   State diagram for a Mealy machine. Output z_1 is asserted whenever $x_1 = 101$; output z_2 is asserted whenever $x_1 = 110$.

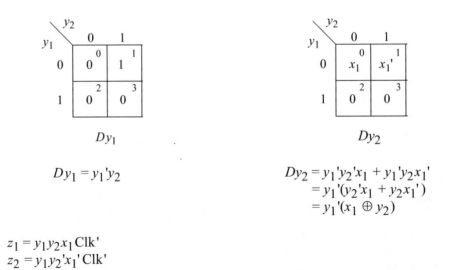

$$Dy_1 = y_1'y_2$$

$$Dy_2 = y_1'y_2'x_1 + y_1'y_2x_1'$$
$$= y_1'(y_2'x_1 + y_2x_1')$$
$$= y_1'(x_1 \oplus y_2)$$

$$z_1 = y_1y_2x_1\,\text{Clk}'$$
$$z_2 = y_1y_2'x_1'\,\text{Clk}'$$

Figure 2.190   Input maps, input equations, and output equations for the Mealy machine of Example 2.25 to detect a sequence on 101 and 110 on input x_1.

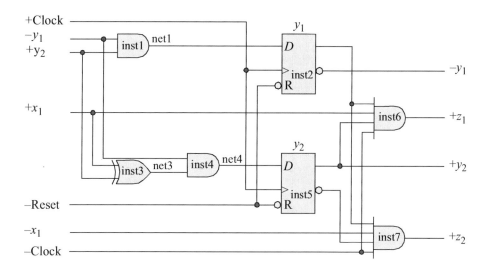

Figure 2.191 Logic diagram for the Mealy machine to detect the input sequences of 101 and 110 on input x_1.

```
//structural mealy ssm to detect 101 and 110
module mealy_ssm16a (rst_n, clk, x1, y, z1, z2);


//define inputs and outputs
input rst_n, clk, x1;
output [1:2] y;
output z1, z2;


//define input nets.  Nets are wire by default
wire net1, net3, net4;


//instantiate the logic for flip-flop y[1]
and2_df inst1 (
   .x1(~y[1]),
   .x2(y[2]),
   .z1(net1)
   );
                              //Continued on next page
```

Figure 2.192 Structural design module for the Mealy machine to detect the input sequences of 101 and 110 on input x_1.

```
d_ff_bh inst2 (
   .rst_n(rst_n),
   .clk(clk),
   .d(net1),
   .q(y[1])
   );

//instantiate the logic for flip-flop y[2]
xor2_df inst3 (
   .x1(x1),
   .x2(y[2]),
   .z1(net3)
   );

and2_df inst4 (
   .x1(~y[1]),
   .x2(net3),
   .z1(net4)
   );

d_ff_bh inst5 (
   .rst_n(rst_n),
   .clk(clk),
   .d(net4),
   .q(y[2])
   );

//instantiate the logic for outputs z1 and z2
and4_df inst6 (
   .x1(y[1]),
   .x2(y[2]),
   .x3(x1),
   .x4(~clk),
   .z1(z1)
   );

and4_df inst7 (
   .x1(y[1]),
   .x2(~y[2]),
   .x3(~x1),
   .x4(~clk),
   .z1(z2)
   );

endmodule
```

Figure 2.192   (Continued)

```
//test bench for mealy machine to detect 101 or 110
module mealy_ssm16a_tb;

reg rst_n, clk, x1;        //inputs are reg for test bench
wire [1:2] y;              //outputs are wire for test bench
wire z1, z2;

//display variables
initial
$monitor ("x1 = %b, state = %b, z1 = %b, z2 = %b",
          x1, y, z1, z2);

//define clock
initial
begin
   clk = 1'b0;
   forever
      #10 clk = ~clk;
end

//define input sequence
initial
begin
   #0    rst_n = 1'b0;                //reset to state_a
         x1 = 1'b0;
   #5    rst_n = 1'b1;                //deassert reset

         x1 = 1'b0;  @ (posedge clk)//go to state_a (00)
   #10   x1 = 1'b1;  @ (posedge clk)//go to state_b (01)
   #15   x1 = 1'b0;  @ (posedge clk)//go to state_c (11)
                     @ (posedge clk)//go to state_a (00)

   #40   x1 = 1'b1;  @ (posedge clk)//go to state_b (01)
   #20   x1 = 1'b0;  @ (posedge clk)//go to state_c (11)
   #20   x1 = 1'b1;  @ (posedge clk)//assert z1 at t2
                                    //then go to state_a (00)
                                    //at posedge clk

                     @ (posedge clk)//go to state_b (01)
                     @ (posedge clk)//go to state_d (10)
   #20   x1 = 1'b0;  @ (posedge clk)//assert z2 at t2
                                    //then go to state_a (00)
                                    //at posedge clk
   #40   $stop;
end
                                    //continued on next page
```

Figure 2.193   Test bench module for the Mealy machine to detect the input sequences of 101 and 110 on input x_1.

```
//instantiate the module into the test bench
mealy_ssm16a inst1 (
    .rst_n(rst_n),
    .clk(clk),
    .x1(x1),
    .y(y),
    .z1(z1),
    .z2(z2)
    );


endmodule
```

Figure 2.193   (Continued)

```
x1 = 0, state = 00, z1 = 0, z2 = 0
x1 = 1, state = 00, z1 = 0, z2 = 0

x1 = 1, state = 01, z1 = 0, z2 = 0
x1 = 0, state = 01, z1 = 0, z2 = 0

x1 = 0, state = 11, z1 = 0, z2 = 0
x1 = 0, state = 00, z1 = 0, z2 = 0

x1 = 1, state = 00, z1 = 0, z2 = 0
x1 = 0, state = 01, z1 = 0, z2 = 0

x1 = 1, state = 11, z1 = 0, z2 = 0
x1 = 1, state = 11, z1 = 1, z2 = 0

x1 = 1, state = 00, z1 = 0, z2 = 0
x1 = 1, state = 01, z1 = 0, z2 = 0

x1 = 0, state = 10, z1 = 0, z2 = 0
x1 = 0, state = 10, z1 = 0, z2 = 1

x1 = 0, state = 00, z1 = 0, z2 = 0
```

Figure 2.194   Outputs for the Mealy machine to detect the input sequences of 101 and 110 on input x_1.

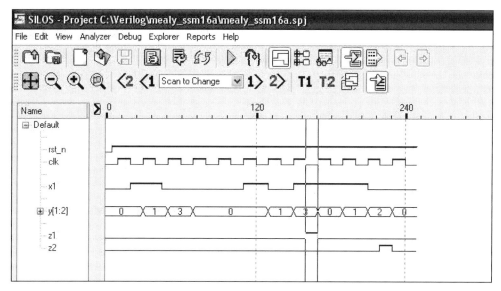

Figure 2.195 Waveforms for the Mealy machine to detect the input sequences of 101 and 110 on input x_1.

Example 2.26 This example designs an equivalent Moore machine to detect a 101 and a 110 sequence using behavioral modeling. Recall that a Moore-type output is a function of the present state only, whereas, a Mealy-type output is a function of both the present state and the present input.

A Mealy machine can be transformed into a corresponding Moore machine, where both machines accept the same set of input vectors and generate the same set of output vectors. To obtain a Moore machine from a Mealy machine, it is necessary to generate two new states for each state of the Mealy model in which different output values occur for different inputs.

The state diagram for the Moore machine that corresponds to the Mealy machine of Figure 2.189 is shown in Figure 2.196. It is possible for the output of the Moore machine to be valid for the entire state time; however, the output is to be asserted at time t_2 and deasserted at time t_3 to maintain compatible output assertion with the Mealy machine. Although both machines detect the same valid input sequence and generate identical output vectors at the same assertion/deassertion times, the output vectors for the Moore machine are delayed by one clock cycle. The delayed outputs are due to the addition of two new states.

The behavioral module using the **case** statement to determine the next state, is shown in Figure 2.197. The test bench module is shown in Figure 2.198, which takes the machine through all possible state transitions. The outputs and waveforms are shown in Figure 2.199 and Figure 2.200, respectively.

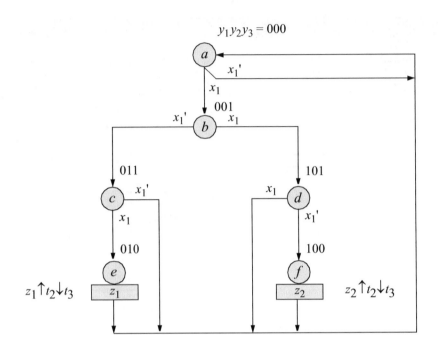

Figure 2.196   Equivalent Moore state diagram for the Mealy machine of Figure 2.189. Output z_1 is asserted whenever $x_1 = 101$; output z_2 is asserted whenever $x_1 = 110$.

```
//behavioral moore synchronous sequential machine
module moore_ssm28_bh (rst_n, clk, x1, y, z1, z2);

input rst_n, clk, x1;           //define inputs and outputs
output [1:3] y;
output z1, z2;

reg [1:3] y, next_state;        //variables are reg in always
reg z1, z2;

//assign state codes.  parameter defines a constant
parameter    state_a = 3'b000,
             state_b = 3'b001,
             state_c = 3'b011,
             state_d = 3'b101,
             state_e = 3'b010,
             state_f = 3'b100;      //continued on next page
```

Figure 2.197   Structural design module for the equivalent Moore machine to detect sequences of 101 and 110.

```
//set next state
always @ (posedge clk)
begin
   if (rst_n == 1'b0)    //if (~rst_n) is true,
      y <= state_a;      //go to state_a
   else
      y <= next_state;   //else go to next_state
end

//----------------------------------------------------
//initialize y and outputs
always @ (rst_n)
begin
   if (~rst_n)
      y = 3'b000;        //initializes state a to 000
      z1 = 1'b0;         //initializes outputs z1 and z2 to 0
      z2 = 1'b0;
end

always @ (y or clk)
begin
   if (y == state_e)     //== specifies logical
      begin              //... equality or compare
         if (~clk)
            z1 = 1'b1;   //assert z1 at t2
         else
            z1 = 1'b0;
      end

   if (y == state_f)
      begin
         if (~clk)
            z2 = 1'b1;
         else
            z2 = 1'b0;
      end

   if (y == state_a)
      z1 = 1'b0;         //assigns a known value to z1

   if (y == state_a)
      z2 = 1'b0;         //assigns a known value to z2

end

                        //continued on next page
```

Figure 2.197   (Continued)

```
//determine next state
always @ (y or x1)
begin
   case (y)                            //case is a multiway
       state_a:                        //... conditional branch
          if (~x1)                     //if y = state_a, then
             next_state = state_a;     //... do if ... else
          else
             next_state = state_b;


       state_b:
          if (~x1)
             next_state = state_c;
          else
             next_state = state_d;


       state_c:
          if (~x1)
             next_state = state_a;
          else
             next_state = state_e;


       state_d:
          if (~x1)
             next_state = state_f;
          else
             next_state = state_a;


       state_e: next_state = state_a;


       state_f: next_state = state_a;


       default: next_state = state_a;

   endcase
end

endmodule
```

Figure 2.197   (Continued)

```
//test bench for moore_ssm28_bh
module moore_ssm28_bh_tb;

reg rst_n, clk, x1;          //inputs are reg for test bench
wire [1:3] y;                //outputs are wire for test bench
wire z1, z2;

//display variables
initial
$monitor ("x1 = %b, state = %b, z1 = %b, z2 = %b",
          x1, y, z1, z2);

//define clock
initial
begin
   clk = 1'b0;
   forever
      #10 clk = ~clk;
end

//define input sequence
initial
begin
   #0    rst_n = 1'b0;             //reset to state_a (000)
         x1 = 1'b0;

   #5    rst_n = 1'b1;             //deassert reset
//--------------------------------------------------------
   x1 = 1'b0;@ (posedge clk)       //go to state_a (000)
   x1 = 1'b1;@ (posedge clk)       //go to state_b (001)
   x1 = 1'b0;@ (posedge clk)       //go to state_c (011)
   x1 = 1'b0;@ (posedge clk)       //go to state_a (000)
//--------------------------------------------------------
   x1 = 1'b1;@ (posedge clk)       //go to state_b (001)
   x1 = 1'b0;@ (posedge clk)       //go to state_c (011)
   x1 = 1'b1;@ (posedge clk)       //go to state_e (010),
                                   //assert z1 at t2
   x1 = $random;@ (posedge clk)    //go to state_a (000)
//--------------------------------------------------------
   x1 = 1'b1;@ (posedge clk)       //go to state_b (001)
   x1 = 1'b1;@ (posedge clk)       //go to state_d (101)
   x1 = 1'b0;@ (posedge clk)       //go to state_f (100),
                                   //assert z2 at t2
   x1 = $random;@ (posedge clk)    //go to state_a (000)
//--------------------------------------------------------
                                   //continued on next page
```

Figure 2.198   Test bench module for the equivalent Moore machine to detect sequences of 101 and 110.

```
//-------------------------------------------------------------
   x1 = 1'b1;@ (posedge clk)//go to state_b (001)
   x1 = 1'b1;@ (posedge clk)//go to state_d (101)
   x1 = 1'b1;@ (posedge clk)//go to state_a (000)
//-------------------------------------------------------------
   #10    $stop;
end


//-------------------------------------------------------------
//instantiate the module into the test bench
moore_ssm28_bh inst1 (
   .rst_n(rst_n),
   .clk(clk),
   .x1(x1),
   .y(y),
   .z1(z1),
   .z2(z2)
   );

endmodule
```

Figure 2.198   (Continued)

```
x1 = 0, state = 000, z1 = 0, z2 = 0
x1 = 1, state = 000, z1 = 0, z2 = 0
x1 = 0, state = 001, z1 = 0, z2 = 0
x1 = 0, state = 011, z1 = 0, z2 = 0
x1 = 1, state = 000, z1 = 0, z2 = 0
x1 = 0, state = 001, z1 = 0, z2 = 0
x1 = 1, state = 011, z1 = 0, z2 = 0
x1 = 0, state = 010, z1 = 0, z2 = 0
x1 = 0, state = 010, z1 = 1, z2 = 0

x1 = 1, state = 000, z1 = 0, z2 = 0
x1 = 1, state = 001, z1 = 0, z2 = 0
x1 = 0, state = 101, z1 = 0, z2 = 0
x1 = 1, state = 100, z1 = 0, z2 = 0
x1 = 1, state = 100, z1 = 0, z2 = 1
x1 = 1, state = 000, z1 = 0, z2 = 0
x1 = 1, state = 001, z1 = 0, z2 = 0
x1 = 1, state = 101, z1 = 0, z2 = 0
x1 = 1, state = 000, z1 = 0, z2 = 0
```

Figure 2.199   Outputs for the equivalent Moore machine to detect sequences of 101 and 110.

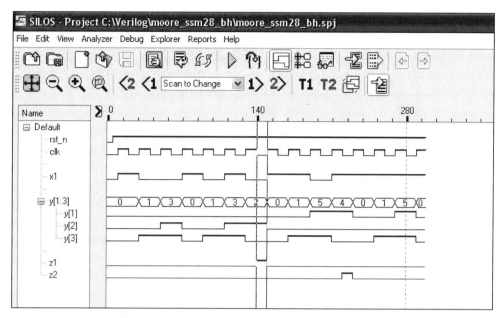

Figure 2.200   Waveforms for the equivalent Moore machine to detect sequences of 101 and 110.

Example 2.27   This example designs an equivalent Moore machine to detect a 101 and a 110 sequence using structural modeling with *JK* flip-flops. The state diagram for the Moore machine that corresponds to the Mealy machine of Figure 2.189 is shown in Figure 2.196. In this example, however, the outputs are asserted at time t_1 and deasserted at time t_3. The excitation table for a *JK* flip-flop is reproduced in Table 2.9 for convenience.

Table 2.9 Excitation Table for a *JK* Flip-Flop

Present state $Y_{j(t)}$	Next state $Y_{k(t+1)}$	Flip-flop inputs JK
0	0	0 –
0	1	1 –
1	0	– 1
1	1	– 0

The next-state table for the equivalent Moore machine using *JK* flip-flops is shown in Table 2.10, as obtained from the state diagram. Using the next-state table, the input maps for the *JK* flip-flops are derived, as shown in Figure 2.201 using x_1 as a map-entered variable. The input equations for the δ next-state logic are also shown in Figure 2.201. The logic diagram is displayed in Figure 2.202 and shows the instantiation names and the net names.

Table 2.10 Next-State Table for the Equivalent Moore Machine Using *JK* Flip-Flops

Present state			Input	Next state			Flip-flop inputs						Outputs	
y_1	y_2	y_3	x_1	y_1	y_2	y_3	Jy_1	Ky_1	Jy_2	Ky_2	Jy_3	Ky_3	z_1	z_2
0	0	0	0	0	0	0	0	–	0	–	0	–	0	0
0	0	0	1	0	0	1	0	–	0	–	1	–	0	0
0	0	1	0	0	1	1	0	–	1	–	–	0	0	0
0	0	1	1	1	0	1	1	–	0	–	–	0	0	0
0	1	0	0	0	0	0	0	–	–	1	0	–	1	0
0	1	0	1	0	0	0	0	–	–	1	0	–	1	0
0	1	1	0	0	0	0	0	–	–	1	–	1	0	0
0	1	1	1	0	1	0	0	–	–	0	–	1	0	0
1	0	0	0	0	0	0	–	1	0	–	0	–	0	1
1	0	0	1	0	0	0	–	1	0	–	0	–	0	1
1	0	1	0	1	0	0	–	0	0	–	–	1	0	0
1	0	1	1	0	0	0	–	1	0	–	–	1	0	0
1	1	0	0	–	–	–	–	–	–	–	–	–	0	0
1	1	0	1	–	–	–	–	–	–	–	–	–	0	0
1	1	1	0	–	–	–	–	–	–	–	–	–	0	0
1	1	1	1	–	–	–	–	–	–	–	–	–	0	0

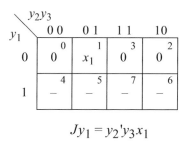

$$Jy_1 = y_2'y_3x_1$$

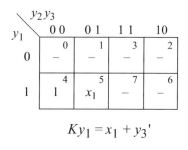

$$Ky_1 = x_1 + y_3'$$

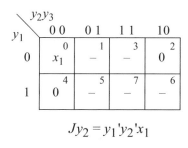

$$Jy_2 = y_1'y_3x_1'$$

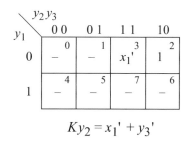

$$Ky_2 = x_1' + y_3'$$

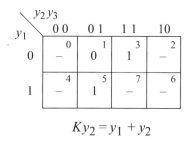

$$Jy_2 = y_1'y_2'x_1$$

$$Ky_2 = y_1 + y_2$$

Figure 2.201   Input maps for the equivalent Moore machine using JK flip-flops to detect sequences of 101 and 110 on input x_1.

The structural design module is shown in Figure 2.203 and the test bench module is shown in Figure 2.204, which takes the machine through the four paths and asserts outputs z_1 and z_2. The outputs and waveforms are shown in Figure 2.205 and Figure 2.206, respectively.

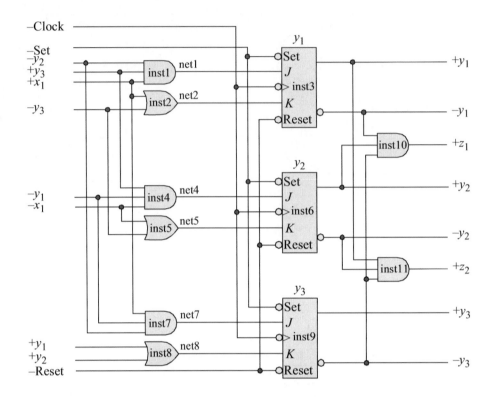

Figure 2.202   Logic diagram for the equivalent Moore machine to detect the sequences 101 and 110 on input x_1.

```
//structural moore synchronous sequential machine
//to detect bit sequences of 101 and 110

module moore_ssm28_jk (rst_n, set_n, clk, x1, y, z1, z2);


//define inputs and outputs
input rst_n, set_n, clk, x1;
output [1:3] y;
output z1, z2;


//define internal nets
wire net1, net2, net4, net5, net7, net8;
                                //continued on next page
```

Figure 2.203   Structural design module for the equivalent Moore machine to detect the sequences 101 and 110 on input x_1.

```
//instantiate the logic for flip-flop y[1]
and3_df inst1 (
    .x1(~y[2]),
    .x2(y[3]),
    .x3(x1),
    .z1(net1)
    );

or2_df inst2 (
    .x1(x1),
    .x2(~y[3]),
    .z1(net2)
    );

jkff_neg_clk inst3 (
    .rst_n(rst_n),
    .set_n(set_n),
    .clk(clk),
    .j(net1),
    .k(net2),
    .q(y[1])
    );

//instantiate the logic for flip-flop y[2]
and3_df inst4 (
    .x1(~y[1]),
    .x2(y[3]),
    .x3(~x1),
    .z1(net4)
    );

or2_df inst5 (
    .x1(~x1),
    .x2(~y[3]),
    .z1(net5)
    );

jkff_neg_clk inst6 (
    .rst_n(rst_n),
    .set_n(set_n),
    .clk(clk),
    .j(net4),
    .k(net5),
    .q(y[2])
    );

                          //continued on next page
```

Figure 2.203   (Continued)

```
//instantiate the logic for flip-flop y[3]
and3_df inst7 (
   .x1(~y[1]),
   .x2(~y[2]),
   .x3(x1),
   .z1(net7)
   );

or2_df inst8 (
   .x1(y[1]),
   .x2(y[2]),
   .z1(net8)
   );

jkff_neg_clk inst9 (
   .rst_n(rst_n),
   .set_n(set_n),
   .clk(clk),
   .j(net7),
   .k(net8),
   .q(y[3])
   );

//instantiate the logic for outputs z1 and z2
and3_df inst10 (
   .x1(~y[1]),
   .x2(y[2]),
   .x3(~y[3]),
   .z1(z1)
   );

and3_df inst11 (
   .x1(y[1]),
   .x2(~y[2]),
   .x3(~y[3]),
   .z1(z2)
   );

endmodule
```

Figure 2.203   (Continued)

```
//test bench for moore_ssm28_jk
module moore_ssm28_jk_tb;

reg set_n, rst_n, clk, x1;  //inputs are reg for test bench
wire [1:3] y;               //outputs are wire for test bench
wire z1, z2;

//display variables
initial
$monitor ("x1 = %b, state = %b, z1 = %b, z2 = %b",
          x1, y, z1, z2);

//define clock
initial
begin
   clk = 1'b0;
   forever
      #10 clk = ~clk;
end

//define input sequence
initial
begin
   #0    rst_n = 1'b0;          //reset to state_a (000)
         set_n = 1'b1;
         x1 = 1'b0;

   #5    rst_n = 1'b1;          //deassert reset

   x1 = 1'b0;@ (negedge clk)  //go to state_a (000)
   x1 = 1'b1;@ (negedge clk)  //go to state_b (001)
   x1 = 1'b0;@ (negedge clk)  //go to state_c (011)
   x1 = 1'b0;@ (negedge clk)  //go to state_a (000)
//------------------------------------------------------------
   x1 = 1'b1;@ (negedge clk)  //go to state_b (001)
   x1 = 1'b0;@ (negedge clk)  //go to state_c (011)
   x1 = 1'b1;@ (negedge clk)  //go to state_e (010), assert z1
   x1 = 1'b0;@ (negedge clk)  //go to state_a (000)
//------------------------------------------------------------
   x1 = 1'b1;@ (negedge clk)  //go to state_b (001)
   x1 = 1'b1;@ (negedge clk)  //go to state_d (101)
   x1 = 1'b0;@ (negedge clk)  //go to state_f (100), assert z2
   x1 = 1'b0;@ (negedge clk)  //go to state_a (000)
//------------------------------------------------------------
                              //continued on next page
```

Figure 2.204   Test bench module for the equivalent Moore machine to detect the sequences 101 and 110 on input x_1.

```
//----------------------------------------------------------
   x1 = 1'b1;@ (negedge clk)   //go to state_b (001)
   x1 = 1'b1;@ (negedge clk)   //go to state_d (101)
   x1 = 1'b1;@ (negedge clk)   //go to state_a (000)
//----------------------------------------------------------
   #10    $stop;
end

//----------------------------------------------------------
//instantiate the module into the test bench
moore_ssm28_jk inst1 (
   .rst_n(rst_n),
   .set_n(set_n),
   .clk(clk),
   .x1(x1),
   .y(y),
   .z1(z1),
   .z2(z2)
   );

endmodule
```

Figure 2.204   (Continued)

```
x1 = 0,  state = 000,  z1 = 0,  z2 = 0
x1 = 1,  state = 000,  z1 = 0,  z2 = 0
x1 = 0,  state = 001,  z1 = 0,  z2 = 0
x1 = 0,  state = 011,  z1 = 0,  z2 = 0
x1 = 1,  state = 000,  z1 = 0,  z2 = 0
x1 = 0,  state = 001,  z1 = 0,  z2 = 0
x1 = 1,  state = 011,  z1 = 0,  z2 = 0
x1 = 0,  state = 010,  z1 = 1,  z2 = 0

x1 = 1,  state = 000,  z1 = 0,  z2 = 0
x1 = 1,  state = 001,  z1 = 0,  z2 = 0
x1 = 0,  state = 101,  z1 = 0,  z2 = 0
x1 = 0,  state = 100,  z1 = 0,  z2 = 1
x1 = 1,  state = 000,  z1 = 0,  z2 = 0
x1 = 1,  state = 001,  z1 = 0,  z2 = 0
x1 = 1,  state = 101,  z1 = 0,  z2 = 0
x1 = 1,  state = 000,  z1 = 0,  z2 = 0
```

Figure 2.205   Outputs for the equivalent Moore machine to detect the sequences 101 and 110 on input x_1.

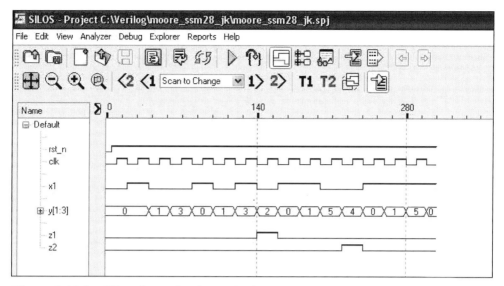

Figure 2.206   Waveforms for the equivalent Moore machine to detect the sequences 101 and 110 on input x_1.

Example 2.28   This example designs an equivalent Moore machine to detect a 101 and a 110 sequence using structural modeling with built-in primitives and D flip-flops. The state diagram for the Moore machine that corresponds to the Mealy machine of Figure 2.189 is shown in Figure 2.196. In this example, however, the outputs are asserted at time t_1 and deasserted at time t_3.

The input maps and input equations for the D flip-flops are shown in Figure 2.207 using x_1 as a map-entered variable. The logic diagram is shown in Figure 2.208 using AND gates and OR gates, together with instantiated D flip-flops.

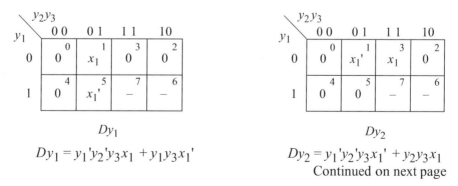

$$Dy_1 = y_1'y_2'y_3x_1 + y_1y_3x_1'$$

$$Dy_2 = y_1'y_2'y_3x_1' + y_2y_3x_1$$

Continued on next page

Figure 2.207   Input maps for the equivalent Moore machine to detect the sequences 101 and 110 on input x_1.

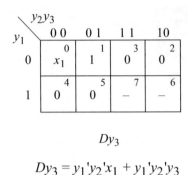

$$Dy_3$$

$$Dy_3 = y_1'y_2'x_1 + y_1'y_2'y_3$$

Figure 2.207   (Continued)

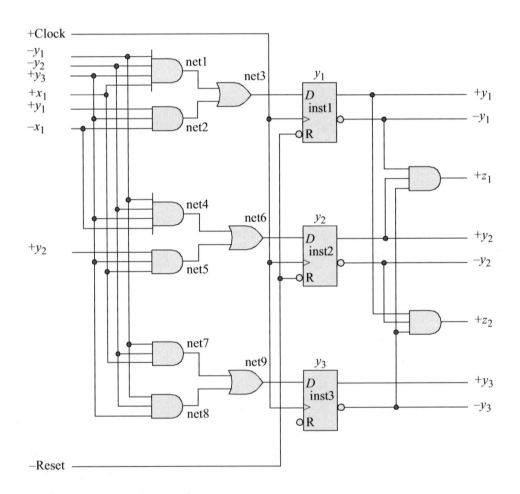

Figure 2.208   Logic diagram for the equivalent Moore machine to detect the sequences 101 and 110 on input x_1.

Verilog uses *built-in primitives* as structural elements that can be instantiated into a larger design to form a more complex structure. Examples are: **and**, **nand**, **or**, **nor**, **xor**, and **xnor**. These built-in primitive gates are used to describe a net and have one or more scalar inputs, but only one scalar output. The output signal is listed first, followed by the inputs in any order. The outputs are declared as **wire;** the inputs can be declared as either **wire** or **reg**. The gates represent combinational logic functions and can be instantiated into a module, as follows, where the instance name is optional:

gate_type inst1 (output, input_1, input_2, . . . , input_n);

The structural design module using built-in primitives and instantiated D flip-flops is shown in Figure 2.209. The test bench module is shown in Figure 2.210, which takes the machine through the four paths and asserts outputs z_1 and z_2. The outputs and waveforms are shown in Figure 2.211 and Figure 2.212, respectively.

```
//structural moore synchronous sequential machine
//using built-in primitives to detect
//bit sequences of 101 and 110
module moore_ssm28_d (rst_n, clk, x1, y, z1, z2);

input rst_n, clk, x1;
output [1:3] y;
output z1, z2;

//define internal nets
wire net1, net2, net3, net4, net5, net6, net7, net8, net9;

//instantiate the logic for flip-flop y[1]
and    (net1, ~y[1], ~y[2], y[3], x1);
and    (net2, y[1], y[3], ~x1);
or     (net3, net1, net2);

d_ff_bh inst1 (
    .rst_n(rst_n),
    .clk(clk),
    .d(net3),
    .q(y[1])
    );

//instantiate the logic for flip-flop y[2]
and    (net4, ~y[1], ~y[2], y[3], ~x1);
and    (net5, y[2], y[3], x1);
or     (net6, net4, net5);
                                //continued on next page
```

Figure 2.209   Structural design module for the equivalent Moore machine to detect the sequences 101 and 110 on input x_1.

```
d_ff_bh inst2 (
   .rst_n(rst_n),
   .clk(clk),
   .d(net6),
   .q(y[2])
   );

//instantiate the logic for flip-flop y[3]
and    (net7, ~y[1], ~y[2], x1);
and    (net8, ~y[1], ~y[2], y[3]);
or     (net9, net7, net8);

d_ff_bh inst3 (
   .rst_n(rst_n),
   .clk(clk),
   .d(net9),
   .q(y[3])
   );

//instantiate the logic for the outputs
and    (z1, ~y[1], y[2], ~y[3]);
and    (z2, y[1], ~y[2], ~y[3]);

endmodule
```

Figure 2.209   (Continued)

```
//test bench for moore_ssm28_d
module moore_ssm28_d_tb;

reg rst_n, clk, x1;       //inputs are reg for test bench
wire [1:3] y;             //outputs are wire for test bench
wire z1, z2;

initial                   //display variables
$monitor ("x1 = %b, state = %b, z1 = %b, z2 = %b",
          x1, y, z1, z2);
initial                   //define clock
begin
   clk = 1'b0;
   forever
      #10 clk = ~clk;
end
                          //continued on next page
```

Figure 2.210   Test bench module for the equivalent Moore machine to detect the sequences 101 and 110 on input x_1.

```
//define input sequence
initial
begin
   #0     rst_n = 1'b0;            //reset to state_a (000)
          x1 = 1'b0;

   #5     rst_n = 1'b1;            //deassert reset

   x1 = 1'b0;@ (negedge clk)   //go to state_a (000)
   x1 = 1'b1;@ (negedge clk)   //go to state_b (001)
   x1 = 1'b0;@ (negedge clk)   //go to state_c (011)
   x1 = 1'b0;@ (negedge clk)   //go to state_a (000)

//--------------------------------------------------------
   x1 = 1'b1;@ (negedge clk)   //go to state_b (001)
   x1 = 1'b0;@ (negedge clk)   //go to state_c (011)
   x1 = 1'b1;@ (negedge clk)   //go to state_e (010), assert z1
   x1 = 1'b0;@ (negedge clk)   //go to state_a (000)

//--------------------------------------------------------
   x1 = 1'b1;@ (negedge clk)   //go to state_b (001)
   x1 = 1'b1;@ (negedge clk)   //go to state_d (101)
   x1 = 1'b0;@ (negedge clk)   //go to state_f (100), assert z2
   x1 = 1'b0;@ (negedge clk)   //go to state_a (000)

//--------------------------------------------------------
   x1 = 1'b1;@ (negedge clk)   //go to state_b (001)
   x1 = 1'b1;@ (negedge clk)   //go to state_d (101)
   x1 = 1'b1;@ (negedge clk)   //go to state_a (000)

//--------------------------------------------------------

   #10     $stop;
end

//--------------------------------------------------------
//instantiate the module into the test bench
moore_ssm28_d inst1 (
   .rst_n(rst_n),
   .clk(clk),
   .x1(x1),
   .y(y),
   .z1(z1),
   .z2(z2)
   );

endmodule
```

Figure 2.210   (Continued)

```
x1 = 0,  state = 000,  z1 = 0,  z2 = 0
x1 = 1,  state = 000,  z1 = 0,  z2 = 0
x1 = 1,  state = 001,  z1 = 0,  z2 = 0
x1 = 0,  state = 001,  z1 = 0,  z2 = 0
x1 = 0,  state = 011,  z1 = 0,  z2 = 0
x1 = 0,  state = 000,  z1 = 0,  z2 = 0
x1 = 1,  state = 000,  z1 = 0,  z2 = 0
x1 = 1,  state = 001,  z1 = 0,  z2 = 0
x1 = 0,  state = 001,  z1 = 0,  z2 = 0
x1 = 0,  state = 011,  z1 = 0,  z2 = 0
x1 = 1,  state = 011,  z1 = 0,  z2 = 0
x1 = 1,  state = 010,  z1 = 1,  z2 = 0
x1 = 0,  state = 000,  z1 = 0,  z2 = 0
x1 = 1,  state = 000,  z1 = 0,  z2 = 0
x1 = 1,  state = 001,  z1 = 0,  z2 = 0
x1 = 1,  state = 101,  z1 = 0,  z2 = 0
x1 = 0,  state = 101,  z1 = 0,  z2 = 0
x1 = 0,  state = 100,  z1 = 0,  z2 = 1
x1 = 0,  state = 000,  z1 = 0,  z2 = 0
x1 = 1,  state = 000,  z1 = 0,  z2 = 0
x1 = 1,  state = 001,  z1 = 0,  z2 = 0
x1 = 1,  state = 101,  z1 = 0,  z2 = 0
x1 = 1,  state = 000,  z1 = 0,  z2 = 0
```

Figure 2.211   Outputs for the equivalent Moore machine to detect the sequences 101 and 110 on input x_1.

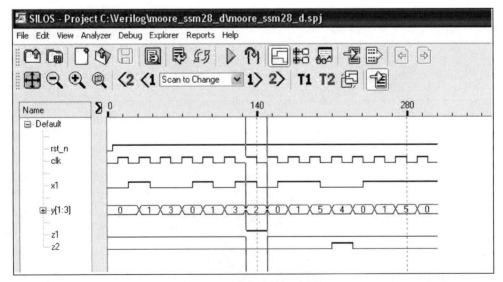

Figure 2.212   Waveforms for the equivalent Moore machine to detect the sequences 101 and 110 on input x_1.

Example 2.29   This example designs an equivalent Moore machine to detect a 101 and a 110 sequence using structural modeling with instantiated logic gates and D flip-flops. The state diagram for the Moore machine that corresponds to the Mealy machine of Figure 2.189 is shown in Figure 2.196. In this example, however, the outputs are asserted at time t_1 and deasserted at time t_3.

This example instantiates the following dataflow modules into the structural module: a 4-input AND gate (*and4_df*), a 3-input AND gate (*and3_df*), and a 2-input OR gate (*or2_df*). Also instantiated is a D flip-flop (*d_ff_bh*) designed using behavioral modeling. The input maps of Figure 2.207 also apply to this example. The logic diagram of Figure 2.208 is redrawn in Figure 2.213 to show the instantiation names of the logic gates and the net names.

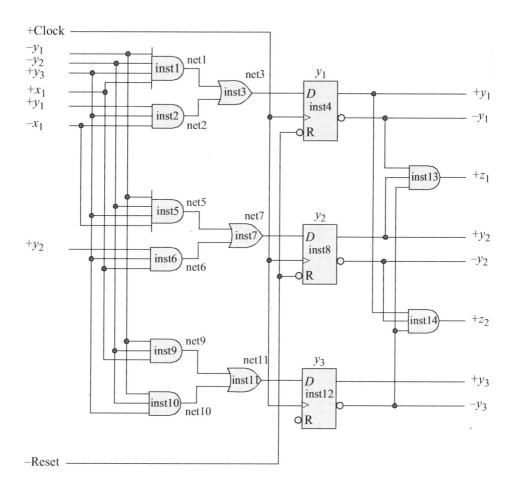

Figure 2.213   Logic diagram for the equivalent Moore machine to detect the sequences 101 and 110 on input x_1.

The structural design module using the instantiated dataflow gates and the behavioral D flip-flops is shown in Figure 2.214. The test bench module is shown in Figure 2.215, which takes the machine through the four paths and asserts outputs z_1 and z_2. The outputs and waveforms are shown in Figure 2.216 and Figure 2.217, respectively.

```verilog
//structural moore synchronous sequential machine
//using module instantiation
//to detect bit sequences of 101 and 110
module moore_ssm28_dff (rst_n, clk, x1, y, z1, z2);

input rst_n, clk, x1;    //define inputs and outputs
output [1:3] y;
output z1, z2;

//define internal nets
wire net1, net2, net3, net5, net6, net7, net9, net10, net11;

//instantiate the logic for flip-flop y[1]
and4_df inst1 (
    .x1(~y[1]),
    .x2(~y[2]),
    .x3(y[3]),
    .x4(x1),
    .z1(net1)
    );

and3_df inst2 (
    .x1(y[1]),
    .x2(y[3]),
    .x3(~x1),
    .z1(net2)
    );

or2_df inst3 (
    .x1(net1),
    .x2(net2),
    .z1(net3)
    );

d_ff_bh inst4 (
    .rst_n(rst_n),
    .clk(clk),
    .d(net3),
    .q(y[1])
    );
                                    //continued on next page
```

Figure 2.214   Structural design module for the equivalent Moore machine to detect the sequences 101 and 110 on input x_1.

```
//instantiate the logic for flip-flop y[2]
and4_df inst5 (
   .x1(~y[1]),
   .x2(~y[2]),
   .x3(y[3]),
   .x4(~x1),
   .z1(net5)
   );

and3_df inst6 (
   .x1(y[2]),
   .x2(y[3]),
   .x3(x1),
   .z1(net6)
   );

or2_df inst7 (
   .x1(net5),
   .x2(net6),
   .z1(net7)
   );

d_ff_bh inst8 (
   .rst_n(rst_n),
   .clk(clk),
   .d(net7),
   .q(y[2])
   );

//instantiate the logic for flip-flop y[3]
and3_df inst9 (
   .x1(~y[1]),
   .x2(~y[2]),
   .x3(x1),
   .z1(net9)
   );

and3_df inst10 (
   .x1(~y[1]),
   .x2(~y[2]),
   .x3(y[3]),
   .z1(net10)
   );

                              //continued on next page
```

Figure 2.214   (Continued)

```
or2_df inst11 (
   .x1(net9),
   .x2(net10),
   .z1(net11)
   );

d_ff_bh inst12 (
   .rst_n(rst_n),
   .clk(clk),
   .d(net11),
   .q(y[3])
   );

//instantiate the logic for outputs z1 and z2
and3_df inst13 (
   .x1(~y[1]),
   .x2(y[2]),
   .x3(~y[3]),
   .z1(z1)
   );

and3_df inst14 (
   .x1(y[1]),
   .x2(~y[2]),
   .x3(~y[3]),
   .z1(z2)
   );

endmodule
```

Figure 2.214   (Continued)

```
//test bench for moore_ssm28_dff ssm
module moore_ssm28_dff_tb;

reg rst_n, clk, x1;   //inputs are reg for test bench
wire [1:3] y;         //outputs are wire for test bench
wire z1, z2;

initial                //display variables
$monitor ("x1 = %b, state = %b, z1 = %b, z2 = %b",
          x1, y, z1, z2);
                               //continued on next page
```

Figure 2.215   Test bench module for the equivalent Moore machine to detect the sequences 101 and 110 on input x_1.

```
initial          //define clock
begin
   clk = 1'b0;
   forever
      #10 clk = ~clk;
end

//define input sequence
initial
begin
   #0    rst_n = 1'b0;         //reset to state_a (000)
         x1 = 1'b0;

   #5    rst_n = 1'b1;         //deassert reset
//------------------------------------------------------------
   x1 = 1'b0;@ (posedge clk)  //go to state_a (000)
   x1 = 1'b1;@ (posedge clk)  //go to state_b (001)
   x1 = 1'b0;@ (posedge clk)  //go to state_c (011)
   x1 = 1'b0;@ (posedge clk)  //go to state_a (000)
//------------------------------------------------------------
   x1 = 1'b1;@ (posedge clk)  //go to state_b (001)
   x1 = 1'b0;@ (posedge clk)  //go to state_c (011)
   x1 = 1'b1;@ (posedge clk)  //go to state_e (010), assert z1
   x1 = $random;@ (posedge clk)//go to state_a (000)
//------------------------------------------------------------
   x1 = 1'b1;@ (posedge clk)  //go to state_b (001)
   x1 = 1'b1;@ (posedge clk)  //go to state_d (101)
   x1 = 1'b0;@ (posedge clk)  //go to state_f (100), assert z2
   x1 = $random;@ (posedge clk)//go to state_a (000)
//------------------------------------------------------------
   x1 = 1'b1;@ (posedge clk)  //go to state_b (001)
   x1 = 1'b1;@ (posedge clk)  //go to state_d (101)
   x1 = 1'b1;@ (posedge clk)  //go to state_a (000)
//------------------------------------------------------------
   #10    $stop;
end

//instantiate the module into the test bench
moore_ssm28_dff inst1 (
   .rst_n(rst_n),
   .clk(clk),
   .x1(x1),
   .y(y),
   .z1(z1),
   .z2(z2)
   );
endmodule
```

Figure 2.215   (Continued)

```
x1 = 0,  state = 000,  z1 = 0,  z2 = 0
x1 = 1,  state = 000,  z1 = 0,  z2 = 0
x1 = 0,  state = 001,  z1 = 0,  z2 = 0
x1 = 0,  state = 011,  z1 = 0,  z2 = 0
x1 = 1,  state = 000,  z1 = 0,  z2 = 0
x1 = 0,  state = 001,  z1 = 0,  z2 = 0
x1 = 1,  state = 011,  z1 = 0,  z2 = 0
x1 = 0,  state = 010,  z1 = 1,  z2 = 0

x1 = 1,  state = 000,  z1 = 0,  z2 = 0
x1 = 1,  state = 001,  z1 = 0,  z2 = 0
x1 = 0,  state = 101,  z1 = 0,  z2 = 0
x1 = 1,  state = 100,  z1 = 0,  z2 = 1

x1 = 1,  state = 000,  z1 = 0,  z2 = 0
x1 = 1,  state = 001,  z1 = 0,  z2 = 0
x1 = 1,  state = 101,  z1 = 0,  z2 = 0
x1 = 1,  state = 000,  z1 = 0,  z2 = 0
```

Figure 2.216   Outputs for the equivalent Moore machine to detect the sequences 101 and 110 on input x_1.

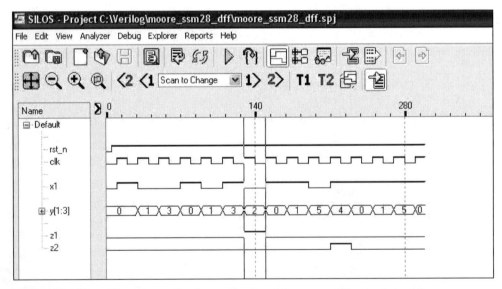

Figure 2.217   Waveforms for the equivalent Moore machine to detect the sequences 101 and 110 on input x_1.

2.6 Output Glitches

A *glitch* in synchronous sequential machines is any false or spurious electronic signal. These narrow, unwanted pulses wreak havoc in digital systems if the glitch occurs on an output signal. Therefore, eliminating output glitches is extremely important, even at the expense of additional logic.

In synchronous sequential machines, glitches can occur in the time period between the active clock transition and circuit stabilization. It is during this time, when the machine is changing states, that the outputs are susceptible to glitches. Although momentary in duration, this transient state can cause an output glitch in both Moore and Mealy machines. If the outputs are enabled at time t_2, then glitches that occur during the period of instability are of no consequence — the machine has long since stabilized. Three methods will be presented, together with the Verilog designs, to eliminate output glitches :

1. State code assignment
2. Complemented clock
3. Delayed clock

2.6.1 Glitch Elimination Using State Code Assignment

This example designs a Moore machine with glitch-free operation using behavioral modeling. The state diagram of Figure 2.218 presents a machine with two Moore-type outputs. State codes were selected, such that there would be no glitches on outputs z_1 or z_2. An incorrect state code assignment would be: state a (000), state b (001), state c (011), state d (101), and state e (111). With this state code assignment, glitches are possible for the state transitions shown in Table 2.11.

Table 2.11 State Transition Sequences that may cause Glitches on z_1 and z_2

Start state	Transient state	End state	
$y_1 y_2 y_3$	$y_1 y_2 y_3$	$y_1 y_2 y_3$	Comments
0 0 1 (b) →	1 0 1 (d) →	1 1 1 (e)	Glitch on z_1
0 1 1 (c) →	1 1 1 (e) →	1 0 1 (d)	Glitch on z_2
1 1 1 (e) →	1 0 1 (d) →	0 0 0 (a)	Glitch on z_1

The path from state b to state e will produce a glitch on output z_1 if flip-flop y_1 sets before y_2 sets. The transition from state c to state d will cause output z_2 to glitch if flip-flop y_1 sets before y_2 resets. The transition from state e to state a results in all flip-flops changing state. Thus, the machine may pass through state d and assert output z_1.

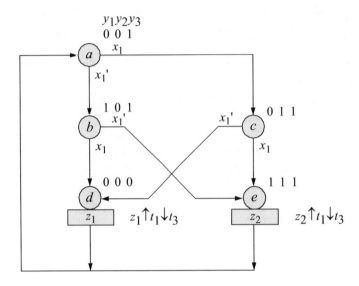

Figure 2.218   State diagram for a Moore machine with correct state code assignment for glitch-free operation. Unused states are $y_1 y_2 y_3 = 010$, 100, and 110.

The behavioral design module is shown in Figure 2.219 using the **case** statement. The test bench module is shown in Figure 2.220, which takes the machine through different state transitions and causes outputs z_1 and z_2 to be asserted. The outputs and waveforms are shown in Figure 2.221 and Figure 2.222, respectively.

```
//behavioral moore synchronous sequential machine
module moore_ssm_29 (rst_n, clk, x1, y, z1, z2);

//define inputs and outputs
input rst_n, clk, x1;
output [1:3] y;
output z1, z2;


reg [1:3] y, next_state;
wire z1, z2;

                              //continued on next page
```

Figure 2.219   Behavioral design module for a Moore machine with glitch-free opeation.

```
//assign state codes
//parameter defines a constant
parameter   state_a = 3'b001,
            state_b = 3'b101,
            state_c = 3'b011,
            state_d = 3'b000,
            state_e = 3'b111;


//-----------------------------------------------------
//set next state
always @ (posedge clk)
begin
   if (~rst_n)          //if (~rst_n) is true
      y <= state_a;     //go to state_a
   else
      y <= next_state;  //else go to next_state
end


//-----------------------------------------------------
//define outputs
assign   z1 = (~y[1] & ~y[2] & ~y[3]),
         z2 = (y[1] & y[2] & y[3]);


//-----------------------------------------------------
//determine next state
always @ (x1 or y)
begin
   case (y)          //case is a multi-way conditional branch
      state_a:       //if y = state_a, do if ... else
         if (~x1)
            next_state = state_b;
         else
            next_state = state_c;

      state_b:
         if (~x1)
            next_state = state_e;
         else
            next_state = state_d;

      state_c:
         if (~x1)
            next_state = state_d;
         else
            next_state = state_e;

                              //continued on next page
```

Figure 2.219   (Continued)

```
        state_d: next_state = state_a;

        state_e: next_state = state_a;

        default: next_state = state_a;
    endcase
end

endmodule
```

Figure 2.219   (Continued)

```
//test bench for moore_ssm_29 synchronous sequential machine
module moore_ssm_29_tb;

reg rst_n, clk, x1;       //inputs are reg for test bench
wire [1:3] y;             //outputs are wire for test bench
wire z1, z2;

initial                   //display variables
$monitor ("x1= %b, state = %b, z1 = %b, z2 = %b", x1, y, z1, z2);

//define clock
initial
begin
   clk = 1'b0;
   forever
      #10 clk = ~clk;
end

//define input sequence
initial
begin
   #0 rst_n = 1'b0;              //reset to state_a (001)
      x1 = 1'b0;

   #5 rst_n = 1'b1;             //deassert reset

//----------------------------------------------------------
               @ (posedge clk)//go to state_a (001)
   x1 = 1'b1;  @ (posedge clk)//go to state_c (011)
   x1 = 1'b0;  @ (posedge clk)//go to state_d (000); assert z1
   x1 = $random;@ (posedge clk)//go to state_a (001)

                              //continued on next page
```

Figure 2.220   Test bench module for a Moore machine with glitch-free operation.

```
//-------------------------------------------------------------
   x1 = 1'b1;@ (posedge clk)   //go to state_c )011)
   x1 = 1'b0;@ (posedge clk)   //go to state_d (000); assert z1
   x1 = $random;@ (posedge clk)//go to state_a (001)
//-------------------------------------------------------------
   x1 = 1'b0;@ (posedge clk)   //go to state_b (101)
   x1 = 1'b0;@ (posedge clk)   //go to state_e (111); assert z2
   x1 = $random;@ (posedge clk)//go to state_a (001)
//-------------------------------------------------------------
   x1 = 1'b1;@ (posedge clk)   //go to state_c (011)
   x1 = 1'b1;@ (posedge clk)   //go to state_e (111); assert z2
   x1 = $random;@ (posedge clk)//go to state_a (001)
//-------------------------------------------------------------
   #20    $stop;
end

//-------------------------------------------------------------
//instantiate the module into the test bench
moore_ssm_29 inst1 (
   .rst_n(rst_n),
   .clk(clk),
   .x1(x1),
   .y(y),
   .z1(z1),
   .z2(z2)
   );
endmodule
```

Figure 2.220   (Continued)

```
x1= 0,  state = xxx,  z1 = x,  z2 = x
x1= 1,  state = 001,  z1 = 0,  z2 = 0
x1= 0,  state = 011,  z1 = 0,  z2 = 0
x1= 0,  state = 000,  z1 = 1,  z2 = 0
x1= 1,  state = 001,  z1 = 0,  z2 = 0
x1= 0,  state = 011,  z1 = 0,  z2 = 0
x1= 1,  state = 000,  z1 = 1,  z2 = 0
x1= 0,  state = 001,  z1 = 0,  z2 = 0
x1= 0,  state = 101,  z1 = 0,  z2 = 0
x1= 1,  state = 111,  z1 = 0,  z2 = 1
x1= 1,  state = 001,  z1 = 0,  z2 = 0
x1= 1,  state = 011,  z1 = 0,  z2 = 0
x1= 1,  state = 111,  z1 = 0,  z2 = 1
x1= 1,  state = 001,  z1 = 0,  z2 = 0
```

Figure 2.221   Outputs for a Moore machine with glitch-free operation.

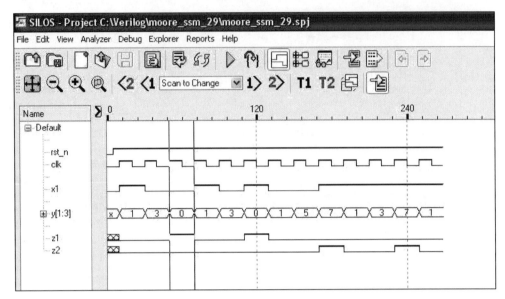

Figure 2.222   Waveforms for a Moore machine with glitch-free operation.

2.6.2 Glitch Elimination Using Complemented Clock

The simplest and most inexpensive method of eliminating output glitches is to include the complement of the machine clock in the implementation of the λ output logic. The output logic will consist of an AND gate which decodes the p-tuple state codes. One input of the AND gate is connected to the complement of the machine clock; that is, the negation of the clock signal which drives the state flip-flops. This will generate an output signal that is only one-half the duration of the clock cycle, but guarantees that the output is free from any erroneous assertions. The output is asserted at time t_2 and deasserted at time t_3.

Example 2.30   This example designs a Moore machine with glitch-free operation using behavioral modeling with complemented clock. The multiple-output Moore machine, depicted by the state diagram of Figure 2.223, contains several possible output glitches if the assertion/deassertion statement for the two outputs z_1 and z_2 is $\uparrow t_1 \downarrow t_3$. Using the complement of the machine clock, however, outputs z_1 and z_2 are asserted at time t_2 and deasserted at time t_3.

Table 2.12 lists the state transition sequences that may cause glitches on the outputs. The four possible output glitches are rendered ineffective, however, by including the machine clock complement in the output decoder for z_1 and z_2. Therefore, outputs z_1 and z_2 are asserted at time t_2 and deasserted at time t_3. If the outputs are enabled at time t_2, then glitches that occur during the period of instability are of no consequence — the machine has long since stabilized.

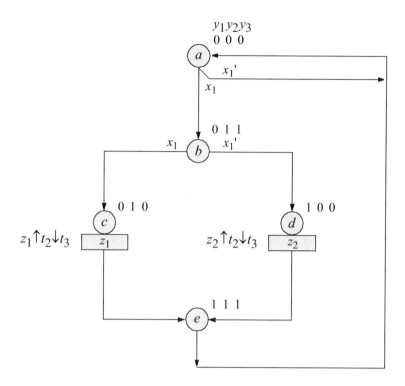

Figure 2.223   State diagram for the Moore machine using the complemented clock to avoid glitches on the outputs.   Unused states are $y_1y_2y_3 = 001$, 101, and 110.

Table 2.12 State Transition Sequences for Figure 2.223 That May Cause Glitches on Outputs z_1 and z_2

Start state $y_1y_2y_3$	Transient state $y_1y_2y_3$	End state $y_1y_2y_3$	Comments
$0\ 0\ 0\ (a) \rightarrow$	$0\ 1\ 0\ (c)$	$\rightarrow\ 0\ 1\ 1\ (b)$	Glitch on z_1
$0\ 1\ 1\ (b) \rightarrow$	$0\ 1\ 0\ (c)$	$\rightarrow\ 1\ 0\ 0\ (d)$	Glitch on z_1
$1\ 1\ 1\ (e) \rightarrow$	$0\ 1\ 0\ (c)$	$\rightarrow\ 0\ 0\ 0\ (a)$	Glitch on z_1
$1\ 1\ 1\ (e) \rightarrow$	$1\ 0\ 0\ (d)$	$\rightarrow\ 0\ 0\ 0\ (a)$	Glitch on z_2

The behavioral design module is shown in Figure 2.224 using the **case** statement. The test bench module is shown in Figure 2.225, which takes the machine through sequences to assert output z_1 and output z_2.   The system task **$random** is used in the test bench to randomly select a value for input x_1 from the values 0 and 1, because the sequences $c \to e$, $e \to a$, and $d \to e$ are independent of the value for x_1. The outputs and waveforms are shown in Figure 2.226 and Figure 2.227, respectively.

```
//behavioral moore machine using
//complemented clock to avoid glitches
module moore_ssm26_bh (rst_n, clk, x1, y, z1, z2);

//define inputs and outputs
input rst_n, clk, x1;
output [1:3] y;
output z1, z2;

//variables are reg in always
reg [1:3] y, next_state;
reg z1, z2;

//assign state codes
//parameter defines a constant
parameter    state_a = 3'b000,
             state_b = 3'b011,
             state_c = 3'b010,
             state_d = 3'b100,
             state_e = 3'b111;


//-----------------------------------------------------------
//determine outputs
always @ (y or clk)
begin
   if (y == state_c)          //== specifies logical
      begin                   //equality or compare
         if (~clk)
            z1 = 1'b1;         //assert z1 at t2
         else
            z1 = 1'b0;
      end

   if (y == state_d)
      begin
         if (~clk)
            z2 = 1'b1;         //assert z2 at t2
         else
            z2 = 1'b0;
      end

                              //continued on next page
```

Figure 2.224   Behavioral design module for the Moore machine using complemented clock to avoid output glitches.

```verilog
   if (y == state_a)
      z1 = 1'b0;

   if (y == state_a)
      z2 = 1'b0;
end

//-------------------------------------------------------
//determine next state
always @ (posedge clk)
begin
   if (~rst_n)
      y <= state_a;
   else
      y <= next_state;
end

//-------------------------------------------------------
//determine next state
always @ (y or x1)
begin
   case (y)                        //case is a multi-way
      state_a:                     //conditional branch
         if (~x1)                  //if y = state_a, then
            next_state = state_a;  //do if ... else
         else
            next_state = state_b;

      state_b:
         if (x1)
            next_state = state_c;
         else
            next_state = state_d;

      state_c: next_state = state_e;

      state_d: next_state = state_e;

      state_e: next_state = state_a;

      default: next_state = state_a;
   endcase
end

endmodule
```

Figure 2.224   (Continued)

```
//test bench for moore machine using complemented clock
module moore_ssm26_bh_tb;

reg rst_n, clk, x1;       //inputs are reg for test bench
wire [1:3] y;             //outputs are wire for test bench
wire z1, z2;

//display variables
initial
$monitor ("x1 = %b, state = %b, z1 = %b, z2 = %b",
            x1, y, z1, z2);

//define clock
initial
begin
   clk = 1'b0;
   forever
      #10 clk = ~clk;
end

//define input sequence
initial
begin
   #0     rst_n = 1'b0;          //reset to state_a (000)
          x1 = 1'b0;

   #5     rst_n = 1'b1;          //deassert reset

   x1 = 1'b0;@ (posedge clk)  //go to state_a (000)
   x1 = 1'b1;@ (posedge clk)  //go to state_b (011)
   x1 = 1'b1;@ (posedge clk)   //go to state_c (010), assert z1
   x1 = $random;@ (posedge clk)//go to state_e (111)
   x1 = $random;@ (posedge clk)//go to state_a (000)
//------------------------------------------------------------
   x1 = 1'b1;@ (posedge clk)  //go to state_b (011)
   x1 = 1'b0;@ (posedge clk)   //go to state_d (100), assert z2
   x1 = $random;@ (posedge clk)//go to state_e (111)
   x1 = $random;@ (posedge clk)//go to state_a (000)
//------------------------------------------------------------

   #10    $stop;
end

                              //continued on next page
```

Figure 2.225   Test bench module for the Moore machine using complemented clock to avoid output glitches.

```
moore_ssm26_bh inst1 (        //instantiate the module
   .rst_n(rst_n),
   .clk(clk),
   .x1(x1),
   .y(y),
   .z1(z1),
   .z2(z2)
   );
endmodule
```

Figure 2.225   (Continued)

```
x1 = 0, state = xxx, z1 = x, z2 = x
x1 = 1, state = 000, z1 = 0, z2 = 0
x1 = 1, state = 011, z1 = 0, z2 = 0
x1 = 0, state = 010, z1 = 0, z2 = 0
x1 = 0, state = 010, z1 = 1, z2 = 0
x1 = 1, state = 111, z1 = 0, z2 = 0
x1 = 1, state = 000, z1 = 0, z2 = 0
x1 = 0, state = 011, z1 = 0, z2 = 0
x1 = 1, state = 100, z1 = 0, z2 = 0
x1 = 1, state = 100, z1 = 0, z2 = 1
x1 = 1, state = 111, z1 = 0, z2 = 0
x1 = 1, state = 000, z1 = 0, z2 = 0
```

Figure 2.226   Outputs for the Moore machine using complemented clock to avoid output glitches.

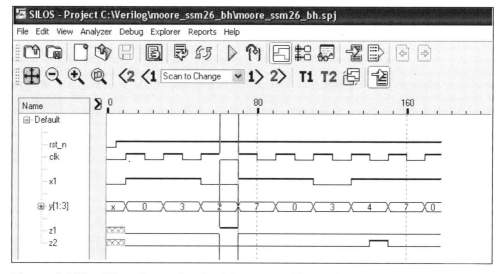

Figure 2.227   Waveforms for the Moore machine using complemented clock to avoid output glitches.

Example 2.31 This example designs a Moore machine for glitch-free operation using structural modeling with *JK* flip-flops and complemented clock. The state diagram of Figure 2.223 will also be used in this example. The excitation table for *JK* flip-flops is reproduced in Table 2.13. Using the state diagram, the input maps for the *JK* flip-flops are shown in Figure 2.228; the output maps are shown in Figure 2.229.

Table 2.13 Excitation Table for a *JK* Flip-Flop

Present state $Y_{j(t)}$	Next state $Y_{k(t+1)}$	Flip-flop inputs $J\,K$
0	0	0 –
0	1	1 –
1	0	– 1
1	1	– 0

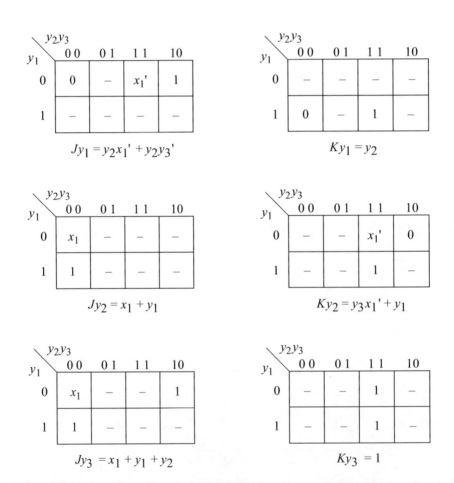

$$Jy_1 = y_2 x_1' + y_2 y_3'$$

$$Ky_1 = y_2$$

$$Jy_2 = x_1 + y_1$$

$$Ky_2 = y_3 x_1' + y_1$$

$$Jy_3 = x_1 + y_1 + y_2$$

$$Ky_3 = 1$$

Figure 2.228 Input maps for the Moore machine using *JK* flip-flops and complemented clock to avoid glitches.

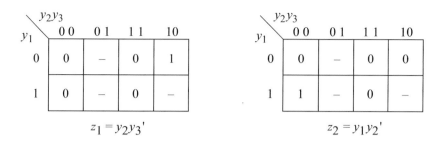

$$z_1 = y_2 y_3{}'$$ $$z_2 = y_1 y_2{}'$$

Figure 2.229   Output maps for the Moore machine using *JK* flip-flops and complemented clock to avoid glitches.

The logic diagram is shown in Figure 2.230 using Boolean equations for the δ input logic. The λ output decoders are implemented with NOR logic functioning as AND gates. It is assumed that the *JK* flip-flops have set and reset inputs. The structural design module is shown in Figure 2.231 using logic gates designed with dataflow modeling and *JK* flip-flops designed with behavioral modeling. The test bench module is shown in Figure 2.232, which asserts outputs z_1 and z_2. The outputs and waveforms are shown in Figure 2.233 and Figure 2.234, respectively.

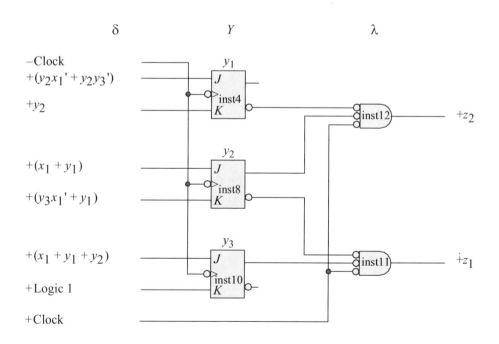

Figure 2.230   Logic diagram for the Moore machine using *JK* flip-flops and complemented clock to avoid glitches.

```
//structural glitch-free moore using jk flip-flops
module mmore_ssm26_jk (set_n, rst_n, clk, x1, y, z1, z2);

input set_n, rst_n, clk, x1;
output [1:3] y;
output z1, z2;

//define internal nets
wire net1, net2, net3, net5, net6, net7, net9;

//design the logic for flip-flop y[1]
and2_df inst1 (
    .x1(y[2]),
    .x2(~x1),
    .z1(net1)
    );

and2_df inst2 (
    .x1(y[2]),
    .x2(~y[3]),
    .z1(net2)
    );

or2_df inst3 (
    .x1(net1),
    .x2(net2),
    .z1(net3)
    );

jkff_neg_clk inst4 (
    .set_n(set_n),
    .rst_n(rst_n),
    .clk(clk),
    .j(net3),
    .k(y[2]),
    .q(y[1])
    );

//design the logic for flip-flop y[2]
or2_df inst5 (
    .x1(x1),
    .x2(y[1]),
    .z1(net5)
    );

                        //continued on next page
```

Figure 2.231   Structural design module for a Moore machine with glitch-free operation.

```
and2_df inst6 (
   .x1(~x1),
   .x2(y[3]),
   .z1(net6)
   );

or2_df inst7 (
   .x1(net6),
   .x2(y[1]),
   .z1(net7)
   );

jkff_neg_clk inst8 (
   .set_n(set_n),
   .rst_n(rst_n),
   .clk(clk),
   .j(net5),
   .k(net7),
   .q(y[2])
   );

//design the logic for flip-flop y[3]
or3_df inst9 (
   .x1(y[2]),
   .x2(y[1]),
   .x3(x1),
   .z1(net9)
   );

jkff_neg_clk inst10 (
   .set_n(set_n),
   .rst_n(rst_n),
   .clk(clk),
   .j(net9),
   .k(1'b1),
   .q(y[3])
   );

//design the logic for outputs z1 and z2
nor3_df inst11 (
   .x1(~y[2]),
   .x2(y[3]),
   .x3(~clk),
   .z1(z1)
   );

                              //continued on next page
```

Figure 2.231   (Continued)

```
nor3_df inst12 (
    .x1(~y[1]),
    .x2(y[2]),
    .x3(~clk),
    .z1(z2)
    );

endmodule
```

Figure 2.231   (Continued)

```
//test bench for glitch-free moore using jk flip-flops
module moore_ssm26_jk_tb;

reg set_n, rst_n, clk, x1; //inputs are reg for test bench
wire [1:3] y;                //outputs are wire for test bench
wire z1, z2;

//display variables
initial
$monitor ("x1 = %b, state = %b, z1 = %b, z2 = %b",
           x1, y, z1, z2);

initial
begin                          //define clock
    clk = 1'b0;
    forever
        #10 clk = ~clk;
end

initial                        //define input sequence
begin
    #0    x1 = 1'b0;
          set_n = 1'b1;
          rst_n = 1'b0;         //reset to state_a (000)

    #5    rst_n = 1'b1;         //deassert reset

    x1 = 1'b0;@ (negedge clk)  //go to state_a (000)
    x1 = 1'b1;@ (negedge clk)  //go to state_b (011)
    x1 = 1'b1;@ (negedge clk)  //go to state_c (010), assert z1
    x1 = $random;@ (negedge clk)//go to state_e (111)
    x1 = $random;@ (negedge clk)//go to state_a (000)
                                //continued on next page
```

Figure 2.232   Test bench module for a Moore machine with glitch-free operation.

```
//---------------------------------------------------------
   x1 = 1'b1;@ (negedge clk)   //go to state_b (011)
   x1 = 1'b0;@ (negedge clk)   //go to state_d (100), assert z2
   x1 = $random;@ (negedge clk)//go to state_e (111)
   x1 = $random;@ (negedge clk)//go to state_a (000)
//---------------------------------------------------------

   #10    $stop;
end

//---------------------------------------------------------
//instantiate the module into the test bench
mmore_ssm26_jk inst1 (
   .set_n(set_n),
   .rst_n(rst_n),
   .clk(clk),
   .x1(x1),
   .y(y),
   .z1(z1),
   .z2(z2)
   );

endmodule
```

Figure 2.232   (Continued)

```
x1 = 0, state = 000, z1 = 0, z2 = 0
x1 = 1, state = 000, z1 = 0, z2· = 0
x1 = 1, state = 011, z1 = 0, z2 = 0
x1 = 0, state = 010, z1 = 0, z2 = 0
x1 = 0, state = 010, z1 = 1, z2 = 0

x1 = 1, state = 111, z1 = 0, z2 = 0
x1 = 1, state = 000, z1 = 0, z2 = 0
x1 = 0, state = 011, z1 = 0, z2 = 0
x1 = 1, state = 100, z1 = 0, z2 = 0
x1 = 1, state = 100, z1 = 0, z2 = 1
x1 = 1, state = 111, z1 = 0, z2 = 0
x1 = 1, state = 000, z1 = 0, z2 = 0
```

Figure 2.233   Outputs for a Moore machine with glitch-free operation.

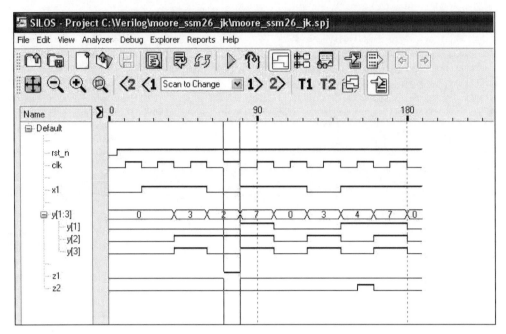

Figure 2.234   Waveforms for a Moore machine with glitch-free operation.

2.6.3 Glitch Elimination Using Delayed Clock

Using the delayed clock technique circumvents the negative effects of glitches. If the machine specifications require that outputs be asserted at time t_1 and deasserted at t_2, then glitches are again possible, because output assertion occurs at the active clock transition. This technique applies to both Moore and Mealy machines.

Figure 2.235 shows a general block diagram for a Mealy machine which uses the active level of the delayed machine clock to enable the λ output logic. The state flip-flops are clocked by the $+Clock$ signal, whereas the λ output logic is asserted by the $+Clock$ delayed signal. The duration of the delay circuit must be equal to or greater than the time Δt — the time when glitches can occur. The machine has stabilized at the termination of the Δt period.

The delay circuit can be either a delay element with a dedicated driver and receiver or simply an even number of inverters. The duration of Δt is quite small in relation to the clock cycle, so that the assertion and deassertion of the output is still considered to be $\uparrow t_1 \downarrow t_2$.

Figure 2.236 shows a state diagram for a Moore machine which will be used to illustrate the delayed clock technique for glitch-free output assertion. The specified assertion and deassertion times for output z_1 are t_1 and t_2, respectively. Since the output logic for z_1 is simply an AND gate to decode $y_1 y_2$, a glitch is possible on z_1 for a state transition from state b to state d if flip-flop y_1 sets before y_2 resets.

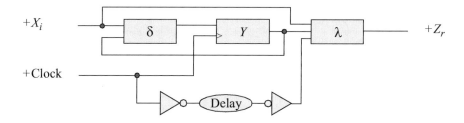

Figure 2.235   A general Mealy machine using clock delayed as an output gating function.

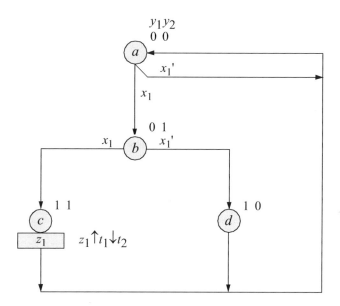

Figure 2.236   Moore machine where a glitch is possible for a transition from state b to state d.

The input maps and output map are shown in Figure 2.237. The logic diagram is shown in Figure 2.238; the D flip-flops have an implied reset input. During a transition from state b to state d, assume that flip-flop y_1 sets before flip-flop y_2 resets. This places a high voltage level on the second and third inputs of the AND gate that generates z_1. If the +Clock delayed signal — which is at a low voltage level at this time — was not connected to the first input of the AND gate, then a glitch would occur on output z_1. Disabling the λ output logic during the time period of Δt by delaying the active clock transition as a gating function, provides an output that is free from spurious signals caused by varying circuit propagation delays.

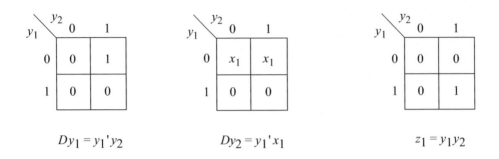

$$Dy_1 = y_1'y_2 \qquad\qquad Dy_2 = y_1'x_1 \qquad\qquad z_1 = y_1y_2$$

Figure 2.237   Input maps and output map for the Moore machine of Figure 2.236.

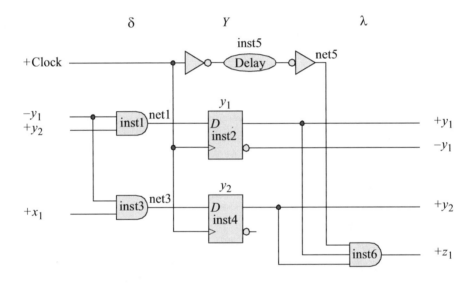

Figure 2.238   Logic diagram for the Moore machine of Figure 2.236.

The structural design module is shown in Figure 2.239, which instantiates logic gates that were designed using dataflow modeling and instantiates two D flip-flops that were designed using behavioral modeling. A built-in primitive, specified as **buf**, is used to delay the machine clock by 5 time units (#5) to prevent output glitches. The delay is variable — depending on the logic family being used — and can be reduced, if necessary. The **buf** gate is a noninverting primitive with one scalar input and one or more scalar outputs. The output terminals are listed first when instantiated; the input is listed last. The time delay of #5 delays the rise and fall times by 5 time units. The test bench module is shown in Figure 2.240. The outputs and waveforms are shown in Figure 2.241 and Figure 2.242, respectively.

```verilog
//structural moore synchronous sequential machine
//using delayed clock to avoid glitches
module moore_ssm_dly (rst_n, clk, x1, y, z1);

input rst_n, clk, x1;
output [1:2]y;
output z1;

wire net1, net3, net5;        //define internal nets

//instantiate the logic for flip-flop y[1]
and2_df inst1 (
    .x1(~y[1]),
    .x2(y[2]),
    .z1(net1)
    );

d_ff_bh inst2 (
    .rst_n(rst_n),
    .clk(clk),
    .d(net1),
    .q(y[1])
    );

//instantiate the logic for flip-flop y[2]
and2_df inst3 (
    .x1(~y[1]),
    .x2(x1),
    .z1(net3)
    );

d_ff_bh inst4 (
    .rst_n(rst_n),
    .clk(clk),
    .d(net3),
    .q(y[2])
    );

//instantiate the logic for output z1
buf #5 inst5 (net5, clk);
and3_df inst6 (
    .x1(net5),
    .x2(y[1]),
    .x3(y[2]),
    .z1(z1)
    );
endmodule
```

Figure 2.239   Structural design module for the Moore machine of Figure 2.238.

```verilog
//test bench for moore_ssm_dly synchronous sequential machine
module moore_ssm_dly_tb;

reg rst_n, clk, x1;        //inputs are reg for test bench
wire [1:2] y;              //outputs are wire for test bench
wire z1;

initial                    //display variables
$monitor ("x1 = %b, state = %b, z1 = %b", x1, y, z1);

initial                    //define clock
begin
   clk = 1'b0;
   forever
      #10 clk = ~clk;
end

//define input sequence
initial
begin
   #0     rst_n = 1'b0;              //reset to state_a (00)
          x1 = 1'b0;

   #5     rst_n = 1'b1;              //deassert reset
//-------------------------------------------------------
   x1 = 1'b0;@ (posedge clk)        //go to state_a (00)
   x1 = 1'b1;@ (posedge clk)        //go to state_b (01)
   x1 = 1'b1;@ (posedge clk)        //go to state_c (11)
                                    //assert z1 (t1 -- t2)
   x1 = $random;@ (posedge clk)     //go to state_a (00)
//-------------------------------------------------------
   x1 = 1'b1;@ (posedge clk)        //go to state_b (01)
   x1 = 1'b0;@ (posedge clk)        //go to state_d (10)
   x1 = $random;@ (posedge clk)     //go to state_a (00)
//-------------------------------------------------------
   #10    $stop;
end

//instantiate the module into the test bench
moore_ssm_dly inst1 (
   .rst_n(rst_n),
   .clk(clk),
   .x1(x1),
   .y(y),
   .z1(z1)
   );
endmodule
```

Figure 2.240   Test bench module for the Moore machine of Figure 2.238.

```
x1 = 0, state = 00, z1 = 0
x1 = 1, state = 00, z1 = 0
x1 = 1, state = 01, z1 = 0
x1 = 0, state = 11, z1 = 0
x1 = 0, state = 11, z1 = 1

x1 = 0, state = 11, z1 = 0
x1 = 1, state = 00, z1 = 0
x1 = 0, state = 01, z1 = 0
x1 = 1, state = 10, z1 = 0
x1 = 1, state = 00, z1 = 0
```

Figure 2.241   Outputs for the Moore machine of Figure 2.238.

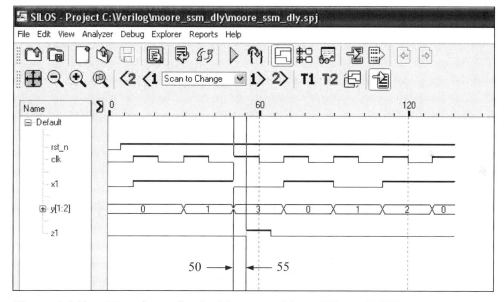

Figure 2.242   Waveforms for the Moore machine of Figure 2.238.

2.7 Problems

2.1   Determine the counting sequence for the counter shown below by designing the counter using structural modeling with built-in primitives and D flip-flops that were designed using behavioral modeling. The D flip-flops have an implied reset input. Obtain the structural design module, the test bench module, the outputs, and the waveforms. The counter is reset initially; that is, $y_1 y_2 = 00$, where y_2 is the low-order flip-flop.

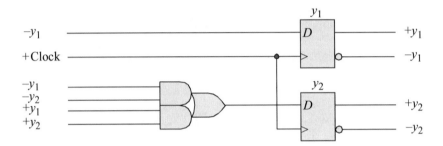

2.2   Determine the counting sequence for the counter shown below by designing the counter using structural modeling with instantiated logic gates that were designed using dataflow modeling and D flip-flops that were designed using behavioral modeling. The D flip-flops have an implied reset input. Obtain the structural design module, the test bench module, the outputs, and the waveforms. The counter is reset initially; that is, $y_1 y_2 = 00$, where y_2 is the low-order flip-flop.

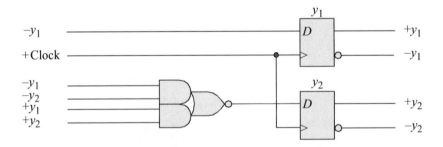

2.3   Using behavioral modeling, design a counter that counts in the sequence shown below. Obtain the design module, the test bench module, the outputs, and the waveforms.

$$y_1 y_2 y_3 y_4 = 0000, 1001, 0001, 1011, 0010, 1100, 0011, 1101, 1000, 1110,$$
$$1010, 1111, 0000, \ldots$$

2.4   Using structural modeling, design a counter that counts in the sequence shown below. Use built-in primitives and D flip-flops that were designed using behavioral modeling. Obtain the design module, the test bench module, the outputs, and the waveforms.

$$y_1 y_2 y_3 y_4 = 0000, 1001, 0001, 1011, 0010, 1100, 0011, 1101, 1000, 1110,$$
$$1010, 1111, 0000, \ldots$$

2.5   Design a modulo-11 counter with no self-starting state using structural modeling with built-in primitives and D flip-flops that were designed using behavioral modeling. Obtain the design module, the test bench module, the outputs, and the waveforms.

2.6   Design a modulo-11 counter with no self-starting state using structural modeling with logic gates that were designed using dataflow modeling and negative-edge triggered JK flip-flops that were designed using behavioral modeling. Obtain the design module, the test bench module, the outputs, and the waveforms.

2.7   Design a modulo-11 counter with no self-starting state using behavioral modeling with the **case** statement. Obtain the design module, the test bench module, the outputs, and the waveforms.

2.8   Design a counter using structural modeling with JK flip-flops that counts in the sequence shown below. The counter is not self-starting. Obtain the design module, the test bench module, the outputs, and the waveforms.

$$y_1 y_2 y_3 = 000, 100, 010, 001, 100, \ldots$$

2.9   Obtain the input equations for flip-flops y_1 and y_4 only, for a BCD counter which counts in the sequence shown below. The equations are to be in minimum form. Use JK flip-flops. There is no self-starting state.

$$y_1 y_2 y_3 y_4 = 0000, 0001, 0010, 0011, 0100, 0101, 0110, 0111, 1000, 1001,$$
$$0000, \cdots .$$

2.10   Design a counter using structural modeling with JK flip-flops which counts in the following decimal sequence: $0, 1, 3, 7, 6, 4, 0, \cdots$. Obtain the design module, the test bench module, the outputs, and the waveforms.

2.11   Design a 4-bit Gray code counter using structural modeling with built-in primitives and JK flip-flops. The counter is initially reset to $y_1 y_2 y_3 y_4 = 0000$. Show the equations for the JK flip-flops, first in a minimum sum-of-products form, then in an exclusive-OR/NOR form, where applicable. Obtain the design module, the test bench module, the outputs, and the waveforms.

2.12   Repeat problem 2.11 using behavioral modeling with the **case** statement. The counting sequence is as follows: 0, 1, 3, 2, 6, 7, 5, 4, 12, 13, 15, 14, 10, 11, 9, 8, 0, $\cdots$. The counter is initially reset to $y_1 y_2 y_3 y_4 = 0000$. Obtain the design module, the test bench module, the outputs, and the waveforms.

2.13   Generate a reduced state diagram for a Moore machine which generates an output z_1 whenever a serial, 4-bit binary word on an input line x_1 is greater than or equal to six. The first bit received in each word is the high-order bit. There is no space between words. Output z_1 is asserted during the fourth bit of a word. Then implement the state diagram in behavioral modeling. Assert output z_1 at time t_2 and deassert z_1 at time t_3. Obtain the design module, the test bench module, the outputs, and the waveforms.

2.14   This problem is similar to Problem 2.13 in that it detects a number that is greater than or equal to six, but uses a user-defined primitive (UDP) that is created by means of a table that defines the functionality of the primitive. UDPs utilizing tables were presented in Section 1.2. The primitive generates an output z_1 whenever a 4-bit binary word x_1, x_2, x_3, x_4 is greater than or equal to six, where x_4 is the low-order bit. Obtain the primitive module, the test bench module, the outputs, and the waveforms.

2.15   A simpler method to detect a number that is greater than or equal to six is to use a Karnaugh map and the continuous assignment statement. This technique models dataflow behavior and is used to design combinational logic. The continuous assignment statement uses the keyword **assign**. The design generates an output z_1 whenever a 4-bit binary word x_1, x_2, x_3, x_4 is greater than or equal to six, where x_4 is the low-order bit. Obtain the design module, the test bench module, the outputs, and the waveforms.

2.16   Generate a reduced state diagram for a Moore machine to detect an input pattern of exactly one group of consecutive 1s on a serial input line x_1. An unconditional output z_1 is asserted when the first 1 occurs and remains asserted for all additional 1s in the group. The output is deasserted for any 0s following the group. If another 1 occurs, then the machine enters and remains in a terminal state which generates an error output.

     For example, a valid input sequence is 000011110000 $\cdots$ 0, because there is a single group of 1s. An invalid sequence is 000010011100 $\cdots$ 0, because there is more than one group of 1s. Show the behavioral design module, the test bench module, the outputs, and the waveforms.

2.17   Generate a reduced state diagram for a Mealy machine which produces a conditional output z_1 whenever a serial data line x_1 contains a sequence of three or more consecutive 1s. Use behavioral modeling with the **case** statement. Show the behavioral design module, the test bench module, the outputs, and the waveforms.

2.18 Repeat Problem 2.17 for a Mealy machine which produces a conditional output z_1 whenever a serial data line x_1 contains a sequence of three or more consecutive 1s. Use structural modeling with built-in primitives and JK flip-flops. Show the Karnaugh maps and equations, the structural design module, the test bench module, the outputs, and the waveforms.

2.19 Generate a reduced state diagram for a Mealy machine which detects a 4-bit word of 1001 on a serial input line x_1. If a correct sequence is detected, then a conditional output z_1 is generated. There is no spacing between words. There is also no overlapping of words. Assert z_1 from time t_2 to time t_3. Obtain the behavioral design module, the test bench module, the outputs, and the waveforms.

2.20 Repeat Problem 2.19 for a Mealy machine which detects a 4-bit word of 1001 on a serial input line x_1. If a correct sequence is detected, then a conditional output z_1 is generated. There is no spacing between words. There is also no overlapping of words. Assert z_1 from time t_2 to time t_3. Obtain the structural design module using instantiated logic gates and D flip-flops, the test bench module, the outputs, and the waveforms.

2.21 Obtain the state diagram for a Mealy synchronous sequential machine to detect a sequence of 0110 on a serial input line x_1. Overlapping sequences are valid. Output z_1 will be asserted during the last half of the clock cycle in the state in which the final 0 is detected. Then obtain the behavioral module, the test bench module, the outputs, and the waveforms. In the test bench, check different paths in the state diagram for correct functional operation and include valid overlapping sequences.

2.22 In the Moore machine shown on the next page, a glitch is possible on output z_1 for a transition from state b to state d if flip-flop y_1 sets before flip-flop y_2 resets. The glitch can be avoided if the output is delayed by a time increment so that the output is asserted only after the machine has stabilized. The delay can be sufficiently small so that the assertion and deassertion of output z_1 can still be considered to be $\uparrow t_1 \downarrow t_2$. Obtain the behavioral design module with an output delay of two time units, the test bench module, the outputs, and the waveforms.

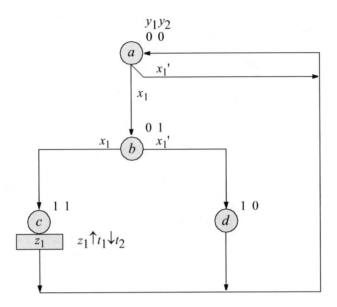

2.23   Select state codes for states d and e for the Moore machine shown below so that there will be no output glitches. Consider all state transitions. Then design a structural module for the Moore machine using built-in primitives and D flip-flops. Obtain the test bench module, the outputs, and the waveforms.

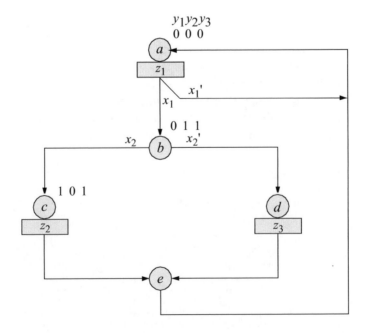

2.24   This repeats Problem 2.23 using the same selected state codes for states *d* and *e* for the Moore machine shown in Problem 2.23 so that there will be no output glitches. Design a structural module for the Moore machine using built-in primitives and *JK* flip-flops. Obtain the test bench module, the outputs, and the waveforms.

2.25   Select state codes for states *c* and *d* for the Moore machine shown below so that there will be no output glitches. Obtain the structural design module using built-in primitives and *D* flip-flops, the test bench module, the outputs, and the waveforms. The outputs are asserted at time t_1 through t_3.

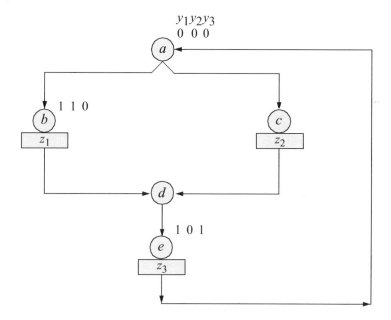

2.26   Obtain the behavioral design module for the Moore machine represented by the state diagram shown below. To avoid any possible output glitches, delay the output assertion by two time units. Obtain the test bench module, the outputs, and the waveforms.

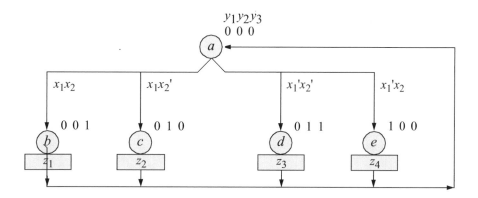

3.1   *Multiplexers for δ Next-State Logic*
3.2   *Decoders for λ Output Logic*
3.3   *Programmable Logic Devices*
3.4   *Iterative Networks*
3.5   *Error Detection in Synchronous*
        Sequential Machines
3.6   *Problems*

3

Synthesis of Synchronous Sequential Machines 2 Using Verilog HDL

This chapter implements synchronous sequential machine designs using Verilog HDL. The designs will be accomplished by utilizing built-in primitives, dataflow modeling, behavioral modeling, structural modeling, or a combination of these modeling techniques. Both linear-select and nonlinear-select multiplexers will be used in various designs for the δ next-state logic. Synchronous sequential machines using decoders for the λ output logic will be synthesized. These include both Moore and Mealy machines.

Programmable logic devices (PLDs) will also be presented in this chapter. This chapter extends the concepts to applications in synthesizing synchronous sequential machines. Different types of PLDs will be presented using Verilog HDL in the design of both Moore and Mealy machines. These include PLDs in the following categories: programmable read-only memory (PROM), programmable array logic (PAL), and programmable logic array (PLA).

Sequential iterative machines will also be covered. An iterative machine is an organization of identical cells (or elements) which are interconnected in an ordered manner. This chapter also describes techniques for error detection for synchronous sequential machines. An error (fault) in a synchronous sequential machine may alter the δ next-state function, the λ output function, or both.

3.1 Multiplexers for δ Next-State Logic

A brief review of multiplexers is appropriate at this time. A multiplexer is a functional logic device that permits two or more data input sources to share a common output transmission circuit, where each data source retains its own independent channel. A multiplexer is essentially a data selector that operates as an electronic switch by connecting one of n inputs to a single output.

There are two general types of multiplexers: linear-select multiplexers and non-linear-select multiplexers. Linear-select multiplexers use all of the data inputs as shown in Figure 3.1(a), for all three flip-flops (y_1, y_2, and y_3), that connect to all three select inputs, yielding an 8:1 multiplexer. Nonlinear-select multiplexers use fewer data inputs (y_1 and y_3) for a smaller 4:1 multiplexer to accomplish the same result, as shown in Figure 3.1(b).

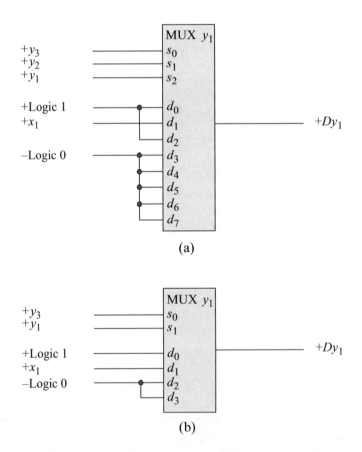

Figure 3.1   The δ next-state logic using multiplexers: (a) linear-select multiplexer; and (b) nonlinear-select multiplexer.

3.1.1 Linear-Select Multiplexers

A general block diagram for a synchronous sequential machine using multiplexers for the δ next-state logic will now be described. The combinational logic which connects to the input of the multiplexer array is either very elementary or nonexistent. In this method, one multiplexer is needed for each state flip-flop and each multiplexer has p select inputs, where p represents the number of storage elements. Therefore, this technique requires $p(2^p:1)$ multiplexers. A machine with 12 states, requiring four storage elements, would have four 16:1 multiplexers. Since most multiplexers have a single output, a design of this type is most easily implemented with D flip-flops.

The active-high output of each state flip-flop is connected to a corresponding select input line; that is, $y_1, y_2, \dots, y_p$ connect to $s_{p-1}, \dots, s_1, s_0$, respectively, where y_p is the low-order flip-flop and s_0 is the low-order select input. Thus, if $p = 3$, then the following state flip-flip-to-select-input connections are necessary: y_3 connects to s_0, y_2 connects to s_1, and y_1 connects to s_2. Since the flip-flop outputs connect to the multiplexer select inputs in a one-to-one mapping, this type of connection can be referred to as *linear selection*.

Example 3.1   A Moore machine will be synthesized that operates according to the state diagram of Figure 3.2. The machine will be synthesized using behavioral modeling with the **case** statement in this example. When input $x_1 = 11$, output z_1 will be asserted at time t_1 and deasserted at time t_2. Output z_1 has an arbitrary delay of two time units.   A structural design will be implemented in Example 3.2.

The behavioral design module is shown in Figure 3.3. The test bench module is shown in Figure 3.4. The outputs and waveforms are shown in Figure 3.5 and Figure 3.6, respectively.

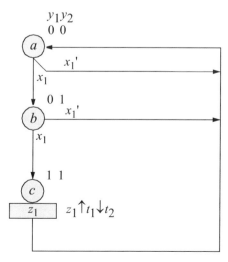

Figure 3.2   Moore machine to assert output z_1 if an input sequence of $x_1 = 11$ is detected. There is one unused state: $y_1 y_2 = 10$.

```verilog
//behavioral moore to assert z1 whenever x1 = 11
module moore_ssm30_bh (rst_n, clk, x1, y, z1);

input rst_n, clk, x1;        //define inputs and outputs
output [1:2] y;
output z1;

reg [1:2] y, next_state;     //variables are reg in always
wire z1;

//assign state codes; parameter defines a constant
parameter    state_a = 2'b00,
             state_b = 2'b01,
             state_c = 2'b11;

//set next state
always @ (posedge clk)
begin
   if (~rst_n)               //if rst_n = 0 (~rst_n is true)
      y <= state_a;          //go to state_a (00)
   else
      y <= next_state;
end

assign #2 z1 = (y[1] && y[2] && clk);   //determine output

always @ (x1 or y)    //determine next state
begin
   case (y)              //csse is a multiway conditional branch
      state_a:          //if y = state_a, do if . . . else
         if (~x1)
            next_state = state_a;
         else
            next_state = state_b;

      state_b:
         if (~x1)
            next_state = state_a;
         else
            next_state = state_c;

      state_c: next_state = state_a;

      default: next_state = state_a;
   endcase
end
endmodule
```

Figure 3.3   Behavioral design module for the Moore machine of Figure 3.2.

```verilog
//test bench for moore to detect x1 = 11
module moore_ssm30_bh_tb;

reg rst_n, clk, x1;           //inputs are reg for test bench
wire [1:2] y;                 //outputs are wire for test bench
wire z1;

initial                       //display variables
$monitor ("x1 = %b, state = %b, z1 = %b", x1, y, z1);

//define clock
initial
begin
   clk = 1'b0;
   forever
      #10 clk = ~clk;
end

//define input sequence
initial
begin
   #0 rst_n = 1'b0;           //reset to state_a (00)
      x1 = 1'b0;

   #5 rst_n = 1'b1;
//-----------------------------------------------------
   x1 = 1'b0; @ (posedge clk)    //go to state_a (00)
   x1 = 1'b1; @ (posedge clk)    //go to state_b (01)

   x1 = 1'b0; @ (posedge clk)    //go to state_a (00)
   x1 = 1'b1; @ (posedge clk)    //go to state_b (01)
   x1 = 1'b1; @ (posedge clk)    //go to state_c (11)
                                 //assert z1 t1 - t2

   x1 = $random; @ (posedge clk) //go to state_a (00)
   #20    $stop;
end
//-----------------------------------------------------
//instantiate the module into the test bench
moore_ssm30_bh inst1 (
   .rst_n(rst_n),
   .clk(clk),
   .x1(x1),
   .y(y),
   .z1(z1)
   );
endmodule
```

Figure 3.4   Test bench module for the Moore machine of Figure 3.2.

```
x1 = 0, state = xx, z1 = 0
x1 = 1, state = 00, z1 = 0
x1 = 0, state = 01, z1 = 0
x1 = 1, state = 00, z1 = 0
x1 = 1, state = 01, z1 = 0
x1 = 0, state = 11, z1 = 0
x1 = 0, state = 11, z1 = 1
x1 = 0, state = 11, z1 = 0
x1 = 0, state = 00, z1 = 0
```

Figure 3.5   Outputs for the Moore machine of Figure 3.2.

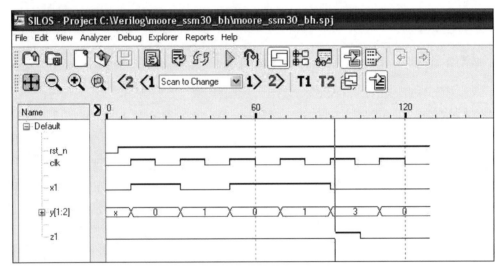

Figure 3.6   Waveforms for the Moore machine of Figure 3.2.

Example 3.2   This example implements the state diagram of Figure 3.2 using structural modeling with 4:1 multiplexers and D flip-flops. The input and output maps are shown in Figure 3.7. The logic diagram using linear-select multiplexers and D flip-flops is shown in Figure 3.8 with instantiation names.

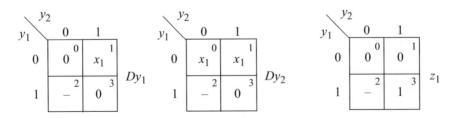

Figure 3.7   Input and output maps for the Moore machine of Figure 3.2.

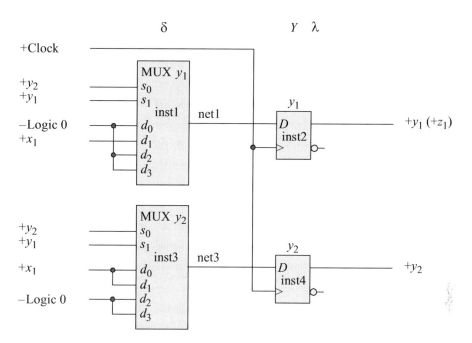

Figure 3.8   Logic diagram for the Moore machine of Figure 3.2 using multiplexers for the δ next-state logic.

The logic diagram for a 4:1 multiplexer is shown in Figure 3.9. The structural design module is shown in Figure 3.10. The test bench module is shown in Figure 3.11. The outputs and waveforms are shown in Figure 3.12 and Figure 3.13, respectively.

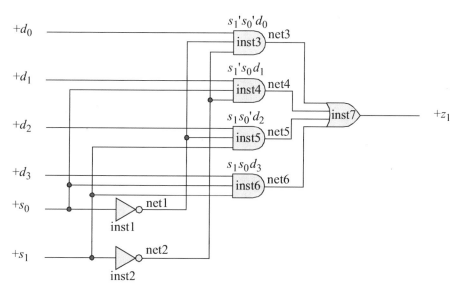

Figure 3.9   Logic diagram for a 4:1 multiplexer.

```
//structural 4:1 multiplexer using built-in primitives
module mux_4to1_struc (d, s, z1);

input [3:0] d;      //data
input [1:0] s;      //select
output z1;

not     inst1 (net1, s[0]),
        inst2 (net2, s[1]);

and     inst3 (net3, d[0], net1, net2),
        inst4 (net4, d[1], s[0], net2),
        inst5 (net5, d[2], net1, s[1]),
        inst6 (net6, d[3], s[0], s[1]);

or      inst7 (z1, net3, net4, net5, net6);

endmodule
```

Figure 3.10   Structural design module for the 4:1 multiplexer.

```
//test bench for 4:1 multiplexer
module mux_4to1_struc_tb;

reg [3:0] d;
reg [1:0] s;
wire z1;

initial
$monitor ("sel = %b, data = %b, z1 = %b", s, d, z1);

initial
begin
   #0    s[0]=1'b0;   s[1]=1'b0;
         d[0]=1'b0;   d[1]=1'b1;   d[2]=1'b0;   d[3]=1'b1;

   #10   s[0]=1'b0;   s[1]=1'b0;
         d[0]=1'b1;   d[1]=1'b1;   d[2]=1'b0;   d[3]=1'b1;

   #10   s[0]=1'b1;   s[1]=1'b0;
         d[0]=1'b1;   d[1]=1'b1;   d[2]=1'b0;   d[3]=1'b1;

   #10   s[0]=1'b0;   s[1]=1'b1;
         d[0]=1'b1;   d[1]=1'b1;   d[2]=1'b0;   d[3]=1'b1;
                                    //continued on next page
```

Figure 3.11   Test bench module for the 4:1 multiplexer.

```
    #10s    [0]=1'b1;   s[1]=1'b0;
            d[0]=1'b1;   d[1]=1'b0;   d[2]=1'b0;   d[3]=1'b1;

    #10     s[0]=1'b1;   s[1]=1'b1;
            d[0]=1'b1;   d[1]=1'b1;   d[2]=1'b0;   d[3]=1'b1;

    #10     s[0]=1'b1;   s[1]=1'b1;
            d[0]=1'b1;   d[1]=1'b1;   d[2]=1'b0;   d[3]=1'b0;

    #10     s[0]=1'b1;   s[1]=1'b1;
            d[0]=1'b1;   d[1]=1'b1;   d[2]=1'b0;   d[3]=1'b0;

    #10     $stop;
end

//instantiate the module into the test bench
mux_4to1_struc inst1 (
    .d(d),
    .s(s),
    .z1(z1)
    );

endmodule
```

Figure 3.11   (Continued)

```
sel = 00, data = 1010, z1 = 0
sel = 00, data = 1011, z1 = 1
sel = 01, data = 1011, z1 = 1
sel = 10, data = 1011, z1 = 0
sel = 01, data = 1001, z1 = 0
sel = 11, data = 1011, z1 = 1
sel = 11, data = 0011, z1 = 0
```

Figure 3.12   Outputs for the 4:1 multiplexer.

The structural design module for the Moore machine is shown in Figure 3.14 which instantiates 4:1 multiplexers and D flip-flops. Note that the data inputs for the multiplexer are labelled *[3:0] data*. Therefore, when the multiplexer is instantiated into the structural module representing the state diagram of Figure 3.2, the data inputs must be listed in the same sequence. The test bench module for the Moore machine is shown in Figure 3.15. The outputs and waveforms are shown in Figure 3.16 and Figure 3.17, respectively.

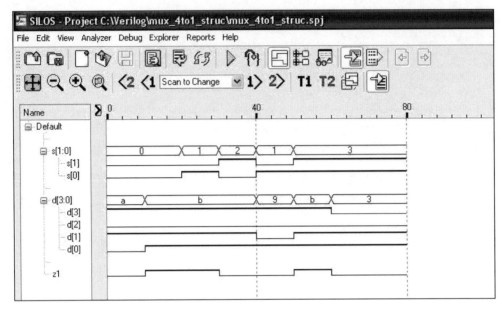

Figure 3.13   Waveforms for the 4:1 multiplexer.

```
//structural moore using muxs and D flip-flops
module moore_mux_dff (rst_n, clk, x1, y, z1);

input rst_n, clk, x1;    //define inputs and outputs
output [1:2] y;
output z1;

wire net1, net3;        //define internal nets
//-----------------------------------------------
//instantiate the logic for flip-flop y[1]
mux_4to1_struc inst1 (
   .d({1'b0, 1'b0, x1, 1'b0}),
   .s({y[1], y[2]}),
   .z1(net1)
   );

d_ff_bh inst2 (
   .rst_n(rst_n),
   .clk(clk),
   .d(net1),
   .q(y[1])
   );
                            //continued on next page
```

Figure 3.14   Structural design module for the Moore machine of Figure 3.2.

```
//------------------------------------------------
//instantiate the logic for flip-flop y[2]
mux_4to1_struc inst3 (
   .d({1'b0, 1'b0, x1, x1}),
   .s({y[1], y[2]}),
   .z1(net3)
   );

d_ff_bh inst4 (
   .rst_n(rst_n),
   .clk(clk),
   .d(net3),
   .q(y[2])
   );

assign z1 = y[1];

endmodule
```

Figure 3.14 (Continued)

```
//test bench for moore machine using muxs
//and D flip-flops to detect x1 = 11
module moore_mux_dff_tb;

//inputs are reg for test bench
//outputs are wire for test bench
reg rst_n, clk, x1;
wire [1:2] y;
wire z1;

//display variables
initial
$monitor ("x1 = %b, state = %b, z1 = %b", x1, y, z1);

//define clock
initial
begin
   clk = 1'b0;
   forever
      #10 clk = ~clk;
end
                              //continued on next page
```

Figure 3.15 Test bench module for the Moore machine of Figure 3.2.

```
//define input sequence
initial
begin
   #0  rst_n = 1'b0;                    //reset to state_a (00)
       x1 = 1'b0;

   #5  rst_n = 1'b1;

//-----------------------------------------------------------
   x1 = 1'b0; @ (posedge clk)      //go to state_a (00)
   x1 = 1'b1; @ (posedge clk)      //go to state_b (01)

   x1 = 1'b0; @ (posedge clk)      //go to state_a (00)
   x1 = 1'b1; @ (posedge clk)      //go to state_b (01)
   x1 = 1'b1; @ (posedge clk)      //go to state_c (11)
                                   //assert z1 t1 - t2

   x1 = $random; @ (posedge clk) //go to state_a (00)

   #20    $stop;
end

//-----------------------------------------------------------
//instantiate the module into the test bench
moore_mux_dff inst1 (
   .rst_n(rst_n),
   .clk(clk),
   .x1(x1),
   .y(y),
   .z1(z1)
   );

endmodule
```

Figure 3.15   (Continued)

```
x1 = 0, state = 00, z1 = 0
x1 = 1, state = 00, z1 = 0
x1 = 0, state = 01, z1 = 0
x1 = 1, state = 00, z1 = 0
x1 = 1, state = 01, z1 = 0
x1 = 0, state = 11, z1 = 1
x1 = 0, state = 00, z1 = 0
```

Figure 3.16   Outputs for the Moore machine of Figure 3.2.

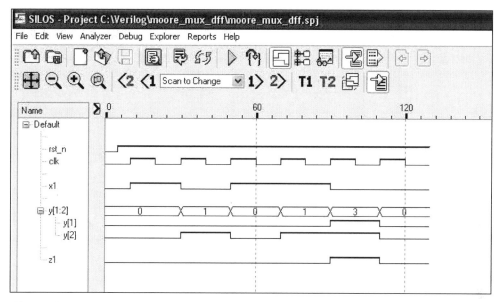

Figure 3.17   Waveforms for the Moore machine of Figure 3.2.

Example 3.3   This example implements the Mealy state diagram of Figure 3.18 using structural modeling with 8:1 multiplexers and D flip-flops. The input and output maps are shown in Figure 3.19. The dataflow design module for an 8:1 multiplexer is shown in Figure 3.20.

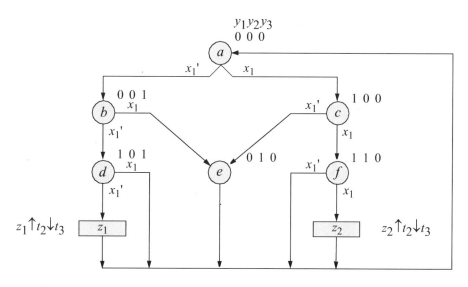

Figure 3.18   Mealy machine that asserts output z_1 if a 3-bit word $x_1 = 000$ is detected and asserts output z_2 if $x_1 = 111$. Unused states are $y_1 y_2 y_3 = 011$ and 111.

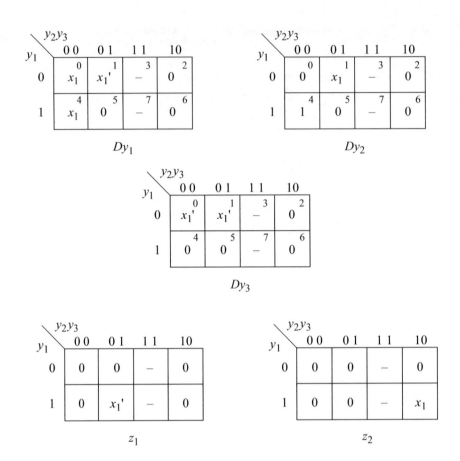

Figure 3.19   Input and output maps for the Mealy machine of Figure 3.18 using input x_1 as a map-entered variable.

```
//dataflow 8:1 multiplexer
module mux_8to1_df (sel, data, out);
input [2:0] sel;
input [7:0] data;
output out;

assign out =(data[0] & ~sel[2] & ~sel[1] & ~sel[0]) |
            (data[1] & ~sel[2] & ~sel[1] & sel[0]) |
            (data[2] & ~sel[2] & sel[1] & ~sel[0]) |
            (data[3] & ~sel[2] & sel[1] & sel[0]) |
            (data[4] & sel[2] & ~sel[1] & ~sel[0]) |
            (data[5] & sel[2] & ~sel[1] & sel[0]) |
            (data[6] & sel[2] & sel[1] & ~sel[0]) |
            (data[7] & sel[2] & sel[1] & sel[0]);
endmodule
```

Figure 3.20   Dataflow design module for an 8:1 multiplexer.

The logic diagram using 8:1 linear-select multiplexers and D flip-flops is shown in Figure 3.21 with instantiation and net names. The D flip-flops have an implied reset input.

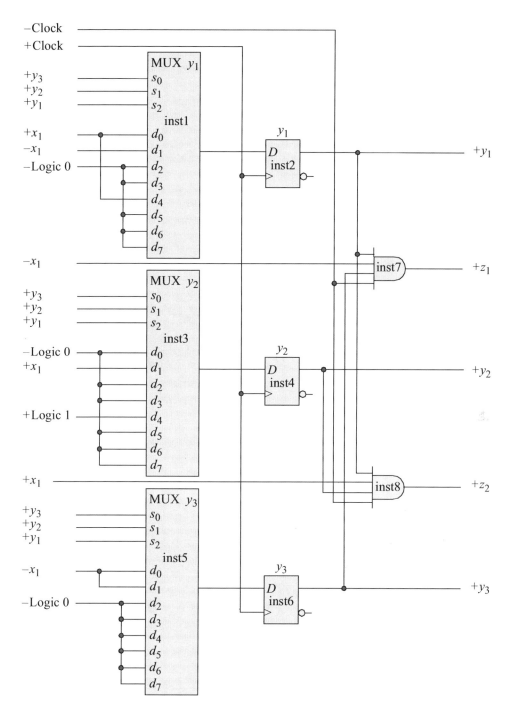

Figure 3.21   Logic diagram for the Mealy machine of Figure 3.18 using multiplexers for the δ next-state logic.

The structural design module is shown in Figure 3.22 using instantiated 8:1 multiplexers designed with dataflow modeling, instantiated D flip-flops designed using behavioral modeling, and instantiated four-input AND gates designed with dataflow modeling. The test bench module is shown in Figure 3.23 which takes the machine through various paths in the state diagram and asserts outputs z_1 and z_2. The outputs and waveforms are shown in Figure 3.24 and Figure 3.25, respectively.

```
//structural mealy using muxs and D flip-flops
//to detect x1 = 000 or x1 = 111

module mealy_mux_dff (rst_n, clk, x1, y, z1, z2);

//define inputs and outputs
input rst_n, clk, x1;
output [1:3] y;
output z1, z2;

//define internal nets
wire net1, net3, net5;

//-------------------------------------------------
//instantiate the logic for flip-flop y[1]
mux_8to1_df inst1 (
   .sel({y[1], y[2], y[3]}),
   .data({{3{1'b0}}, x1, 1'b0, 1'b0, ~x1, x1}),
   .out(net1)
   );

d_ff_bh inst2 (
   .rst_n(rst_n),
   .clk(clk),
   .d(net1),
   .q(y[1])
   );

//-------------------------------------------------
//instantiate the logic for flip-flop y[2]
mux_8to1_df inst3 (
   .sel({y[1], y[2], y[3]}),
   .data({{3{1'b0}}, 1'b1, 1'b0, 1'b0, x1, 1'b0}),
   .out(net3)
   );

                                    //continued on next page
```

Figure 3.22   Structural design module for the Mealy machine of Figure 3.18, which asserts output z_1 if $x_1 = 000$ and asserts output z_2 if x1 = 111.

```
d_ff_bh inst4 (
    .rst_n(rst_n),
    .clk(clk),
    .d(net3),
    .q(y[2])
    );


//-------------------------------------------------
//instantiate the logic for flip-flop y[3]
mux_8to1_df inst5 (
    .sel({y[1], y[2], y[3]}),
    .data({{6{1'b0}}, ~x1, ~x1}),
    .out(net5)
    );

d_ff_bh inst6 (
    .rst_n(rst_n),
    .clk(clk),
    .d(net5),
    .q(y[3])
    );


//-------------------------------------------------
//instantiate the and gates for the outputs
and4_df inst7 (
    .x1(y[1]),
    .x2(~x1),
    .x3(y[3]),
    .x4(~clk),
    .z1(z1)
    );

and4_df inst8 (
    .x1(y[1]),
    .x2(x1),
    .x3(y[2]),
    .x4(~clk),
    .z1(z2)
    );

endmodule
```

Figure 3.22   (Continued)

```
//test bench for mealy using muxs and D flip-flops
//to detect x1 = 000 or x1 = 111

module mealy_mux_dff_tb;

//inputs are reg for test bench
//outputs are wire for test bench
reg rst_n, clk, x1;
wire [1:3] y;
wire z1, z2;

//display variables
initial
$monitor ("x1 = %b, state = %b, z1 = %b, z2 = %b",
          x1, y, z1, z2);

//define clock
initial
begin
   clk = 1'b0;
   forever
      #10 clk = ~clk;
end

//define input sequence
initial
begin
   #0 rst_n = 1'b0;              //reset to state_a (000)
      x1 = 1'b0;

   #5 rst_n = 1'b1;

//----------------------------------------------------------
              @ (posedge clk)
   x1 = 1'b0;@ (posedge clk)    //go to state_b (001)
   x1 = 1'b0;@ (posedge clk)    //go to state_d (101)
   x1 = 1'b0;@ (posedge clk)    //assert z1, go to stat_a (000)

   x1 = 1'b1;@ (posedge clk)    //go to state_c (100)
   x1 = 1'b1;@ (posedge clk)    //go to state_f (110)
   x1 = 1'b1;@ (posedge clk)    //assert z2, go to state_a (000)

                                //continued on next page
```

Figure 3.23   Test bench module for the Mealy machine of Figure 3.18, which asserts output z_1 if $x_1 = 000$ and asserts output z_2 if $x_1 = 111$.

```
      x1 = 1'b1;@ (posedge clk)   //go to state_c (100)
      x1 = 1'b0;@ (posedge clk)   //go to state_e (010)
      x1 = 1'b1;@ (posedge clk)   //go to state_a (000)
      x1 = 1'b0;@ (posedge clk)   //go to state_c (100)

      #20    $stop;

end

//-----------------------------------------------------------
//instantiate the module into the test bench
mealy_mux_dff inst1 (
   .rst_n(rst_n),
   .clk(clk),
   .x1(x1),
   .y(y),
   .z1(z1),
   .z2(z2)
   );

endmodule
```

Figure 3.23   (Continued)

```
x1 = 0, state = 000, z1 = 0, z2 = 0
x1 = 0, state = 001, z1 = 0, z2 = 0
x1 = 0, state = 101, z1 = 0, z2 = 0
x1 = 0, state = 101, z1 = 1, z2 = 0
x1 = 0, state = 000, z1 = 0, z2 = 0
x1 = 0, state = 001, z1 = 0, z2 = 0
x1 = 1, state = 101, z1 = 0, z2 = 0
x1 = 1, state = 000, z1 = 0, z2 = 0
x1 = 1, state = 100, z1 = 0, z2 = 0
x1 = 1, state = 110, z1 = 0, z2 = 0
x1 = 1, state = 110, z1 = 0, z2 = 1
x1 = 0, state = 000, z1 = 0, z2 = 0
x1 = 1, state = 001, z1 = 0, z2 = 0
x1 = 0, state = 010, z1 = 0, z2 = 0
x1 = 0, state = 000, z1 = 0, z2 = 0
```

Figure 3.24   Outputs for the Mealy machine of Figure 3.18, which asserts output z_1 if $x_1 = 000$ and asserts output z_2 if $x_1 = 111$.

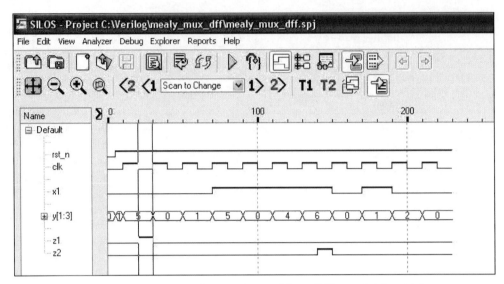

Figure 3.25 Waveforms for the Mealy machine of Figure 3.18, which asserts output z_1 if $x_1 = 000$ and asserts output z_2 if $x_1 = 111$.

Example 3.4 This example designs a Moore machine which operates according to the state diagram of Figure 3.26. The machine will be designed using structural modeling with multiplexers and built-in primitives for the δ next-state logic. The input maps are shown in Figure 3.27; the output maps are shown in Figure 3.28.

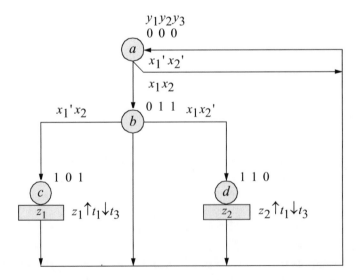

Figure 3.26 State diagram of a Moore machine for Example 3.4. Unused states are: $y_1y_2y_3 = 001, 010, 100,$ and 111.

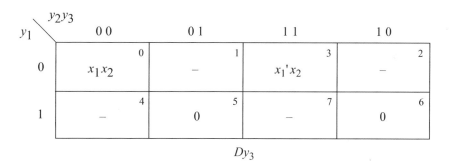

$$Dy_1$$

$$Dy_2$$

$$Dy_3$$

Figure 3.27   Input maps for the Moore machine of Figure 3.26.

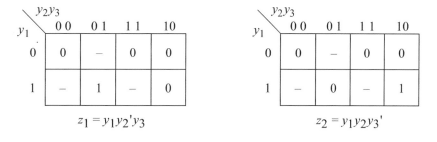

$$z_1 = y_1 y_2' y_3$$

$$z_2 = y_1 y_2 y_3'$$

Figure 3.28   Output maps for the Moore machine of Figure 3.26.

The logic diagram is shown in Figure 3.29. The structural design module is shown in Figure 3.30 using built-in primitives, 8:1 multiplexers, and D flip-flops. The test bench module is shown in Figure 3.31. The outputs and waveforms are shown in Figure 3.32 and Figure 3.33, respectively.

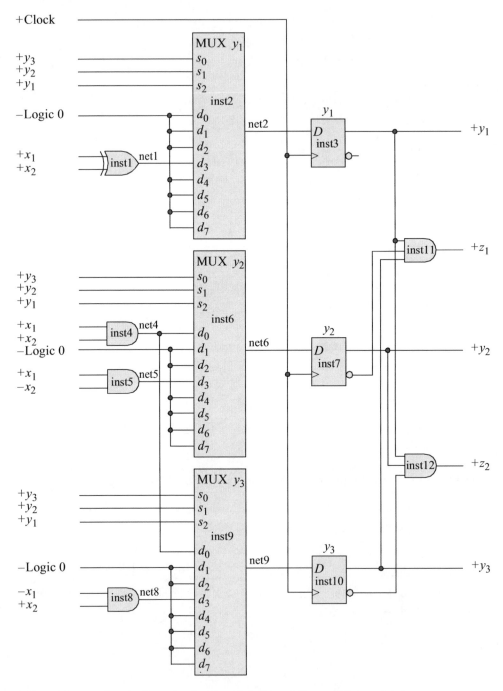

Figure 3.29   Logic diagram for the Moore machine of Figure 3.26.

```
//structural moore to assert z1 for a sequence: x1x2 --> ~x1x2
//and assert z2 for a sequence: x1x2 --> x1~x2

module moore_mux2_dff (rst_n, clk, x1, x2, y, z1, z2);

//define inputs and outputs
input rst_n, clk, x1, x2;
output [1:3] y;
output z1, z2;

//define internal nets
wire net1, net2, net4, net5, net6, net8, net9;

//-----------------------------------------------------
//instantiate the logic for flip-flop y[1]
xor inst1 (net1, x1, x2);

mux_8to1_df inst2 (
   .sel({y[1], y[2], y[3]}),
   .data({{4{1'b0}}, net1, {3{1'b0}}}),
   .out(net2)
   );

d_ff_bh inst3 (
   .rst_n(rst_n),
   .clk(clk),
   .d(net2),
   .q(y[1])
   );

//-----------------------------------------------------
//instantiate the logic for flip-flop y[2]
and   inst4 (net4, x1, x2);
and   inst5 (net5, x1, ~x2);

mux_8to1_df inst6 (
   .sel({y[1], y[2], y[3]}),
   .data({{4{1'b0}}, net5, 1'b0, 1'b0, net4}),
   .out(net6)
   );

d_ff_bh inst7 (
   .rst_n(rst_n),
   .clk(clk),
   .d(net6),
   .q(y[2])
   );                              //continued on next page
```

Figure 3.30   Structural design module for the Moore machine of Figure 3.26.

```
//---------------------------------------------------------
//instantiate the logic for flip-flop y[3]
and inst8 (net8, ~x1, x2);

mux_8to1_df inst9 (
    .sel({y[1], y[2], y[3]}),
    .data({{4{1'b0}}, net8, 1'b0, 1'b0, net4}),
    .out(net9)
    );

d_ff_bh inst10 (
    .rst_n(rst_n),
    .clk(clk),
    .d(net9),
    .q(y[3])
    );

//---------------------------------------------------------
//instantiate the logic for the outputs
and inst11 (z1, y[1], ~y[2], y[3]);
and inst12 (z2, y[1], y[2], ~y[3]);

endmodule
```

Figure 3.30   (Continued)

```
//test bench for moore using muxs and D flip-flops
//to assert z1 in state_c (101)
//and assert z2 in state_d (110)

module moore_mux2_dff_tb;

//inputs are reg for test bench
//outputs are wire for test bench
reg rst_n, clk, x1, x2;
wire [1:3] y;
wire z1, z2;

//display variables
initial
$monitor ("x1 x2 = %b, state = %b, z1 z2 = %b",
            {x1, x2}, y, {z1,z2});

                              //continued on next page
```

Figure 3.31   Test bench module for the Moore machine of Figure 3.26.

```verilog
//define clock
initial
begin
   clk = 1'b0;
   forever
      #10 clk = ~clk;
end


//define input sequence
initial
begin
   #0 rst_n = 1'b0;             //reset to state_a (000)
      x1 = 1'b0;
      x2 = 1'b0;

   #5 rst_n = 1'b1;

//----------------------------------------------------------
   x1 = 1'b1; x2 = 1'b1; @ (posedge clk)//go to state_b (011)
   x1 = 1'b0; x2 = 1'b1; @ (posedge clk)//go to state_c (101)
                                        //assert z1
   x1 = 1'b0; x2 = 1'b0; @ (posedge clk)//go to state_a (000)

//----------------------------------------------------------
   x1 = 1'b1; x2 = 1'b1; @ (posedge clk)//go to state_b (011)
   x1 = 1'b1; x2 = 1'b0; @ (posedge clk)//go to state_d (110)
                                        //assert z2

//----------------------------------------------------------
   x1 = 1'b0; x2 = 1'b0; @ (posedge clk)//go to state_a (000)
   x1 = 1'b1; x2 = 1'b1; @ (posedge clk)//go to state_b (011)
   x1 = 1'b0; x2 = 1'b0; @ (posedge clk)//go to state_a (000)

   #20   $stop;

end

//----------------------------------------------------------
//instantiate the module into the test bench
//as a single line
moore_mux2_dff inst1 (rst_n, clk, x1, x2, y, z1, z2);

endmodule
```

Figure 3.31   (Continued)

```
x1 x2 = 00, state = 000, z1 z2 = 00
x1 x2 = 11, state = 000, z1 z2 = 00
x1 x2 = 01, state = 011, z1 z2 = 00
x1 x2 = 00, state = 101, z1 z2 = 10

x1 x2 = 11, state = 000, z1 z2 = 00
x1 x2 = 10, state = 011, z1 z2 = 00
x1 x2 = 00, state = 110, z1 z2 = 01

x1 x2 = 11, state = 000, z1 z2 = 00
x1 x2 = 00, state = 011, z1 z2 = 00
x1 x2 = 00, state = 000, z1 z2 = 00
```

Figure 3.32   Outputs for the Moore machine of Figure 3.26.

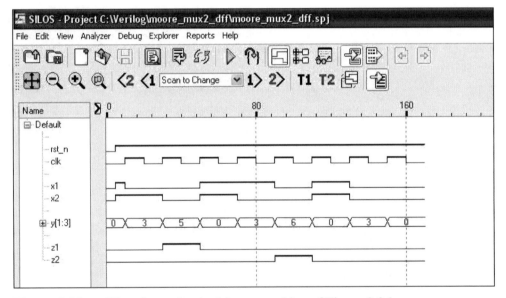

Figure 3.33   Waveforms for the Moore machine of Figure 3.26.

Example 3.5   This example is similar to the Moore machine design of Example 3.4, but with an additional state added to the state diagram, as shown in Figure 3.34. Also, the Moore machine will be designed using behavioral modeling; therefore, multiplexers will not be utilized in the design. Behavioral modeling describes the *behavior* of a digital system and is not concerned with the direct implementation of logic gates but more on the architecture of the system. This is an algorithmic approach to hardware implementation and represents a higher level of abstraction than either dataflow modeling or structural modeling.

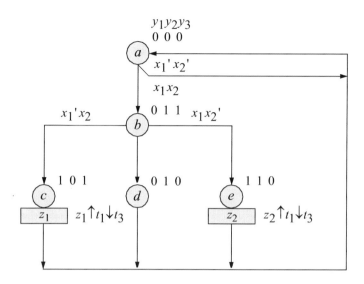

Figure 3.34 State diagram for the Moore machine of Example 3.5. Unused states are: $y_1 y_2 y_3 = 001$, 100, and 111.

Since behavioral modeling is used in this design, there is no need for input maps, output maps, or a logic diagram. The behavioral design module is shown in Figure 3.35. The test bench module is shown in Figure 3.36, which takes the machine through all the paths in the state diagram. The outputs and waveforms are shown in Figure 3.37 and Figure 3.38, respectively.

```verilog
//behavioral moore with 2 inputs and 2 outputs
module moore_bh (rst_n, clk, x1, x2, y, z1, z2);

input rst_n, clk, x1, x2;      //define inputs and outputs
output [1:3] y;
output z1, z2;

reg [1:3] y, next_state;       //must be reg for always
wire z1, z2;

//assign state codes
parameter   state_a = 3'b000,
            state_b = 3'b011,
            state_c = 3'b101,
            state_d = 3'b010,
            state_e = 3'b110;           //continued on next page
```

Figure 3.35 Behavioral design module for the Moore machine of Figure 3.34.

```verilog
//set next state
always @ (posedge clk)
begin
        if (~rst_n)
            y <= state_a;
        else
            y <= next_state;
end

//determine outputs
assign z1 = (y[1] & ~y[2] & y[3]);
assign z2 = (y[1] & y[2] & ~y[3]);

//determine next state
always @ (y or x1 or x2)
begin
   case (y)
        state_a:
            if (x1==0 & x2==0)
                next_state = state_a;

            else if (x1==1 & x2==1)
                next_state = state_b;

        state_b:
            if (x1==0 & x2==1)
                next_state = state_c;

            else if (x1==1 & x2==0)
                next_state = state_e;

            else next_state = state_d;

        state_c: next_state = state_a;

        state_d: next_state = state_a;

        state_e: next_state = state_a;

        default: next_state = state_a;
   endcase

end

endmodule
```

Figure 3.35   (Continued)

```
//test bench for moore with 2 inputs and 2 outputs

module moore_bh_tb;

//inputs are reg for test bench
//outputs are wire for test bench
reg rst_n, clk, x1, x2;
wire [1:3] y;
wire z1, z2;

//display variables
initial
$monitor ("x1 x2 = %b, state = %b, z1 z2 = %b",
           {x1, x2}, y, {z1, z2});

//define clock
initial
begin
    clk = 1'b0;
    forever
        #10 clk = ~clk;
end

//define input sequence
initial
begin
    #0 rst_n = 1'b0;          //reset to state_a (000)
       x1 = 1'b0;
       x2 = 1'b0;

    #5 rst_n = 1'b1;

//------------------------------------------------------------
                        @ (posedge clk)
    x1 = 1'b1; x2 = 1'b1; @ (posedge clk)//go to state_b (011)
    x1 = 1'b0; x2 = 1'b1; @ (posedge clk)//go to state_c (101)
                                         //assert z1
    x1 = 1'b0; x2 = 1'b0; @ (posedge clk)//go to state_a (000)

    x1 = 1'b1; x2 = 1'b1; @ (posedge clk)//go to state_b (011)
    x1 = 1'b1; x2 = 1'b0; @ (posedge clk)//go to state_e (110)
                                         //assert z2
    x1 = 1'b0; x2 = 1'b0; @ (posedge clk)//go to state_a (000)

                        //continued on next page
```

Figure 3.36   Test bench module for the Moore machine of Figure 3.34.

```
   x1 = 1'b1; x2 = 1'b1; @ (posedge clk)//go to state_b (011)
   x1 = 1'b0; x2 = 1'b0; @ (posedge clk)//go to state_d (010)

   x1 = 1'b0; x2 = 1'b0; @ (posedge clk)//go to state_a (000)

   #20     $stop;

end
//------------------------------------------------------------
//instantiate the module into the test bench
//as a single line
moore_bh inst1 (rst_n, clk, x1, x2, y, z1, z2);

endmodule
```

Figure 3.36   (Continued)

```
x1 x2 = 00, state = xxx, z1 z2 = xx
x1 x2 = 11, state = 000, z1 z2 = 00
x1 x2 = 01, state = 011, z1 z2 = 00
x1 x2 = 00, state = 101, z1 z2 = 10

x1 x2 = 11, state = 000, z1 z2 = 00
x1 x2 = 10, state = 011, z1 z2 = 00
x1 x2 = 00, state = 110, z1 z2 = 01

x1 x2 = 11, state = 000, z1 z2 = 00
x1 x2 = 00, state = 011, z1 z2 = 00
x1 x2 = 00, state = 010, z1 z2 = 00
x1 x2 = 00, state = 000, z1 z2 = 00
```

Figure 3.37   Outputs for the Moore machine of Figure 3.34.

3.1.2 Nonlinear-Select Multiplexers

If the number of unique entries in an input map for flip-flop y_i satisfies the expression of Equation 3.1, where u is the number of unique entries and p is the number of storage elements, then at most a $(2^p \div 2){:}1$ multiplexer will satisfy the requirements of Dy_i.

$$1 < u \geq (2^p \div 2) \tag{3.1}$$

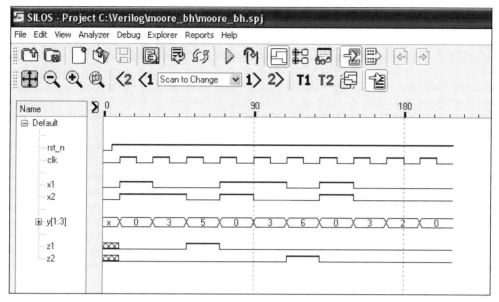

Figure 3.38   Waveforms for the Moore machine of Figure 3.34.

If, however, $u > 2^p \div 2$, then a 2^p:1 multiplexer is necessary. The largest multiplexer with which to economically implement the input logic is a 16:1 multiplexer, and then only if the number of distinct entries in the input map warrants a multiplexer of this size. Other techniques, such as a PLD implementation, would make more efficient use of current technology.

If a multiplexer has unused data inputs — corresponding to unused states in the input map — then these unused inputs can be connected to logically adjacent multiplexer inputs. The resulting linked set of inputs can be addressed by a common select variable. Thus, in a 4:1 multiplexer, if data input $d_2 = 1$ and $d_3 =$ "don't care," then d_2 and d_3 can both be connected to a logic 1. The two inputs can now be selected by $s_1 s_0 = 10$ or 11; that is, $s_1 s_0 = 1-$. Also, multiplexers containing the same number of data inputs should be addressed by the same select input variables, if possible. This permits the utilization of noncustom technology, where multiplexers in the same integrated circuit share common select inputs.

Example 3.6   Given the input map for Dy_1 in Figure 3.39(a), the δ next-state logic for flip-flop y_1 will be obtained using a nonlinear-select multiplexer. If linear selection is used, then an 8:1 multiplexer is required. However, the input map contains only three distinct entries: 1, 0, and x_1 — the "don't care" entry can be set to any value. Therefore, a 4:1 multiplexer is sufficient for this implementation, as shown in the input map of Figure 3.39(b).

Figure 3.40(a) and Figure 3.40(b) show the logic macros for both the linear-select and the nonlinear-select implementations. Since a 4:1 multiplexer is required, only two select inputs are necessary, either $y_1 y_2$, $y_1 y_3$, or $y_2 y_3$, where the low-order flip-flop y_2 or y_3 connects to the low-order multiplexer select input s_0. The two multiplexers will be designed using structural modeling with instantiated 8:1 and 4:1 multiplexers.

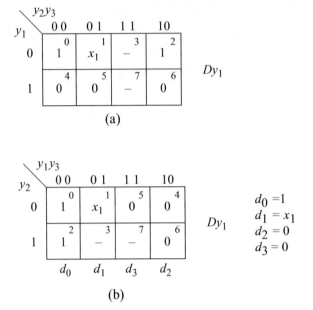

(a)

(b)

Figure 3.39   Input maps for Example 3.6: (a) using a linear-select multiplexer and (b) using a nonlinear-select multiplexer with $y_1 y_3$ as select inputs.

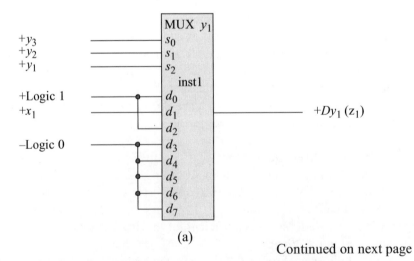

(a)

Continued on next page

Figure 3.40   The δ next-state logic for the input maps of Figure 3.39 using multiplexers: (a) linear-select multiplexer; and (b) nonlinear-select multiplexer.

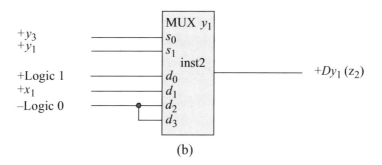

(b)

Figure 3.40   (Continued)

The structural design module is shown in Figure 3.41 for both multiplexers. The test bench module is shown in Figure 3.42, which takes the design through all combinations of the concatenated values for y_1, y_2, y_3, and x_1, then displays the resulting values for the outputs, labelled z_1 and z_2. The outputs are shown in Figure 3.43.

```
//structural linear-select and nonlinear-select multiplexers

module mux_nonlinear6 (x1, y, z1, z2);

//define inputs and outputs
input x1;
input [1:3] y;
output z1, z2;

//instantiate the 8:1 multiplexer
mux_8to1_df inst1 (
   .sel({y[1], y[2], y[3]}),
   .data({{5{1'b0}}, 1'b1, x1, 1'b1}),
   .out(z1)
   );

//instantiate the 4:1 multiplexer
mux_4to1_struc inst2 (
   .s({y[1], y[3]}),
   .d({1'b0, 1'b0, x1, 1'b1}),
   .z1(z2)
   );

endmodule
```

Figure 3.41   Structural design module for the linear-select and nonlinear-select multiplexers of Example 3.6.

```
//test bench for linear-select and nonlinear-select muxs
module mux_nonlinear6_tb;

//inputs are reg for test bench
//outputs are wire for test bench
reg x1;
reg [1:3] y;
wire z1, z2;

initial           //define input sequence
begin: apply_stimulus
   reg [4:0] invect;
   for (invect = 0; invect < 16; invect = invect + 1)
      begin
         {y, x1} = invect [4:0];
         #10 $display ("y = %b, x1 = %b, z1 z2 = %b",
                        y, x1, {z1, z2});
      end
end

//instantiate the module into the test bench
mux_nonlinear6 inst1 (
   .x1(x1),
   .y(y),
   .z1(z1),
   .z2(z2)
   );

endmodule
```

Figure 3.42   Test bench module for the linear-select and nonlinear-select multiplexers of Example 3.6.

```
y = 000, x1 = 0, z1 z2 = 11        y = 100, x1 = 0, z1 z2 = 00
y = 000, x1 = 1, z1 z2 = 11        y = 100, x1 = 1, z1 z2 = 00
y = 001, x1 = 0, z1 z2 = 00        y = 101, x1 = 0, z1 z2 = 00
y = 001, x1 = 1, z1 z2 = 11        y = 101, x1 = 1, z1 z2 = 00

y = 010, x1 = 0, z1 z2 = 11        y = 110, x1 = 0, z1 z2 = 00
y = 010, x1 = 1, z1 z2 = 11        y = 110, x1 = 1, z1 z2 = 00
y = 011, x1 = 0, z1 z2 = 00        y = 111, x1 = 0, z1 z2 = 00
y = 011, x1 = 1, z1 z2 = 01        y = 111, x1 = 1, z1 z2 = 00
```

Figure 3.43   Outputs for the linear-select and nonlinear-select multiplexers of Example 3.6.

Example 3.7   Given the input map for Dy_1 in Figure 3.44(a), the δ next-state logic for flip-flop y_1 will be obtained using two nonlinear-select multiplexer designs. Figure 3.44(a) can be permuted, as shown in Figure 3.44(b) and Figure 3.44(c). Figure 3.44(b) requires an exclusive-OR function; whereas, Figure 3.44(c) yields a solution that requires no gates.

The designs of the multiplexer logic for the input maps of Figure 3.44(b) and Figure 3.44(c) are shown in Figure 3.45(a) and Figure 3.45(b), respectively. Both implementation methods are presented to illustrate different techniques for synthesizing with nonlinear-select multiplexers.

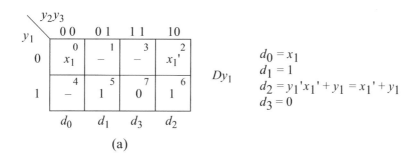

$d_0 = x_1$
$d_1 = 1$
$d_2 = y_1'x_1' + y_1 = x_1' + y_1$
$d_3 = 0$

(a)

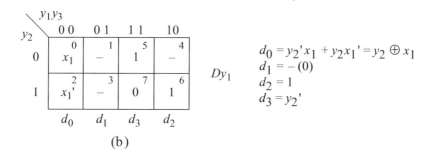

$d_0 = y_2'x_1 + y_2x_1' = y_2 \oplus x_1$
$d_1 = -(0)$
$d_2 = 1$
$d_3 = y_2'$

(b)

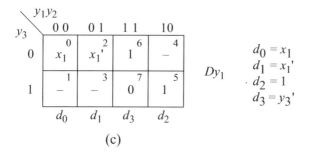

$d_0 = x_1$
$d_1 = x_1'$
$d_2 = 1$
$d_3 = y_3'$

(c)

Figure 3.44   Input maps for flip-flop y_1 for Example 3.7: (a) using y_2y_3 as select inputs; (b) using y_1y_3 as select inputs; and (c) using y_1y_2 as select inputs.

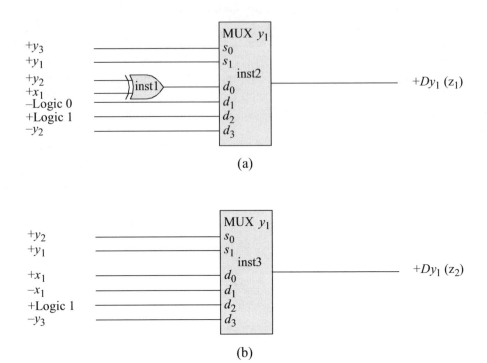

$+y_3$
$+y_1$
$+y_2$
$+x_1$
$-$Logic 0
$+$Logic 1
$-y_2$

MUX y_1
s_0
s_1
inst2
d_0
d_1
d_2
d_3

inst1

$+Dy_1$ (z_1)

(a)

$+y_2$
$+y_1$
$+x_1$
$-x_1$
$+$Logic 1
$-y_3$

MUX y_1
s_0
s_1
inst3
d_0
d_1
d_2
d_3

$+Dy_1$ (z_2)

(b)

Figure 3.45   The δ next-state logic for the input maps of Figure 3.44(b) and Figure 3.44(c): (a) non minimized configuration; and (b) minimized configuration.

The structural design module for the nonlinear-select multiplexers is shown in Figure 3.46. The test bench module is shown in Figure 3.47, which takes the design through all combinations of the concatenated values for y_1, y_2, y_3, and x_1, then displays the resulting values for the outputs, labelled z_1 and z_2. The outputs are shown in Figure 3.48.

```
//structural nonlinear-select multiplexers
module mux_nonlinear7 (x1, y, z1, z2);

//define inputs and outputs
input x1;
input [1:3] y;
output z1, z2;
                            //continued on next page
```

Figure 3.46   Structural design module for the nonlinear-select multiplexers of Example 3.7.

```
//instantiate the logic and 4:1 multiplexer for z1
xor inst1 (net1, y[2], x1);

mux_4to1_struc inst2 (
    .s({y[1], y[3]}),
    .d({~y[2], 1'b1, 1'b0, net1}),
    .z1(z1)
    );

//instantiate the logic and 4:1 multiplexer for z2
mux_4to1_struc inst3 (
    .s({y[1], y[2]}),
    .d({~y[3], 1'b1, ~x1, x1}),
    .z1(z2)
    );

endmodule
```

Figure 3.46   (Continued)

```
//test bench for the nonlinear-select multiplexers

module mux_nonlinear7_tb;

//inputs are reg for test bench
//outputs are wire for test bench
reg x1;
reg [1:3] y;
wire z1, z2;

//define input sequence
initial
begin: apply_stimulus
    reg [4:0] invect;
    for (invect = 0; invect < 16; invect =  invect + 1)
       begin
           {y, x1} = invect [4:0];
           #10 $display ("y = %b, x1 = %b, z1 z2 = %b",
                         y, x1, {z1, z2});
    end
end
```
                                //continued on next page

Figure 3.47   Test bench module for the nonlinear-select multiplexers of Example 3.7.

```
//instantiate the module into the test bench
mux_nonlinear7 inst1 (
    .x1(x1),
    .y(y),
    .z1(z1),
    .z2(z2)
    );

endmodule
```

Figure 3.47   (Continued)

```
y = 000,  x1 = 0,  z1 z2 = 00
y = 000,  x1 = 1,  z1 z2 = 11
y = 001,  x1 = 0,  z1 z2 = 00
y = 001,  x1 = 1,  z1 z2 = 01

y = 010,  x1 = 0,  z1 z2 = 11
y = 010,  x1 = 1,  z1 z2 = 00
y = 011,  x1 = 0,  z1 z2 = 01
y = 011,  x1 = 1,  z1 z2 = 00

y = 100,  x1 = 0,  z1 z2 = 11
y = 100,  x1 = 1,  z1 z2 = 11
y = 101,  x1 = 0,  z1 z2 = 11
y = 101,  x1 = 1,  z1 z2 = 11

y = 110,  x1 = 0,  z1 z2 = 11
y = 110,  x1 = 1,  z1 z2 = 11
y = 111,  x1 = 0,  z1 z2 = 00
y = 111,  x1 = 1,  z1 z2 = 00
```

Figure 3.48   Outputs for the nonlinear-select multiplexers of Example 3.7.

Example 3.8   Example 3.8 designs a Moore machine using nonlinear-select multiplexers that operates in accordance with the machine specifications described below.

$$\text{If } x_1 = 110, \text{ then } z_1 \uparrow t_2 \downarrow t_3$$
$$\text{If } x_1 = 101, \text{ then } z_2 \uparrow t_2 \downarrow t_3$$

The input data on x_1 consists of a bit stream of binary data. Valid overlapping sequences are possible, such that the input sequence that asserts output z_1 may be part of a valid sequence that asserts outputs z_2. The converse is true for a z_2, z_1 sequence. The outputs are asserted at time t_2 and deasserted at time t_3.

The state diagram of Figure 3.49 portrays the machine's behavior according to the machine specifications. The machine remains in state a until the first 1 appears on input x_1. State c is the state that is entered whenever two consecutive 1s occur anywhere in the bit stream, signifying the beginning bit configuration of a valid input sequence to assert output z_1. State f is the state that is entered whenever the sequence $x_1 = \ldots 10$ occurs, signifying the beginning bit configuration of a valid input sequence to assert output z_2.

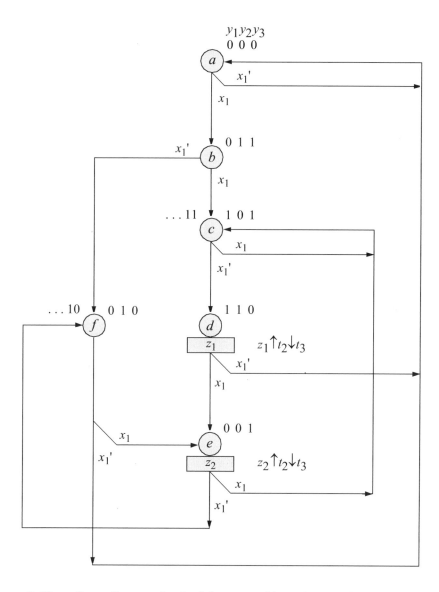

Figure 3.49 State diagram for the Moore machine of Example 3.8 to detect input sequences of $x_1 = 110$ to assert output z_1 and $x_1 = 101$ to assert output z_2. Unused states are: $y_1 y_2 y_3 = 100$ and 111.

The input maps are shown in Figure 3.50, which shows the permuted map for Dy_1 and the unchanged map for Dy_2, together with the multiplexer data input equations for both flip-flops. The input map for Dy_3 needs no equations, because every minterm location contains the value of x_1, including the "don't care" minterms. The output maps are obtained from the state diagram and are shown in Figure 3.51.

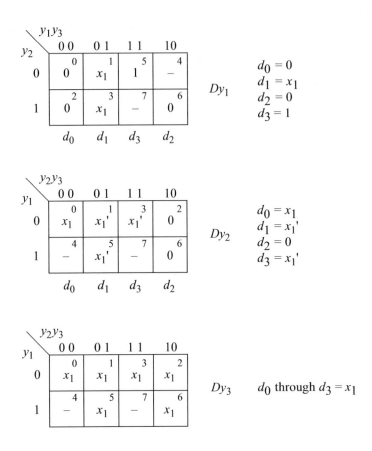

Figure 3.50   Input maps for the Moore machine of Figure 3.49 using D flip-flops.

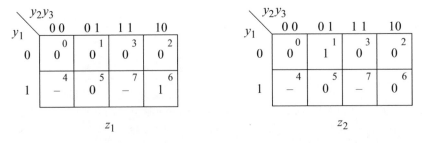

Figure 3.51   Output maps for the Moore machine of Figure 3.49.

The logic diagram is shown in Figure 3.52 using two 4:1 nonlinear-select multiplexers for the δ next-state input logic, D flip-flops with implied reset inputs, and AND gates for the λ output logic.

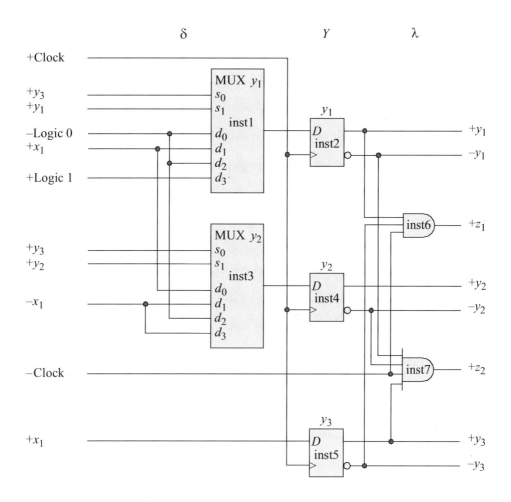

Figure 3.52   Logic diagram for the Moore machine of Figure 3.49 using multiplexers for the δ next-state logic.

The structural design module is shown in Figure 3.53, which instantiates 4:1 multiplexers designed using structural modeling, instantiates D flip-flops designed using behavioral modeling, and instantiates built-in primitives to decode the outputs. The test bench module is shown in Figure 3.54, which takes the machine through various paths in the state diagram and asserts outputs z_1 and z_2 from t_2 to t_3. The outputs and waveforms are shown in Figure 3.55 and Figure 3.56, respectively.

```
//structural moore using muxs to assert
//z1 if x1 = 110 and to assert z2 if x1 = 101

module moore_mux3_dff (rst_n, clk, x1, y, z1, z2);

//define inputs and outputs
input rst_n, clk, x1;
output [1:3] y;
output z1, z2;

//define internal nets
wire net1, net3;

//-----------------------------------------------
//instantiate the logic for flip-flop y[1]
mux_4to1_struc inst1 (
    .s({y[1], y[3]}),
    .d({1'b1, 1'b0, x1, 1'b0}),
    .z1(net1)
    );

d_ff_bh inst2 (
    .rst_n(rst_n),
    .clk(clk),
    .d(net1),
    .q(y[1])
    );

//-----------------------------------------------
//instantiate the logic for flip-flop y[2]
mux_4to1_struc inst3 (
    .s({y[2], y[3]}),
    .d({~x1, 1'b0, ~x1, x1}),
    .z1(net3)
    );

d_ff_bh inst4 (
    .rst_n(rst_n),
    .clk(clk),
    .d(net3),
    .q(y[2])
    );

//-----------------------------------------------
                          //continued on next page
```

Figure 3.53   Structural design module for the logic diagram of Figure 3.52, which represents the state diagram of Figure 3.49.

```
//-----------------------------------------------
//instantiate the logic for flip-flop y[3]
d_ff_bh inst5 (
   .rst_n(rst_n),
   .clk(clk),
   .d(x1),
   .q(y[3])
   );


//-----------------------------------------------
//instantiate the logic for the outputs
and inst6 (z1, y[1], ~y[3], ~clk);
and inst7 (z2, ~y[1], ~y[2], y[3], ~clk);

endmodule
```

Figure 3.53   (Continued)

```
//test bench for moore to detect two sequences.
//assert z1 if x1 = 110; assert z2 if x1 = 101

module moore_mux3_dff_tb;


//inputs are reg for test bench
//outputs are wire for test bench
reg rst_n, clk, x1;
wire [1:3] y;
wire z1, z2;


//display variables
initial
$monitor ("x1 = %b, state = %b, z1 z2 = %b", x1, y, {z1, z2});


//define clock
initial
begin
   clk = 1'b0;
   forever
      #10 clk = ~clk;
end

                                   //continued on next page
```

Figure 3.54   Test bench module for the logic diagram of Figure 3.52.

```
//define input sequence
initial
begin
   #0 rst_n = 1'b0;              //reset to state_a (000)
      x1 = 1'b0;
   #5 rst_n = 1'b1;

//-----------------------------------------------------------
   x1 = 1'b0;@ (posedge clk)  //go to state_a (000)
   x1 = 1'b1;@ (posedge clk)  //go to state_b (011)
   x1 = 1'b1;@ (posedge clk)  //go to state_c (101)
   x1 = 1'b0;@ (posedge clk)  //go to state_d (110)
                              //assert z1
   x1 = 1'b1;@ (posedge clk)  //go to state_e (001)
                              //assert z2
   x1 = 1'b0;@ (posedge clk)  //go to state_f (010)
   x1 = 1'b0;@ (posedge clk)  //go to state_a (000)

//-----------------------------------------------------------
   x1 = 1'b1;@ (posedge clk)  //go to state_b (011)
   x1 = 1'b1;@ (posedge clk)  //go to state_c (101)
   x1 = 1'b0;@ (posedge clk)  //go to state_d (110)
                              //assert z1
   x1 = 1'b1;@ (posedge clk)  //go to state_e (001)
                              //assert z2

//-----------------------------------------------------------
   x1 = 1'b1;@ (posedge clk)  //go to state_c (101)
   x1 = 1'b0;@ (posedge clk)  //go to state_d (110)
                              //assert z1
   x1 = 1'b0;@ (posedge clk)  //go to state_a (000)
   x1 = 1'b1;@ (posedge clk)  //go to state_b (011)

   x1 = 1'b0;@ (posedge clk)  //go to state_f (010)
   x1 = 1'b1;@ (posedge clk)  //go to state_e (001)
                              //assert z2
   x1 = 1'b0;@ (posedge clk)  //go to state_f (010)
   x1 = 1'b0;@ (posedge clk)  //go to state_a (000)
   #20    $stop;
end

//-----------------------------------------------------------
//instantiate the module into the test bench
//as a single line
moore_mux3_dff inst1 (rst_n, clk, x1, y, z1, z2);

endmodule
```

Figure 3.54   (Continued)

```
x1 = 1, state = 000, z1 z2 = 00
x1 = 1, state = 011, z1 z2 = 00
x1 = 0, state = 101, z1 z2 = 00
x1 = 1, state = 110, z1 z2 = 00
x1 = 1, state = 110, z1 z2 = 10
x1 = 0, state = 001, z1 z2 = 00
x1 = 0, state = 001, z1 z2 = 01
x1 = 0, state = 010, z1 z2 = 00
x1 = 1, state = 000, z1 z2 = 00
x1 = 1, state = 011, z1 z2 = 00
x1 = 0, state = 101, z1 z2 = 00
x1 = 1, state = 110, z1 z2 = 00 .
x1 = 1, state = 110, z1 z2 = 10
x1 = 1, state = 001, z1 z2 = 00
x1 = 1, state = 001, z1 z2 = 01
x1 = 0, state = 101, z1 z2 = 00
x1 = 0, state = 110, z1 z2 = 00
x1 = 0, state = 110, z1 z2 = 10
x1 = 1, state = 000, z1 z2 = 00
x1 = 0, state = 011, z1 z2 = 00
x1 = 1, state = 010, z1 z2 = 00
x1 = 0, state = 001, z1 z2 = 00
x1 = 0, state = 001, z1 z2 = 01
x1 = 0, state = 010, z1 z2 = 00
x1 = 0, state = 000, z1 z2 = 00
```

Figure 3.55   Outputs for the logic diagram of Figure 3.52.

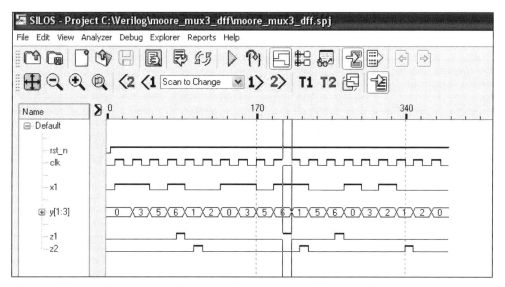

Figure 3.56   Waveforms for the logic diagram of Figure 3.52.

3.2 Decoders for λ Output Logic

Decoders are reviewed briefly in this section. A *decoder* is a combinational macro logic circuit that translates a binary input number to an equivalent output number for a specific radix. In general, there are n input lines and m output lines, where $m = 2^n$. For each combination of the 2^n input values, only one unique output signal is active — all other outputs are inactive. Thus, a fundamental characteristic of a decoder is the mutual exclusiveness of the outputs.

The multiplexer data inputs correspond to the minterm locations in an input map; whereas, the outputs of a decoder correspond to the minterms locations in an output map. The m output signals, labeled $f_0, f_1, \ldots, f_{m-1}$ correspond to the 2^n minterm values represented by the binary number on inputs $x_{n-1}, \ldots, x_1, x_0$, where x_0 is the low-order input variable. A decoder consisting of n inputs and 2^n outputs can be used to implement any Boolean switching function, because each output represents a unique minterm. By connecting the appropriate decoder outputs to an OR gate, the desired function can be realized.

Example 3.9   A Moore machine will be designed using structural modeling, which generates a sequence of six contiguous, nonoverlapping pulses. The six pulses are mutually exclusive and each pulse is active for one clock period. The δ next-state logic consists of logic gates and D flip-flops; the λ output logic consists of a 3:8 decoder. The assertion/deassertion statement for all outputs is $z_i \uparrow t_1 \downarrow t_3$. A typical application of this type of Moore machine is as a ring counter, which functions as a finite-state machine to control the operation of an external digital system. Each output of the machine represents a different state.

The state diagram is illustrated in Figure 3.57. In order to generate glitch-free outputs, a Gray code arrangement is used for the state code assignment. The state codes adhere to the definition of a Gray code, in which each code word differs from the preceding word in only one bit position. However, since only six out of the eight possible combinations of three variables are used, the code represented in Figure 3.57 differs slightly from the traditional Gray code in order to maintain logical adjacency between states e and f and between states f and a. Each state generates an output and each output is asserted for the entire state time.

The input maps are derived from the state diagram and are shown in Figure 3.58. Since there are only two values (0 and 1) in any input map, multiplexers are not appropriate — logic gates will suffice for each flip-flop input. Since a decoder is used for the λ output logic, the output maps are not shown. Each Moore output corresponds to a distinct decoder minterm and can be implemented directly from the state diagram. A 3:8 decoder is ideal for the λ output logic for this multiple output machine. The decoder has two unused outputs that correspond to state codes $y_1 y_2 y_3 = 101$ and 111.

The logic diagram is shown in Figure 3.59, using logic gates for the δ next-state logic and a modified 3:8 decoder for the λ output logic. The active-high output of each state flip-flop is connected to the appropriate decoder address input. Thus, y_1, y_2, and y_3 connect to x_2, x_1, and x_0, respectively, where y_3 and x_0 are the low-order variables. The uncommitted decoder outputs represent the unused states in the state diagram and input maps. An enabling function is not necessary for the decoder outputs.

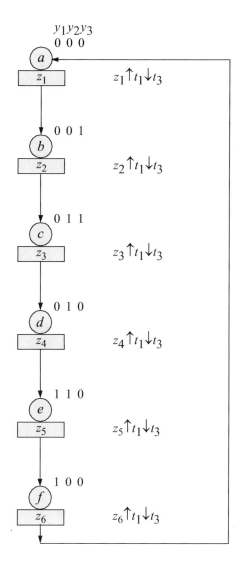

Figure 3.57   State diagram for the Moore machine of Example 3.9. Unused states are: $y_1 y_2 y_3 = 101$ and 111.

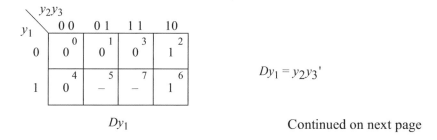

Figure 3.58   Input maps for Example 3.9.

Continued on next page

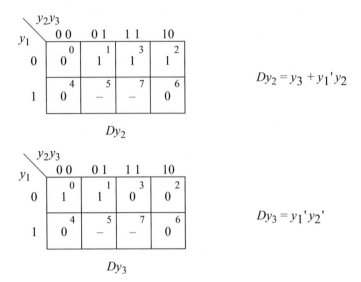

$$Dy_2 = y_3 + y_1'y_2$$

$$Dy_3 = y_1'y_2'$$

Figure 3.58   (Continued)

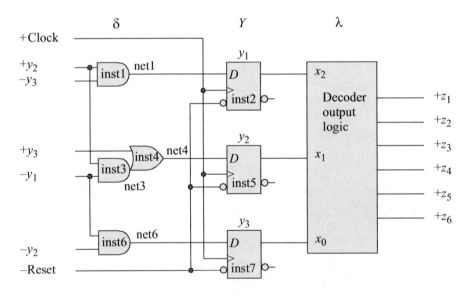

Figure 3.59   Logic diagram for the Moore machine of Figure 3.57.

Figure 3.60 shows the structural design module for the Moore machine to generate six contiguous nonoverlapping pulses. The test bench is shown in Figure 3.61. The outputs and waveforms are shown in Figure 3.62 and Figure 3.63, respectively.

```
//structural moore to generate 6 contiguous pulses
module moore_6_pulses (rst_n, clk, y, z);
input rst_n, clk;                //define inputs and outputs
output [1:3] y;
output [1:6] z;

wire net1, net3, net4, net6;  //define internal nets

//instantiate the logic for flip-flop y[1]
and inst1 (net1, y[2], ~y[3]);

d_ff_bh inst2 (
   .rst_n(rst_n),
   .clk(clk),
   .d(net1),
   .q(y[1])
   );

//instantiate the logic for flip-flop y[2]
and inst3 (net3, ~y[1], y[2]);
or  inst4 (net4, y[3], net3);

d_ff_bh inst5 (
   .rst_n(rst_n),
   .clk(clk),
   .d(net4),
   .q(y[2])
   );

//instantiate the logic for flip-flop y[3]
and inst6 (net6, ~y[1], ~y[2]);

d_ff_bh inst7 (
   .rst_n(rst_n),
   .clk(clk),
   .d(net6),
   .q(y[3])
   );

//define the output logic
assign z[1] = ~y[1] & ~y[2] & ~y[3];
assign z[2] = ~y[1] & ~y[2] & y[3];
assign z[3] = ~y[1] & y[2] & y[3];
assign z[4] = ~y[1] & y[2] & ~y[3];
assign z[5] = y[1] & y[2] & ~y[3];
assign z[6] = y[1] & ~y[2] & ~y[3];
endmodule
```

Figure 3.60   Structural design module for the Moore machine of Example 3.9.

```verilog
//test bench for moore to generate 6 contiguous pulses

module moore_6_pulses_tb;

//inputs are reg for test bench
//outputs are wire for test bench
reg rst_n, clk;
wire [1:3] y;
wire [1:6] z;

//display variables
initial
$monitor ("output = %b", z);

//define reset
initial
begin
    #0      rst_n = 1'b0;
    #20     rst_n = 1'b1;
end

//define clock
initial
begin
    clk = 1'b0;
    #10 forever
        #10 clk = ~clk;
end

//define length of simulation
initial
begin
    #140 $finish;
end

//instantiate the module into the test bench
moore_6_pulses inst1 (
    .rst_n(rst_n),
    .clk(clk),
    .y(y),
    .z(z)
    );

endmodule
```

Figure 3.61   Test bench module for the Moore machine of Example 3.9.

```
output = 100000
output = 010000
output = 001000
output = 000100
output = 000010
output = 000001
output = 100000
```

Figure 3.62   Outputs for the Moore machine of Example 3.9.

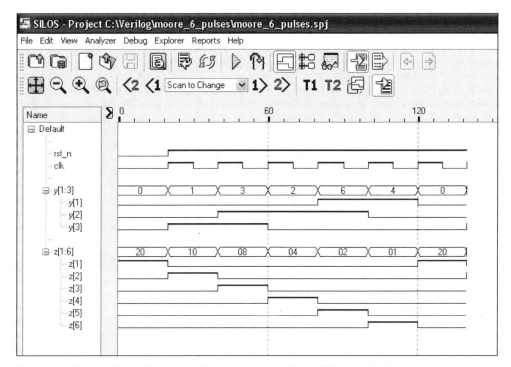

Figure 3.63   Waveforms for the Moore machine of Example 3.9.

Example 3.10   This example repeats Example 3.9 using the same state diagram and *D* flip-flops. However, discrete logic gates are used to implement the decoder outputs, as shown in the logic diagram of Figure 3.64. The structural design module is shown in Figure 3.65. The test bench module is shown in Figure 3.66. The outputs and waveforms are show in Figure 3.67 and Figure 3.68, respectively.

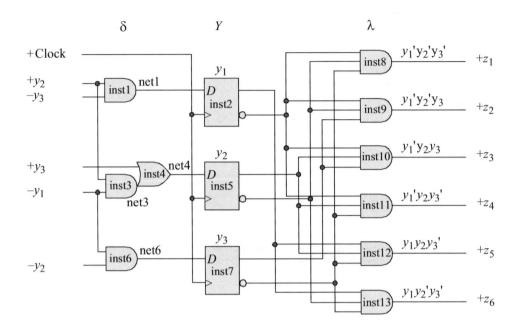

Figure 3.64   Logic diagram for the Moore machine of Example 3.10 to generate six consecutive nonoverlapping pulses.

```
//structural moore to generate 6 contiguous pulses
module moore_decoder_3to8c (rst_n, clk, y, z);
input rst_n, clk;              //define inputs and outputs
output [1:3] y;
output [1:6] z;

wire net1, net3, net4, net6;  //define internal nets

//---------------------------------------------
//instantiate the logic for flip-flop y[1]
and inst1 (net1, y[2], ~y[3]);

d_ff_bh inst2 (
   .rst_n(rst_n),
   .clk(clk),
   .d(net1),
   .q(y[1])
   );                          //continued on next page
```

Figure 3.65   Structural design module for the Moore machine of Example 3.10 to generate six consecutive nonoverlapping pulses.

```
//-----------------------------------------------
//instantiate the logic for flip-flop y[2]
and inst3 (net3, ~y[1], y[2]);
or  inst4 (net4, y[3], net3);

d_ff_bh inst5 (
   .rst_n(rst_n),
   .clk(clk),
   .d(net4),
   .q(y[2])
   );

//-----------------------------------------------
//instantiate the logic for flip-flop y[3]
and inst6 (net6, ~y[1], ~y[2]);

d_ff_bh inst7 (
   .rst_n(rst_n),
   .clk(clk),
   .d(net6),
   .q(y[3])
   );

//-----------------------------------------------
//define the logic for the outputs
and inst8  (z[1], ~y[1], ~y[2], ~y[3]);
and inst9  (z[2], ~y[1], ~y[2],  y[3]);
and inst10 (z[3], ~y[1],  y[2],  y[3]);
and inst11 (z[4], ~y[1],  y[2], ~y[3]);
and inst12 (z[5],  y[1],  y[2], ~y[3]);
and inst13 (z[6],  y[1], ~y[2], ~y[3]);

endmodule
```

Figure 3.65   (Continued)

```
//test bench for moore to generate 6 contiguous pulses
module moore_decoder_3to8c_tb;

reg rst_n, clk;       //inputs are reg for test bench
wire [1:3] y;         //outputs are wire for test bench
wire [1:6] z;                        //continued on next page
```

Figure 3.66   Test bench module for the Moore machine of Example 3.10 to generate six consecutive nonoverlapping pulses.

```
//display variables
initial
$monitor ("output = %b", z);

initial                //define reset
begin
    #0     rst_n = 1'b0;
    #20    rst_n = 1'b1;
end

initial                //define clock
begin
    clk = 1'b0;
    #10 forever
        #10 clk = ~clk;
end

//define length of simulation
initial
begin
    #140 $finish;
end

//instantiate the module into the test bench
moore_decoder_3to8c inst1 (
    .rst_n(rst_n),
    .clk(clk),
    .y(y),
    .z(z)
    );

endmodule
```

Figure 3.66   (Continued)

```
output = 100000
output = 010000
output = 001000
output = 000100
output = 000010
output = 000001
output = 100000
```

Figure 3.67   Outputs for the Moore machine of Example 3.10 to generate six consecutive nonoverlapping pulses.

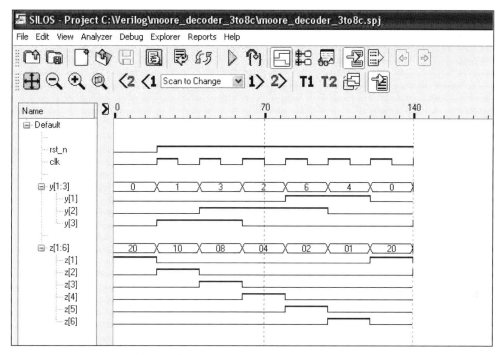

Figure 3.68   Waveforms for the Moore machine of Example 3.10 to generate six consecutive nonoverlapping pulses.

Example 3.11   This example uses the state diagram of Figure 3.57 to generate six contiguous nonoverlapping pulses. The structural module will be designed using instantiated negative-edge triggered *JK* flip-flops that are designed with behavioral modeling. The instantiated logic elements will be designed using dataflow modeling.

Recall that the *JK* flip-flop is an edge-triggered storage device in which the active clock transition can be either the positive or negative edge. The functional characteristics of the *JK* data inputs are defined as shown in Table 3.1. Table 3.2 shows an excitation table in which a particular state transition predicates a set of values for *J* and *K*. This table is especially useful in the synthesis of synchronous sequential machines.

Table 3.1 *JK* Functional Characteristic Table

J K	Function
0 0	No change
0 1	Reset
1 0	Set
1 1	Toggle

Table 3.2 Excitation Table for a *JK* Flip-Flop

Present state $Y_{j(t)}$	Next state $Y_{k(t+1)}$	Data inputs *J K*
0	0	0 –
0	1	1 –
1	0	– 1
1	1	– 0

The state diagram for the Moore machine to generate six contiguous nonoverlapping pulses is reproduced in Figure 3.69 to assist in generating the JK input maps, which are shown in Figure 3.70 together with the input equations.

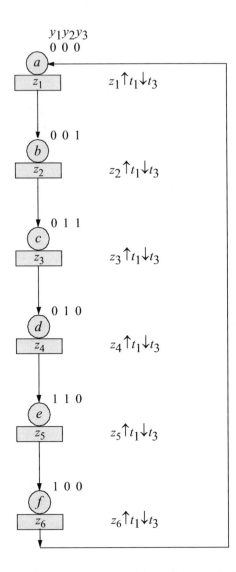

Figure 3.69   State diagram for the Moore machine of Example 3.11 to generate six contiguous nonoverlapping pulses. Unused states are: $y_1y_2y_3 = 101$ and 111.

The logic diagram is shown in Figure 3.71. The structural design module is shown in Figure 3.72. The test bench module is shown in Figure 3.73. The outputs and waveforms are shown in Figure 3.74 and Figure 3.75, respectively. Since NAND gates are used to implement the outputs, the six contiguous nonoverlapping pulses are asserted as negative pulses.

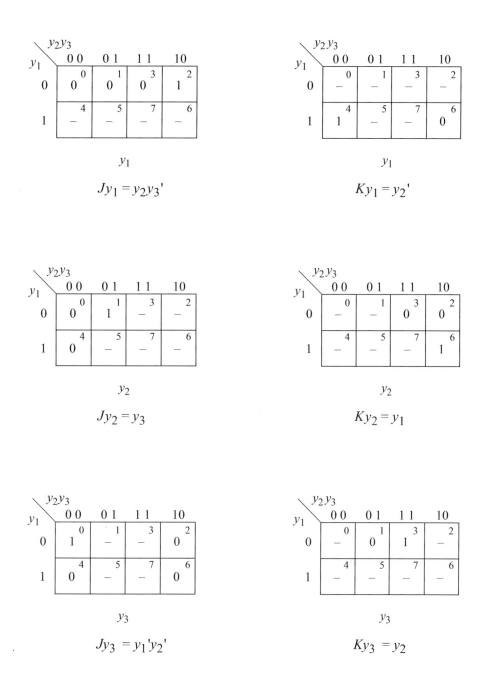

Figure 3.70   Input maps for the Moore machine of Example 3.11 to generate six contiguous nonoverlapping pulses.

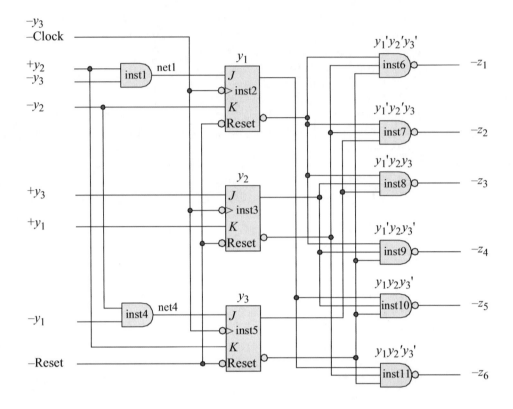

Figure 3.71   Logic diagram for the Moore machine of Example 3.11 to generate six contiguous nonoverlapping pulses.

```
//structural moore to generate 6 contiguous pulses

module moore_decoder_jk (rst_n, clk, y, z);

//define inputs and outputs
input rst_n, clk;
output [1:3] y;

output [1:6] z;

//define internal nets
wire net1, net4;

                              //continued on next page
```

Figure 3.72   Structural design module for the Moore machine of Example 3.11 to generate six contiguous nonoverlapping pulses.

```
//------------------------------------------------
//instantiate the logic for flip-flop y[1]
and2_df inst1 (
   .x1(y[2]),
   .x2(~y[3]),
   .z1(net1)
   );

jkff inst2 (
   .set_n(1'b1),
   .rst_n(rst_n),
   .clk(clk),
   .j(net1),
   .k(~y[2]),
   .q(y[1])
   );

//------------------------------------------------
//instantiate the logic for flip-flop y[2]
jkff inst3 (
   .set_n(1'b1),
   .rst_n(rst_n),
   .clk(clk),
   .j(y[3]),
   .k(y[1]),
   .q(y[2])
   );

//------------------------------------------------
//instantiate the logic for flip-flop y[3]
and2_df inst4 (
   .x1(~y[2]),
   .x2(~y[1]),
   .z1(net4)
   );

jkff inst5 (
   .set_n(1'b1),
   .rst_n(rst_n),
   .clk(clk),
   .j(net4),
   .k(y[2]),
   .q(y[3])
   );

//------------------------------------------------
                           //continued on next page
```

Figure 3.72   (Continued)

```verilog
//-----------------------------------------------
//instantiate the logic for the outputs
nand3_df inst6 (
    .x1(~y[1]),
    .x2(~y[2]),
    .x3(~y[3]),
    .z1(z[1])
    );

nand3_df inst7 (
    .x1(~y[1]),
    .x2(~y[2]),
    .x3(y[3]),
    .z1(z[2])
    );

nand3_df inst8 (
    .x1(~y[1]),
    .x2(y[2]),
    .x3(y[3]),
    .z1(z[3])
    );

nand3_df inst9 (
    .x1(~y[1]),
    .x2(y[2]),
    .x3(~y[3]),
    .z1(z[4])
    );

nand3_df inst10 (
    .x1(y[1]),
    .x2(y[2]),
    .x3(~y[3]),
    .z1(z[5])
    );

nand3_df inst11 (
    .x1(y[1]),
    .x2(~y[2]),
    .x3(~y[3]),
    .z1(z[6])
    );

endmodule
```

Figure 3.72   (Continued)

```
//test bench for moore to generate 6 contiguous pulses

module moore_decoder_jk_tb;

//define inputs and outputs
reg rst_n, clk;          //inputs are reg for test bench
wire [1:3] y;            //outputs are wire for test bench
wire [1:6] z;

//display outputs
initial
$monitor ("output = %b", z);

//define reset
initial
begin
   #0    rst_n = 1'b0;
   #20   rst_n = 1'b1;
end

//define clock
initial
begin
   clk = 1'b0;
   #10 forever
      #10 clk = ~clk;
end

//define length of simulation
initial
begin
   #140 $finish;
end

//instantiate the module into the test bench
moore_decoder_jk inst1 (
   .rst_n(rst_n),
   .clk(clk),
   .y(y),
   .z(z)
   );

endmodule
```

Figure 3.73   Test bench for the Moore machine of Example 3.11 to generate six contiguous nonoverlapping pulses.

```
output  =  011111
output  =  101111
output  =  110111
output  =  111011
output  =  111101
output  =  111110
output  =  011111
```

Figure 3.74   Waveforms for the Moore machine of Example 3.11 to generate six contiguous nonoverlapping pulses.

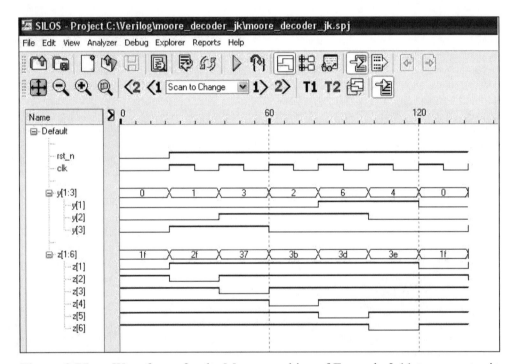

Figure 3.75   Waveforms for the Moore machine of Example 3.11 to generate six contiguous nonoverlapping pulses.

Example 3.12   This example designs a Mealy machine that examines 3-bit words on a serial input line x_1. The format for the serial data is shown below, where $b_i = 0$ or 1. The serial words are contiguous with no space between words.

$$x_1 = \quad \cdots \quad \left| \, b_1 b_2 b_3 \, \right| \, b_1 b_2 b_3 \, \left| \, b_1 b_2 b_3 \, \right| \quad \cdots$$

The machine generates four different outputs, depending upon the bit configuration of the received data:

If $x_1 = 001$, then assert output z_1
If $x_1 = 011$, then assert output z_2
If $x_1 = 101$, then assert output z_3
If $x_1 = 111$, then assert output z_4

For all other bit patterns, the four outputs are inactive. The assertion/deassertion statement for all outputs is: $z_i \uparrow t_2 \downarrow t_3$. Logic gates will be used for the δ next-state logic, D flip-flops will be used for the storage elements, and continuous assignment statements will be used to generate a decoder for the λ output logic. The state diagram is developed by generating a path for each of the eight bit sequences (000 through 111) while maintaining three state levels for each word, as shown in Figure 3.76.

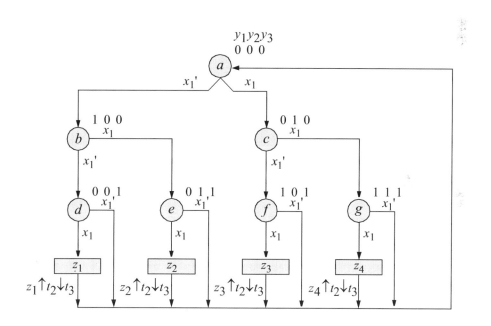

Figure 3.76   State diagram for the Mealy machine of Example 3.12. There is one unused state: $y_1 y_2 y_3 = 110$.

Notice that all four outputs have one thing in common: All outputs occur in a state where input $x_1 = 1$. Therefore, input x_1 can be used as an input variable for the decoder output logic, together with the negative assertion of the clock signal to assert the outputs from t_2 to t_3. The input maps, obtained from the state diagram, are shown in Figure 3.77 together with the input equations.

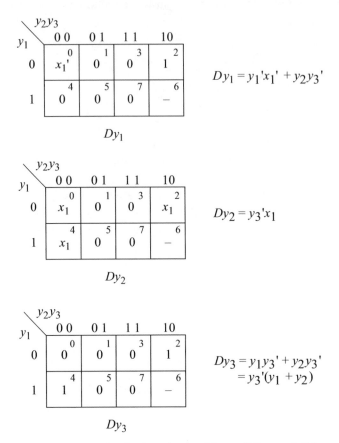

$$Dy_1 = y_1'x_1' + y_2y_3'$$

$$Dy_2 = y_3'x_1$$

$$Dy_3 = y_1y_3' + y_2y_3'$$
$$= y_3'(y_1 + y_2)$$

Figure 3.77   Input maps for the Mealy machine of Figure 3.76.

The output maps and equations are shown in Figure 3.78, as obtained from the state diagram. The logic diagram is generated from the input maps and is illustrated in Figure 3.79. The enable gate for the output decoder logic contains two inputs: the $+x_1$ signal to generate the Mealy-type outputs and the $-$Clock signal to provide assertion at time t_2 and deassertion at t_3.

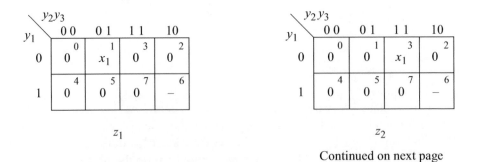

Continued on next page

Figure 3.78   Output maps for the Mealy machine of Figure 3.76.

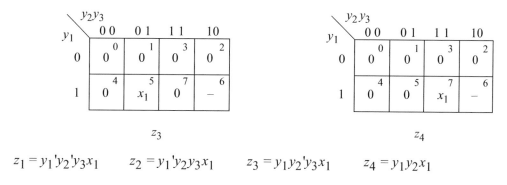

$$z_1 = y_1'y_2'y_3x_1 \qquad z_2 = y_1'y_2y_3x_1 \qquad z_3 = y_1y_2'y_3x_1 \qquad z_4 = y_1y_2x_1$$

Figure 3.78 (Continued)

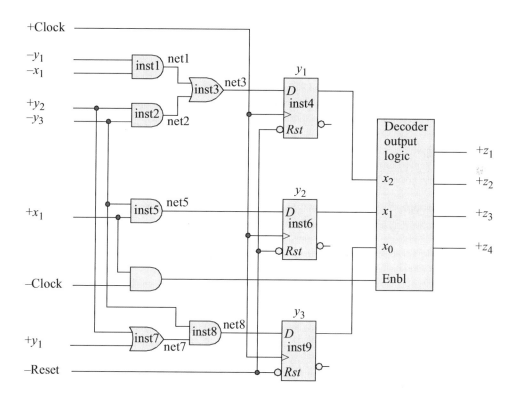

Figure 3.79 Logic diagram for the Mealy machine of Figure 3.76. All outputs are asserted at time t_2 and deasserted at t_3.

The structural design module is shown in Figure 3.80 using instantiated *D* flip-flops and built-in primitives. The test bench module is shown in Figure 3.81, which takes the machine through the four paths to assert the four outputs. The outputs and waveforms are shown in Figure 3.82 and Figure 3.83, respectively.

```
//structural mealy using decoder to generate four outputs

module mealy_dcdr (rst_n, clk, x1, y, z);

//define inputs and outputs
input rst_n, clk, x1;
output [1:3] y;
output [1:4] z;

//define internal nets
wire net1, net2, net3, net5, net7, net8;


//----------------------------------------
//instantiate the logic for flip-flop y[1]
and inst1 (net1, ~x1, ~y[1]);
and inst2 (net2, y[2], ~y[3]);
or  inst3 (net3, net1, net2);

d_ff_bh inst4 (
   .rst_n(rst_n),
   .clk(clk),
   .d(net3),
   .q(y[1])
   );


//----------------------------------------
//instantiate the logic for flip-flop y[2]
and inst5 (net5, ~y[3], x1);

d_ff_bh inst6 (
   .rst_n(rst_n),
   .clk(clk),
   .d(net5),
   .q(y[2])
   );

                        //continued on next page
```

Figure 3.80   Structural design module for the Mealy machine of Example 3.12 to assert four outputs.

```
//--------------------------------------------
//instantiate the logic for flip-flop y[3]
or   inst7 (net7, y[2], y[1]);
and inst8 (net8, ~y[3], net7);

d_ff_bh inst9 (
    .rst_n(rst_n),
    .clk(clk),
    .d(net8),
    .q(y[3])
    );

//--------------------------------------------
//generate the decoder logic for the outputs
assign z[1] = ~y[1] & ~y[2] & y[3] & x1 & ~clk;
assign z[2] = ~y[1] & y[2] & y[3] & x1 & ~clk;
assign z[3] = y[1] & ~y[2] & y[3] & x1 & ~clk;
assign z[4] = y[1] & y[2] & y[3] & x1 & ~clk;

endmodule
```

Figure 3.80 (Continued)

```
//test bench for mealy to generate four outputs
module mealy_dcdr_tb;

reg rst_n, clk, x1;        //inputs are reg for test bench
wire [1:3] y;              //outputs are wire for test bench
wire [1:4] z;

//display variables
initial
$monitor ("x1 = %b, state = %b, z1 z2 z3 z4 = %b", x1, y, z);

//define clock
initial
begin
    clk = 1'b0;
    forever
        #10 clk = ~clk;
end
                                //continued on next page
```

Figure 3.81 Test bench module for the Mealy machine of Example 3.12 to assert four outputs.

```
//define input sequence
initial
begin
   #0 rst_n = 1'b0;
      x1 = 1'b0;

   #5 rst_n = 1'b1;

//------------------------------------------------------------
                  @ (posedge clk)
   x1 = 1'b0;   @ (posedge clk)     //go to state_b (100)
   x1 = 1'b0;   @ (posedge clk)     //go to state_d (001)
   x1 = 1'b1;   @ (posedge clk)     //assert z1, then
                                    //go to state_a (000)

   x1 = 1'b0;   @ (posedge clk)     //go to state_b (100)
   x1 = 1'b1;   @ (posedge clk)     //go to state_e (011)
   x1 = 1'b1;   @ (posedge clk)     //assert z2, then
                                    //go to state_a (000)

   x1 = 1'b1;   @ (posedge clk)     //go to state_c (010)
   x1 = 1'b0;   @ (posedge clk)     //go to state_f (101)
   x1 = 1'b1;   @ (posedge clk)     //assert z3, then
                                    //go to state_a (000)

   x1 = 1'b1;   @ (posedge clk)     //go to state_c (010)
   x1 = 1'b1;   @ (posedge clk)     //go to state_g (111)
   x1 = 1'b1;   @ (posedge clk)     //assert z4, then
                                    //go to state_a (000)

//------------------------------------------------------------
   #20     $stop;

end

//------------------------------------------------------------
//instantiate the module into the test bench
mealy_dcdr inst1 (
   .rst_n(rst_n),
   .clk(clk),
   .x1(x1),
   .y(y),
   .z(z)
   );

endmodule
```

Figure 3.81   (Continued)

```
x1 = 0, state = 000, z1 z2 z3 z4 = 0000
x1 = 0, state = 100, z1 z2 z3 z4 = 0000
x1 = 0, state = 001, z1 z2 z3 z4 = 0000
x1 = 0, state = 100, z1 z2 z3 z4 = 0000
x1 = 1, state = 001, z1 z2 z3 z4 = 0000
x1 = 1, state = 001, z1 z2 z3 z4 = 1000
x1 = 0, state = 000, z1 z2 z3 z4 = 0000
x1 = 1, state = 100, z1 z2 z3 z4 = 0000
x1 = 1, state = 011, z1 z2 z3 z4 = 0000
x1 = 1, state = 011, z1 z2 z3 z4 = 0100
x1 = 1, state = 000, z1 z2 z3 z4 = 0000
x1 = 0, state = 010, z1 z2 z3 z4 = 0000
x1 = 1, state = 101, z1 z2 z3 z4 = 0000
x1 = 1, state = 101, z1 z2 z3 z4 = 0010
x1 = 1, state = 000, z1 z2 z3 z4 = 0000
x1 = 1, state = 010, z1 z2 z3 z4 = 0000
x1 = 1, state = 111, z1 z2 z3 z4 = 0000
x1 = 1, state = 111, z1 z2 z3 z4 = 0001
x1 = 1, state = 000, z1 z2 z3 z4 = 0000
```

Figure 3.82   Outputs for the Mealy machine of Example 3.12 to assert four outputs.

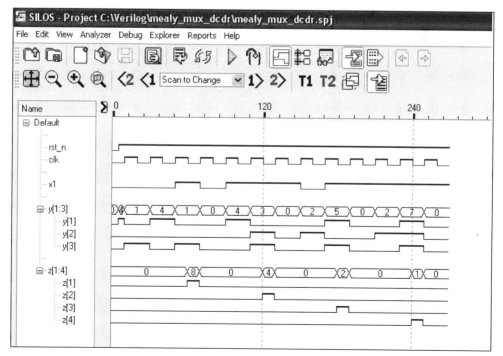

Figure 3.83   Waveforms Outputs for the Mealy machine of Example 3.12.

Example 3.13   To complete this section on decoders, a 3:8 decoder will be designed using the continuous assignment statement, then used to implement a Boolean function. As stated before, a decoder is a combinational logic macro that has n binary inputs and m mutually exclusive outputs, where $2^n \geq m$. An $n{:}m$ (n-to-m) decoder is also classified as a demultiplexer (DX). Each output represents a minterm that corresponds to the binary representation of the input vector. Thus, $z_i = m_i$, where m_i is the ith minterm of the n input variables. For example, if $n = 3$ and $x_1 x_2 x_3 = 101$, then output z_5 is asserted.

A decoder with n inputs, therefore, has a maximum of 2^n outputs. Because the outputs are mutually exclusive, only one output is active for each different combination of the inputs. The decoder outputs may be asserted high or low. Decoders have many applications in digital engineering, ranging from instruction decoding to memory addressing to code conversion.

A 3:8 (binary-to-octal) decoder is shown in Figure 3.84 which decodes a binary number into the corresponding octal number. The three inputs are $x[2{:}0]$ with binary weights of 2^2, 2^1, and 2^0, respectively. Thus, an input of $x[2]\,x[1]\,x[0] = 110$ will assert output $f[6]$. A decoder may also have an enable function which allows the selected output to be asserted. The enable function may be a single input or an AND gate with two or more inputs. A 3:8 decoder generates all eight minterms z_0 through z_7 of three binary variables x_2, x_1, and x_0.

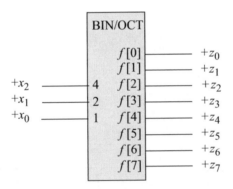

Figure 3.84   A binary-to-octal decoder.

The structural design module of a 3:8 decoder is shown in Figure 3.85 using the continuous assignment statement. The test bench module is shown in Figure 3.86, which applies input vectors for all eight combinations of three variables. The outputs and waveforms are shown in Figure 3.87 and Figure 3.88, respectively.

```
//dataflow 3:8 decoder
module decoder_3to8_df2 (x, f);

input [2:0] x;
output [7:0] f;

assign    f[0] = ~x[2] & ~x[1] & ~x[0],
          f[1] = ~x[2] & ~x[1] &  x[0],
          f[2] = ~x[2] &  x[1] & ~x[0],
          f[3] = ~x[2] &  x[1] &  x[0],
          f[4] =  x[2] & ~x[1] & ~x[0],
          f[5] =  x[2] & ~x[1] &  x[0],
          f[6] =  x[2] &  x[1] & ~x[0],
          f[7] =  x[2] &  x[1] &  x[0];
endmodule
```

Figure 3.85   Structural design module for a 3:8 decoder.

```
//test bench for 3:8 decoder
module decoder_3to8_df2_tb;

reg [2:0] x;
wire [7:0] f;

initial         //display variables
$monitor ("data in = %b, f = %b", x, f);

initial         //apply input vectors
begin
   #0    x = 3'b000;
   #10   x = 3'b001;
   #10   x = 3'b010;
   #10   x = 3'b011;
   #10   x = 3'b100;
   #10   x = 3'b101;
   #10   x = 3'b110;
   #10   x = 3'b111;
   #10   $stop;
end

//instantiate the module into the test bench
decoder_3to8_df2 inst1 (
   .x(x),
   .f(f)
   );
endmodule
```

Figure 3.86   Test bench module for a 3:8 decoder.

```
                    f[7] -- f[0]

data in = 000, f = 00000001
data in = 001, f = 00000010
data in = 010, f = 00000100
data in = 011, f = 00001000
data in = 100, f = 00010000
data in = 101, f = 00100000
data in = 110, f = 01000000
data in = 111, f = 10000000
```

Figure 3.87   Outputs for a 3:8 decoder.

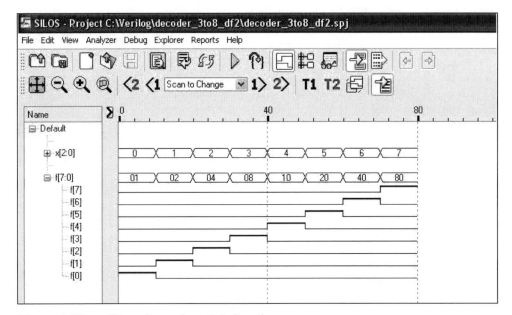

Figure 3.88   Waveforms for a 3:8 decoder.

Using the 3:8 decoder of this example, a Boolean function will be synthesized using the Karnaugh map shown in Figure 3.89, which yields the equation for output z_1 as shown in Equation 3.2. The logic diagram for the implementation is shown in Figure 3.90. The structural design module is shown in Figure 3.91. The test bench module is shown in Figure 3.92. The outputs and waveforms are shown in Figure 3.93 and Figure 3.94, respectively.

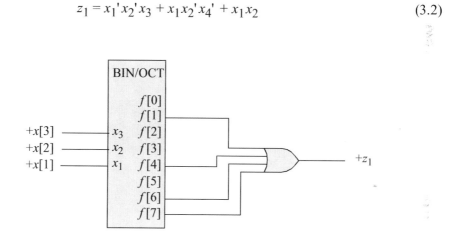

Figure 3.89   Karnaugh map to be implemented with a 3:8 decoder.

$$z_1 = x_1'x_2'x_3 + x_1x_2'x_4' + x_1x_2 \tag{3.2}$$

Figure 3.90   Logic diagram for the 3:8 decoder implementation to design Equation 3.2.

```
//structural design using decoder to
//implement a Boolean function
module decoder_3to8_boolean (x, z1);

//define input and output
input [1:3] x;
output z1;
                              //continued on next page
```

Figure 3.91   Structural design module to implement Figure 3.90.

```
//define internal nets
wire net0, net1, net2, net3, net4, net5, net6, net7;

//instantiate the logic to implement the function
decoder_3to8_df2 inst1 (
    .x(x),
    .f({net7, net6, net5, net4, net3, net2, net1, net0})
    );

or4_df inst2 (
    .x1(net1),
    .x2(net4),
    .x3(net6),
    .x4(net7),
    .z1(z1)
    );
endmodule
```

Figure 3.91   (Continued)

```
//test bench for decoder to implement a Boolean function
module decoder_3to8_boolean_tb;

reg [1:3] x;           //inputs are reg for test bench
wire z1;               //outputs are wire for test bench

//apply input vectors and display variables
initial
begin: apply_stimulus
    reg [3:0] invect;
    for (invect = 0; invect < 8; invect = invect + 1)
        begin
            x = invect [3:0];
            #10 $display ("x = %b, z1 = %b", x, z1);
        end
end

//instantiate the module into the test bench
decoder_3to8_boolean inst1 (
    .x(x),
    .z1(z1)
    );
endmodule
```

Figure 3.92   Test bench module for Figure 3.90.

```
x = 000, z1 = 0
x = 001, z1 = 1      z_1 = x_1'x_2'x_3
x = 010, z1 = 0
x = 011, z1 = 0
x = 100, z1 = 1      z_1 = x_1 x_2' x_4'
x = 101, z1 = 0
x = 110, z1 = 1      z_1 = x_1 x_2
x = 111, z1 = 1      z_1 = x_1 x_2
```

$$z_1 = x_1'x_2'x_3$$

$$z_1 = x_1 x_2' x_4'$$

$$z_1 = x_1 x_2$$

$$z_1 = x_1 x_2$$

Figure 3.93   Outputs for Figure 3.90.

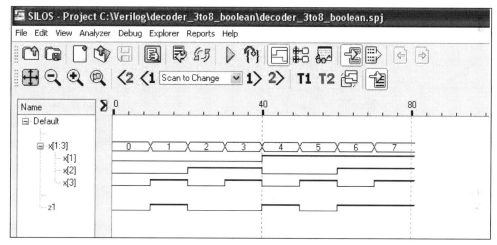

Figure 3.94   Waveforms for Figure 3.90.

3.3 Programmable Logic Devices

Programmable logic devices (PLDs) can be used in applications involving the synthesize of synchronous sequential machines using traditional techniques. This section synthesizes PLDs using Verilog HDL. There are three main types of programmable logic devices (PLDs): *programmable read-only memories* (PROMs), *programmable array logic* (PAL) devices, and *programmable logic array* (PLA) devices.

Current processors and memory are designed using a hardware description language (HDL), such as Verilog HDL. These HDL software systems simplify the task of logic design using PLDs and also perform logic minimization and test vector generation for system simulation. The ultimate goal is to specify the input/output characteristics of a machine in a high-level language. The hardware-software system then synthesizes the machine to yield a minimized, functionally tested unit.

Programmable logic devices can be used in the synthesis (design) of both combinational and sequential logic networks. PLDs are prefabricated integrated circuits (ICs) in which fused and hard-wired interconnections are used and implement 2-level switching functions by means of an AND array and an OR array.

The basic organization of a PLD consists of an AND array driving an OR array as shown in Figure 3.95. There is a set of inputs X_i containing n input signals and a set of outputs Z_i containing m output signals. The amount of programming capability depends upon the type of PLD that is used. For example, a PROM contains a fixed AND array and a programmable OR array; a PAL contains a programmable AND array and a fixed OR array; a PLA contains both a programmable AND array and a programmable OR array. Both PAL and PLA architectures have versions which contain storage elements in conjunction with combinational logic.

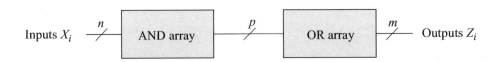

Figure 3.95   Basic organization of a programmable logic device.

The following sections illustrate the use of PLDs in the synthesis of combinational logic and synchronous sequential machines. The PLDs that will be presented are programmable read-only memories (PROMs), programmable array logic (PAL) devices, programmable logic arrays (PLAs).

3.3.1 Programmable Read-Only Memory

A PROM is a storage device in which the information is permanently stored; that is, the data remains valid even after power is turned off. PROMs are used for application programs, tables, code conversion, control store for microprogram sequencers, and other functions in which the stored data is not changed. The organization of a PROM is essentially the same as that for other PLDs: an input vector (an address) connects to an AND array which in turn connects to an OR array which generates the output vector (or word) for the PROM.

The concept of read-only memories (ROMs) for sequential machine synthesis and processor control is quite common. ROMs are also used extensively in developing new microprogram-controlled systems. In general, a PROM contains n inputs and m outputs. Because the inputs function as an address, there are 2^n unique addresses to select one of 2^n words. The AND array decodes the address to select a specific word in memory.

Thus, the interconnections in the AND array are fixed by the manufacturer and cannot be programmed by the user. The OR array, however, is programmable. The interconnections in the OR array are programmed by the user using special internal circuitry and a programming device to indicate the bit configuration of each word in memory. Each interconnection functions as a fuse; thus, the fuse can be left intact (indicating a logic 1) or opened (indicating a logic 0).

Example 3.14   Equation 3.3 will be implemented using a PROM. Figure 3.96 illustrates the internal organization of a representative PROM to implement the four equations of Equation 3.3. There are two address inputs x_1 and x_2 and four outputs $f_1, f_2,$ $f_3,$ and f_4. Inputs x_1 and x_2 select one of four words using the AND array decoder: word 0, 1, 2, or 3 that corresponds to $x_1 x_2 = 00, 01, 10,$ or 11, respectively.

Thus, the AND array cannot be programmed, as indicated by the "hardwired" connection symbol "● ." The OR array, however, is programmable. The symbol "×" indicates an intact fuse at the intersection of the AND gate product term and the OR gate input and provides a logic 1 to the specified OR gate input. The absence of an × indicates an open fuse, which provides a logic 0 to the OR gate input.

$$f_1(x_1, x_2) = \Sigma_m(0, 2) = x_1'x_2' + x_1 x_2'$$

$$f_2(x_1, x_2) = \Sigma_m(1, 2) = x_1'x_2 + x_1 x_2'$$

$$f_3(x_1, x_2) = \Sigma_m(0, 3) = x_1'x_2' + x_1 x_2$$

$$f_4(x_1, x_2) = \Sigma_m(1, 3) = x_1'x_2 + x_1 x_2 \qquad (3.3)$$

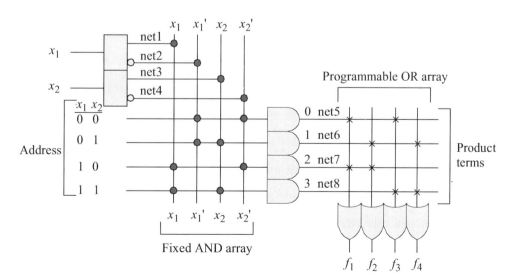

Figure 3.96   PROM organization with two address inputs: x_2 and x_2 and four outputs: $f_1, f_2, f_3,$ and f_4 to implement Equation 3.3.

The structural design module is shown in Figure 3.97 using built-in primitives that represent the logic of a typical PROM to implement the combinational logic equations of Equation 3.3. The test bench module is shown in Figure 3.98. The outputs and waveforms are shown in Figure 3.99 and Figure 3.100, respectively.

```verilog
//structural prom to generate four equations

module prom2 (x1, x2, f1, f2, f3, f4);


//define inputs and outputs
input x1, x2;
output f1, f2, f3, f4;


//define internal nets
wire net1, net2, net3, net4, net5, net6, net7, net8;


//define the input logic
buf (net1, x1);
not (net2, x1);


buf (net3, x2);
not (net4, x2);


//define the logic for the and array
and (net5, net2, net4);
and (net6, net2, net3);
and (net7, net1, net4);
and (net8, net1, net3);


//define the logic for the or array
or  (f1, net5, net7);
or  (f2, net6, net7);
or  (f3, net5, net8);
or  (f4, net6, net8);


endmodule
```

Figure 3.97   Structural design module to implement the equations of Equation 3.3.

```
//test bench for the structural prom module

module prom2_tb;

reg x1, x2;                //inputs are reg for test bench
wire f1, f2,f3, f4;        //outputs are wire for test bench

//display variables
initial
$monitor ("x1 x2 = %b, f1 f2 f3 f4 = %b",
          {x1, x2}, {f1, f2, f3, f4});

//apply input vectors
initial
begin
    #0     x1 = 1'b0;   x2 = 1'b0;
    #10    x1 = 1'b0;   x2 = 1'b1;
    #10    x1 = 1'b1;   x2 = 1'b0;
    #10    x1 = 1'b1;   x2 = 1'b1;

    #10    $stop;
end

//instantiate the module into the test bench
prom2 inst1 (
    .x1(x1),
    .x2(x2),
    .f1(f1),
    .f2(f2),
    .f3(f3),
    .f4(f4)
    );

endmodule
```

Figure 3.98   Test bench for the PROM of Figure 3.97 to implement the equations of Equation 3.3.

```
x1 x2 = 00, f1 f2 f3 f4 = 1010
x1 x2 = 01, f1 f2 f3 f4 = 0101
x1 x2 = 10, f1 f2 f3 f4 = 1100
x1 x2 = 11, f1 f2 f3 f4 = 0011
```

$$f_1 = \Sigma_m(0, 2) = x_1' x_2' + x_1 x_2'$$
$$f_2 = \Sigma_m(1, 2) = x_1' x_2 + x_1 x_2'$$
$$f_3 = \Sigma_m(0, 3) = x_1' x_2' + x_1 x_2$$
$$f_4 = \Sigma_m(1, 3) = x_1' x_2 + x_1 x_2$$

Figure 3.99   Outputs for the logic to implement Equation 3.3.

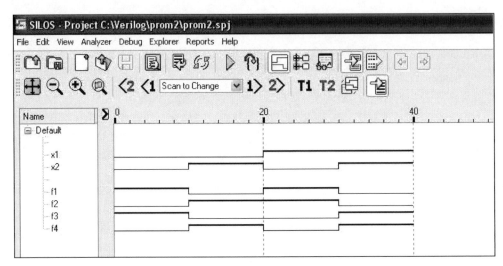

Figure 3.100   Waveforms for the logic to implement Equation 3.3.

3.3.2 Programmable Array Logic

A programmable array logic (PAL) device is structured with a programmable AND array and a non-programmable (fixed) OR array. The number of gates in the AND array is not a function of the number of inputs, as for PROMs. The AND array allows product terms to be programmed which then connect to a predefined OR array. The restriction of the prewired OR array is compensated by the wide variety of available PAL configurations. Specifying the configuration of the OR array, therefore, is simply a matter of device selection.

Example 3.15   The equations of Equation 3.4 will be implemented in the PAL device shown in Figure 3.101, which is the organization of a basic PAL. Each AND gate has $2n$ inputs, where n is the number of device inputs. The symbol "$\times$" indicates an intact fuse, which connects a unique variable — either true or complemented — to one of the six AND gate inputs. The absence of an $\times$ indicates an open fuse, which supplies a logic 1 to the AND gate. Thus, the product terms consist only of the input variables specified by an $\times$.

$$z_1 = x_1 x_2' + x_1' x_2$$

$$z_2 = x_1 x_2 x_3 + x_1' x_2' x_3'$$

$$z_3 = x_1 x_2' x_3 \tag{3.4}$$

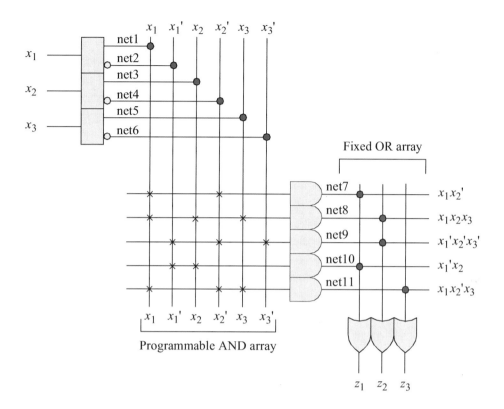

Figure 3.101 Organization of a PAL to implement the following equations: $z_1 = x_1x_2' + x_1'x_2$; $z_2 = x_1x_2x_3 + x_1'x_2'x_3'$; $z_3 = x_1x_2'x_3$.

All programmed terms of the three input variables x_1, x_2, and x_3 are available at the three outputs z_1, z_2, and z_3 in a sum-of-minterms or sum-of-products form. Each hardwired connection in the OR array indicates that the corresponding product term is a component of the appropriate output function.

The structural design module is shown in Figure 3.102 using built-in primitives that represent the logic of a typical PAL device to implement the equations of Equation 3.4. The test bench module is shown in Figure 3.103. The outputs and waveforms are shown in Figure 3.104 and Figure 3.105, respectively.

More complex PALs contain not only the basic AND-OR array organization but also additional circuitry for feedback signals and output registers specified by flip-flops. The basic organization of a PAL conforms to the general structure of a synchronous sequential machine. The input drivers, together with the AND-OR arrays, constitute the δ next-state logic; the flip-flops are the storage elements; and the drivers represent the λ output logic.

The input equations obtained from input maps are in a sum-of-products form, which is the requisite format for programming the AND array. Thus, PALs and other PLDs are ideally suited for synthesizing synchronous sequential machines.

```
//structural pal to generate equations for z1, z2, and z3

module pal (x1, x2, x3, z1,z2, z3);

//define inputs and outputs
input x1, x2, x3;
output z1, z2, z3;

//define internal nets
wire net1, net2, net3, net4, net5, net6,
     net7, net8, net9, net10, net11;

//define the input logic
buf (net1, x1);
not (net2, x1);

buf (net3, x2);
not (net4, x2);

buf (net5, x3);
not (net6, x3);

//define the logic for the and array
and (net7, net1, net4);
and (net8, net1, net3, net5);
and (net9, net2, net4, net6);
and (net10, net2, net3);
and (net11, net1, net4, net5);

//define the logic for the or array
or  (z1, net7, net10);
or  (z2, net8, net9);
or  (z3, net11);

endmodule
```

Figure 3.102   Structural design module to implement the equations of Equation 3.4.

```
//test bench pal to generate equations for z1, z2, and z3

module pal_tb;

reg x1, x2, x3;         //inputs are reg for test bench
wire z1, z2, z3;        //outputs are wire for test bench
                                    //continued on next page
```

Figure 3.103   Test bench module to implement the equations of Equation 3.4.

```
//display variables
initial
$monitor ("x1 x2 x3 = %b, z1, z2, z3 = %b",
          {x1, x2, x3}, {z1, z2, z3});

initial      //apply input vectors
begin
    #0     x1 = 1'b0;   x2 = 1'b0;   x3 = 1'b0;
    #10    x1 = 1'b0;   x2 = 1'b0;   x3 = 1'b1;
    #10    x1 = 1'b0;   x2 = 1'b1;   x3 = 1'b0;
    #10    x1 = 1'b0;   x2 = 1'b1;   x3 = 1'b1;

    #10    x1 = 1'b1;   x2 = 1'b0;   x3 = 1'b0;
    #10    x1 = 1'b1;   x2 = 1'b0;   x3 = 1'b1;
    #10    x1 = 1'b1;   x2 = 1'b1;   x3 = 1'b0;
    #10    x1 = 1'b1;   x2 = 1'b1;   x3 = 1'b1;

    #10    $stop;
end

//instantiate the module into the test bench
pal inst1 (
    .x1(x1),
    .x2(x2),
    .x3(x3),
    .z1(z1),
    .z2(z2),
    .z3(z3)
    );
endmodule
```

Figure 3.103   (Continued)

$$z_1 = x_1 x_2' + x_1' x_2; \qquad z_2 = x_1 x_2 x_3 + x_1' x_2' x_3'; \qquad z_3 = x_1 x_2' x_3$$

```
x1 x2 x3 = 000, z1, z2, z3 = 010
x1 x2 x3 = 001, z1, z2, z3 = 000
x1 x2 x3 = 010, z1, z2, z3 = 100
x1 x2 x3 = 011, z1, z2, z3 = 100
x1 x2 x3 = 100, z1, z2, z3 = 100
x1 x2 x3 = 101, z1, z2, z3 = 101
x1 x2 x3 = 110, z1, z2, z3 = 000
x1 x2 x3 = 111, z1, z2, z3 = 010
```

Figure 3.104   Outputs for the PAL device of Figure 3.101 to implement the equations of Equation 3.4.

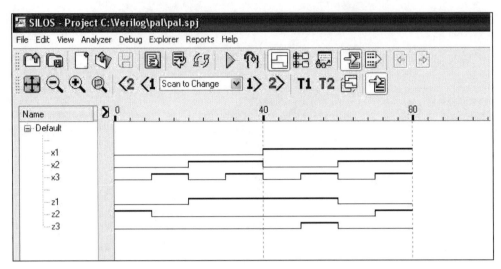

Figure 3.105   Waveforms for the PAL device of Figure 3.101 to implement the equations of Equation 3.4.

Example 3.16   A 3-bit Gray code counter will be synthesized that counts in the following sequence: 000, 001, 011, 010, 110, 111, 101, 100, 000, ... A single PAL device will be used for both the δ next-state logic and the storage elements, which consist of positive-edge-triggered D flip-flops. The input maps are shown in Figure 3.106. The equations for flip-flops y_1, y_2, and y_3 are listed in Equation 3.5 in a sum-of-products form. The logic diagram using PAL technology is shown in Figure 3.107 and is programmed directly from the sum-of-products input equations of Equation 3.5. The only inputs to the counter are the Clock signal and an implied Reset for the flip-flops.

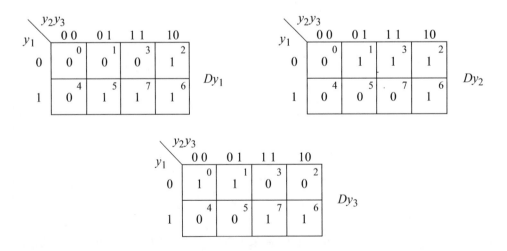

Figure 3.106   Input maps for the Gray code counter.

$$Dy_1 = y_1y_3 + y_2y_3'$$
$$Dy_2 = y_1'y_3 + y_2y_3'$$
$$Dy_3 = y_1'y_2' + y_1y_2$$

(3.5)

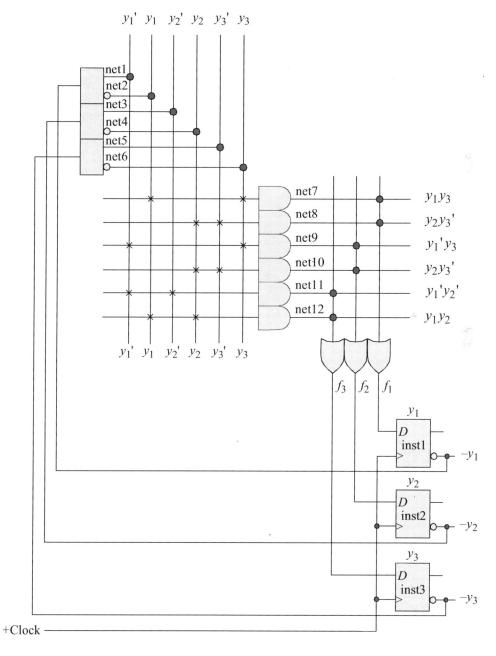

Figure 3.107   Logic diagram for the Gray code counter using a PAL device.

All AND array inputs are feedback signals from the state flip-flops through drivers with inverting and noninverting outputs. The complemented output of flip-flop y_i is fed back to an input driver. Thus, when flip-flop y_i is set (=1), the complemented output of the driver generates a positive voltage level which is available to all AND gate inputs.

If the input equation for y_i contains a flip-flop variable in its true form, then the complemented output of the driver is programmed to connect to the appropriate positive-input AND gate in the AND array. The δ next-state logic consists of the feedback drivers, the programmable AND array, and the fixed OR array. Outputs f_1, f_2, and f_3 of the OR gates connect to Dy_1, Dy_2, and Dy_3, respectively. The present state $Y_{j(t)}$ is fed back through drivers to the AND array and, after an appropriate propagation delay, appears at the flip-flop D inputs as the next state $Y_{k(t+1)}$.

The structural design module that represents the logic of a typical PAL device to implement the counter is shown in Figure 3.108 using built-in primitives and instantiated D flip-flops. The test bench module is shown in Figure 3.109. The outputs and waveforms are shown in Figure 3.110 and Figure 3.111, respectively.

```verilog
//structural pal to design a Gray code counter
module pal2 (rst_n, clk, y);

input rst_n, clk;      //define inputs and outputs
output [1:3] y;

//define internal nets
wire net1, net2, net3, net4, net5, net6, net7, net8,
     net9, net10, net11, net12, f1, f2, f3;

//define the input drivers
buf (net1, ~y[1]);
not (net2, ~y[1]);

buf (net3, ~y[2]);
not (net4, ~y[2]);

buf (net5, ~y[3]);
not (net6, ~y[3]);


//define the logic for and array, or array, and y[1]
and (net7, net2, net6);
and (net8, net4, net5);
or  (f1, net7, net8);
d_ff_bh inst1 (
   .rst_n(rst_n),
   .clk(clk),
   .d(f1),
   .q(y[1])
   );                              //continued on next page
```

Figure 3.108   Structural design module for the 3-bit Gray code counter.

```
//define the logic for and array, or array, and y[2]
and (net9, net1, net6);
and (net10, net4, net5);
or  (f2, net9, net10);
d_ff_bh inst2 (
   .rst_n(rst_n),
   .clk(clk),
   .d(f2),
   .q(y[2])
   );

//define the logic for and array, or array, and y[3]  .
and (net11, net1, net3);
and (net12, net2, net4);
or  (f3, net11, net12);
d_ff_bh inst3 (
   .rst_n(rst_n),
   .clk(clk),
   .d(f3),
   .q(y[3])
   );

endmodule
```

Figure 3.108   (Continued)

```
//test bench for Gray code counter

module pal2_tb;

reg rst_n, clk;       //inputs are reg for test bench
wire [1:3] y;         //outputs are wire for test bench

//display outputs
initial
$monitor ("count = %b", y);

//define reset
initial
begin
   #0 rst_n = 1'b0;
   #5 rst_n = 1'b1;
end                              //continued on next page
```

Figure 3.109   Test bench module for the 3-bit Gray code counter.

```
initial              //define clock
begin
   clk = 1'b0;
   forever
      #10 clk = ~clk;
end

initial              //define length of simulation
begin
   #145 $finish;
end

//instantiate the module into the test bench
pal2 inst1 (
   .rst_n(rst_n),
   .clk(clk),
   .y(y)
   );
endmodule
```

Figure 3.109   (Continued)

```
count = 000      count = 110
count = 001      count = 111
count = 011      count = 101
count = 010      count = 100
                 count = 000
```

Figure 3.110   Outputs for the 3-bit Gray code counter.

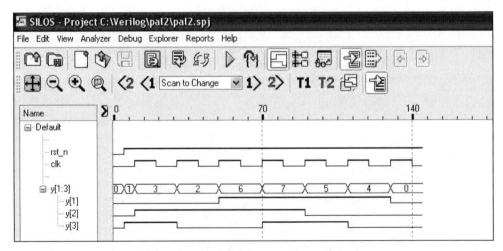

Figure 3.111   Waveforms for the 3-bit Gray code counter.

Example 3.17 PAL devices can be used to implement any synchronous sequential machine, including a sequence detector. A Mealy machine will be synthesized that checks for the sequence 01111110 on a serial input line x_1. Input x_1 remains at a high voltage level until transmission is to begin, at which time x_1 assumes a low voltage level for one bit period, providing a negative transition.

The state diagram for the Mealy machine is shown in Figure 3.112. The machine remains in state a until the start of transmission is indicated by a high-to-low transition on input x_1. The bit sequence is then received beginning with bit 0, one bit per clock period. When the first 1 bit has been detected in state b, any subsequent 0 bit that occurs before six consecutive 1s returns the machine to state b to begin checking for a new valid sequence. Similarly, seven consecutive 1s returns the machine to state a to begin checking for a new valid sequence. Only when $x_1 = 01111110$ is output z_1 asserted.

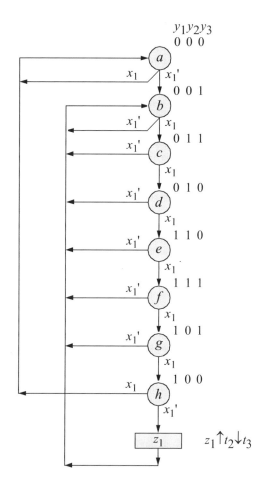

Figure 3.112 State diagram for the Mealy machine of Example 3.17.

The input maps are shown in Figure 3.113. In state a ($y_1 y_2 y_3 = 000$), the next states for flip-flops y_1 and y_2 are 0 and 0, respectively, regardless of the value of x_1. The next state for y_3, however, is determined by the value of x_1; if $x_1 = 0$, then the next state for y_3 is 1, otherwise the next state is 0. In states d, e, f, and g, flip-flop y_1 has a next state of 1 only if $x_1 = 1$. Therefore, x_1 is entered in the map as a map-entered variable.

Similarly, in states b, c, d, and e, flip-flop y_2 has a next value of 1 only if $x_1 = 1$. In the input map for flip-flop y_3, the next state is never an unconditional 0, irrespective of the path taken. The next state will be either an unconditional 1 or a value dependent upon x_1; that is, if $x_1 = 0$, then $y_3 = 1$. Since a logic $1 = x_1 + x_1'$, therefore, every minterm location in the map can be given a value of x_1'. This accounts for the x_1' term in the equation for Dy_3 of the input equations shown in Equation 3.6. The x_1 term of the logic 1 expression must now be taken into account. This is very easily accomplished by reverting to the minterm value of 1, which generates the two remaining terms for Dy_3. The PAL logic diagram is illustrated in Figure 3.114.

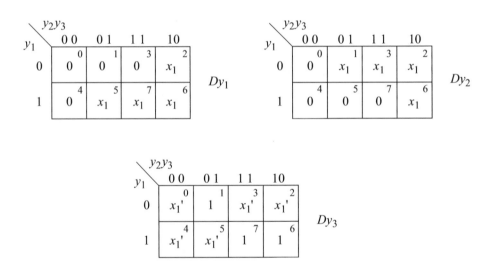

Figure 3.113   Input maps for the Mealy machine of Figure 3.112.

$$Dy_1 = y_1 y_3 x_1 + y_2 y_3' x_1$$

$$Dy_2 = y_1' y_3 x_1 + y_2 y_3' x_1$$

$$Dy_3 = x_1' + y_1' y_2' y_3 + y_1 y_2 \qquad (3.6)$$

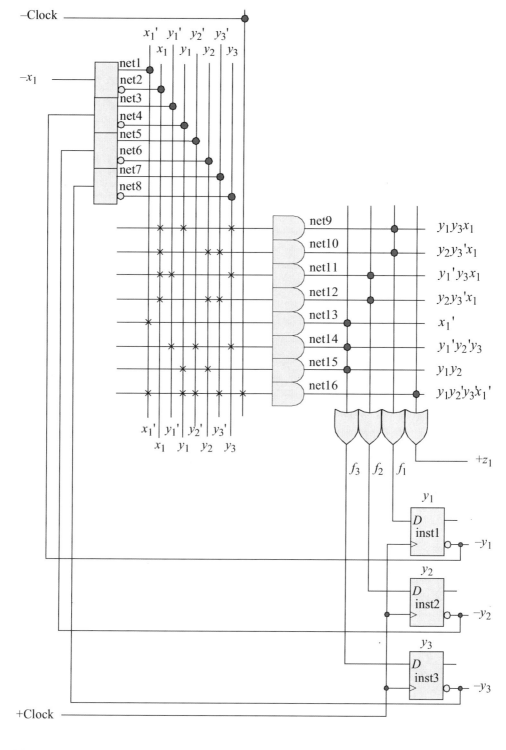

Figure 3.114   Logic diagram for the Mealy machine of Figure 3.112 using a PAL device.

The structural design module that represents the logic of a typical PAL device to implement the design is shown in Figure 3.115 using built-in primitives and instantiated D flip-flops. The test bench module is shown in Figure 3.116. The outputs and waveforms are shown in Figure 3.117 and Figure 3.118, respectively.

```
//structural pal to detect the sequence x1 = 01111110

module pal3 (rst_n, clk, x1, y, z1);

//define inputs and outputs
input rst_n, clk, x1;
output [1:3] y;
output z1;

//define internal nets
wire net1, net2, net3, net4, net5, net6, net7, net8, net9,
      net10, net11, net12, net13, net14, net15, net16,
      f1, f2, f3;

//define the input drivers
buf (net1, ~x1);
not (net2, ~x1);

buf (net3, ~y[1]);
not (net4, ~y[1]);

buf (net5, ~y[2]);
not (net6, ~y[2]);

buf (net7, ~y[3]);
not (net8, ~y[3]);

//define the logic for and array, or array, and y[1]
and (net9, net2, net4, net8);
and (net10, net2, net6, net7);
or  (f1, net9, net10);

d_ff_bh inst1 (
   .rst_n(rst_n),
   .clk(clk),
   .d(f1),
   .q(y[1])
   );
                                //continued on next page
```

Figure 3.115   Structural design module for the PAL device of Example 3.17 to detect the sequence $x_1 = 01111110$.

```
//define the logic for and array, or array, and y[2]
and (net11, net2, net3, net8);
and (net12, net2, net6, net7);
or  (f2, net11, net12);

d_ff_bh inst2 (
   .rst_n(rst_n),
   .clk(clk),
   .d(f2),
   .q(y[2])
   );

//define the logic for and array, or array, and y[3]
and (net13, net1);
and (net14, net3, net5, net8);
and (net15, net4, net6);
or  (f3, net13, net14, net15);

d_ff_bh inst3 (
   .rst_n(rst_n),
   .clk(clk),
   .d(f3),
   .q(y[3])
   );

//define the logic for output z1
and (z1, net1, net4, net5, net7, ~clk);

endmodule
```

Figure 3.115   (Continued)

```
//test bench to detect the sequence x1 = 01111110
module pal3_tb;

reg rst_n, clk,x1;    //inputs are reg for test bench
wire [1:3] y;         //outputs are wire for test bench
wire z1;

//display variables
initial
$monitor ("x1 = %b, state = %b, z1 = %b", x1, y, z1);
                                //continued on next page
```

Figure 3.116   Test bench module for the PAL device of Example 3.17.

```
initial                  //define clock
begin
   clk = 1'b0;
   forever
      #10 clk = ~clk;
end

initial                  //define input sequence
begin
   #0 rst_n = 1'b0;
      x1 = 1'b0;

   #5 rst_n = 1'b1;
//-------------------------------------------------
   #25   x1 = 1'b1;
   #120  x1 = 1'b0;

   #40    $stop;
end

//instantiate the module into the test bench
pal3 inst1 (
   .rst_n(rst_n),
   .clk(clk),
   .x1(x1),
   .y(y),
   .z1(z1)
   );

endmodule
```

Figure 3.116   (Continued)

```
x1 = 0, state = 000, z1 = 0
x1 = 0, state = 001, z1 = 0
x1 = 1, state = 001, z1 = 0
x1 = 1, state = 011, z1 = 0
x1 = 1, state = 010, z1 = 0
x1 = 1, state = 110, z1 = 0
x1 = 1, state = 111, z1 = 0
x1 = 1, state = 101, z1 = 0
x1 = 0, state = 100, z1 = 0
x1 = 0, state = 100, z1 = 1
x1 = 0, state = 001, z1 = 0
```

Figure 3.117   Outputs for the PAL device to detect the sequence $x_1 = 01111110$.

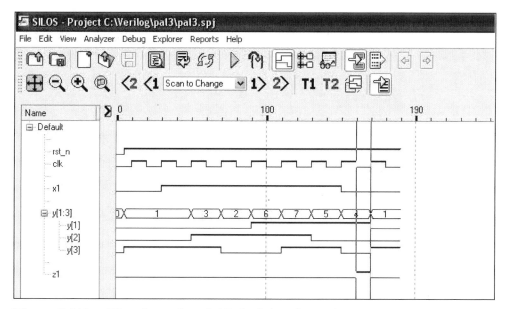

Figure 3.118   Waveforms for a PAL device to detect the sequence $x_1 = 01111110$.

Example 3.18   As a final example utilizing programmable array logic, a PAL device will be used to implement the Boolean equations specified in Equation 3.7. When designing with a PAL, the Boolean expressions should be minimized, if necessary, to reduce the number of product terms in each expression of the AND-OR structure.

$$z_1(x_1, x_2, x_3) = \Sigma_m(1, 2, 6)$$

$$z_2(x_1, x_2, x_3) = \Sigma_m(0, 1, 5, 6, 7)$$

$$z_3(x_1, x_2, x_3) = \Sigma_m(1, 2, 4, 6, 7) \tag{3.7}$$

Using Boolean algebra or Karnaugh maps, the above functions convert to the sum-of-products forms shown in Equation 3.8. The PAL logic diagram is illustrated in Figure 3.119.

$$z_1 = x_1'x_2'x_3 + x_2x_3'$$

$$z_2 = x_1'x_2' + x_1x_2 + x_2'x_3$$

$$z_3 = x_1'x_2'x_3 + x_2x_3' + x_1x_2 + x_1x_3'$$

$$= z_1 + x_1x_2 + x_1x_3' \tag{3.8}$$

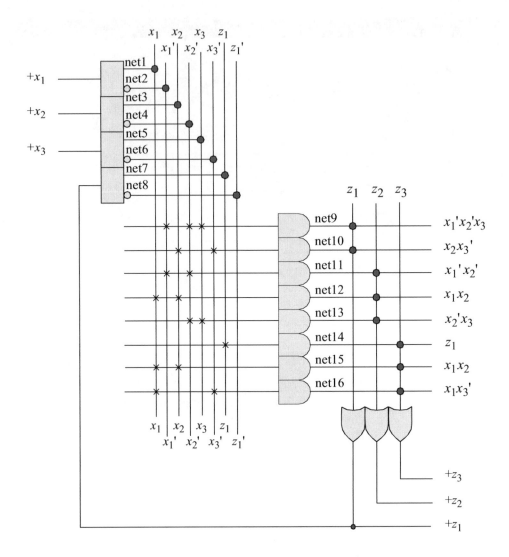

Figure 3.119   A PAL device using three inputs and three outputs to implement the Boolean expressions of Equation 3.8.

Each input of Figure 3.119 is connected to a buffer-driver which generates both true and complemented outputs of the corresponding input. The device consists of eight AND gates forming the programmable AND array and three OR gates which form the fixed OR array. Each AND gate has eight fused programmable inputs as shown by the eight vertical lines intersecting each horizontal line. The horizontal line is called the *product line* and symbolizes the multiple-input configuration of the AND gate. The output of each AND gate is the corresponding *product term*.

The structural design module is shown in Figure 3.120 using built-in primitives. The test bench module is shown in Figure 3.121, which applies all eight combinations of the three variables $x_1 x_2 x_3$. The outputs and waveforms are shown in Figure 3.122 and Figure 3.123, respectively.

```
//structural pal to implement three equations:
//z1 = x1'x2'x3 + x2x3'
//z2 = x1'x2' + x1x2 + x2'x3
//z3 = z1 + x1x2 + x1x3'
module pal4 (x1, x2, x3, z1, z2, z3);

input x1, x2, x3;        //define inputs and outputs
output z1, z2, z3;

//define internal nets
wire net1, net2, net3, net4, net5, net6, net7, net8, net9;
wire net10, net11, net12, net13, net14, net15, net16;

//define the input drivers
buf (net1, x1);
not (net2, x1);

buf (net3, x2);
not (net4, x2);

buf (net5, x3);
not (net6, x3);

buf (net7, z1);
not (net8, z1);

//define the logic for the and array and the or array for z1
and (net9, net2, net4, net5);
and (net10, net3, net6);
or  (z1, net9, net10);

//define the logic for the and array and the or array for z2
and (net11, net2, net4);
and (net12, net1, net3);
and (net13, net4, net5);
or  (z2, net11, net12, net13);

//define the logic for the and array and the or array for z3
and (net14, net7);
and (net15, net1, net3);
and (net16, net1, net6);
or  (z3, net14, net15, net16);

endmodule
```

Figure 3.120 Structural design module to implement the Boolean equations of Equation 3.8.

```
//test bench to implement the three equations of pal4
//z1 = x1'x2'x3 + x2x3'
//z2 = x1'x2' + x1x2 + x2'x3
//z3 = z1 + x1x2 + x1x3'

module pal4_tb;

//inputs are reg for test bench
//outputs are wire for test bench
reg   x1, x2, x3;
wire  z1, z2, z3;

//display variables
initial
$monitor ("x1 x2 x3 = %b, z1 z2 z3 = %b",
          {x1, x2, x3}, {z1, z2, z3});


//apply input vectors
initial
begin
    #0    x1 = 1'b0;  x2 = 1'b0;  x3 = 1'b0;
    #10   x1 = 1'b0;  x2 = 1'b0;  x3 = 1'b1;
    #10   x1 = 1'b0;  x2 = 1'b1;  x3 = 1'b0;
    #10   x1 = 1'b0;  x2 = 1'b1;  x3 = 1'b1;

    #10   x1 = 1'b1;  x2 = 1'b0;  x3 = 1'b0;
    #10   x1 = 1'b1;  x2 = 1'b0;  x3 = 1'b1;
    #10   x1 = 1'b1;  x2 = 1'b1;  x3 = 1'b0;
    #10   x1 = 1'b1;  x2 = 1'b1;  x3 = 1'b1;

    #10    $stop;
end

//instantiate the module into the test bench
pal4 inst1 (
    .x1(x1),
    .x2(x2),
    .x3(x3),
    .z1(z1),
    .z2(z2),
    .z3(z3)
    );

endmodule
```

Figure 3.121   Test bench module to implement the Boolean equations of Equation 3.8.

```
x1 x2 x3 = 000, z1 z2 z3 = 010
x1 x2 x3 = 001, z1 z2 z3 = 111
x1 x2 x3 = 010, z1 z2 z3 = 101
x1 x2 x3 = 011, z1 z2 z3 = 000

x1 x2 x3 = 100, z1 z2 z3 = 001
x1 x2 x3 = 101, z1 z2 z3 = 010
x1 x2 x3 = 110, z1 z2 z3 = 111
x1 x2 x3 = 111, z1 z2 z3 = 011
```

Figure 3.122   Outputs for the Boolean equations of Equation 3.8.

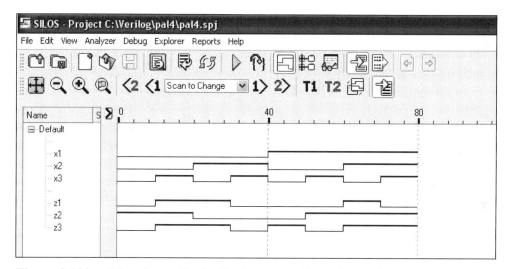

Figure 3.123   Waveforms for the Boolean equations of Equation 3.8.

3.3.3 Programmable Logic Array

Programmable logic arrays (PLAs) offer a high degree of flexibility, because both the AND array and the OR array are programmable. Unlike the AND array for a PROM, the AND array for a PLA does not require 2^n AND gates to accommodate all combinations of n inputs. A PLA has n input variables, $x_1, x_2, \ldots, x_n$ and m output functions, $z_1, z_2, \ldots, z_m$. The OR array permits each OR gate to access any product term. Thus, the programmable OR array allows all OR gates to access the same product terms simultaneously. Each output z_i is generated from a sum-of-product expression which is a function of the input variables.

The output function in a PLA is limited only by the number of AND gates in the AND array, since all AND gates can be programmed to connect to all OR gates. The output function in a PAL, however, is restricted not only by the number of AND gates in the AND array, but also by the fixed connections from the AND array outputs to the OR array.

Example 3.19   The basic organization of a PLA is shown in Figure 3.124 and will be used in this example to implement the Boolean functions of Equation 3.9. Both the AND array and the OR array are programmable. Since both arrays are programmable, the PLA has more programming capability and thus, more flexibility than the PROM or PAL.

$$z_1 = x_1 x_2' + x_1' x_2$$

$$z_2 = x_1 x_3 + x_1' x_3'$$

$$z_3 = x_1 x_2' + x_1 x_3 + x_1' x_2' x_3' \qquad (3.9)$$

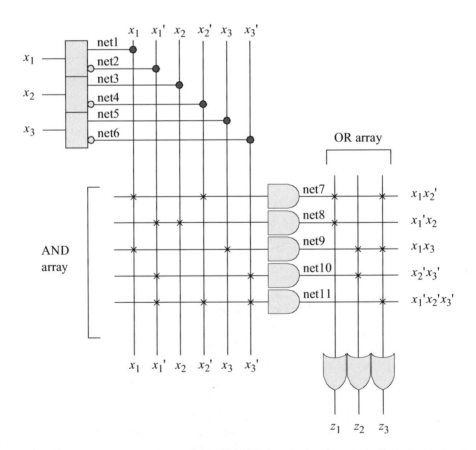

Figure 3.124   Basic organization of a PLA implementation using three inputs and three outputs.

The structural design module is shown in Figure 3.125 using built-in primitives. The test bench module is shown in Figure 3.126, which applies all eight combinations of the three variables $x_1 x_2 x_3$. The outputs and waveforms are shown in Figure 3.127 and Figure 3.128, respectively.

```verilog
//structural pla to implement three equations:
//z1 = x1x2' + x1'x2
//z2 = x1x3 + x1'x3'
//z3 = x1x2' + x1x3 + x1'x2'x3'

module pla (x1, x2, x3, z1, z2, z3);

//define inputs and outputs
input x1, x2, x3;
output z1, z2, z3;

//define internal nets
wire net1, net2, net3, net4, net5, net6, net7,
     net8, net9, net10, net11;

//define the input drivers
buf (net1, x1);
not (net2, x1);

buf (net3, x2);
not (net4, x2);

buf (net5, x3);
not (net6, x3);

//define the logic for the and array and the or array for z1
and (net7, net1, net4);
and (net8, net2, net3);
or  (z1, net7, net8);

//define the logic for the and array and the or array for z2
and (net9, net1, net5);
and (net10, net2, net6);
or  (z2, net9, net10);

//define the logic for the and array and the or array for z3
and (net11, net2, net4, net6);
or  (z3, net7, net9, net11);

endmodule
```

Figure 3.125 Structural design module to implement the three equations of Equation 3.9.

```
//test bench to implement the three equations of pla

module pla_tb;

//inputs are reg for test bench
//outputs are wire for test bench
reg   x1, x2, x3;
wire z1, z2, z3;


//display variables
initial
$monitor ("x1 x2 x3 = %b,  z1 z2 z3 = %b",
            {x1, x2, x3}, {z1, z2, z3});


//apply input vectors
initial
begin
   #0     x1 = 1'b0;   x2 = 1'b0;   x3 = 1'b0;
   #10    x1 = 1'b0;   x2 = 1'b0;   x3 = 1'b1;
   #10    x1 = 1'b0;   x2 = 1'b1;   x3 = 1'b0;
   #10    x1 = 1'b0;   x2 = 1'b1;   x3 = 1'b1;

   #10    x1 = 1'b1;   x2 = 1'b0;   x3 = 1'b0;
   #10    x1 = 1'b1;   x2 = 1'b0;   x3 = 1'b1;
   #10    x1 = 1'b1;   x2 = 1'b1;   x3 = 1'b0;
   #10    x1 = 1'b1;   x2 = 1'b1;   x3 = 1'b1;

   #10    $stop;
end


//instantiate the module into the test bench
pla inst1 (
   .x1(x1),
   .x2(x2),
   .x3(x3),
   .z1(z1),
   .z2(z2),
   .z3(z3)
   );

endmodule
```

Figure 3.126   Test bench module to implement the three equations of Example 3.19.

```
x1 x2 x3 = 000,  z1 z2 z3 = 011
x1 x2 x3 = 001,  z1 z2 z3 = 000
x1 x2 x3 = 010,  z1 z2 z3 = 110
x1 x2 x3 = 011,  z1 z2 z3 = 100

x1 x2 x3 = 100,  z1 z2 z3 = 101
x1 x2 x3 = 101,  z1 z2 z3 = 111
x1 x2 x3 = 110,  z1 z2 z3 = 000
x1 x2 x3 = 111,  z1 z2 z3 = 011
```

Figure 3.127   Outputs for the Boolean functions of Equation 3.9.

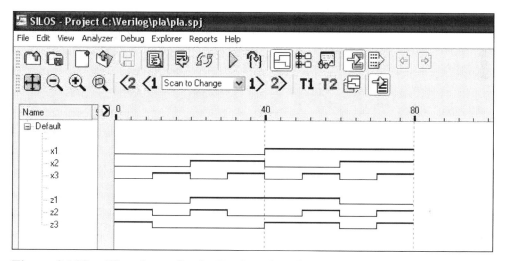

Figure 3.128   Waveforms for the Boolean functions of Equation 3.9.

Example 3.20   PLAs are characterized by the number of input variables, the number of product terms, and the number of output functions. All Boolean expressions can be decomposed into a sum-of-products representation. For example, the exclusive-OR function $x_1 \oplus x_2$ equates to $x_1 x_2' + x_1' x_2$. The equations shown in Equation 3.10 will be implemented in a programmable representation using the PLA device shown in Figure 3.129. The symbol × indicates an intact fuse; the absence of an × indicates an open fuse, where the unconnected input assumes a logic 1 voltage level for an AND gate input and a logic 0 voltage level for an OR gate input.

$$z_1 = x_1 x_2' + x_1' x_2$$

$$z_2 = x_1' x_2 x_3 + x_1 x_3' + x_4' \qquad (3.10)$$

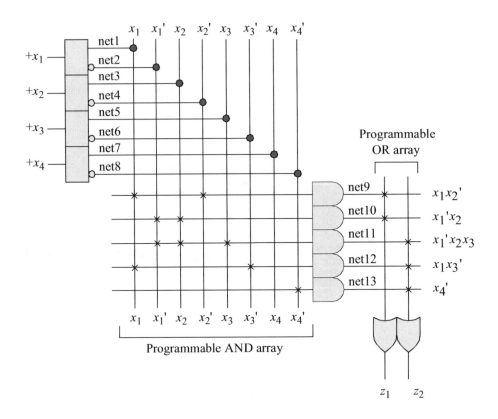

Figure 3.129 Implementation of a multiple-input, multiple-output, 2-level logic circuit using a PLA device. Output $z_1 = x_1 x_2' + x_1' x_2$. Output $z_2 = x_1' x_2 x_3 + x_1 x_3' + x_4'$.

The structural design module is shown in Figure 3.130 using built-in primitives. The test bench module is shown in Figure 3.131, which applies all 16 combinations of the four variables $x_1 x_2 x_3 x_4$. The outputs and waveforms are shown in Figure 3.132 and Figure 3.133, respectively.

```
//structural pla to implement two equations:
//z1 = x1x2' + x1'x2
//z2 = x1'x2x3 + x1x3' + x4'

module pla2 (x1, x2, x3, x4, z1, z2);
                                        //continued on next page
```

Figure 3.130 Structural design module to implement the two equations of Equation 3.10.

```
//define inputs and outputs
input x1, x2, x3, x4;
output z1, z2;

//define internal nets
wire net1, net2, net3, net4, net5, net6, net7, net8,
     net9, net10, net11, net12, net13;

//define the input drivers
buf (net1, x1);
not (net2, x1);

buf (net3, x2);
not (net4, x2);

buf (net5, x3);
not (net6, x3);

buf (net7, x4);
not (net8, x4);

//define the logic for the and array and the or array for z1
and (net9, net1, net4);
and (net10, net2, net3);
or  (z1, net9, net10);

//define the logic for the and array and the or array for z2
and (net11, net2, net3, net5);
and (net12, net1, net6);
and (net13, net8);
or  (z2, net11, net12, net13);

endmodule
```

Figure 3.130   (Continued)

```
//test bench to implement the two equations of pla2

module pla2_tb;

//inputs are reg for test bench
//outputs are wire for test bench
reg x1, x2, x3, x4;
wire z1, z2;
                                //continued on next page
```

Figure 3.131   Test bench module to implement the two equations of Equation 3.10.

```
//apply input vectors and display variables
initial
begin: apply_stimulus
   reg [4:0] invect;
   for (invect = 0; invect < 16; invect = invect + 1)
      begin
         {x1, x2, x3, x4} = invect [4:0];
         #10 $display ("x1 x2 x3 x4 = %b, z1 z2 = %b",
                       {x1, x2, x3, x4}, {z1, z2});
   end
end

//instantiate the module into the test bench as a single line
pla2 inst1 (x1, x2, x3, x4, z1, z2);
endmodule
```

Figure 3.131   (Continued)

```
x1 x2 x3 x4 = 0000, z1 z2 = 01     x1 x2 x3 x4 = 1000, z1 z2 = 11
x1 x2 x3 x4 = 0001, z1 z2 = 00     x1 x2 x3 x4 = 1001, z1 z2 = 11
x1 x2 x3 x4 = 0010, z1 z2 = 01     x1 x2 x3 x4 = 1010, z1 z2 = 11
x1 x2 x3 x4 = 0011, z1 z2 = 00     x1 x2 x3 x4 = 1011, z1 z2 = 10
x1 x2 x3 x4 = 0100, z1 z2 = 11     x1 x2 x3 x4 = 1100, z1 z2 = 01
x1 x2 x3 x4 = 0101, z1 z2 = 10     x1 x2 x3 x4 = 1101, z1 z2 = 01
x1 x2 x3 x4 = 0110, z1 z2 = 11     x1 x2 x3 x4 = 1110, z1 z2 = 01
x1 x2 x3 x4 = 0111, z1 z2 = 11     x1 x2 x3 x4 = 1111, z1 z2 = 00
```

Figure 3.132   Outputs to display the results of Equation 3.10.

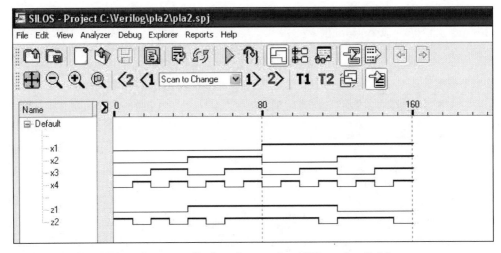

Figure 3.133   Waveforms to display the results of Equation 3.10.

Example 3.21 The synthesis of synchronous sequential machines can be realized using PLAs in a manner analogous to that used for PROMs and PALs. The assignment of state codes, however, is more crucial, since a judicious choice of state codes can reduce the number of product terms required, and thus, reduce the size of the PLA device.

Figure 3.134 illustrates a state diagram for a Moore machine with one input x_1 and two outputs z_1 and z_2. The machine will be implemented using a PLA and three positive-edge-triggered D flip-flops. Because the outputs are asserted at time t_1 and deasserted at t_3, the λ output logic simply decodes states c ($y_1y_2y_3 = 111$) and f ($y_1y_2y_3 = 100$), asserting outputs z_1 and z_2, respectively. Glitches may occur on these Moore outputs, however, unless state codes are assigned, such that no state transition will generate a transient state equal to state c or state f.

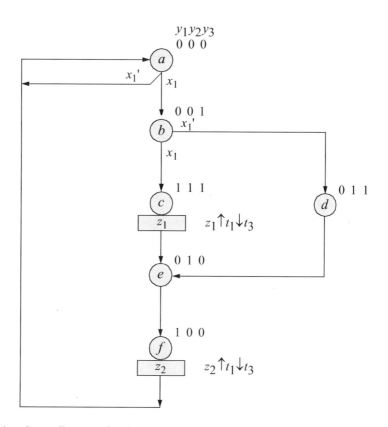

Figure 3.134 State diagram for the Moore machine of Example 3.21. Unused states are: $y_1y_2y_3 = 101$ and 110.

Using the rules for state code adjacency, states a and f should be adjacent, because the next state for both is state a. Also, states c and d should be adjacent, because they are both possible next states for state b and have the same next state, e. The state assignment shown in Figure 3.134 precludes the possibility of glitches on outputs z_1 and

z_2. This can be verified by checking all paths to determine if any state transition produces a transient state that is identical to the state codes for states c or f. This condition can occur only if two or more flip-flops change state for a particular state transition.

The path from state a to state b produces a change of state for only one flip-flop (y_3). Similarly, the path from state b to state d produces only one change — y_2 changes from 0 to 1. Although the transition from state b to state c produces two changes, flip-flop y_3 remains set — y_3 must be reset for the machine to enter state f and assert output z_2.

The path from state d to state e results in only one change of flip-flop variable (y_3). Both y_1 and y_3 change state when the machine proceeds from state c to state e; however, y_2 remains set, thus negating a glitch on z_2 in state f. The path from state e to state f produces a change to both y_1 and y_2, but flip-flop y_3 remains reset — y_3 must be set for the machine to enter state c and assert output z_1. Finally, the path from state f to state a occurs when only y_1 changes state. Thus, no state transition will produce a glitch on either output z_1 or z_2.

The input maps are shown in Figure 3.135 as obtained from the state diagram using input x_1 as a map-entered variable. A Karnaugh map yields a minimum sum-of-products expression, which is a requirement for generating output functions for a PLA device. Five product terms are required as shown in Equation 3.11.

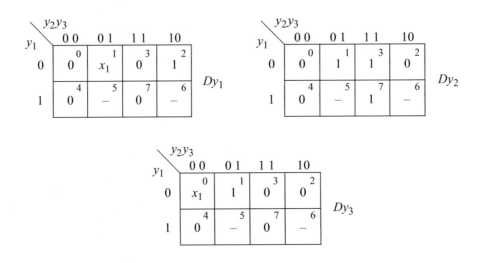

Figure 3.135   Input maps for the Moore machine of Figure 3.134.

$$Dy_1 = y_2'y_3x_1 + y_2y_3'$$

$$Dy_2 = y_3$$

$$Dy_3 = y_1'y_2'x_1 + y_2'y_3 \qquad (3.11)$$

The output maps are shown in Figure 3.136. The output equations can be minimized if the minterms for z_1 and z_2 are combined with unused minterm $y_1y_2y_3 = 101$ or 110. If these unused states are to be used for minimization, however, they must not function as transient states for any state sequence that does not include the corresponding output.

The only transition that may pass through unused state $y_1y_2y_3 = 101$ is the path from state b to state c. This presents no hazard for output z_1, however, because this sequence includes z_1. Output z_1 may be asserted slightly early, but no glitch will be generated. Therefore, a 1 can be inserted in $y_1y_2y_3 = 101$ in order to minimize the equation for z_1.

This is not true for z_2, however. The path from state b to state c does not include output z_2 in either the initial state or the destination state; therefore, a 0 must be inserted in state $y_1y_2y_3 = 101$ in the output map for z_2. The output equations are shown in Equation 3.12.

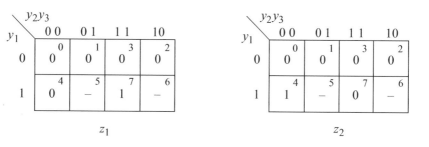

Figure 3.136   Output maps for the Moore machine of Figure 3.134.

$$z_1 = y_1y_3$$
$$z_2 = y_1y_2'y_3' \tag{3.12}$$

The logic diagram is shown in Figure 3.137 using a PLA with positive-edge-triggered D flip-flops. To obtain the logic function for Dy_1, Dy_2, and Dy_3, the AND array is programmed according to the product terms of Equation 3.11 and the OR array is programmed to obtain the appropriate sum-of-products for the respective Dy_i input. In the same manner, the AND and OR arrays are programmed to generate outputs z_1 and z_2 according to Equation 3.12.

The structural design module that represents the logic of a typical PAL device to implement the design is shown in Figure 3.138 using built-in primitives and instantiated D flip-flops. The test bench module is shown in Figure 3.139. The outputs and waveforms are shown in Figure 3.140 and Figure 3.141, respectively.

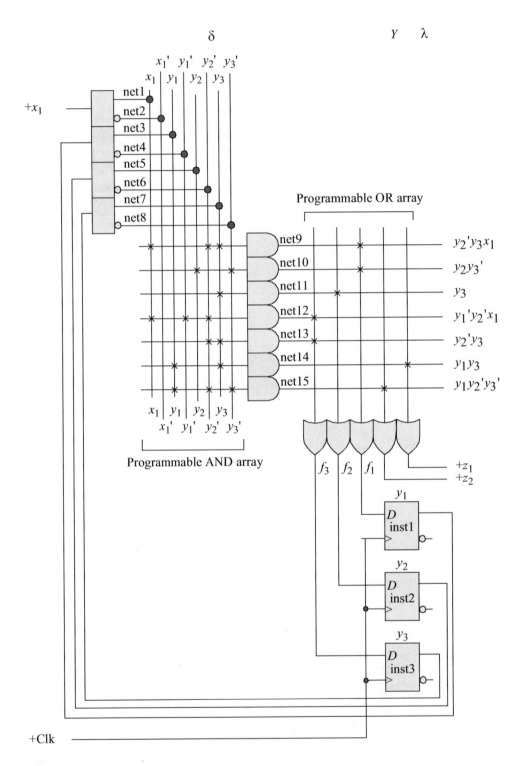

Figure 3.137   Implementation of the Moore machine of Figure 3.134 using a PLA device.

```
//structural pla to implement the three equations:
//Dy1 = y2'y3x1 + y2y3'
//Dy2 = y3
//Dy3 = y1'y2'x1 + y2'y3

module pla3 (rst_n, clk, x1, y, z1, z2);


//define inputs and outputs
input rst_n, clk, x1;
output [1:3] y;
output z1, z2;


//define internal nets
wire net1, net2, net3, net4, net5, net6, net7, net8, net9;
wire net10, net11, net12, net13, net14, net15, f1, f2, f3;


//define the input drivers
buf (net1, x1);
not (net2, x1);

buf (net3, y[1]);
not (net4, y[1]);

buf (net5, y[2]);
not (net6, y[2]);

buf (net7, y[3]);
not (net8, y[3]);


//define the logic for the and array, the or array, and y[1]
and (net9, net1, net6, net7);
and (net10, net5, net8);
or  (f1, net9, net10);

d_ff_bh inst1 (
   .rst_n(rst_n),
   .clk(clk),
   .d(f1),
   .q(y[1])
   );

                              //continued on next page
```

Figure 3.138   Structural design module for the Moore machine of Figure 3.134 using a PLA device.

```
//define the logic for the and array, the or array, and y[2]
and (net11, net7);
or  (f2, net11);

d_ff_bh inst2 (
   .rst_n(rst_n),
   .clk(clk),
   .d(f2),
   .q(y[2])
   );

//define the logic for the and array, the or array, and y[3]
and (net12, net1, net4, net6);
and (net13, net6, net7);
or  (f3, net12, net13);

d_ff_bh inst3 (
   .rst_n(rst_n),
   .clk(clk),
   .d(f3),
   .q(y[3])
   );

//define the logic for output z1
and (net14, net3, net7);
or  (z1, net14);

//define the logic for output z2
and (net15, net3, net6, net8);
or  (z2, net15);

endmodule
```

Figure 3.138   (Continued)

```
//test bench to implement the three equations of pla3
module pla3_tb;

//inputs are reg for test bench
//outputs are wire for test bench
reg rst_n, clk, x1;
wire [1:3] y;
wire z1, z2;                          //continued on next page
```

Figure 3.139   Test bench module for the Moore machine of Figure 3.134 using a PLA device.

```
//display variables
initial
$monitor ("x1 = %b, state = %b, z1 z2 = %b", x1, y, {z1, z2});

//define clock
initial
begin
   clk = 1'b0;
   forever
      #10 clk = ~clk;
end

//define input sequence
initial
begin
   #0 rst_n = 1'b0;//reset to state_a (000)
      x1 = 1'b0;

   #5 rst_n = 1'b1;

//--------------------------------------------------------
            @ (posedge clk)
   x1 = 1'b0;@ (posedge clk)      //go to state_a (000)
   x1 = 1'b1;@ (posedge clk)      //go to state_b (001)
   x1 = 1'b1;@ (posedge clk)      //go to state_c (111)
                                  //assert z1
   x1 = $random;@ (posedge clk)   //go to state_e (010)
   x1 = $random;@ (posedge clk)   //go to state_f (100)
                                  //assert z2
   x1 = $random;@ (posedge clk)   //go to state_a (000)

//--------------------------------------------------------
   x1 = 1'b1;@ (posedge clk)      //go to state_b (001)
   x1 = 1'b0;@ (posedge clk)      //go to state_d (011)
   x1 = $random;@ (posedge clk)   //go to state_e (010)
   x1 = $random;@ (posedge clk)   //go to state_f (100)
                                  //assert z2
   x1 = $random;@ (posedge clk)   //go to state_a (000)

   #10    $stop;
end

//--------------------------------------------------------
//instantiate the module into the test bench as a single line
pla3 inst1 (rst_n, clk, x1, y, z1, z2);

endmodule
```

Figure 3.139   (Continued)

```
x1 = 0, state = 000, z1 z2 = 00
x1 = 1, state = 000, z1 z2 = 00
x1 = 1, state = 001, z1 z2 = 00
x1 = 0, state = 111, z1 z2 = 10

x1 = 1, state = 010, z1 z2 = 00
x1 = 1, state = 100, z1 z2 = 01
x1 = 1, state = 000, z1 z2 = 00
x1 = 0, state = 001, z1 z2 = 00
x1 = 1, state = 011, z1 z2 = 00
x1 = 1, state = 010, z1 z2 = 00
x1 = 1, state = 100, z1 z2 = 01
x1 = 1, state = 000, z1 z2 = 00
```

Figure 3.140   Outputs for the Moore machine of Figure 3.134 using a PLA device.

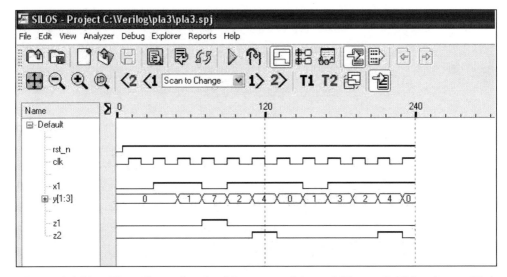

Figure 3.141   Waveforms for the Moore machine of Figure 3.134 using a PLA device.

3.4 Iterative Networks

An *iterative network* is a logical structure composed of identical cells. It is a cascade of identical combinational or sequential circuits (cells) in which the first or last cells may be different than the other cells in the network. Since an iterative network consists of identical cells, it is only necessary to design a typical cell, and then to replicate that cell for the entire network.

An iterative machine (or network) may consist of combinational logic arranged in a linear array in which signals between cells propagate in one direction only. A parity checker and comparator are examples of combinational iterative networks. Or, the iterative network may consist of sequential cells, such as found in shift registers and simple binary counters.

Example 3.22   This example demonstrates a method to design a single-bit detection circuit. In this example, a typical cell will be designed, then instantiated four times into a higher-level module to detect a single bit in a 4-bit input vector $x[1:4]$. Figure 3.142 shows the block diagram of a typical cell and Figure 3.143 shows the internal logic of the cell, which will be instantiated four times into the higher-level circuit of Figure 3.144.

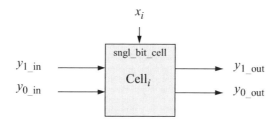

Figure 3.142   Typical cell for a single-bit detection circuit that will be instantiated four times into a higher-level structural module.

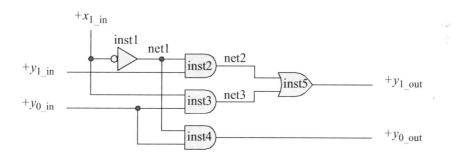

Figure 3.143   Internal logic for a typical cell for the single-bit detection circuit of Example 3.22.

In a combinational network, the operation is complete after an appropriate propagation delay, at which time the outputs are stable. In a functionally equivalent sequential machine, the operation is complete only when all inputs have been sequenced through the machine and the outputs have stabilized. Thus, k clock cycles are required to establish the final machine outputs, where k is the number of input sets. That is, one clock is necessary to process each set of inputs x_{ij}, where $1 \leq i \leq k$ and $1 \leq j \leq n$.

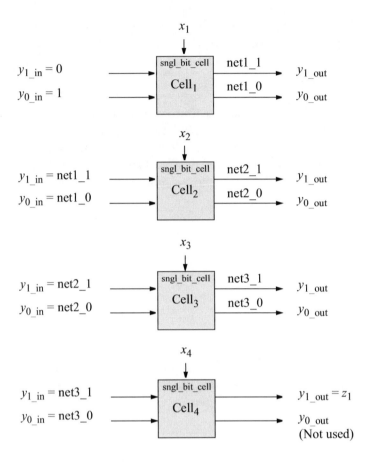

Figure 3.144   Block diagram to detect a single 1 bit in a 4-bit input vector $x[1:4]$.

In Figure 3.144, the input and output lines are defined as follows:

- y_{1_in} is an active-high input line indicating that a single 1 bit was detected up to that cell.
- y_{0_in} is an active-high input line indicating that no 1 bits were detected up to that cell.
- y_{1_out} is an active-high output line indicating that a single 1 bit was detected up to and including that cell.
- y_{0_out} is an active-high output line indicating that no 1 bits were detected up to and including that cell.

The structural design module for the typical cell is shown in Figure 3.145 using built-in logic primitives. The structural design module for the single-bit detection circuit and the test bench module are shown in Figure 3.146 and Figure 3.147, respectively. The outputs are shown in Figure 3.148.

```
//typical cell for single-bit detection
module sngl_bit_cell (x1_in, y1_in, y0_in, y1_out, y0_out);


input x1_in, y1_in, y0_in;
output y1_out, y0_out;


not  inst1 (net1, x1_in);
and  inst2 (net2, net1, y1_in);
and  inst3 (net3, x1_in, y0_in);
and  inst4 (y0_out, net1, y0_in);
or   inst5 (y1_out, net2, net3);


endmodule
```

Figure 3.145   Typical cell that is instantiated four times into a structural module to detect a single 1 bit in an input vector *x[1:4]*.

```
//structural single-bit detection module
module sngl_bit_detect2 (x1, x2, x3, x4, z1);

input x1, x2, x3, x4;
output z1;

//instantiate the single-bit cell modules
//cell 1
sngl_bit_cell  inst1(
   .x1_in(x1),
   .y1_in(1'b0),
   .y0_in(1'b1),
   .y1_out(net1_1),
   .y0_out(net1_0)
   );

//cell 2
sngl_bit_cell  inst2(
   .x1_in(x2),
   .y1_in(net1_1),
   .y0_in(net1_0),
   .y1_out(net2_1),
   .y0_out(net2_0)
   );
                              //continued on next page
```

Figure 3.146   Structural design module to detect a single 1 bit in a 4-bit input vector *x[1:4]* in which the typical cell of Figure 3.145 is instantiated four times.

```
//cell 3
sngl_bit_cell   inst3(
   .x1_in(x3),
   .y1_in(net2_1),
   .y0_in(net2_0),
   .y1_out(net3_1),
   .y0_out(net3_0)
   );

//cell 4
sngl_bit_cell   inst4(
   .x1_in(x4),
   .y1_in(net3_1),
   .y0_in(net3_0),
   .y1_out(z1)
   );
endmodule
```

Figure 3.146   (Continued)

```
//test bench for the single-bit detection module
module sngl_bit_detect2_tb;

reg x1, x2, x3, x4;
wire z1;

initial      //apply input vectors
begin: apply_stimulus
   reg [4:0] invect;
   for (invect=0; invect<16; invect=invect+1)
      begin
         {x1, x2, x3, x4} = invect [4:0];
         #10 $display ("x1x2x3x4 = %b, z1 = %b",
                       {x1, x2, x3, x4}, z1);
      end
end
//instantiate the module into the test bench
sngl_bit_detect2 inst1 (
   .x1(x1),
   .x2(x2),
   .x3(x3),
   .x4(x4),
   .z1(z1)
   );
endmodule
```

Figure 3.147   Test bench module for the single-bit detector.

```
x1x2x3x4  =  0000,  z1  =  0          x1x2x3x4  =  1000,  z1  =  1
x1x2x3x4  =  0001,  z1  =  1          x1x2x3x4  =  1001,  z1  =  0
x1x2x3x4  =  0010,  z1  =  1          x1x2x3x4  =  1010,  z1  =  0
x1x2x3x4  =  0011,  z1  =  0          x1x2x3x4  =  1011,  z1  =  0
x1x2x3x4  =  0100,  z1  =  1          x1x2x3x4  =  1100,  z1  =  0
x1x2x3x4  =  0101,  z1  =  0          x1x2x3x4  =  1101,  z1  =  0
x1x2x3x4  =  0110,  z1  =  0          x1x2x3x4  =  1110,  z1  =  0
x1x2x3x4  =  0111,  z1  =  0          x1x2x3x4  =  1111,  z1  =  0
```

Figure 3.148   Outputs for the single-bit detection module of Figure 3.146.

Example 3.23   This example repeats Example 3.22 for a single-bit detection; however, built-in primitives will be used to design the four cells using the logic diagram of Figure 3.143 for each cell. The general equations for $cell_i$ are shown in Equation 3.13. The input vector consists of four bits: $x_1 x_2 x_3 x_4$. The equations for the single-bit detector are shown in Equation 3.14 for each cell. As stated previously, the first and last cell of the network may be different than the other cells. This is evident in the logic diagram of the iterative network shown in Figure 3.149.

$$y_{i(1)} = x_i' y_{i-1(1)} + x_i y_{i-1(0)}$$

$$y_{i(0)} = x_i' y_{i-1(0)} \tag{3.13}$$

Bit 1 cell $y_{1(1)} = x_1$ One 1

$\quad\quad\quad\quad\quad y_{1(0)} = x_1'$ No 1s

Bit 2 cell $y_{2(1)} = y_{1(1)} x_2' + y_{1(0)} x_2$ One 1

$\quad\quad\quad\quad\quad y_{2(0)} = y_{1(0)} x_2'$ No 1s

Bit 3 cell $y_{3(1)} = y_{2(1)} x_3' + y_{2(0)} x_3$ One 1

$\quad\quad\quad\quad\quad y_{3(0)} = y_{2(0)} x_3'$ No 1s

Bit 4 cell $y_{4(1)} = y_{3(1)} x_4' + y_{3(0)} x_4$ One 1 to assert output z_1

$\quad\quad\quad\quad\quad y_{4(0)} = y_{3(0)} x_4'$ No 1s (3.14)

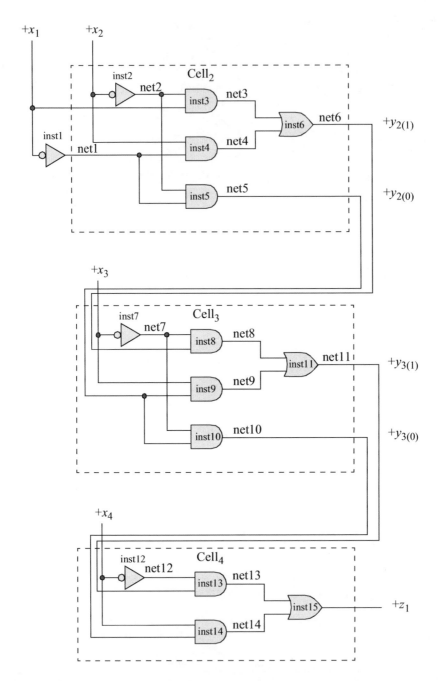

Figure 3.149   Iterative network to detect a single 1 bit in an input vector of $x_1 x_2 x_3 x_4$.

Referring to the iterative network logic diagram, the leftmost cell ($cell_1$) is simply an inverter. $Cell_2$ and $cell_3$ are identical, but $cell_4$ is different than the other cells. The output of $cell_4$, indicating that the input vector contained a single 1 bit, is z_1 from instantiation *inst15*. Output z_1 will be a logical 1 if the previous cells detected a single 1 bit and $x_4 = 0$ or if the previous cells detected all 0s and $x_4 = 1$.

The structural design module is shown in Figure 3.150 using built-in primitives. The test bench, outputs, and waveforms are shown in Figure 3.151, Figure 3.152, and Figure 3.153, respectively.

```
//single bit detector
module sngl_bit_detect (x1, x2, x3, x4, z1);

input x1, x2, x3, x4;
output z1;

//cell 1 **********************************************
not    inst1    (net1, x1);

//cell 2 **********************************************
not    inst2    (net2, x2);
and    inst3    (net3, net2, x1);
and    inst4    (net4, x2, net1);
and    inst5    (net5, net2, net1);
or     inst6    (net6, net3, net4);

//cell 3 **********************************************
not    inst7    (net7, x3);
and    inst8    (net8, net7, net6);
and    inst9    (net9, x3, net5);
and    inst10   (net10, net7, net5);
or     inst11   (net11, net8, net9);

//cell 4 **********************************************
not    inst12   (net12, x4);
and    inst13   (net13, net12, net11);
and    inst14   (net14, x4, net10);
or #1  inst15   (z1, net13, net14);

endmodule
```

Figure 3.150 Structural design module for the single-bit detector iterative network of Figure 3.149.

```
//test bench for single bit detection
module sngl_bit_detect_tb;

reg x1, x2, x3, x4;
wire z1;

initial          //apply input vectors
begin: apply_stimulus
   reg [4:0] invect;
   for (invect=0; invect<16; invect=invect+1)
      begin
         {x1, x2, x3, x4} = invect [4:0];
         #10 $display ("x1x2x3x4 = %b, z1 = %b",
                        {x1, x2, x3, x4}, z1);
      end
end


//instantiate the module into the test bench
sngl_bit_detect inst1 (
   .x1(x1),
   .x2(x2),
   .x3(x3),
   .x4(x4),
   .z1(z1)
   );

endmodule
```

Figure 3.151   Test bench module for the single-bit detector iterative network of Figure 3.149.

```
x1x2x3x4 = 0000, z1 = 0        x1x2x3x4 = 1000, z1 = 1
x1x2x3x4 = 0001, z1 = 1        x1x2x3x4 = 1001, z1 = 0
x1x2x3x4 = 0010, z1 = 1        x1x2x3x4 = 1010, z1 = 0
x1x2x3x4 = 0011, z1 = 0        x1x2x3x4 = 1011, z1 = 0
x1x2x3x4 = 0100, z1 = 1        x1x2x3x4 = 1100, z1 = 0
x1x2x3x4 = 0101, z1 = 0        x1x2x3x4 = 1101, z1 = 0
x1x2x3x4 = 0110, z1 = 0        x1x2x3x4 = 1110, z1 = 0
x1x2x3x4 = 0111, z1 = 0        x1x2x3x4 = 1111, z1 = 0
```

Figure 3.152   Outputs for the single-bit detector iterative network of Figure 3.149.

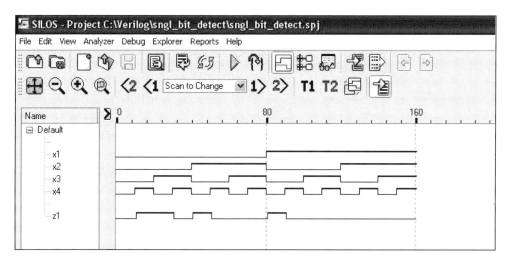

Figure 3.153   Waveforms for the single-bit detector iterative network of Figure 3.149.

Example 3.24   As a final example for iterative networks, Example 3.23 will be repeated to illustrate an alternative method to design a single-bit detector. In this implementation, all four cells are identical, including the first cell and the fourth cell, as shown in Figure 3.154. The logic diagram is shown in Figure 3.155.

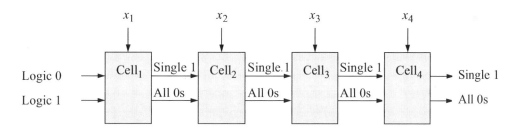

Figure 3.154   Single-bit detector using identical cells.

The structural design module is shown in Figure 3.156 using built-in primitives. The OR built-in primitive for cell 4 delays output z_1 by one time unit to avoid possible glitches. The test bench module is shown in Figure 3.157 using a **for** loop, which executes a procedural statement or a block of procedural statements a specified number of times. The outputs and waveforms are shown in Figure 3.158 and Figure 3.159, respectively.

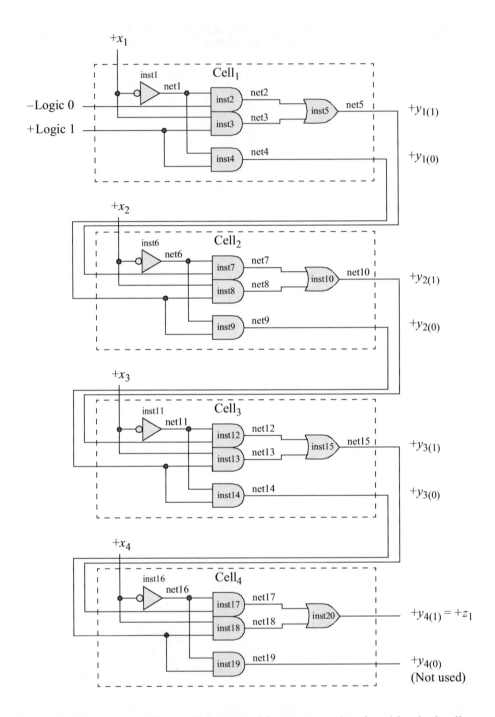

Figure 3.155   Logic diagram for a single-bit detector using four identical cells.

```
//single bit detector
module sngl_bit_detect1 (x1, x2, x3, x4, z1);


input x1, x2, x3, x4;
output z1;


//cell 1 *********************************************
not   inst1    (net1, x1);
and   inst2    (net2, net1, 1'b0);
and   inst3    (net3, x1, 1'b1);
and   inst4    (net4, net1, 1'b1);
or    inst5    (net5, net2, net3);


//cell 2 *********************************************
not   inst6    (net6, x2);
and   inst7    (net7, net6, net5);
and   inst8    (net8, x2, net4);
and   inst9    (net9, net6, net4);
or    inst10   (net10, net7, net8);


//cell 3 *********************************************
not   inst11   (net11, x3);
and   inst12   (net12, net11, net10);
and   inst13   (net13, x3, net9);
and   inst14   (net14, net11, net9);
or    inst15   (net15, net12, net13);


//cell 4 *********************************************
not   inst16   (net16, x4);
and   inst17   (net17, net16, net15);
and   inst18   (net18, x4, net14);
and   inst19   (net19, net16, net14);
or #1 inst20   (z1, net17, net18);

endmodule
```

Figure 3.156   Design module for the single-bit detection circuit of Figure 3.155 using four identical cells.

```
//test bench for single bit detection
module sngl_bit_detect1_tb;

reg x1, x2, x3, x4;
wire z1;


//apply input vectors
initial
begin: apply_stimulus
   reg [4:0] invect;
   for (invect=0; invect<16; invect=invect+1)
      begin
         {x1, x2, x3, x4} = invect [4:0];
         #10 $display ("x1x2x3x4 = %b, z1 = %b",
                       {x1, x2, x3, x4}, z1);
      end
end


//instantiate the module into the test bench
sngl_bit_detect1 inst1 (
   .x1(x1),
   .x2(x2),
   .x3(x3),
   .x4(x4),
   .z1(z1)
   );

endmodule
```

Figure 3.157   Test bench for the single-bit detection module of Figure 3.156.

```
x1x2x3x4 = 0000, z1 = 0        x1x2x3x4 = 1000, z1 = 1
x1x2x3x4 = 0001, z1 = 1        x1x2x3x4 = 1001, z1 = 0
x1x2x3x4 = 0010, z1 = 1        x1x2x3x4 = 1010, z1 = 0
x1x2x3x4 = 0011, z1 = 0        x1x2x3x4 = 1011, z1 = 0
x1x2x3x4 = 0100, z1 = 1        x1x2x3x4 = 1100, z1 = 0
x1x2x3x4 = 0101, z1 = 0        x1x2x3x4 = 1101, z1 = 0
x1x2x3x4 = 0110, z1 = 0        x1x2x3x4 = 1110, z1 = 0
x1x2x3x4 = 0111, z1 = 0        x1x2x3x4 = 1111, z1 = 0
```

Figure 3.158   Outputs for the single-bit detection module of Figure 3.156.

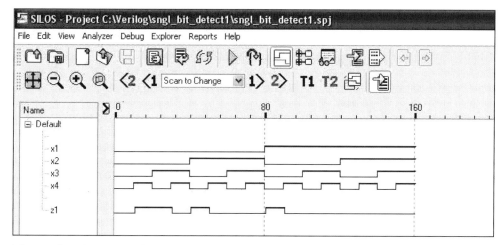

Figure 3.159   Waveforms for the single-bit detection module of Figure 3.156.

3.5   Error Detection in Synchronous Sequential Machines

Before discussing error detection in synchronous sequential machines, a brief presentation will be given on general error detection and correction. Transferring data within a computer or between computers is subject to error, either permanent or transient. Permanent errors can be caused by hardware malfunctions; transient errors can be caused by transmission errors due to noise. In either case, the data error must at least be detected and preferably corrected.

3.5.1   Overview of Error Detection and Correction

Parity   An extra bit can be added to a message to make the overall parity of the code word either odd or even; that is, the number of 1s in the code word — message bits plus parity bit — will be either odd or even. The parity bit to maintain even parity for a 4-bit message $x_1 x_2 x_3 x_4$ can be generated by modulo-2 addition, as shown in Equation 3.15. The parity bit for odd parity generation is the complement of Equation 3.15, which is shown in Equation 3.16 and designed in Figure 3.160.

$$\text{Parity bit (even)} = x_1 \oplus x_2 \oplus x_3 \oplus x_4 \qquad (3.15)$$

$$\text{Parity bit (odd)} = (x_1 \oplus x_2 \oplus x_3 \oplus x_4)' \qquad (3.16)$$

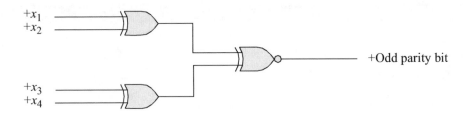

Figure 3.160   Odd parity generator shown in Equation 3.16.

Parity implementation can detect an odd number of errors, but cannot correct the errors, because the bits in error cannot be determined. If a single error occurred, then an incorrect code word would be generated and the error would be detected. If two errors occurred, then parity would be unchanged and still correct. Every adjacent pair of code words — message plus parity — has a minimum distance of two; that is, they differ in two bit positions. This means that two bits must change to still maintain a correct code word.

Hamming code   Richard W. Hamming developed a code in 1950 that resolves the problem associated with parity implementation. The *Hamming code* can be considered as an extension of the parity code, because multiple parity bits provide parity for subsets of the message bits. The subsets overlap, such that each message bit is contained in at least two subsets. The basic Hamming code can detect single or double errors and can correct a single error.

A *code word* contains n bits consisting of m message bits plus k parity check bits as shown in Figure 3.161. The m bits represent the information or message part of the code word; the k bits are used for detecting and correcting errors, where $k = n - m$.

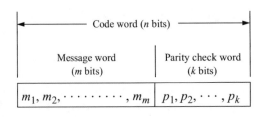

Figure 3.161   Code word of n bits containing m message bits and k parity check bits.

Since there can be an error in *any* bit position, including the parity check bits, there must be a sufficient number of k parity check bits to identify any of the $m + k$ bit positions. The parity check bits are normally embedded in the code word and are positioned in columns with column numbers that are a power of two, as shown below for a code word containing four message bits (m_3, m_5, m_6, m_7) and three parity bits (p_1, p_2, p_4).

Column number	1	2	3	4	5	6	7
Code word =	p_1	p_2	m_3	p_4	m_5	m_6	m_7

Each parity bit maintains odd parity over a unique group of bits as shown below for a code word of four message bits.

$$E_1 = \quad p_1 \quad m_3 \quad m_5 \quad m_7$$
$$E_2 = \quad p_2 \quad m_3 \quad m_6 \quad m_7$$
$$E_4 = \quad p_4 \quad m_5 \quad m_6 \quad m_7$$

The placement of the parity bits in certain columns is not arbitrary. Each of the variables in group E_1 contains a 1 in column 1 (2^0) of the binary representation of the column number as shown below.

	8	4	2	1
Group E_1	2^3	2^2	2^1	2^0
p_1	0	0	0	1
m_3	0	0	1	1
m_5	0	1	0	1
m_7	0	1	1	1
. . .				

Since p_1 has only a single 1 in the binary representation of column 1, p_1 can therefore be used as a parity check bit for a message bit in *any* column in which the binary representation of the column number has a 1 in column 1 (2^0). Thus, group E_1 can be expanded to include other message bits, as shown below.

$$p_1, m_3, m_5, m_7, m_9, m_{11}, m_{13}, m_{15}, m_{17}, \cdots$$

In a similar manner, each of the variables in group E_2 contains a 1 in column 2 (2^1) of the binary representation of the column number. Thus, group E_2 can be expanded to include other message bits, as shown below.

$$p_2, m_3, m_6, m_7, m_{10}, m_{11}, m_{14}, m_{15}, m_{18}, \cdots$$

Each of the variables in group E_4 contains a 1 in column 4 (2^2) of the binary representation of the column number. Since p_4 has only a single 1 in the binary representation of column 4, p_4 can therefore be used as a parity check bit for a message bit in *any* column in which the binary representation of the column number has a 1 in column 4 (2^2). Thus, group E_4 can be expanded to include other message bits, as shown below.

$$p_4, m_5, m_6, m_7, m_{12}, m_{13}, m_{14}, m_{15}, m_{20}, \cdots$$

The format for embedding parity bits in the code word can be extended easily to any size message. For example, the code word for an 8-bit message is encoded as follows:

$$p_1, p_2, m_3, p_4, m_5, m_6, m_7, p_8, m_9, m_{10}, m_{11}, m_{12}$$

where $m_3, m_5, m_6, m_7, m_9, m_{10}, m_{11}, m_{12}$ are the message bits and p_1, p_2, p_4, p_8 are the parity check bits for groups E_1, E_2, E_4, E_8, respectively.

Example 3.25   A 4-bit message (0110) will be encoded using the Hamming code then transmitted. The message, transmitted code word, and received code word are shown below.

	p_1	p_2	m_3	p_4	m_5	m_6	m_7
Message to be sent			0		1	1	0
Code word sent	0	0	0	1	1	1	0
Code word received	0	0	0	1	0	1	0

From the received code word, it is seen that bit 5 is in error. When the code word is received, the parity of each group is checked using the bits assigned to that group, as shown below.

Group $E_1 =$	$p_1\ m_3\ m_5\ m_7 =$	$0\ 0\ 0\ 0 =$	Error $=$ 1
Group $E_2 =$	$p_2\ m_3\ m_6\ m_7 =$	$0\ 0\ 1\ 0 =$	No error $=$ 0
Group $E_4 =$	$p_4\ m_5\ m_6\ m_7 =$	$1\ 0\ 1\ 0 =$	Error $=$ 1

A parity error is assigned a value of 1; no parity error is assigned a value of 0. The groups are then listed according to their binary weight. The resulting binary number is called the *syndrome word* and indicates the bit in error; in this case, bit 5. The bit in error is then complemented to yield a correct message of 0110.

Cyclic redundancy check code A class of codes has been developed specifically for serial data transfer called cyclic redundancy check (CRC) codes. This section will provide a brief introduction to the CRC codes. Cyclic redundancy check codes can detect both single-bit errors and multiple-bit errors and are especially useful for large strings of serial binary data found on single-track storage devices, such as disk drives. They are also used in serial data transmission networks and in 9-track magnetic tape systems, where each track is treated as a serial bit stream.

The generation of a CRC character uses modulo-2 addition, which is a linear operation; therefore, a linear feedback shift register is used in its implementation. The CRC character that is generated is placed at or near the end of the message.

A possible track format for a disk drive is shown in Figure 3.162, which has separate address and data fields. There is a CRC character for each of the two fields; the CRC character in the address field checks the cylinder, head, and sector addresses; the CRC character in the data field checks the data stream.

The address field and data field both have a *preamble* and a *postamble*. The preamble consists of fifteen 0s followed by a single 1 and is used to synchronize the clock to the data and to differentiate between 0s and 1s. The postamble consists of sixteen 0s and is used to separate the address and data fields.

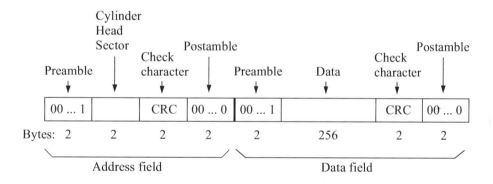

Figure 3.162 Possible track format for a disk drive.

Checksum The *checksum* is the sum derived from the application of an algorithm that is calculated before and after transmission to ensure that the data is free from errors. The checksum character is a numerical value that is based on the number of asserted bits in the message and is appended to the end of the message. The receiving unit then applies the same algorithm to the message and compares the results with the appended checksum character.

There are many versions of checksum algorithms — one version is described next. If the information consists of n bytes of data, then a simple checksum algorithm is to perform modulo-256 addition on the bytes in the message. The sum thus obtained is the checksum byte and is appended to the last byte creating a message of $n + 1$ bytes.

The receiving unit then regenerates the checksum by obtaining the sum of the first n bytes and compares that sum to byte $n + 1$. This method can detect a single byte error.

Alternatively, the sum that is obtained by modulo-256 addition in the transmitting unit is 2s complemented and becomes the checksum character which is appended to the end of the transmitted message. The receiving unit uses the same algorithm and adds the recalculated uncomplemented checksum character to the transmitted checksum character, resulting in a sum of zero if the message had no errors. An example of this algorithm is shown in Figure 3.163 for a 4-byte message.

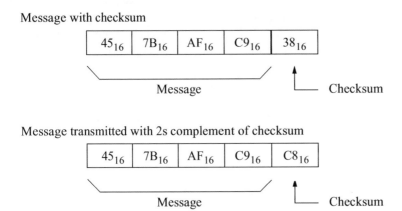

Figure 3.163   Checksum generated for a 4-byte message.

Two-out-of five code   The two-out-of-five code is 5-bit nonweighted code that is characterized by having exactly two 1s and three 0s in any code word — $0 = 00011$, 00101, 01001, 10001, 00110, 01010, 10010, 01100, 10100, $9 = 11000$. This code has a minimum distance of two and it is relatively easy to provide error detection by counting the number of 1s in a code word.

An error is detected whenever the number of 1s in a code word is not equal to two. This can result from a change of one or more bits which cause the total number of 1s to differ from two. However, an error will be undetected if there are two simultaneous bit changes which result in a valid code word with two 1s. For example, if code word 01100 (7) were changed during transmission to 01010 (5), then the error would be undetected. The two-out-of-five code is representative of m-out-of-n codes.

3.5.2 Examples of Error Detection in Synchronous Sequential Machines

This section describes techniques of error detection for synchronous sequential machines, such as counters and Moore machines. Where applicable, the designs will include the state diagram and the logic diagram. All designs will be implemented in Verilog. It is assumed that the design procedure resulted in a correctly synthesized machine which operates reliably according to the machine specifications and has no output glitches.

Example 3.26   The synthesis of a parity-checked counter will now be presented. If each state code differs by only one bit between contiguous state codes, then parity checking is easily implemented. The Gray code meets this requirement. The state diagram for a 3-bit Gray code counter is shown in Figure 3.164 together with the state code assignment and the corresponding state of the parity flip-flop.

The flip-flop y_p is the parity flip-flop which maintains odd parity for the four flip-flops $y_1 y_2 y_3 y_p$. The parity flip-flop is set to a value of 1 initially and will change state with each clock pulse. If an incorrect state transition occurs, then the parity of $y_1 y_2 y_3 y_p$ will be even and an error will be indicated. Parity is checked by means of a parity checker $2k + 1$, which is an odd parity circuit, as shown in the logic diagram of Figure 3.165.

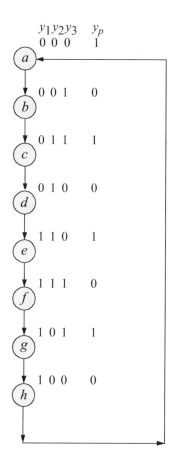

Figure 3.164   State diagram for the parity-checked 3-bit Gray code counter.

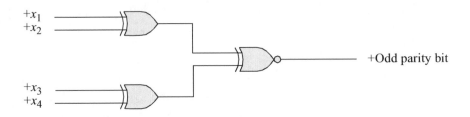

Figure 3.165   Logic diagram for a 4-bit odd parity generator.

The input maps are derived from the state diagram and are shown in Figure 3.166 using D flip-flops. The input equations are listed in Equation 3.17. The logic diagram is shown in Figure 3.167. The error signal (+error) can be stored in a flip-flop to indicate that an error has occurred. The counter logic can then be examined in order to correct the error. In order to minimize the possibility of double errors, the term y_2y_3' in the equations for Dy_1 and Dy_2 will be duplicated. Thus, a fault in the logic represented by the term y_2y_3' in flip-flop y_1 or y_2, but not both, will result in an incorrect state transition for only one flip-flop.

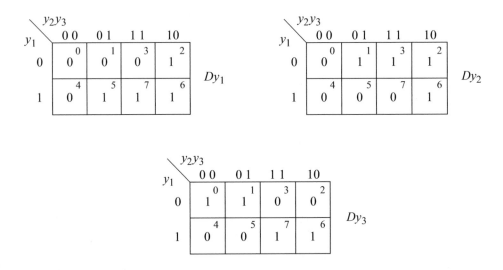

Figure 3.166   Input maps for the parity-checked Gray code counter.

$$Dy_1 = y_1y_3 + y_2y_3'$$
$$Dy_2 = y_1'y_3 + y_2y_3'$$
$$Dy_3 = y_1'y_2' + y_1y_2 \qquad\qquad (3.17)$$

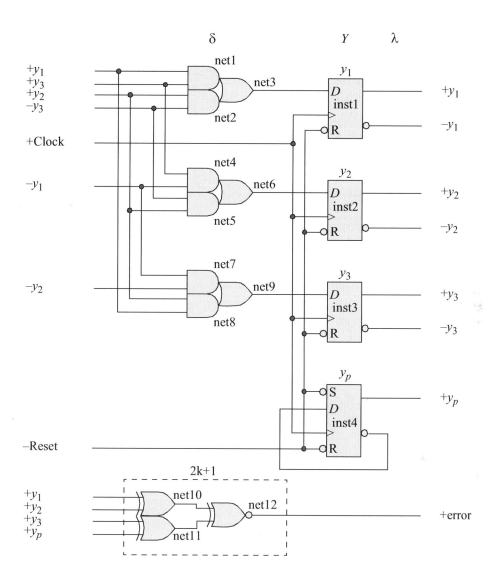

Figure 3.167   Logic diagram for the parity-checked 3-bit Gray code counter.

The structural design module is shown in Figure 3.168 using built-in primitives and instantiated D flip-flops. The test bench module is shown in Figure 3.169, which takes the counter through the Gray code sequence for three bits. The outputs are shown in Figure 3.170. The initial reset signal resets all flip-flops to a 0000 state — causing an error signal to be asserted. However, when the counter begins a new sequence at $y_1 y_2 y_3 = 000$, the parity flip-flop is asserted, resulting in no parity error. The waveforms are shown in Figure 3.171.

```
//structural parity checked gray code counter

module gray_code_ctr_par_chk (rst_n, clk, y, yp, error);

//define inputs and outputs
input rst_n, clk;
output [1:3] y;
output yp, error;

//define internal nets
wire net1, net2, net3, net4, net5, net6, net7,
     net8, net9, net10, net11, net12;

//---------------------------------------------
//define the logic for flip-flop y[1]
and (net1, y[1], y[3]);
and (net2, y[2], ~y[3]);
or  (net3, net1, net2);

d_ff_bh inst1 (
   .rst_n(rst_n),
   .clk(clk),
   .d(net3),
   .q(y[1])
   );

//---------------------------------------------
//define the logic for flip-flop y[2]
and (net4, y[3], ~y[1]);
and (net5, ~y[3], y[2]);
or  (net6, net4, net5);

d_ff_bh inst2 (
   .rst_n(rst_n),
   .clk(clk),
   .d(net6),
   .q(y[2])
   );

//---------------------------------------------
//define the logic for flip-flop y[3]
and (net7, ~y[1], ~y[2]);
and (net8, y[1], y[2]);
or  (net9, net7, net8);

                          //continued on next page
```

Figure 3.168   Structural design module for a 3-bit, parity-checked Gray code counter.

```
d_ff_bh inst3 (
    .rst_n(rst_n),
    .clk(clk),
    .d(net9),
    .q(y[3])
    );

//-------------------------------------------
//define the logic for the parity flip-flop yp
d_ff inst4 (
    .rst_n(rst_n),
    .set_n(rst_n),
    .clk(clk),
    .d(~yp),
    .q(yp)
    );

//-------------------------------------------
//define the logic (2k+1) to detect a parity error
xor (net10, y[1], y[2]);
xor (net11, y[3], yp);
xnor (error, net10, net11);

endmodule
```

Figure 3.168 (Continued)

```
//test bench for the gray code parity-checked counter
module gray_code_ctr_par_chk_tb;

reg rst_n, clk;        //inputs are reg for test bench
wire [1:3] y;          //outputs are wire for test bench
wire yp, error;

//display outputs
initial
$monitor ("count = %b, parity = %b, error = %b", y, yp, error);

initial                //define reset
begin
    #0     rst_n = 1'b0;
    #5     rst_n = 1'b1;
end                                    //continued on next page
```

Figure 3.169 Test bench module for a 3-bit, parity-checked Gray code counter.

```
//define clock
initial
begin
   clk = 1'b0;
   forever
      #10  clk = ~clk;
end


//define length of simulation
initial
begin
   #150   $finish;
end


//instantiate the module into the test bench
gray_code_ctr_par_chk inst1 (
   .rst_n(rst_n),
   .clk(clk),
   .y(y),
   .yp(yp),
   .error(error)
   );

endmodule
```

Figure 3.169   (Continued)

```
count = 000, parity = 0, error = 1
count = 001, parity = 0, error = 0
count = 011, parity = 1, error = 0
count = 010, parity = 0, error = 0
count = 110, parity = 1, error = 0
count = 111, parity = 0, error = 0
count = 101, parity = 1, error = 0
count = 100, parity = 0, error = 0

count = 000, parity = 1, error = 0
```

Figure 3.170   Outputs for a 3-bit, parity-checked Gray code counter.

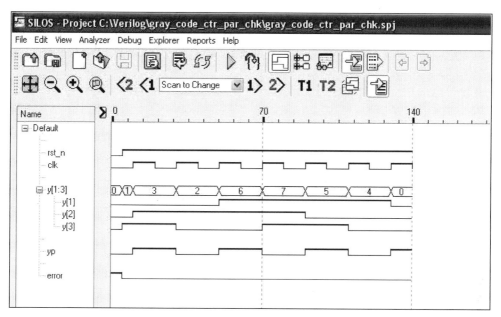

Figure 3.171   Waveforms for a 3-bit, parity-checked Gray code counter.

Example 3.27 This example synthesizes a Moore machine using structural modeling that accepts serial data in the form of 3-bit words on an input line x_1. There is one bit space between contiguous words, as shown below, where $b_i = 0$ or 1. Whenever a word contains the bit pattern $b_1 b_2 b_3 = 111$, the machine will assert output z_1 during the bit time between words. The assertion/deassertion will be as follows: $z_1 \uparrow t_2 \downarrow t_3$.

$$x_1 = \quad \cdots \quad \left| b_1 b_2 b_3 \right| \quad \left| b_1 b_2 b_3 \right| \quad \left| b_1 b_2 b_3 \right| \quad \cdots$$

An example of a valid word in a series of words is shown below. Notice that the output signal is displaced in time with respect to the input sequence and occurs one state time later.

$$x_1 = \quad \cdots \quad \left| 0\ 0\ 1 \right| \quad \left| 1\ 0\ 1 \right| \quad \left| 0\ 1\ 1 \right| \quad \left| 1\ 1\ 1 \right| \quad \left| 0\ 1\ 0 \right| \quad \cdots$$

$$\text{Output } z_1 \uparrow t_2 \downarrow t_3 \ \underline{\hspace{1cm}} \rule{0pt}{1em} \underline{\hspace{0.5cm}}$$

The state diagram is shown in Figure 3.172 and includes the parity flip-flop y_p which maintains odd parity over the four flip-flops y_1, y_2, y_3, and y_p. The input maps for all four flip-flops are shown in Figure 3.173 using input x_1 as a map-entered variable. The input equations and output equation are shown in Equation 3.18.

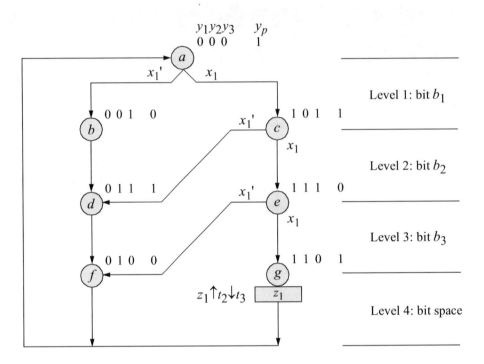

Figure 3.172   State diagram for the Moore machine using D flip-flops with adjacent state codes. Output z_1 is asserted whenever a 3-bit word $x_1 = 111$. There is one unused state (100).

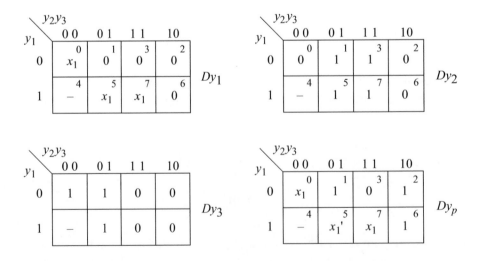

Figure 3.173   Input maps for the Moore machine of Figure 3.172.

$$Dy_1 = y_2'y_3'x_1 + y_1y_3x_1$$

$$Dy_2 = y_3$$

$$Dy_3 = y_2'$$

$$Dy_p = y_1'y_2'x_1 + y_2'y_3x_1' + y_1y_2x_1 + y_2y_3'$$

$$z_1 = y_1y_3'\text{clk}' \qquad (3.18)$$

The logic diagram is shown in Figure 3.174. The structural design module is shown in Figure 3.175 using built-in primitives and instantiated D flip-flops. The $2k+1$ logic uses two exclusive-OR circuits and one exclusive-NOR circuit. The test bench is shown in Figure 3.176, which takes the machine through two sequences. The outputs and waveforms are shown in Figure 3.177 and Figure 3.178, respectively.

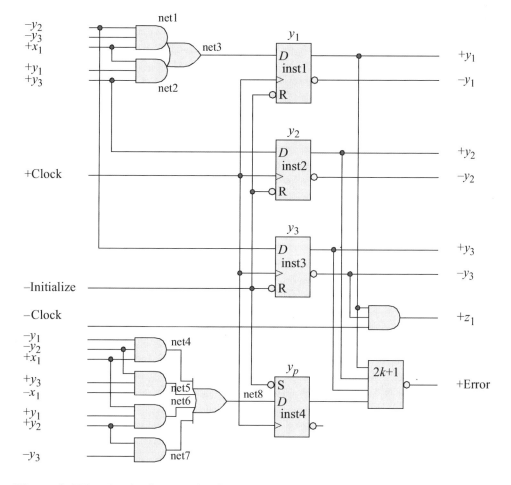

Figure 3.174   Logic diagram for the Moore machine of Figure 3.172.

```
//structural parity-checked moore machine

module moore_par_chk (rst_n, clk, x1, y, yp, z1, error);

//define inputs and outputs
input rst_n, clk, x1;
output [1:3] y;
output z1, yp, error;

//define internal nets
wire net1, net2, net3, net4, net5, net6, net7, net8;

//------------------------------------------------
//define the logic for flip-flop y[1]
and (net1, ~y[2], ~y[3], x1);
and (net2, x1, y[1], y[3]);
or  (net3, net1, net2);

d_ff_bh inst1 (
   .rst_n(rst_n),
   .clk(clk),
   .d(net3),
   .q(y[1])
   );

//------------------------------------------------
//define the logic for flip-flop y[2]
d_ff_bh inst2 (
   .rst_n(rst_n),
   .clk(clk),
   .d(y[3]),
   .q(y[2])
   );

//------------------------------------------------
//define the logic for flip-flop y[3]
d_ff_bh inst3 (
   .rst_n(rst_n),
   .clk(clk),
   .d(~y[2]),
   .q(y[3])
   );

//------------------------------------------------
                              //continued on next page
```

Figure 3.175   Structural design module for the Moore machine of Figure 3.172 including error checking logic using a parity flip-flop y_p.

```
//-------------------------------------------------
//define the logic for flip-flop yp
and (net4, ~y[1], ~y[2], x1);
and (net5, ~y[2], y[3], ~x1);
and (net6, x1, y[1], y[2]);
and (net7, y[2], ~y[3]);
or  (net8, net4, net5, net6, net7);

d_ff inst4 (
   .rst_n(rst_n),
   .set_n(rst_n),
   .clk(clk),
   .d(net8),
   .q(yp)
   );

//-------------------------------------------------
//define the logic (2k+1) to detect a parity error
xor  (net10, y[1], y[2]);
xor  (net11, y[3], yp);
xnor (error, net10, net11);

//-------------------------------------------------
//define the logic for output z1
and (z1, y[1], ~y[3], ~clk);

endmodule
```

Figure 3.175   (Continued)

```
//test bench for the parity-checked moore machine
module moore_par_chk_tb;

reg rst_n, clk, x1;      //inputs are reg for test bench
wire [1:3] y;            //outputs are wire for test bench
wire z1, yp, error;

//display variables
initial
$monitor ("x1 = %b, state = %b, par = %b, z1 = %b",
          x1, y, yp, z1);
                                //continued on next page
```

Figure 3.176   Test bench for the Moore machine of Figure 3.172.

```
//define clock
initial
begin
   clk = 1'b0;
   forever
      #10 clk = ~clk;
end


//define input sequence
initial
begin
   #0 rst_n = 1'b0;  //reset to state_a (000)
      x1 = 1'b0;

   #5 rst_n = 1'b1;


//-----------------------------------------------------------
   x1 = 1'b0;  @ (posedge clk)    //go to state_b (001)
   x1 = 1'b0;  @ (posedge clk)    //go to state_d (011)
   x1 = 1'b0;  @ (posedge clk)    //go to state_f (010)
   x1 = 1'b1;  @ (posedge clk)    //go to state_a (000)


//-----------------------------------------------------------
   x1 = 1'b1;  @ (posedge clk)    //go to state_c (101)
   x1 = 1'b1;  @ (posedge clk)    //go to state_e (111)
   x1 = 1'b1;  @ (posedge clk)    //go to state_g (110)
                                  //assert z1 (t2 - t3)
   x1 = 1'b1;  @ (posedge clk)    //go to state_a (000)


//-----------------------------------------------------------
   $stop;

end

//-----------------------------------------------------------
//instantiate the module into the test bench as a single line
moore_par_chk inst1 (rst_n, clk, x1, y, yp, z1, error);

endmodule
```

Figure 3.176   (Continued)

```
x1 = 0, state = 000, par = 0, z1 = 0
x1 = 0, state = 001, par = 0, z1 = 0
x1 = 0, state = 011, par = 1, z1 = 0
x1 = 0, state = 010, par = 0, z1 = 0
x1 = 1, state = 000, par = 1, z1 = 0
x1 = 1, state = 101, par = 1, z1 = 0
x1 = 1, state = 111, par = 0, z1 = 0
x1 = 1, state = 110, par = 1, z1 = 0
x1 = 1, state = 110, par = 1, z1 = 1
x1 = 1, state = 000, par = 1, z1 = 0
```

Figure 3.177   Outputs for the Moore machine of Figure 3.172.

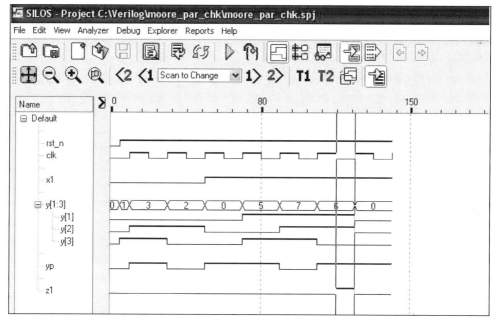

Figure 3.178   Waveforms for the Moore machine of Figure 3.172.

3.6 Problems

3.1   Given the input map shown below for flip-flop z_1, design the logic using a linear-select multiplexer. Use the least amount of logic. Obtain the design module, the test bench module, the outputs, and the waveforms.

$y_1 \backslash y_2 y_3$	0 0	0 1	1 1	1 0
0	0 ⁰	0 ¹	1 ³	1 ²
1	$x_1 x_2$ ⁴	x_3 ⁵	$x_1 + x_2$ ⁷	x_1 ⁶

$$z_1$$

3.2   Given the state diagram shown below for a synchronous sequential machine containing Moore- and Mealy-type outputs, synthesize the machine using linear-select multiplexers for the δ next-state logic. Use inputs x_1 and x_2 as map-entered variables. Use the least amount of logic. Obtain the structural design module, the test bench module, the outputs, and the waveforms.

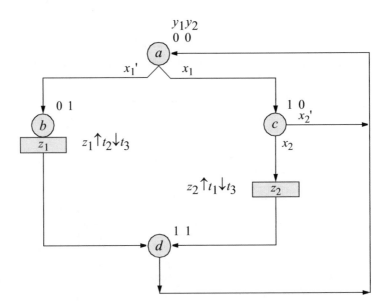

3.3   The state diagram for a Moore synchronous sequential machine is shown below. Implement the machine using linear-select multiplexers for the δ next-state logic, D flip-flops for the storage elements, and any additional logic for the λ output logic. Use x_1 and x_2 as map-entered variables. Obtain the structural design module, the test bench module, the outputs, and the waveforms.

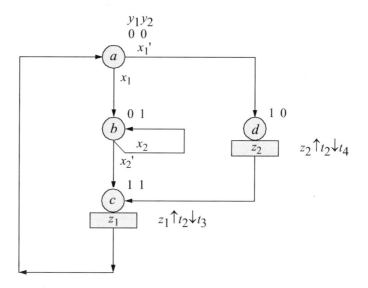

3.4   Given the state diagram shown below for a Moore synchronous sequential machine, derive the input maps for flip-flops y_1, y_2, and y_3 and the corresponding input equations. Synthesize the machine using linear-select multiplexers for the δ next-state logic, D flip-flops for the storage elements, and built-in primitives for any additional logic. Obtain the structural design module, the test bench module, the outputs, and the waveforms.

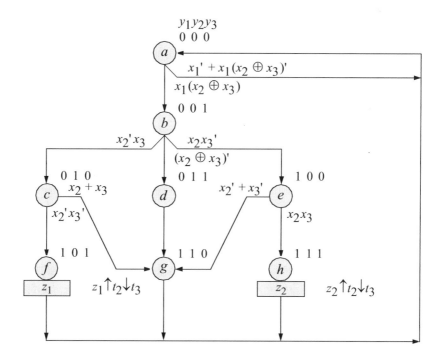

3.5   Given the Karnaugh map shown below for output z_1 of a synchronous sequential machine, implement the δ next-state logic using a nonlinear-select multiplexer and additional logic gates, if necessary. Use the least amount of logic. The Karnaugh map can be permuted, allowing only y_1 and y_2 to be used as the multiplexer select inputs. Obtain the structural design module, the test bench module, the outputs, and the waveforms.

y_1 \ $y_2 y_3$	0 0	0 1	1 1	1 0
0	1 0	1 1	$x_1 x_2$ 3	$-$ 2
1	$-$ 4	1 5	$x_1' + x_2'$ 7	$(x_1 x_2)'$ 6

z_1

3.6   Given the output map shown below for output z_1 of a synchronous sequential machine, implement the logic using a nonlinear-select multiplexer and additional logic gates, if necessary. Use the least amount of logic. Obtain the structural design module, the test bench module, the outputs, and the waveforms.

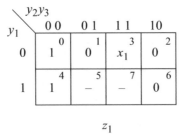

y_1 \ $y_2 y_3$	0 0	0 1	1 1	1 0
0	1 0	0 1	x_1 3	0 2
1	1 4	$-$ 5	$-$ 7	0 6

z_1

3.7   Given the state diagram shown below for a Moore machine, implement the design using nonlinear-select multiplexers for the δ next-state logic, D flip-flops for the storage elements, and continuous assignment statements for the λ output logic. Outputs z_1, z_2, and z_3 are asserted at time t_2 and deasserted at t_3. Obtain the structural design module, the test bench module, the outputs, and the waveforms.

3.8   Design a 3-bit Johnson counter using a PAL device for the δ next-state logic, the storage elements using D flip-flops, and the λ output logic. The counter counts in the following sequence: 000, 100, 110, 111, 011, 001, 000, ... Obtain the structural design module, the test bench module, the outputs, and the waveforms.

3.9   Design a 4-bit Moore sequential machine that operates according to the following sequence: $y_1 y_2 y_3 y_4 = 0000, 1000, 1100, 1110, 1111, 0000, ...$. Output z_1 is asserted unconditionally in state $y_1 y_2 y_3 y_4 = 1111$. The assertion/deassertion statement for z_1 is: $\uparrow t_2 \downarrow t_3$. Use a PAL device for the δ next-state logic, the storage elements using D flip-flops, and the λ output logic. Obtain the structural design module, the test bench module, the outputs, and the waveforms.

3.10   Design a synchronous count-down counter that counts according to the following sequence: $y_1 y_2 y_3 = 111, 110, 101, 100, 011, 010, 001, 000$. The counter has an initial state of $y_1 y_2 y_3 = 111$. Use a PLA device for the δ next-state logic and the D flip-flops, which specify the state of the counter. Obtain the structural design module, the test bench module, the outputs, and the waveforms. Review Section 3.3.3 for the architecture of a typical programmable logic array device.

3.11   Design a 4-bit Johnson counter using D flip-flops. The counter counts in the following sequence: 0000, 1000, 1100, 1110, 1111, 0111, 0011, 0001, 0000. Obtain the structural design module, the test bench module, the outputs, and the waveforms.

3.12   Design a 4-bit Gray code counter using NOR, exclusive-OR, exclusive-NOR, AND, and OR logic built-in primitives for the δ next-state logic. Use negative-edge-triggered JK flip-flops for the storage elements. Obtain the structural design module, the test bench module, the outputs, and the waveforms. The counting sequence is as follows:

$0000, 0001, 0011, 0010, 0110, 0111, 0101, 0100, 1100, 1101, 1111, 1110,$
$1010, 1011, 1001, 1000, 0000, \dots .$

3.13   Design a parity-checked Mealy synchronous sequential machine that generates an output z_1 whenever the sequence 1001 is detected on a serial data input line x_1. Overlapping sequences are valid. Output z_1 is asserted at time t_2 and deasserted at t_3. The parity flip-flop maintains odd parity over the state flip-flops and the parity flip-flop itself. Use built-in primitives for the δ next-state logic and the output logic. Use D flip-flops as the storage elements. Obtain the structural design module, the test bench module, the outputs, and the waveforms.

3.14   Design a parity-checked Mealy synchronous sequential machine that generates an output z_1 whenever the sequence 1001 is detected on a serial data input line x_1. Overlapping sequences are valid. Output z_1 is asserted at time t_2 and deasserted at t_3. The parity flip-flop maintains odd parity over the state flip-flops and the parity flip-flop itself. Use built-in primitives for the δ next-state logic and the output logic. This problem repeats Problem 3.13, but uses JK flip-flops as the storage elements. Obtain the structural design module, the test bench module, the outputs, and the waveforms.

3.15   Design a parity-checked Mealy synchronous sequential machine that generates an output z_1 whenever the sequence 1001 is detected on a serial data input line x_1. Overlapping sequences are valid. Output z_1 is asserted at time t_2 and deasserted at t_3. The parity flip-flop maintains odd parity over the state flip-flops and the parity flip-flop itself. This problem repeats Problem 3.13, but uses behavioral modeling. Obtain the structural design module, the test bench module, the outputs, and the waveforms.

4.1   Introduction
4.2   Synthesis Examples
4.3   Problems

4

Synthesis of Asynchronous Sequential Machines Using Verilog HDL

This chapter implements asynchronous sequential machine designs using Verilog HDL. The designs will be accomplished by utilizing one or more of the following modeling methods for each design: built-in primitive gates, dataflow modeling, behavioral modeling, structural modeling. These four modeling methods are briefly reviewed in the following section.

4.1   Introduction

This section briefly describes the four modeling methods of the Verilog hardware description language that will be used to design asynchronous sequential machines. Different types of asynchronous sequential machines will be designed using Verilog.

4.1.1   Built-In Primitive Gates

These gates describe a net and have one or more scalar inputs, but only one scalar output. The multiple-input gates are **and**, **nand**, **or**, **nor**, **xor**, and **xnor**. The output signal is listed first, followed by the inputs in any order. The outputs are declared as

wire; the inputs can be declared as either **wire** or **reg**. The gates represent combinational logic functions and can be instantiated into a module, as follows, where the instance name is optional:

> **gate_type** inst1 (output, input_1, input_2, . . . , input_*n*);

Two or more instances of the same type of gate can be specified in the same construct, as shown below. Note that only the last instantiation has a semicolon terminating the line. All previous lines are terminated by a comma.

> **gate_type** inst1 (output_1, input_11, input_12, . . . , input_1*n*),
> inst2 (output_2, input_21, input_22, . . . , input_2*n*),
>
> .
>
> .
>
> inst*m* (output_*m*, input_*m*1, input_*m*2, . . . , input_*mn*);

4.1.2   Dataflow Modeling

This method is at a higher level of abstraction than gate-level modeling using built-in primitives. Design automation tools are used to create gate-level logic from dataflow modeling by a process called *logic synthesis*. Register transfer level (RTL) is a combination of dataflow modeling and behavioral modeling and characterizes the flow of data through logic circuits. The *continuous assignment* statement models dataflow behavior and is used to design combinational logic without using gates and interconnecting nets. Continuous assignment statements provide a Boolean correspondence between the right-hand side expression and the left-hand side target. The continuous assignment statement uses the keyword **assign** and has the following syntax with optional drive strength and delay:

> **assign** [drive_strength] [delay] left-hand side target = right-hand side expression

The **assign** statement continuously monitors the right-hand side expression. If a variable changes value, then the expression is evaluated and the result is assigned to the target after any specified delay. If no delay is specified, then the default delay is zero. The continuous assignment statement can be considered to be a form of behavioral modeling, because the behavior of the circuit is specified, not the implementation.

4.1.3   Behavioral Modeling

This method describes the behavior of a digital system and is not concerned with the direct implementation of logic gates but more on the architecture of the system. This is an algorithmic approach to hardware implementation and represents a higher level of abstraction than the previous modeling methods. A Verilog module may contain a mixture of the four modeling constructs. The constructs in behavioral modeling closely resemble those used in the C programming language.

A *procedure* is a series of operations taken to design a module. A Verilog module that is designed using behavioral modeling contains no internal structural details; it simply defines the behavior of the hardware in an abstract, algorithmic description. Verilog contains two structured procedure statements or behaviors: **initial** and **always**. A behavior may consist of a single statement or a block of statements delimited by the keywords **begin . . . end**. A module may contain multiple **initial** and **always** statements. These statements are the basic statements used in behavioral modeling and execute concurrently starting at time zero in which the order of execution is not important. All other behavioral statements are contained inside these structured procedure statements.

Initial statement   All statements within an **initial** statement comprise an **initial** block. An **initial** statement executes only once beginning at time zero, then suspends execution. An **initial** statement provides a method to initialize and monitor variables before the variables are used in a module; it is also used to generate waveforms. For a given time unit, all statements within the **initial** block execute sequentially. The syntax for an **initial** statement is shown below.

> **initial** [optional timing control] procedural statement or
> block of procedural statements

Always statement   The **always** statement executes the behavioral statements within the **always** block repeatedly in a looping manner and begins execution at time zero. Execution of the statements continues indefinitely until the simulation is terminated. The keywords **initial** and **always** specify a behavior and the statements within a behavior are classified as *behavioral* or *procedural*. The syntax for the **always** statement is shown below.

> **always** [optional timing control] procedural statement or
> block of procedural statements

Conditional statements   Conditional statements alter the flow within a behavior based upon certain conditions. The choice among alternative statements depends on the Boolean value of an expression. The alternative statements can be a single statement or a block of statements delimited by the keywords **begin . . . end**. The keywords **if** and **else** are used in conditional statements. There are three categories of the conditional statement as shown below. A true value is 1 or any nonzero value; a false value is 0, **x**, or **z**. If the evaluation is false, then the next expression in the activity flow is evaluated.

//no **else** statement
if (expression) statement1;   //if expression is true, then statement1 is executed.

//one **else** statement	//choice of two statements. Only one is executed.
if (expression) statement1;	//if expression is true, then statement1 is executed.
else statement2;	//if expression is false, then statement2 is executed.

//nested **if-else if**	//choice of multiple statements. Only one is executed.
if (expression1) statement1;	//if expression1 is true, then statement1 is executed.
else if (expression2) statement2;	//if expression2 is true, then statement2 is executed.
else if (expression3) statement3;	//if expression3 is true, then statement3 is executed.
else default statement;	

Case statement   The **case** statement is an alternative to the **if** . . . **else if** construct and may simplify the readability of the Verilog code. The **case** statement is a multiple-way conditional branch. It executes one of several different procedural statements depending on the comparison of an expression with a case item. The expression and the case item are compared bit-by-bit and must match exactly. The statement that is associated with a case item may be a single procedural statement or a block of statements delimited by the keywords **begin** . . . **end.** The **case** statement has the following syntax:

```
case (expression)
    case_item1 : procedural_statement1;
    case_item2 : procedural_statement2;
    case_item3 : procedural_statement3;
            .
            .
            .
    case_itemn : procedural_statementn;
    default : default_statement;
endcase
```

While loop   The **while** loop executes a procedural statement or a block of procedural statements as long as a Boolean expression returns a value of true. When the procedural statements are executed, the Boolean expression is reevaluated. The loop is executed until the expression returns a value of false. If the evaluation of the expression is false, then the **while** loop is terminated and control is passed to the next statement in the module. If the expression is false before the loop is initially entered, then the **while** loop is never executed. The syntax for a **while** statement is as follows:

```
while (expression)
    procedural statement or block of procedural statements
```

4.1.4   Structural Modeling

Structural modeling consists of instantiating one or more of the following design objects:

- Built-in primitives
- User-defined primitives (UDPs)
- Design modules

Instantiation means to use one or more lower-level modules — including built-in primitives — that are interconnected in the construction of a higher-level structural module. A module can be a logic gate, an adder, a multiplexer, a counter, or some other logical function. The objects that are instantiated are called *instances*. Structural modeling is described by the interconnection of these lower-level logic primitives or modules. The interconnections are made by wires that connect primitive terminals or module ports. Ports provide a means for the module to communicate with its external environment.

4.2   Synthesis Examples

The examples which follow illustrate the synthesis procedure for asynchronous sequential machines using a timing diagram and/or a verbal specification. The traditional synthesis procedure is used in designing the asynchronous sequential machines. In order to prevent possible race conditions and associated timing problems when two or more inputs change value simultaneously, it will be assumed that only one input variable will change state at a time. This is referred to as a *fundamental-mode model*, further defined with the following characteristics:

1. Only one input will change at a time.
2. No other input will change until the machine has sequenced to a stable state.

Example 4.1   A Mealy asynchronous sequential machine has two inputs x_1 and x_2 and one output z_1. The initial conditions are: $x_1 x_2 z_1 = 000$. Output z_1 is asserted whenever $x_1 x_2 = 11$ if and only if the input sequence was $x_1 x_2 = 00, 01, 11$. The output is to change as fast as possible. The waveforms of Figure 4.1 depict some typical input sequences. Many other input sequences are possible and must be considered in order to thoroughly construct a primitive flow table.

The primitive flow table is the initial step in the design process and is constructed by plotting the sequence of state changes as specified by the timing diagram, while allowing only one stable state per row. Beginning at the leftmost section of the timing diagram, where the initial conditions are specified, a stable state is assigned to each different combination of the input vector. Figure 4.2 shows the complete primitive flow table. The column headings represent the input vector. The table entries specify the stable states, transient states, invalid state transitions which are represented as dashes, and outputs.

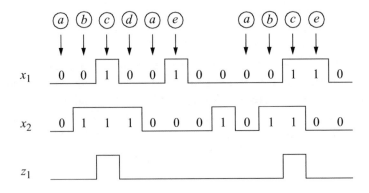

Figure 4.1   Timing diagram for the asynchronous sequential machine of Example 4.1.

x_1x_2	00	01	11	10	z_1
	ⓐ	b	–	e	0
	a	ⓑ	c	–	0
	–	d	ⓒ	e	1
	a	ⓓ	f	–	0
	a	–	f	ⓔ	0
	–	d	ⓕ	e	0

Figure 4.2   Complete (reduced) primitive flow table for the asynchronous sequential machine of Example 4.1.

The merger diagram graphically portrays the result of the merging process in which an attempt is made to combine two or more rows of the reduced primitive flow table into a single row. The merger diagram is shown in Figure 4.3.

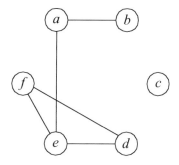

Figure 4.3   Merger diagram for the asynchronous sequential machine of Example 4.1.

The merged flow table is constructed from the reduced primitive flow table and the merger diagram as shown in Figure 4.4. The merged flow table is obtained by transferring the individual rows from the primitive flow table to the merged flow table in accordance with the partition assignment of Figure 4.3.

x_1x_2		00	01	11	10
1	$(a),(b)$	(a)	(b)	c	e
2	(c)	$-$	d	(c)	e
3	$(d),(e),(f)$	a	(d)	(f)	(e)
4		$-$	$-$	$-$	$-$

Figure 4.4   Merged flow table obtained from the merger diagram of Figure 4.3.

The transition diagram graphically portrays the information displayed in the merged flow table and is shown in Figure 4.5. The arrows in the transition diagram are added to visualize the sequence of transitions. As mentioned previously, a transition diagram with only three vertices (rows) cannot all have adjacent contiguous rows; therefore, a fourth row must be added, as shown in Figure 4.5.

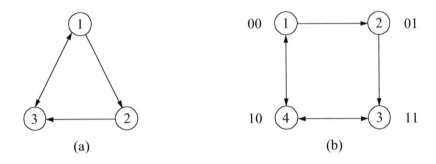

(a)   (b)

Figure 4.5   Transition diagram for the merged flow table of Figure 4.4.

The combined excitation map for excitation variables Y_{1e} and Y_{2e} is shown in Figure 4.6. This is a combined excitation map for excitation variables Y_{1e} and Y_{2e}, in which the feedback variables are y_{1f} and y_{2f}, respectively. The stable states in each row are assigned excitation values that are equal to the feedback values of the corresponding row. The unstable states are assigned excitation values that direct the machine to the destination stable state.

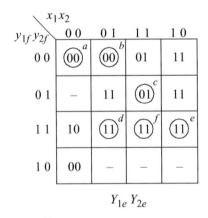

Figure 4.6   Combined excitation map for the merged flow table of Figure 4.5.

The combined excitation map of Figure 4.6 is now separated into its constituent parts to obtain individual excitation maps for Y_{1e} and Y_{2e}, as shown in Figure 4.7. To obtain the excitation map for Y_{1e}, simply transfer the values for Y_{1e} from the minterm locations in the combined map to the same squares in the individual map. Repeat the process to obtain the excitation map for Y_{2e}.

$y_{1f} y_{2f}$ \ $x_1 x_2$	0 0	0 1	1 1	1 0
0 0	0	0	0	1
0 1	–	1	0	1
1 1	1	1	1	1
1 0	0	–	–	–

Y_{1e}

$y_{1f} y_{2f}$ \ $x_1 x_2$	0 0	0 1	1 1	1 0
0 0	0	0	1	1
0 1	–	1	1	1
1 1	0	1	1	1
1 0	0	–	–	–

Y_{2e}

Figure 4.7 Individual excitation maps obtained from Figure 4.6 for the asynchronous sequential machine of Example 4.1.

The equations for the excitation variables are derived directly from the individual excitation maps. The equations can be specified in either a sum-of-products form or a product-of-sums form. Regardless of the form used, the Boolean equations must be free of static-1 and static-0 hazards. A course in asynchronous sequential machines specifies that if an input variable changes value causing an output to be deasserted momentarily when the output should remain asserted, then this is classified as a *static-1 hazard*. Conversely, when an input change causes an output to become asserted momentarily when it should remain deasserted, a *static-0 hazard* results. The sum-of-products notation is shown in Equation 4.1 for Y_{1e} and Y_{2e}. All equations are free of static hazards.

$$Y_{1e} = x_1 x_2' + y_{1f} y_{2f} + y_{2f} x_1' + y_{2f} x_2' \text{ (Hazard cover)}$$
$$Y_{2e} = x_1 + y_{2f} x_2 \tag{4.1}$$

The map for output z_1 is shown in Figure 4.8. Output values are assigned for all nonstable states so that no transient signals appear on the output. In this step, the speed of the asynchronous sequential machine can also be determined. The values entered in the unstable states can permit fast or slow output changes. In this example, output z_1 is to change as fast as possible; therefore, a value of 1 is placed in minterm location $y_{1f} y_{2f} x_1 x_2 = 0011$. Then the output equation is derived from the output map assuring that the output will be free of static-1 hazards, as shown in Equation 4.2.

$y_{1f}y_{2f}$ \ x_1x_2	0 0	0 1	1 1	1 0
0 0	0 *a*	0 *b*	1	0
0 1	–	0	1 *c*	0
1 1	0	0 *d*	0 *f*	0 *e*
1 0	–	–	–	–

z_1

Figure 4.8 Output map for the asynchronous sequential machine of Example 4.1.

$$z_1 = x_1x_2y_{1f}' \tag{4.2}$$

The Verilog design module is shown in Figure 4.9 using built-in primitives. The test bench module is shown in Figure 4.10 and takes the machine through the sequences of the timing diagram. The outputs and waveforms are shown in Figure 4.11 and Figure 4.12, respectively. The waveforms clearly show the sequences displayed in the timing diagram where output z_1 is asserted whenever $x_1x_2 = 11$ if and only if the input sequence was $x_1x_2 = 00, 01, 11$.

```
//built-in primitive design for asm

module asm25_bip (rst_n, x1, x2, y1e, y2e, z1);

//define inputs and outputs
input rst_n, x1, x2;
output y1e, y2e, z1;

//define internal nets
wire net1, net2, net3, net4, net5;

                                    //continued on next page
```

Figure 4.9 Output z_1 is asserted whenever $x_1x_2 = 11$ if and only if the input sequence was $x_1x_2 = 00, 01, 11$.

```
//design the logic for y1e
and     (net1, x1, ~x2),
        (net2, y1e, y2e, rst_n),
        (net3, y2e, ~x1, rst_n),
        (net4, y2e, ~x2, rst_n);
or      (y1e, net1, net2, net3, net4);

//design the logic for y2e
and     (net5, y2e, x2, rst_n);
or      (y2e, x1, net5);

//design the logic for output z1
and     (z1, x1, x2, ~y1e);

endmodule
```

Figure 4.9   (Continued)

```
//test bench for the asm using built-in primitives

module asm25_bip_tb;

//inputs are reg for test bench
//outputs are wire for test bench
reg rst_n, x1, x2;
wire y1e, y2e, z1;

//display variables
initial
$monitor ("x1x2 = %b, state = %b, z1 = %b",
          {x1, x2}, {y1e, y2e}, z1);

//apply input vectors
initial
begin
    #0      rst_n = 1'b0;
            x1 = 1'b0;
            x2 = 1'b0;

    #5      rst_n = 1'b1;
                                //continued on next page
```

Figure 4.10   Test bench for the asynchronous sequential machine of Example 4.1.

```
    #10     x1 = 1'b0;   x2 = 1'b0;
    #10     x1 = 1'b0;   x2 = 1'b1;
    #10     x1 = 1'b1;   x2 = 1'b1;   //assert z1
    #10     x1 = 1'b0;   x2 = 1'b1;
    #10     x1 = 1'b0;   x2 = 1'b0;


    #10     x1 = 1'b1;   x2 = 1'b0;
    #10     x1 = 1'b0;   x2 = 1'b0;
    #10     x1 = 1'b0;   x2 = 1'b1;
    #10     x1 = 1'b0;   x2 = 1'b0;
    #10     x1 = 1'b0;   x2 = 1'b1;
    #10     x1 = 1'b1;   x2 = 1'b1;   //assert z1
    #10     x1 = 1'b1;   x2 = 1'b0;
    #10     x1 = 1'b0;   x2 = 1'b0;


    #10     $stop;

end


//instantiate the module into the test bench as a single line
asm25_bip inst1 (rst_n, x1, x2, y1e, y2e, z1);

endmodule
```

Figure 4.10   (Continued)

```
x1x2 = 00, state = 00, z1 = 0
x1x2 = 01, state = 00, z1 = 0
x1x2 = 11, state = 01, z1 = 1
x1x2 = 01, state = 11, z1 = 0
x1x2 = 00, state = 00, z1 = 0

x1x2 = 10, state = 11, z1 = 0
x1x2 = 00, state = 00, z1 = 0
x1x2 = 01, state = 00, z1 = 0
x1x2 = 00, state = 00, z1 = 0
x1x2 = 01, state = 00, z1 = 0
x1x2 = 11, state = 01, z1 = 1
x1x2 = 10, state = 11, z1 = 0
x1x2 = 00, state = 00, z1 = 0
```

Figure 4.11   Outputs for the asynchronous sequential machine of Example 4.1.

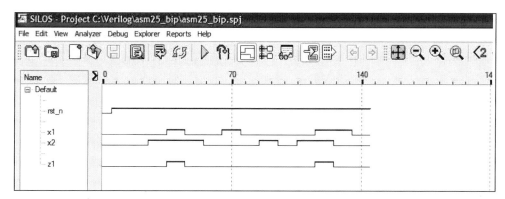

Figure 4.12   Waveforms for the asynchronous sequential machine of Example 4.1.

Example 4.2   This example repeats Example 4.1; however, the continuous assignment statement of dataflow modeling is used rather than built-in primitives. The excitation equations of Equation 4.1 and the output equation of Equation 4.2 will be used in this design. Recall that the asynchronous sequential machine has two inputs x_1 and x_2 and one output z_1. The initial conditions are: $x_1 x_2 z_1 = 000$. Output z_1 is asserted whenever $x_1 x_2 = 11$ if and only if the input sequence was $x_1 x_2 = 00, 01, 11$. The output is to change as fast as possible. The excitation equations and the output equation are reproduced in Equation 4.3 for Y_{1e}, Y_{2e}, and z_1.

$$Y_{1e} = x_1 x_2' + y_{1f} y_{2f} + y_{2f} x_1' + y_{2f} x_2' \text{ (Hazard cover)}$$

$$Y_{2e} = x_1 + y_{2f} x_2$$

$$(4.3)$$

$$z_1 = x_1 x_2 y_{1f}'$$

The design module is shown in Figure 4.13 in a sum-of-products form using the **assign** statement of dataflow modeling. When a variable on the right-hand side changes value, the right-hand side expression is evaluated and the value is assigned to the left-hand side net after the specified delay. The continuous assignment is used to place a value on a net.

The test bench module is shown in Figure 4.14 and takes the machine through the sequences of the timing diagram of Figure 4.1. The outputs and waveforms are shown in Figure 4.15 and Figure 4.16, respectively. The waveforms indicate the sequences displayed in the timing diagram of Figure 4.1, where output z_1 is asserted whenever $x_1 x_2 = 11$ if and only if the input sequence was $x_1 x_2 = 00, 01, 11$.

```
//dataflow for sum-of-products asm

module asm25_df (rst_n, x1, x2, y1e, y2e, z1);

//define inputs and outputs
input rst_n, x1, x2;
output y1e, y2e, z1;

//define internal nets
wire net1, net2, net3, net4, net5;

//design the logic for y1e
assign    net1 = x1 & ~x2,
          net2 = y1e & y2e & rst_n,
          net3 = ~x1 & y2e & rst_n,
          net4 = ~x2 & y2e & rst_n,
          y1e = net1 | net2 | net3 | net4;

//design the logic for y2e
assign    net5 = x2 & y2e,
          y2e = x1 | net5;

//design the logic for output z1
assign    z1 = x1 & x2 & ~y1e;

endmodule
```

Figure 4.13   Design module for the asynchronous sequential machine of Example 4.2 using dataflow modeling.

```
module asm25_df_tb;

//inputs are reg for test bench
//outputs are wire for test bench
reg rst_n, x1, x2;
wire y1e, y2e, z1;

//display variables
initial
$monitor ("x1x2 = %b, state = %b, z1 = %b",
          {x1, x2}, {y1e, y2e}, z1);
                                        //continued on next page
```

Figure 4.14   Test bench for Example 4.2.

```
//apply input vectors
initial
begin
   #0     rst_n = 1'b0;
          x1 = 1'b0;
          x2 = 1'b0;
   #5     rst_n = 1'b1;

   #10    x1 = 1'b0;   x2 = 1'b0;
   #10    x1 = 1'b0;   x2 = 1'b1;
   #10    x1 = 1'b1;   x2 = 1'b1;   //assert z1
   #10    x1 = 1'b0;   x2 = 1'b1;
   #10    x1 = 1'b0;   x2 = 1'b0;

   #10    x1 = 1'b1;   x2 = 1'b0;
   #10    x1 = 1'b0;   x2 = 1'b0;
   #10    x1 = 1'b0;   x2 = 1'b1;
   #10    x1 = 1'b0;   x2 = 1'b0;
   #10    x1 = 1'b0;   x2 = 1'b1;
   #10    x1 = 1'b1;   x2 = 1'b1;   //assert z1
   #10    x1 = 1'b1;   x2 = 1'b0;
   #10    x1 = 1'b0;   x2 = 1'b0;
   #10    $stop;
end

//instantiate the module into the test bench as a single line
asm25_df inst1 (rst_n, x1, x2, y1e, y2e, z1);
endmodule
```

Figure 4.14   (Continued)

```
x1x2 = 00, state = 00, z1 = 0
x1x2 = 01, state = 00, z1 = 0
x1x2 = 11, state = 01, z1 = 1
x1x2 = 01, state = 11, z1 = 0
x1x2 = 00, state = 00, z1 = 0

x1x2 = 10, state = 11, z1 = 0
x1x2 = 00, state = 00, z1 = 0
x1x2 = 01, state = 00, z1 = 0
x1x2 = 00, state = 00, z1 = 0
x1x2 = 01, state = 00, z1 = 0
x1x2 = 11, state = 01, z1 = 1
x1x2 = 10, state = 11, z1 = 0
x1x2 = 00, state = 00, z1 = 0
```

Figure 4.15   Outputs for Example 4.2.

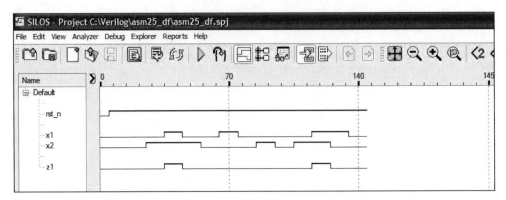

Figure 4.16   Waveforms for Example 4.2.

Example 4.3   This example repeats Example 4.1; however, behavioral modeling is used with the **case** statement rather than built-in primitives. The waveforms are reproduced in Figure 4.17 for convenience and depict some typical input sequences. Recall that the asynchronous sequential machine has two inputs x_1 and x_2 and one output z_1. The initial conditions are: $x_1 x_2 z_1 = 000$. Output z_1 is asserted whenever $x_1 x_2 = 11$ if and only if the input sequence was $x_1 x_2 = 00, 01, 11$. The output is to change as fast as possible.

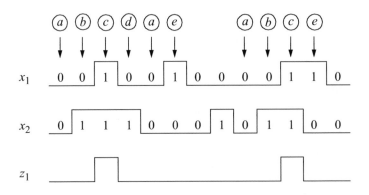

Figure 4.17   Waveforms for the asynchronous sequential machine of Example 4.3.

The primitive flow table is shown in Figure 4.18 indicating the possible transition sequences for the machine. Note that state $\textcircled{e}$ sequences to state $\textcircled{f}$ if $x_1 x_2 = 11$, but does not assert output z_1 because the input sequence was not $x_1 x_2 = 00, 01, 11$. Six states require three excitation variables.

x_1x_2	00	01	11	10	z_1	$[1:3]\, y_e$
	ⓐ	b	–	e	0	ⓐ = 000
	a	ⓑ	c	–	0	ⓑ = 001
	–	d	ⓒ	e	1	ⓒ = 011
	a	ⓓ	f	–	0	ⓓ = 010
	a	–	f	ⓔ	0	ⓔ = 110
	–	d	ⓕ	e	0	ⓕ = 111

Figure 4.18   Primitive flow table for the asynchronous sequential machine of Example 4.1 that is used in Example 4.3.

The design module is shown in Figure 4.19. The test bench module is shown in Figure 4.20 and takes the machine through the sequences of the timing diagram of Figure 4.17. The outputs and waveforms are shown in Figure 4.21 and Figure 4.22, respectively. The waveforms indicate the sequences displayed in the timing diagram of Figure 4.17, where output z_1 is asserted whenever $x_1x_2 = 11$ if and only if the input sequence was $x_1x_2 = 00, 01, 11$.

```
//behavioral asynchronous sequential machine

module asm25_bh (rst_n, x1, x2, ye, z1);

//define inputs and outputs
//do not have to declare inputs as wire
//they are wire by default

input rst_n, x1, x2;
output [1:3] ye;
output z1;

                              //continued on next page
```

Figure 4.19   Behavioral design module for the asynchronous sequential machine of Example 4.3.

```verilog
reg [1:3] ye, next_state;       //variables are reg in always
reg z1;

//assign state codes; parameter defines a constant
//state names must have at least 2 characters
parameter    state_a = 3'b000,
             state_b = 3'b001,
             state_c = 3'b011,
             state_d = 3'b010,
             state_e = 3'b110,
             state_f = 3'b111;

always @ (rst_n or x1 or x2)   //set next state
begin
   if (~rst_n)
      ye <= state_a;
   else
      ye <= next_state;
end

always @ (x1 or x2 or ye)       //define output z1
begin
   if (ye == state_c)           //== is a logical equality
      z1 = 1'b1;
   else
      z1 = 1'b0;
end

//determine next state
always @ (x1 or x2)
begin
   case (ye)
      state_a:
         if (x1==1'b0 &   x2==1'b1)
            next_state = state_b;
         else if (x1==1'b1 & x2==1'b0)
            next_state = state_e;
         else
            next_state = state_a;

      state_b:
         if (x1==1'b0 &   x2==1'b0)
            next_state = state_a;
         else if (x1==1'b1 & x2==1'b1)
            next_state = state_c;
         else
            next_state = state_b;     //continued on next page
```

Figure 4.19   (Continued)

```
    state_c:
         if (x1==1'b0 &   x2==1'b1)
            next_state = state_d;
         else if (x1==1'b1 & x2==1'b0)
            next_state = state_e;
         else
            next_state = state_c;


    state_d:
         if (x1==1'b0 &   x2==1'b0)
            next_state = state_a;
         else if (x1==1'b1 & x2==1'b1)
            next_state = state_f;
         else
            next_state = state_d;


    state_e:
         if (x1==1'b0 &   x2==1'b0)
            next_state = state_a;
         else if (x1==1'b1 & x2==1'b1)
            next_state = state_f;
         else
            next_state = state_e;


    state_f:
         if (x1==1'b0 &   x2==1'b1)
            next_state = state_d;
         else if (x1==1'b1 & x2==1'b0)
            next_state = state_e;
         else
            next_state = state_f;


    default: next_state = state_a;

  endcase

end

endmodule
```

Figure 4.19   (Continued)

```verilog
//test bench for the asynchronous sequential machine

module asm25_bh_tb;

//inputs are reg for test bench
//outputs are wire for test bench
reg rst_n, x1, x2;
wire [1:3] ye;
wire z1;

//display variables
initial
$monitor ("x1x2 = %b, state = %b, z1 = %b", {x1, x2}, ye, z1);

//define input vectors
initial
begin
   #0    rst_n = 1'b0;
         x1 = 1'b0;
         x2 = 1'b0;

   #5    rst_n = 1'b1;

   #10   x1 = 1'b0;  x2 = 1'b0;  //go to state_a
   #10   x1 = 1'b0;  x2 = 1'b1;  //go to state_b
   #10   x1 = 1'b1;  x2 = 1'b1;  //go to state_c, assert z1
   #10   x1 = 1'b0;  x2 = 1'b1;  //go to state_d
   #10   x1 = 1'b0;  x2 = 1'b0;  //go to state_a

   #10   x1 = 1'b1;  x2 = 1'b0;  //go to state_e
   #10   x1 = 1'b0;  x2 = 1'b0;  //go to state_a
   #10   x1 = 1'b0;  x2 = 1'b1;  //go to state_b
   #10   x1 = 1'b0;  x2 = 1'b0;  //go to state_a
   #10   x1 = 1'b0;  x2 = 1'b1;  //go to state_b
   #10   x1 = 1'b1;  x2 = 1'b1;  //go to state_c, assert z1
   #10   x1 = 1'b1;  x2 = 1'b0;  //go to state_e
   #10   x1 = 1'b0;  x2 = 1'b0;  //go to state_a

   #10   $stop;
end

//instantiate the module into the test bench as a single line
asm25_bh inst1 (rst_n, x1, x2, ye, z1);

endmodule
```

Figure 4.20   Test bench module for the asynchronous sequential machine of Example 4.3.

```
x1x2 = 00, state = 000, z1 = 0
x1x2 = 01, state = 001, z1 = 0
x1x2 = 11, state = 011, z1 = 1
x1x2 = 01, state = 010, z1 = 0
x1x2 = 00, state = 000, z1 = 0

x1x2 = 10, state = 110, z1 = 0
x1x2 = 00, state = 000, z1 = 0
x1x2 = 01, state = 001, z1 = 0
x1x2 = 00, state = 000, z1 = 0
x1x2 = 01, state = 001, z1 = 0
x1x2 = 11, state = 011, z1 = 1
x1x2 = 10, state = 110, z1 = 0
x1x2 = 00, state = 000, z1 = 0
```

Figure 4.21   Outputs for the asynchronous sequential machine of Example 4.3.

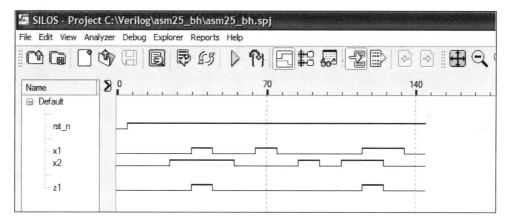

Figure 4.22   Waveforms for the asynchronous sequential machine of Example 4.3.

Example 4.4   This example repeats Example 4.1; however, structural modeling is used with instantiated AND gates and OR gates that are designed with dataflow modeling. The waveforms are reproduced in Figure 4.23 for convenience and depict some typical input sequences. The excitation and output equations are reproduced in Equation 4.4 for Y_{1e}, Y_{2e}, and z_1. The logic diagram for the asynchronous sequential machine is shown in Figure 4.24. Recall that the initial conditions are: $x_1x_2z_1 = 000$. Therefore, there will be a reset input to initialize the machine. Output z_1 is asserted whenever $x_1x_2 = 11$ if and only if the input sequence was $x_1x_2 = 00, 01, 11$. The output is to change as fast as possible.

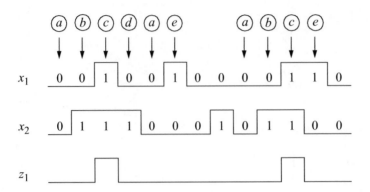

Figure 4.23   Waveforms for the asynchronous sequential machine of Example 4.4.

$$Y_{1e} = x_1 x_2' + y_{1f} y_{2f} + y_{2f} x_1' + y_{2f} x_2' \text{ (Hazard cover)}$$

$$Y_{2e} = x_1 + y_{2f} x_2$$

$$(4.4)$$

$$z_1 = x_1 x_2 y_{1f}'$$

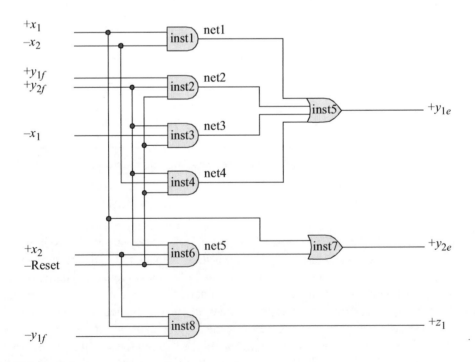

Figure 4.24   Logic diagram for Example 4.4.

The structural design module is shown in Figure 4.25. The test bench module is shown in Figure 4.26 and takes the machine through the sequences of the timing diagram of Figure 4.23. The outputs and waveforms are shown in Figure 4.27 and Figure 4.28, respectively. The waveforms indicate the sequences displayed in the timing diagram of Figure 4.23, where output z_1 is asserted whenever $x_1 x_2 = 11$ if and only if the input sequence was $x_1 x_2 = 00, 01, 11$.

```verilog
//structural asynchronous sequential machine

module asm25_struc (rst_n, x1, x2, y1e, y2e, z1);

//define inputs and outputs
input rst_n, x1, x2;
output y1e, y2e, z1;

//define internal nets
wire net1, net2, net3, net4, net6;

//instantiate the logic for y1e
and2_df inst1 (
   .x1(x1),
   .x2(~x2),
   .z1(net1)
   );

and3_df inst2 (
   .x1(y1e),
   .x2(y2e),
   .x3(rst_n),
   .z1(net2)
   );

and3_df inst3 (
   .x1(y2e),
   .x2(~x1),
   .x3(rst_n),
   .z1(net3)
   );

and3_df inst4 (
   .x1(y2e),
   .x2(~x2),
   .x3(rst_n),
   .z1(net4)
   );                          //continued on next page
```

Figure 4.25   Structural design module for the asynchronous sequential machine of Example 4.4.

```
or4_df inst5 (
   .x1(net1),
   .x2(net2),
   .x3(net3),
   .x4(net4),
   .z1(y1e)
   );

//instantiate the logic for y2e
and3_df inst6 (
   .x1(y2e),
   .x2(x2),
   .x3(rst_n),
   .z1(net6)
   );

or2_df inst7 (
   .x1(x1),
   .x2(net6),
   .z1(y2e)
   );

//instantiate the logic for output z1
and3_df inst8 (
   .x1(x2),
   .x2(x1),
   .x3(~y1e),
   .z1(z1)
   );

endmodule
```

Figure 4.25   (Continued)

```
//test bench for the structural asm

module asm25_struc_tb;

//inputs are reg for test bench
//outputs are wire for test bench
reg rst_n, x1, x2;
wire y1e, y2e, z1;
                                 //continued on next page
```

Figure 4.26   Test bench module for the asynchronous sequential machine of Example 4.4.

```
//display variables
initial
$monitor ("x1x2 = %b, state = %b, z1 = %b",
            {x1, x2}, {y1e, y2e}, z1);


//apply input vectors
initial
begin
    #0     rst_n = 1'b0;
           x1 = 1'b0;
           x2 = 1'b0;

    #5     rst_n = 1'b1;


    #10    x1 = 1'b0;   x2 = 1'b0;
    #10    x1 = 1'b0;   x2 = 1'b1;
    #10    x1 = 1'b1;   x2 = 1'b1;   //assert z1
    #10    x1 = 1'b0;   x2 = 1'b1;
    #10    x1 = 1'b0;   x2 = 1'b0;


    #10    x1 = 1'b1;   x2 = 1'b0;
    #10    x1 = 1'b0;   x2 = 1'b0;
    #10    x1 = 1'b0;   x2 = 1'b1;
    #10    x1 = 1'b0;   x2 = 1'b0;
    #10    x1 = 1'b0;   x2 = 1'b1;
    #10    x1 = 1'b1;   x2 = 1'b1;   //assert z1
    #10    x1 = 1'b1;   x2 = 1'b0;
    #10    x1 = 1'b0;   x2 = 1'b0;

    #10    $stop;

end

//instantiate the module into the test bench
asm25_struc inst1 (
    .rst_n(rst_n),
    .x1(x1),
    .x2(x2),
    .y1e(y1e),
    .y2e(y2e),
    .z1(z1)
    );

endmodule
```

Figure 4.26   (Continued)

```
x1x2 = 00, state = 00, z1 = 0
x1x2 = 01, state = 00, z1 = 0
x1x2 = 11, state = 01, z1 = 1
x1x2 = 01, state = 11, z1 = 0
x1x2 = 00, state = 00, z1 = 0

x1x2 = 10, state = 11, z1 = 0
x1x2 = 00, state = 00, z1 = 0
x1x2 = 01, state = 00, z1 = 0
x1x2 = 00, state = 00, z1 = 0
x1x2 = 01, state = 00, z1 = 0
x1x2 = 11, state = 01, z1 = 1
x1x2 = 10, state = 11, z1 = 0
x1x2 = 00, state = 00, z1 = 0
```

Figure 4.27   Outputs for the asynchronous sequential machine of Example 4.4.

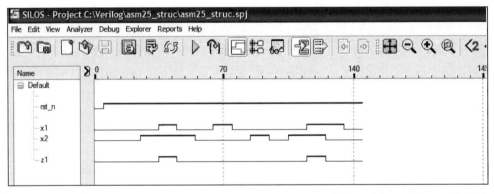

Figure 4.28   Waveforms for the asynchronous sequential machine of Example 4.4.

Example 4.5   This example designs a Mealy machine with two inputs x_1 and x_2 and one output z_1. The machine specifications state that z_1 is to be asserted coincident with the first assertion of x_2 if and only if x_1 is already asserted. Output z_1 is to be deasserted simultaneously with the deassertion of x_1. Output z_1 is to change value as fast as possible. The excitation equations will be in a sum-of-products form. The machine will be implemented with a programmable logic array (PLA) device using built-in primitives. A representative timing diagram depicting one possible sequence is shown in Figure 4.29.

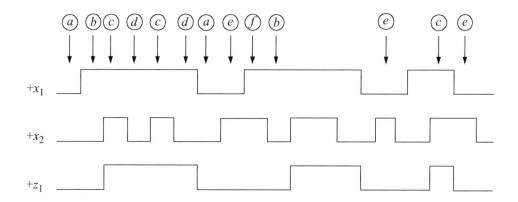

Figure 4.29 A representative timing diagram for the asynchronous sequential machine of Example 4.5.

The primitive flow table is shown in Figure 4.30 and is constructed from the sequences depicted in the timing diagram. Proceeding through the design procedure, the combined excitation map for Y_{1e} and Y_{2e} is obtained as shown in Figure 4.31. The individual excitation maps are shown in Figure 4.32. The excitation equations are shown in Equation 4.5 in a sum-of-products form. The output map for the fastest possible change is shown in Figure 4.33 and the corresponding output equation is shown in Equation 4.6.

x_1x_2 00	01	11	10	z_1
$\textcircled{a}$	e	$-$	b	0
a	$-$	c	$\textcircled{b}$	0
$-$	e	$\textcircled{c}$	d	1
a	$-$	c	$\textcircled{d}$	1
a	$\textcircled{e}$	f	$-$	0
$-$	e	$\textcircled{f}$	b	0

Figure 4.30 Primitive flow table for the asynchronous sequential machine of Example 4.5.

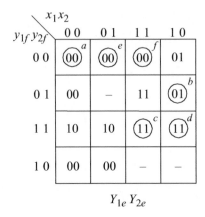

Figure 4.31   Combined excitation map for Example 4.5.

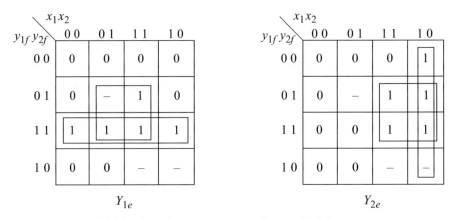

Figure 4.32   Individual excitation maps for Example 4.5.

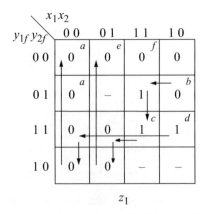

Figure 4.33   Output map for Example 4.5 indicating the fastest change.

$$Y_{1e} = y_{1f}y_{2f} + y_{2f}x_2$$

$$Y_{2e} = y_{2f}x_1 + x_1x_2' \tag{4.5}$$

$$z_1 = y_{1f}x_1 + y_{2f}x_1x_2 \quad \text{Fastest operation} \tag{4.6}$$

The logic diagram using a PLA is shown in Figure 4.34. The structural design module is shown in Figure 4.35 and the test bench module is shown in Figure 4.36. The outputs and waveforms are shown in Figure 4.37 and Figure 4.38, respectively.

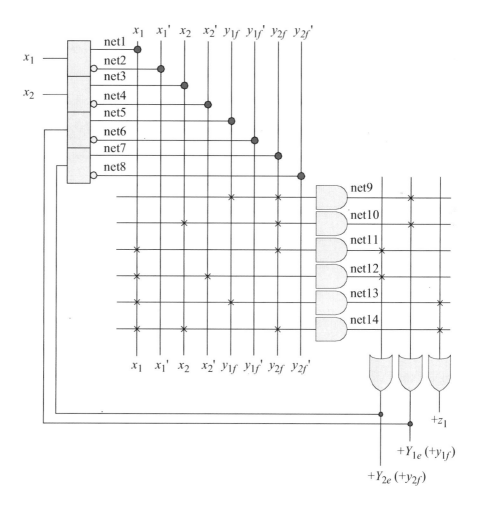

Figure 4.34   Logic diagram for Example 4.5 using a PLA implementation, which represents the fastest possible output changes.

```
//structural pla for an asynchronous sequential machine

module asm_pla (x1, x2, y1e, y2e, z1);

//define inputs and outputs
input x1, x2;
output y1e, y2e, z1;

//define internal nets
wire net1, net2, net3, net4, net5, net6, net7, net8;
wire net9, net10, net11, net12, net13, net14;


//define the input drivers
buf (net1, x1);
not (net2, x1);

buf (net3, x2);
not (net4, x2);

buf (net5, y1e);
not (net6, y1e);

buf (net7, y2e);
not (net8, y2e);


//define the logic for the and array and the or array for y1e
and (net9, net5, net7);
and (net10, net3, net7);
or  (y1e, net9, net10);


//define the logic for the and array and the or array for y2e
and (net11, net1, net7);
and (net12, net1, net4);
or  (y2e, net11, net12);


//define the logic for z1
and (net13, net1, net5);
and (net14, net1, net3, net7);
or  (z1, net13, net14);

endmodule
```

Figure 4.35   Structural design module for the asynchronous sequential machine of Example 4.5.

```verilog
//test bench for the asynchronous sequential machine

module asm_pla_tb;

//inputs are reg for test bench
//outputs are wire for test bench
reg x1, x2;
wire y1e, y2e, z1;

//display variables
initial
$monitor("x1x2 = %b, z1 = %b", {x1, x2}, z1);

//apply input vectors
initial
begin
   #0    x1 = 1'b0;  x2 = 1'b0;
   #10   x1 = 1'b1;  x2 = 1'b0;
   #10   x1 = 1'b1;  x2 = 1'b1;    //assert z1
   #10   x1 = 1'b1;  x2 = 1'b0;    //assert z1
   #10   x1 = 1'b1;  x2 = 1'b1;    //assert z1
   #10   x1 = 1'b1;  x2 = 1'b0;    //assert z1
   #10   x1 = 1'b0;  x2 = 1'b0;

   #10   x1 = 1'b0;  x2 = 1'b1;
   #10   x1 = 1'b1;  x2 = 1'b1;
   #10   x1 = 1'b1;  x2 = 1'b0;
   #10   x1 = 1'b1;  x2 = 1'b1;    //assert z1
   #10   x1 = 1'b1;  x2 = 1'b0;    //assert z1
   #10   x1 = 1'b0;  x2 = 1'b0;

   #10   x1 = 1'b0;  x2 = 1'b1;
   #10   x1 = 1'b0;  x2 = 1'b0;

   #10   x1 = 1'b1;  x2 = 1'b0;
   #10   x1 = 1'b1;  x2 = 1'b1;    //assert z1
   #10   x1 = 1'b0;  x2 = 1'b1;
   #10   x1 = 1'b0;  x2 = 1'b0;

   #10   $stop;
end

//instantiate the module into the test bench as a single line
asm_pla inst1 (x1, x2, y1e, y2e, z1);

endmodule
```

Figure 4.36   Test bench module for the asynchronous sequential machine of Example 4.5.

```
x1x2 = 00, z1 = 0
x1x2 = 10, z1 = 0
x1x2 = 11, z1 = 1
x1x2 = 10, z1 = 1
x1x2 = 11, z1 = 1
x1x2 = 10, z1 = 1
x1x2 = 00, z1 = 0

x1x2 = 01, z1 = 0
x1x2 = 11, z1 = 0
x1x2 = 10, z1 = 0
x1x2 = 11, z1 = 1
x1x2 = 10, z1 = 1

x1x2 = 00, z1 = 0
x1x2 = 01, z1 = 0
x1x2 = 00, z1 = 0
x1x2 = 10, z1 = 0
x1x2 = 11, z1 = 1
x1x2 = 01, z1 = 0
x1x2 = 00, z1 = 0
```

Figure 4.37   Outputs for the asynchronous sequential machine of Example 4.5.

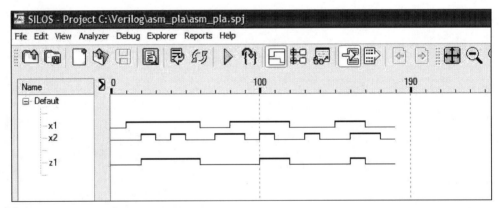

Figure 4.38   Waveforms for the asynchronous sequential machine of Example 4.5.

Example 4.6   This example repeats the Mealy machine of Example 4.5 without using a PLA. Also, this example uses dataflow modeling instead of built-in primitives. Recall that the machine has two inputs x_1 and x_2 and one output z_1. The machine specifications state that z_1 is to be asserted coincident with the first assertion of x_2 if and only if x_1 is already asserted. Output z_1 is to be deasserted simultaneously with the deassertion of x_1. Output z_1 is to change value as fast as possible. The excitation equations will be in a sum-of-products form. The timing diagram is reproduced in Figure 4.39.

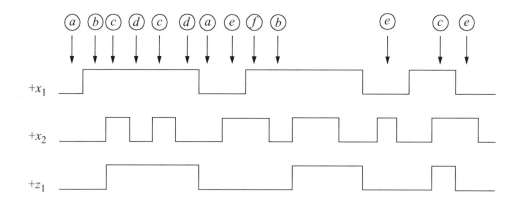

Figure 4.39   A representative timing diagram for the asynchronous sequential machine of Example 4.6.

Dataflow modeling using the continuous assignment statement is used to describe combinational logic where the output of the circuit is evaluated whenever an input changes value; that is, the value of the right-hand side expression is *continuously assigned* to the left-hand side net. Continuous assignments are similar to Boolean algebra, which is a systematic treatment of logical operations. Continuous assignments can be used only for nets, not for register variables.

Dataflow modeling allows implementation of a logical function at a higher level of abstraction than gate level modeling using built-in primitives. The fundamental method of designing in dataflow is the continuous assignment statement **assign**, an example of which is shown below. The excitation and output equations are reproduced in Equation 4.7 and Equation 4.8, respectively.

$$\textbf{assign}\ \ sum = a \wedge b \wedge cin$$

$$Y_{1e} = y_{1f}y_{2f} + y_{2f}x_2$$
$$Y_{2e} = y_{2f}x_1 + x_1x_2'   \tag{4.7}$$

$$z_1 = y_{1f}x_1 + y_{2f}x_1x_2 \quad \text{Fastest operation} \tag{4.8}$$

The dataflow design module is shown in Figure 4.40. The test bench module is shown in Figure 4.41. The outputs and waveforms are shown in Figure 4.42 and Figure 4.43, respectively.

```verilog
//dataflow for asm pla using dataflow modeling

module asm26_df (rst_n, x1, x2, y1e, y2e, z1);

//define inputs and outputs
input rst_n, x1, x2;
output y1e, y2e, z1;

//define internal nets
wire net1, net2, net3, net4, net5, net6;

//define the logic for y1e
assign    net1 = y1e & y2e & rst_n,
          net2 = y2e & x2 & rst_n,
          y1e = net1 | net2;

//define the logic for y2e
assign    net3 = y2e & x1 & rst_n,
          net4 = x1 & ~x2,
          y2e = net3 | net4;

//define the logic for output z1
assign    net5 = y1e & x1,
          net6 = y2e & x1 & x2,
          z1 = net5 | net6;

endmodule
```

Figure 4.40   Dataflow design module for the asynchronous sequential machine of Example 4.6.

```verilog
//test bench for the pla asm using dataflow modeling

module am26_df_tb;

//inputs are reg for test bench
//outputs are wire for test bench
reg rst_n, x1, x2;
wire y1e, y2e, z1;                        //continued on next page
```

Figure 4.41   Test bench module for Example 4.6.

```
//display variables
initial
$monitor ("x1x2 = %b,  state = %b,  z1 = %b",
          {x1, x2}, {y1e, y2e}, z1);

//apply input vectors
initial
begin
   #0    rst_n = 1'b0;
         x1 = 1'b0;
         x2 = 1'b0;

   #5    rst_n = 1'b1;

   #10   x1 = 1'b0;   x2 = 1'b0;
   #10   x1 = 1'b1;   x2 = 1'b0;
   #10   x1 = 1'b1;   x2 = 1'b1;    //assert z1
   #10   x1 = 1'b1;   x2 = 1'b0;    //assert z1
   #10   x1 = 1'b1;   x2 = 1'b1;    //assert z1
   #10   x1 = 1'b1;   x2 = 1'b0;    //assert z1
   #10   x1 = 1'b0;   x2 = 1'b0;

   #10   x1 = 1'b0;   x2 = 1'b1;
   #10   x1 = 1'b1;   x2 = 1'b1;
   #10   x1 = 1'b1;   x2 = 1'b0;
   #10   x1 = 1'b1;   x2 = 1'b1;    //assert z1
   #10   x1 = 1'b1;   x2 = 1'b0;    //assert z1
   #10   x1 = 1'b0;   x2 = 1'b0;

   #10   x1 = 1'b0;   x2 = 1'b1;
   #10   x1 = 1'b0;   x2 = 1'b0;

   #10   x1 = 1'b1;   x2 = 1'b0;
   #10   x1 = 1'b1;   x2 = 1'b1;    //assert z1
   #10   x1 = 1'b0;   x2 = 1'b1;
   #10   x1 = 1'b0;   x2 = 1'b0;

   #10   $stop;

end

//instantiate the module into the test bench as a single line
asm26_df inst1 (rst_n, x1, x2, y1e, y2e, z1);

endmodule
```

Figure 4.41   (Continued)

```
x1x2 = 00, state = 00, z1 = 0
x1x2 = 10, state = 01, z1 = 0
x1x2 = 11, state = 11, z1 = 1
x1x2 = 10, state = 11, z1 = 1
x1x2 = 11, state = 11, z1 = 1
x1x2 = 10, state = 11, z1 = 1
x1x2 = 00, state = 00, z1 = 0


x1x2 = 01, state = 00, z1 = 0
x1x2 = 11, state = 00, z1 = 0
x1x2 = 10, state = 01, z1 = 0
x1x2 = 11, state = 11, z1 = 1
x1x2 = 10, state = 11, z1 = 1


x1x2 = 00, state = 00, z1 = 0
x1x2 = 01, state = 00, z1 = 0
x1x2 = 00, state = 00, z1 = 0
x1x2 = 10, state = 01, z1 = 0
x1x2 = 11, state = 11, z1 = 1
x1x2 = 01, state = 00, z1 = 0
x1x2 = 00, state = 00, z1 = 0
```

Figure 4.42   Outputs for Example 4.6.

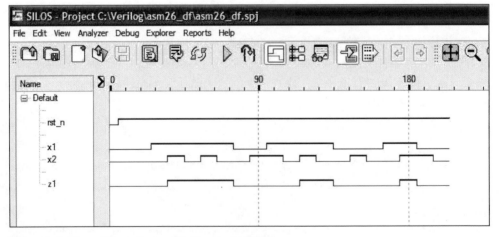

Figure 4.43   Waveforms for Example 4.6.

Example 4.7   This example repeats the Mealy machine of Example 4.5, but uses behavioral modeling with the **case** statement, which is a convenient way to achieve multiple-way branching. The keywords **case, endcase,** and **default** are used in the **case** statement as previously indicated in Section 4.1.3. The **case** statement executes one of several different procedural statements depending on the comparison of an expression with a case item. For convenience, the timing diagram and the primitive flow table are reproduced in Figure 4.44 and Figure 4.45, respectively.

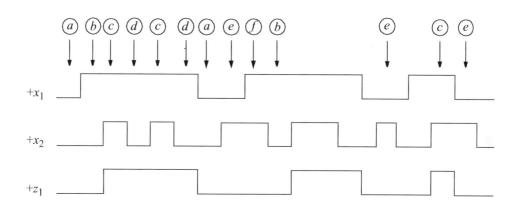

Figure 4.44   A representative timing diagram for the asynchronous sequential machine of Example 4.7.

$[1:3]\, y_e$ $\overset{x_1 x_2}{}$	00	01	11	10	z_1
$(a) = 000$	(a)	e	$-$	b	0
$(b) = 001$	a	$-$	c	(b)	0
$(c) = 011$	$-$	e	(c)	d	1
$(d) = 010$	a	$-$	c	(d)	1
$(e) = 110$	a	(e)	f	$-$	0
$(f) = 111$	$-$	e	(f)	b	0

Figure 4.45   Primitive flow table for the asynchronous sequential machine of Example 4.7.

The behavioral design module is shown in Figure 4.46 and the test bench module is shown in Figure 4.47. The outputs and waveforms are shown in Figure 4.48 and Figure 4.49, respectively.

```
//behavioral for asynchronous sequential machine

module asm26_bh (rst_n, x1, x2, ye, z1);

//define inputs and outputs
input rst_n, x1, x2;
output [1:3] ye;
output z1;

//inputs are wire by default

//variables are reg in always
reg [1:3] ye, next_state;
reg z1;

//assign state codes; parameter defines a constant
//state names must have at least 2 characters
parameter    state_a = 3'b000,
             state_b = 3'b001,
             state_c = 3'b011,
             state_d = 3'b010,
             state_e = 3'b110,
             state_f = 3'b111;

//set next state
always @ (rst_n or x1 or x2)
begin
   if (~rst_n)      //if (~rst_n) is true
      ye <= state_a;
   else
      ye <= next_state;
end

//define output z1 for each state
always @ (ye)
begin
   if (ye == state_a)
      z1 = 1'b0;

   if (ye == state_b)
      z1 = 1'b0;                     //continued on next page
```

Figure 4.46   Behavioral design module for the asynchronous sequential machine of Example 4.7.

```
   if (ye == state_c)
      z1 = 1'b1;

   if (ye == state_d)
      z1 = 1'b1;

   if (ye == state_e)
      z1 = 1'b0;

   if (ye == state_f)
      z1 = 1'b0;
end

always @ (x1 or x2)       //determine next state
begin
   case (ye)
      state_a:
         if (x1==1'b0 & x2==1'b1)
            next_state = state_e;
         else if (x1==1'b1 & x2==1'b0)
            next_state = state_b;
         else
            next_state = state_a;

      state_b:
         if (x1==1'b0 & x2==1'b0)
            next_state = state_a;
         else if (x1==1'b1 & x2==1'b1)
            next_state = state_c;
         else
            next_state = state_b;

   state_c:
         if (x1==1'b0 & x2==1'b1)
            next_state = state_e;
         else if (x1==1'b1 & x2==1'b0)
            next_state = state_d;
         else
            next_state = state_c;

      state_d:
         if (x1==1'b0 & x2==1'b0)
            next_state = state_a;
         else if (x1==1'b1 & x2==1'b1)
            next_state = state_c;
         else
            next_state = state_d;   //continued on next page
```

Figure 4.46   (Continued)

```
      state_e:
         if (x1==1'b0 & x2==1'b0)
            next_state = state_a;
         else if (x1==1'b1 & x2==1'b1)
            next_state = state_f;
         else
            next_state = state_e;

      state_f:
         if (x1==1'b0 & x2==1'b1)
            next_state = state_e;
         else if (x1==1'b1 & x2==1'b0)
            next_state = state_b;
         else
            next_state = state_f;

      default: next_state = state_a;

   endcase
end

endmodule
```

Figure 4.46   (Continued)

```
//test bench for the asm using behavioral modeling

module asm26_bh_tb;

//inputs are reg for test bench
//outputs are wire for test bench
reg rst_n, x1, x2;
wire [1:3] ye;
wire z1;

//display variables
initial
$monitor ("x1x2 = %b, state = %b, z1 = %b", {x1, x2}, ye, z1);

                        //continued on next page
```

Figure 4.47   Test bench module for the asynchronous sequential machine of Example 4.7.

```
//apply input vectors
initial
begin
   #0    rst_n = 1'b0;
         x1 = 1'b0;
         x2 = 1'b0;
   #5    rst_n = 1'b1;


   #10   x1 = 1'b0;   x2 = 1'b0;
   #10   x1 = 1'b1;   x2 = 1'b0;
   #10   x1 = 1'b1;   x2 = 1'b1;   //assert z1
   #10   x1 = 1'b1;   x2 = 1'b0;   //assert z1
   #10   x1 = 1'b1;   x2 = 1'b1;   //assert z1
   #10   x1 = 1'b1;   x2 = 1'b0;   //assert z1
   #10   x1 = 1'b0;   x2 = 1'b0;


   #10   x1 = 1'b0;   x2 = 1'b1;
   #10   x1 = 1'b1;   x2 = 1'b1;
   #10   x1 = 1'b1;   x2 = 1'b0;
   #10   x1 = 1'b1;   x2 = 1'b1;   //assert z1
   #10   x1 = 1'b1;   x2 = 1'b0;   //assert z1
   #10   x1 = 1'b0;   x2 = 1'b0;


   #10   x1 = 1'b0;   x2 = 1'b1;
   #10   x1 = 1'b0;   x2 = 1'b0;


   #10   x1 = 1'b1;   x2 = 1'b0;
   #10   x1 = 1'b1;   x2 = 1'b1;   //assert z1
   #10   x1 = 1'b0;   x2 = 1'b1;
   #10   x1 = 1'b0;   x2 = 1'b0;


   #10   $stop;
end

//instantiate the module into the test bench as a single line
asm26_bh inst1 (rst_n, x1, x2, ye, z1);

endmodule
```

Figure 4.47   (Continued)

```
x1x2 = 00, state = 000, z1 = 0
x1x2 = 10, state = 001, z1 = 0
x1x2 = 11, state = 011, z1 = 1
x1x2 = 10, state = 010, z1 = 1
x1x2 = 11, state = 011, z1 = 1
x1x2 = 10, state = 010, z1 = 1
x1x2 = 00, state = 000, z1 = 0


x1x2 = 01, state = 110, z1 = 0
x1x2 = 11, state = 111, z1 = 0
x1x2 = 10, state = 001, z1 = 0
x1x2 = 11, state = 011, z1 = 1
x1x2 = 10, state = 010, z1 = 1


x1x2 = 00, state = 000, z1 = 0
x1x2 = 01, state = 110, z1 = 0
x1x2 = 00, state = 000, z1 = 0
x1x2 = 10, state = 001, z1 = 0
x1x2 = 11, state = 011, z1 = 1
x1x2 = 01, state = 110, z1 = 0
x1x2 = 00, state = 000, z1 = 0
```

Figure 4.48   Outputs for the asynchronous sequential machine of Example 4.7.

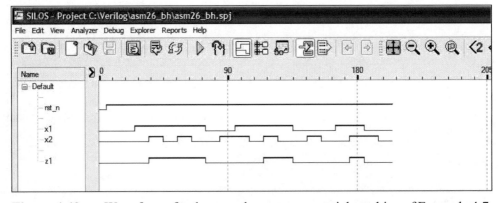

Figure 4.49   Waveforms for the asynchronous sequential machine of Example 4.7.

Example 4.8 A Mealy asynchronous sequential machine has two inputs x_1 and x_2 and one output z_1. The merged flow table is shown in Figure 4.50. The transition diagram is shown in Figure 4.51 in which the rows of the merged flow table are arranged so that all state transitions can be realized without regard for race conditions. Thus, all contiguous rows in a cycle are adjacent.

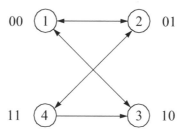

Figure 4.50 Merged flow table for Example 4.8..

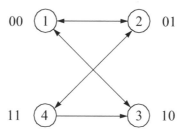

Figure 4.51 Transition diagram for Example 4.8.

Figure 4.52 illustrates the combined excitation map in which the rows are reordered and assigned the appropriate state variable codes. The individual excitation maps are shown in Figure 4.53. The excitation equations are listed in Equation 4.9 and in Equation 4.10 in a sum-of-products form and in a product-of-sums form. Both forms are free of static-1 and static-0 hazards, respectively. The sum-of-products form will be used in the implementation of the asynchronous sequential machine, since the product-of-sums equation requires additional logic gates.

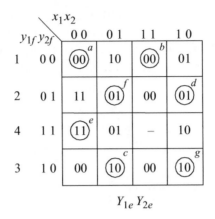

Figure 4.52   Combined excitation map for Example 4.8.

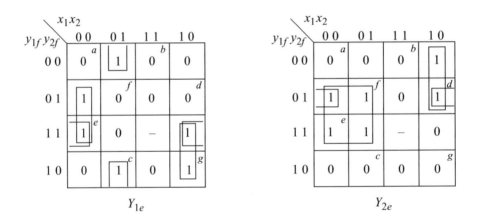

Figure 4.53   Individual excitation maps for Example 4.8.

$$Y_{1e} = y_{2f}x_1'x_2' + y_{2f}'x_1'x_2 + y_{1f}x_1x_2' + y_{1f}y_{2f}x_2' \quad (net1 \dots net4)$$

$$Y_{1e} = (y_{1f} + x_1')(x_1' + x_2')(y_{2f}' + x_2')(y_{2f} + x_1 + x_2)(y_{1f} + y_{2f} + x_2) \quad (4.9)$$

$$Y_{2e} = y_{2f}x_1' + y_{1f}'x_1x_2' + y_{1f}'y_{2f}x_2' \quad (net5 \dots net7)$$

$$Y_{2e} = (y_{1f}' + x_1')(y_{2f} + x_1)(x_1' + x_2')(y_{1f}' + y_{2f})(y_{2f} + x_2') \quad (4.10)$$

Assume that output z_1 has the values in the corresponding stable states, as shown below. The output map is shown in Figure 4.54. All state transitions preclude the possibility of momentary false outputs. This is achieved by assigning appropriate output values to the intermediate transient states of a cycle, in which the initial and final stable states contain identical output values. Also, output response time is devised to be as fast as possible. Thus, if the initial and final stable states have different output values, then the output variable changes value in the first unstable state of a cycle. The output equation is shown in Equation 4.11.

Stable state	Output z_1
a	0
b	1
c	0
d	1
e	0
f	1
g	1

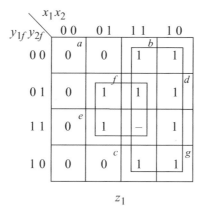

Figure 4.54 Output map for the asynchronous sequential machine of Example 4.8. Output values are assigned such that there will be no momentary false outputs and output changes will be as fast as possible.

$$z_1 = x_1 + y_{2f}x_2 \quad \text{(net8)} \qquad (4.11)$$

The design module using **nand** built-in primitive logic gates is shown in Figure 4.55. A reset input is applied so that all excitation variables are initialized to a value of logic zero. The test bench module is shown in Figure 4.56. The outputs and waveforms are shown in Figure 4.57 and Figure 4.58, respectively.

```verilog
//built-in primitive design for a Mealy asm

module asm29_bip (rst_n, x1, x2, y1e, y2e, z1);

//define inputs and output
input rst_n, x1, x2;
output y1e, y2e, z1;

//define internal nets
wire net1, net2, net3, net4, net5, net6, net7, net8;

//design the logic for y1e
nand    (net1, y2e, ~x1, ~x2, rst_n),
        (net2, ~y2e, ~x1, x2, rst_n),
        (net3, y1e, x1, ~x2, rst_n),
        (net4, y1e, y2e, ~x2, rst_n),
        (y1e, net1, net2, net3, net4);

//design the logic for y2e
nand    (net5, y2e, ~x1, rst_n),
        (net6, ~y1e, x1, ~x2, rst_n),
        (net7, ~y1e, y2e, ~x2, rst_n),
        (y2e, net5, net6, net7);

//design the logic for output z1
nand    (net8, y2e, x2),
        (z1, ~x1, net8);

endmodule
```

Figure 4.55   Design module for the Mealy asynchronous sequential machine of Example 4.8.

```verilog
//test bench for the Mealy asm
module ams_bip_tb;

//inputs are reg for test bench
//outputs are wire for test bench
reg rst_n, x1, x2;
wire y1e, y2e, z1;

//display variables
initial
$monitor ("x1x1 = %b, state = %b, z1= %b",
            {x1, x2}, {y1e, y2e}, z1);

//apply input vectors
initial
begin
   #0      rst_n = 1'b0;
           x1 = 1'b0;
           x2 = 1'b0;
   #5      rst_n = 1'b1;

   #10     x1 = 1'b0;   x2 = 1'b0;
   #10     x1 = 1'b1;   x2 = 1'b0;
   #10     x1 = 1'b1;   x2 = 1'b1;
   #10     x1 = 1'b0;   x2 = 1'b1;

   #10     x1 = 1'b0;   x2 = 1'b0;
   #10     x1 = 1'b1;   x2 = 1'b0;
   #10     x1 = 1'b0;   x2 = 1'b0;
   #10     x1 = 1'b0;   x2 = 1'b1;
   #10     x1 = 1'b1;   x2 = 1'b1;
   #10     x1 = 1'b1;   x2 = 1'b0;
   #10     x1 = 1'b0;   x2 = 1'b0;
   #10     $stop;
end

//instantiate the module into the test bench
asm29_bip inst1 (
   .rst_n(rst_n),
   .x1(x1),
   .x2(x2),
   .y1e(y1e),
   .y2e(y2e),
   .z1(z1)
   );
endmodule
```

Figure 4.56   Test bench for the Mealy machine of Example 4.8.

```
x1x1 = 00, state = 00, z1= 0
x1x1 = 10, state = 01, z1= 1
x1x1 = 11, state = 00, z1= 1
x1x1 = 01, state = 10, z1= 0
x1x1 = 00, state = 00, z1= 0
x1x1 = 10, state = 01, z1= 1
x1x1 = 00, state = 11, z1= 0
x1x1 = 01, state = 01, z1= 1
x1x1 = 11, state = 00, z1= 1
x1x1 = 10, state = 01, z1= 1
x1x1 = 00, state = 11, z1= 0
```

Figure 4.57   Outputs for the Mealy machine of Example 4.8.

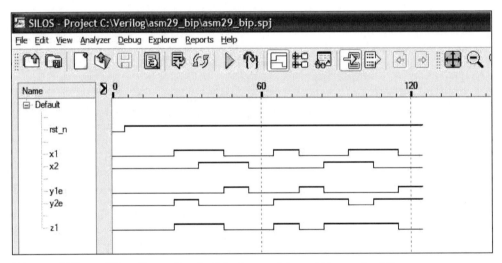

Figure 4.58   Waveforms for the Mealy machine of Example 4.8.

Example 4.9   This example designs a Mealy asynchronous sequential machine using dataflow modeling with the continuous assignment statement. The continuous assignment statement uses the keyword **assign** and has the following syntax with optional drive strength and delay:

assign [drive_strength] [delay] left-hand side target = right-hand side expression

The excitation equations and the output equation are shown in Equation 4.12 and Equation 4.13 in a sum-of-products form, respectively. The logic diagram is shown in Figure 4.59 using AND, OR, and NAND gates displaying the net names. There is an implied reset input to initialize the machine to a value of logic zero.

$$Y_{1e} = y_{2f}x_1'x_2' + y_{2f}'x_1'x_2 + y_{1f}x_1x_2' + y_{1f}y_{2f}x_2'$$

(4.12)

$$Y_{2e} = y_{2f}x_1' + y_{1f}'x_1x_2' + y_{1f}'y_{2f}x_2'$$

$$z_1 = x_1 + y_{2f}x_2$$

(4.13)

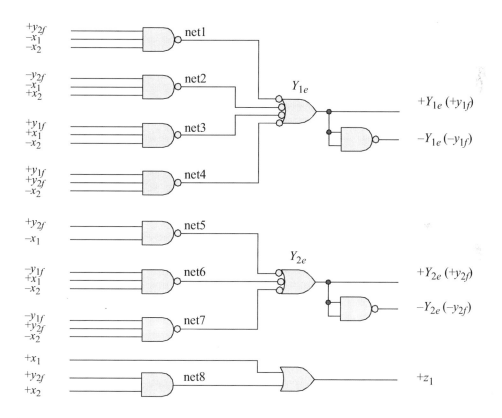

Figure 4.59   Logic diagram for the Mealy asynchronous sequential machine of Example 4.9.

The dataflow design module is shown in Figure 4.60 and the test bench module is shown in Figure 4.61. The outputs and waveforms are shown in Figure 4.62 and Figure 4.63, respectively.

```
//dataflow for Mealy asm

module asm29_df (rst_n, x1, x2, y1e, y2e, z1);

//define inputs and output
input rst_n, x1, x2;
output y1e, y2e, z1;

//define internal nets
wire net1, net2, net3, net4, net6, net7, net8, net10;

//define the logic for y1e
assign    net1 = ~(y2e & ~x1 & ~x2 & rst_n),
          net2 = ~(~y2e & ~x1 & x2 & rst_n),
          net3 = ~(y1e & x1 & ~x2 & rst_n),
          net4 = ~(y1e & y2e & ~x2 & rst_n),
          y1e  = ~(net1 & net2 & net3 & net4);

//define the logic for y2e
assign    net5 = ~(y2e & ~x1 & rst_n),
          net6 = ~(~y1e & x1 & ~x2 & rst_n),
          net7 = ~(~y1e & y2e & ~x2 & rst_n),
          y2e  = ~(net5 & net6 & net7);

//define the logic for output z1
assign    net8 = (y2e & x2);
assign    z1 = x1 | net8;

endmodule
```

Figure 4.60   Dataflow design module for the Mealy machine of Example 4.9.

```
//test bench for the Mealy asm

module asm29_df_tb;

//inputs are reg for test bench
//outputs are wire for test bench
reg rst_n, x1, x2;
wire y1e, y2e, z1;

                              //continued on next page
```

Figure 4.61   Test bench module for the Mealy machine of Example 4.9.

```
//display variables
initial
$monitor ("x1x1 = %b, state = %b, z1= %b",
          {x1, x2}, {y1e, y2e}, z1);


//apply input vectors
initial
begin
   #0     rst_n = 1'b0;
          x1 = 1'b0;
          x2 = 1'b0;

   #5     rst_n = 1'b1;


   #10    x1 = 1'b0;   x2 = 1'b0;
   #10    x1 = 1'b1;   x2 = 1'b0;
   #10    x1 = 1'b1;   x2 = 1'b1;
   #10    x1 = 1'b0;   x2 = 1'b1;


   #10    x1 = 1'b0;   x2 = 1'b0;
   #10    x1 = 1'b1;   x2 = 1'b0;
   #10    x1 = 1'b0;   x2 = 1'b0;
   #10    x1 = 1'b0;   x2 = 1'b1;
   #10    x1 = 1'b1;   x2 = 1'b1;
   #10    x1 = 1'b1;   x2 = 1'b0;
   #10    x1 = 1'b0;   x2 = 1'b0;


   #10    $stop;

end


//instantiate the module into the test bench
asm29_df inst1 (
   .rst_n(rst_n),
   .x1(x1),
   .x2(x2),
   .y1e(y1e),
   .y2e(y2e),
   .z1(z1)
   );

endmodule
```

Figure 4.61   (Continued)

```
x1x1 = 00, state = 00, z1= 0
x1x1 = 10, state = 01, z1= 1
x1x1 = 11, state = 00, z1= 1
x1x1 = 01, state = 10, z1= 0
x1x1 = 00, state = 00, z1= 0
x1x1 = 10, state = 01, z1= 1
x1x1 = 00, state = 11, z1= 0
x1x1 = 01, state = 01, z1= 1
x1x1 = 11, state = 00, z1= 1
x1x1 = 10, state = 01, z1= 1
x1x1 = 00, state = 11, z1= 0
```

Figure 4.62   Outputs for the Mealy machine of Example 4.9.

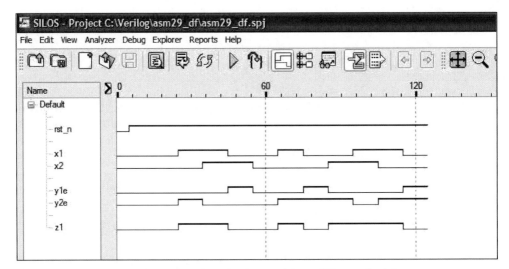

Figure 4.63   Waveforms for the Mealy machine of Example 4.9.

Example 4.10   A Mealy asynchronous sequential machine has two inputs x_1 and x_2 and one output z_1. The machine is reset initially; that is, $x_1 x_2 z_1 = 000$. A specific condition of the operational characteristics is that input x_1 must envelop all occurrences of the x_2 pulse. Thus, the allowable input vectors are $x_1 x_2 = 00$, 10, or 11; the input combination of $x_1 x_2 = 01$ will never occur. Output z_1 is to be asserted coincident with the assertion of every second x_2 pulse and is to remain asserted until the deassertion of x_2. Output assertion is to be as fast as possible.

A representative timing diagram is shown in Figure 4.64. Although the timing diagram illustrates a valid input sequence to generate an output, other variations are possible and must be considered to adequately represent the operation of the machine for all valid input sequences. The primitive flow table, shown in Figure 4.65, is obtained by assigning a unique stable state name to each different combination of the input vector and the associated output z_1 in the timing diagram.

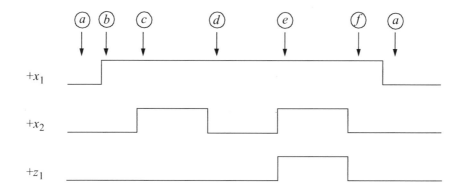

Figure 4.64   Timing diagram for the asynchronous sequential machine of Example 4.10.

x_1x_2	00	01	11	10	z_1
a	a	–	–	b	0
b	a	–	c	b	0
c	–	–	c	d	0
d	a	–	e	d	0
e	–	–	e	f	1
f	a	–	c	f	0

Figure 4.65   Primitive flow table for the asynchronous sequential machine of Example 4.10.

The reduced primitive flow table is shown in Figure 4.66. Rows a and b can merge, because there is no conflict in any column of the two rows. The only other row with which row a can merge is row e — all other rows have a conflict in at least one column. The merged flow table for partition {ⓐ, ⓔ}, {ⓑ}, {ⓒ}, {ⓓ} is shown in Figure 4.67.

x_1x_2 00	01	11	10	z_1
ⓐ	–	–	b	0
a	–	c	ⓑ	0
–	–	ⓒ	d	0
a	–	e	ⓓ	0
–	–	ⓔ	b	1

Figure 4.66 Reduced primitive flow table obtained from the primitive flow table of Figure 4.65.

$y_{1f}\,y_{2f}$		x_1x_2 00	01	11	10
00	ⓐ,ⓔ	ⓐ	–	ⓔ	b
01	ⓑ	a	–	c	ⓑ
11	ⓒ	–	–	ⓒ	d
10	ⓓ	a	–	e	ⓓ

Figure 4.67 Merged flow table for partition {ⓐ, ⓔ}, {ⓑ}, {ⓒ}, {ⓓ}.

The combined excitation map for excitation variables Y_{1e} and Y_{2e} is shown in Figure 4.68. The stable states are assigned excitation values that are the same as the feedback values of the corresponding rows. It is important to not inadvertently assign excitation values to the "don't care" states that would generate a stable state. The individual excitation maps are shown in Figure 4.69 and the resulting hazard-free excitation equations in Equation 4.14 in a sum-of-products form. The rightmost term in each equation is the hazard cover.

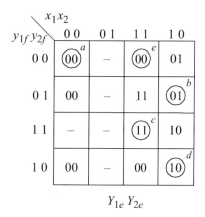

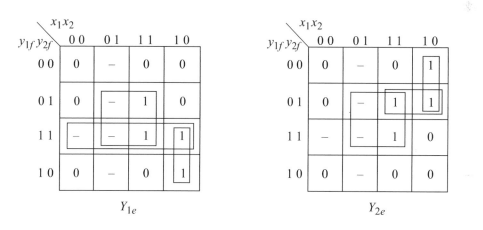

Figure 4.68   Combined excitation map for the merged flow table of Figure 4.67.

Figure 4.69   Individual excitation maps for Y_{1e} and Y_{2e} obtained from the combined excitation map of Figure 4.68.

$$Y_{1e} = y_{2f}x_2 + y_{1f}x_1x_2' + y_{1f}y_{2f}$$

$$Y_{2e} = y_{2f}x_2 + y_{1f}'x_1x_2' + y_{1f}'y_{2f}x_1 \qquad (4.14)$$

The output map is constructed from the merged flow table of Figure 4.67 and the reduced primitive flow table of Figure 4.66. The merged flow table indicates the location of the stable states and the reduced primitive flow table specifies the output values of the stable states. The output map is shown in Figure 4.70 and the output equation is shown in Equation 4.15.

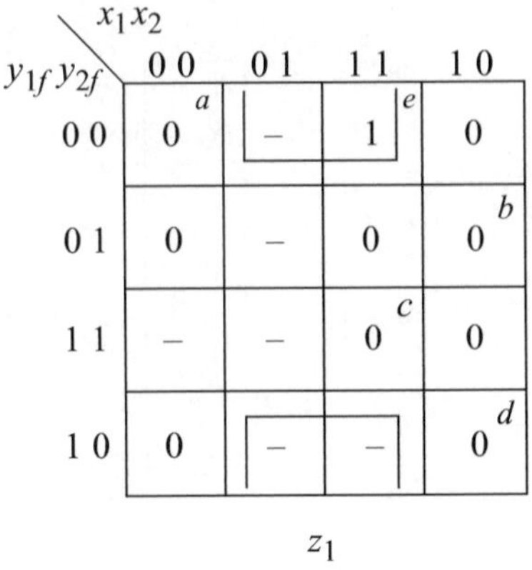

Figure 4.70   Output map for Example 4.10.

$$z_1 = y_{2f}{}' x_2 \tag{4.15}$$

The design module using dataflow modeling is shown in Figure 4.71 using the excitation and output equations of Equation 4.14 and Equation 4.15, respectively. The test bench module is shown in Figure 4.72. The outputs and waveforms are shown in Figure 4.73 and Figure 4.74, respectively.

```
//dataflow asynchronous sequential machine
module asm24_df (rst_n, x1, x2, y1e, y2e, z1);

//define inputs and outputs
input rst_n, x1, x2;
output y1e, y2e, z1;

//define internal nets
wire net1, net2, net3, net4, net5;

//design the logic for y1e
assign    net1 = y2e & x2 & rst_n,
          net2 = y1e & x1 & ~x2 & rst_n,
          net3 = y1e & y2e & rst_n,
          y1e = net1 | net2 | net3;   //continued on next page
```

Figure 4.71   Dataflow design module for Mealy machine of Example 4.10.

```
//design the logic for y2e
assign    net4 = ~y1e & x1 & ~x2 & rst_n,
          net5 = ~y1e & y2e & x1 & rst_n,
          y2e = net1 | net4 | net5;

//design the logic for output z1
assign    z1 = ~y2e & x2;

endmodule
```

Figure 4.71   (Continued)

```
//test bench for the asynchronous sequential machine

module asm24_df_tb;

//inputs are reg for test bench
//outputs are wire for test bench
reg rst_n, x1, x2;
wire y1e, y2e, z1;

//display variables
initial
$monitor ("x1x2 = %b, state = %b, z1 = %b",
          {x1, x2}, {y1e, y2e}, z1);

//apply input vectors
initial
begin
   #0     rst_n = 1'b0;      //reset to state_a (000)
          x1 = 1'b0;
          x2 = 1'b0;

   #5     rst_n = 1'b1;      //deassert reset

   #10    x1=1'b1; x2=1'b0; //go to state_b (001)
   #10    x1=1'b0; x2=1'b0; //go to state_a (000)
   #10    x1=1'b1; x2=1'b0; //go to state_b (001)
   #10    x1=1'b1; x2=1'b1; //go to state_c (011)

   #10    x1=1'b1; x2=1'b0; //go to state_d (010)
   #10    x1=1'b0; x2=1'b0; //go to state_a (000)
                                     //continued on next page
```

Figure 4.72   Test bench module for the Mealy machine of Example 4.10.

```
    #10    x1=1'b1; x2=1'b0; //go to state_b (001)
    #10    x1=1'b1; x2=1'b1; //go to state_c (011)

    #10    x1=1'b1; x2=1'b0; //go to state_d (010)
    #10    x1=1'b0; x2=1'b0; //go to state_a (000)

    #10    x1=1'b1; x2=1'b0; //go to state_b (001)
    #10    x1=1'b1; x2=1'b1; //go to state_c (011)
    #10    x1=1'b1; x2=1'b0; //go to state_d (010)
    #10    x1=1'b1; x2=1'b1; //go to state_e (110)    //assert z1

    #10    x1=1'b1; x2=1'b0; //go to state_f (111)
    #10    x1=1'b1; x2=1'b1; //go to state_c (011)
    #10    x1=1'b1; x2=1'b0; //go to state_d (010)
    #10    x1=1'b1; x2=1'b1; //go to state_e (110)    //assert z1

    #10    x1=1'b1; x2=1'b0; //go to state_f (111)
    #10    x1=1'b0; x2=1'b0; //go to state_a (000)

    #10    $stop;

end

//instantiate the module into the test bench as a single line
asm24_df inst1 (rst_n, x1, x2, y1e, y2e, z1);

endmodule
```

Figure 4.72   (Continued)

```
x1x2 = 00, state = 00, z1 = 0        x1x2 = 10, state = 01, z1 = 0
x1x2 = 10, state = 01, z1 = 0        x1x2 = 11, state = 11, z1 = 0
x1x2 = 00, state = 00, z1 = 0        x1x2 = 10, state = 10, z1 = 0
x1x2 = 10, state = 01, z1 = 0        x1x2 = 11, state = 00, z1 = 1
x1x2 = 11, state = 11, z1 = 0
x1x2 = 10, state = 10, z1 = 0        x1x2 = 10, state = 01, z1 = 0
x1x2 = 00, state = 00, z1 = 0        x1x2 = 11, state = 11, z1 = 0
                                     x1x2 = 10, state = 10, z1 = 0
x1x2 = 10, state = 01, z1 = 0        x1x2 = 11, state = 00, z1 = 1
x1x2 = 11, state = 11, z1 = 0
x1x2 = 10, state = 10, z1 = 0        x1x2 = 10, state = 01, z1 = 0
x1x2 = 00, state = 00, z1 = 0        x1x2 = 00, state = 00, z1 = 0
```

Figure 4.73   Outputs for the Mealy machine of Example 4.10.

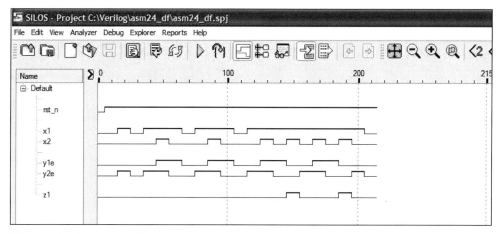

Figure 4.74   Waveforms for the Mealy machine of Example 4.10.

Example 4.11   This example repeats Example 4.10, but uses behavioral modeling with the **case** statement in the design process. The **case** statement is an alternative to the **if** ... **else** construct and may simplify the readability of the Verilog code. The **case** statement is a multiple-way conditional branch.

Behavioral modeling describes the behavior of the machine and does not require direct implementation with logic gates. This is an algorithmic approach which describes the architecture of the machine. Therefore, only a primitive flow table is required for this design. The timing diagram is reproduced in Figure 4.75 and the primitive flow table is reproduced in Figure 4.76 for convenience.

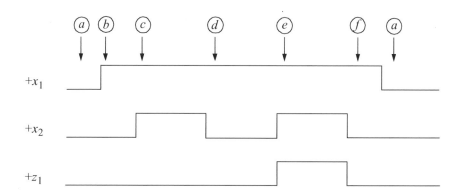

Figure 4.75   Timing diagram for the asynchronous sequential machine of Example 4.11.

$[1:3]\, y_e$ $\backslash$ x_1x_2	00	01	11	10	z_1
$\textcircled{a}$ = 000	$\textcircled{a}$	–	–	b	0
$\textcircled{b}$ = 001	a	–	c	$\textcircled{b}$	0
$\textcircled{c}$ = 011	–	–	$\textcircled{c}$	d	0
$\textcircled{d}$ = 010	a	–	e	$\textcircled{d}$	0
$\textcircled{e}$ = 110	–	–	$\textcircled{e}$	f	1
$\textcircled{f}$ = 111	a	–	c	$\textcircled{f}$	0

Figure 4.76   Primitive flow table for the asynchronous sequential machine of Example 4.11.

The behavioral design module is shown in Figure 4.77. The test bench module is shown in Figure 4.78, which takes the Mealy machine through a variety of paths in the primitive flow table. The outputs and the waveforms are shown in Figure 4.79 and Figure 4.80, respectively.

```
//behavioral Mealy asynchronous sequential machine

module asm24_bh (rst_n, x1, x2, ye, z1);

//define inputs and outputs
input rst_n, x1, x2;
output [1:3] ye;
output z1;

wire rst_n, x1, x2;      //alternatively do not declare wires
                         //because inputs are wire by default

reg [1:3] ye, next_state;   //variables are reg in always
reg z1;
                            //continued on next page
```

Figure 4.77   Behavioral design module for the Mealy asynchronous sequential machine of Example 4.11.

```
//assign state codes; parameter defines a constant
//state names must have at least 2 characters
parameter    state_a = 3'b000,
             state_b = 3'b001,
             state_c = 3'b011,
             state_d = 3'b010,
             state_e = 3'b110,
             state_f = 3'b111;

//set next state
always @ (rst_n or x1 or x2)
begin
      if (~rst_n)
   ye <= state_a;
   else
      ye <= next_state;
end

//define output z1
always @ (x1 or x2 or ye)
begin
   if (ye == state_e)        //== is a logical equality
      z1 = 1'b1;
   else
      z1 = 1'b0;
end

//determine next state
always @ (x1 or x2)
begin
   case (ye)
      state_a:
         if (x1==1'b1 & x2==1'b0)
            next_state = state_b;
         else
            next_state = state_a;

      state_b:
         if (x1==1'b0 & x2==1'b0)
            next_state = state_a;
         else if (x1==1'b1 & x2==1'b1)
            next_state = state_c;
         else
            next_state = state_b;

                              //continued on next page
```

Figure 4.77   (Continued)

```
        state_c:
            if (x1==1'b1 & x2==1'b0)
                next_state = state_d;
            else
                next_state = state_c;

        state_d:
            if (x1==1'b0 & x2==1'b0)
                next_state = state_a;
            else if (x1==1'b1 & x2==1'b1)
                next_state = state_e;
            else
                next_state = state_d;

        state_e:
            if (x1==1'b1 & x2==1'b0)
                next_state = state_f;
            else
                next_state = state_e;

        state_f:
            if (x1==1'b0 & x2==1'b0)
                next_state = state_a;
            else if (x1==1'b1 & x2==1'b1)
                next_state = state_c;
            else
                next_state = state_f;

        default: next_state = state_a;
    endcase
end
endmodule
```

Figure 4.77   (Continued)

```
//test bench for asynchronous sequential machine
module asm24_bh_tb;

//inputs are reg for test bench
//outputs are wire for test bench
reg rst_n, x1, x2;
wire [1:3] ye;                       //continued on next page
```

Figure 4.78   Test bench module for the Mealy asynchronous sequential machine of Example 4.11.

```
//display variables
initial
$monitor ("x1x2 = %b, state = %b, z1 = %b", {x1, x2}, ye, z1);

//define input vectors
initial
begin
   #0    rst_n = 1'b0;      //reset to state_a (000)
         x1 = 1'b0;
         x2 = 1'b0;

   #5    rst_n = 1'b1;      //deassert reset

   #10   x1=1'b1; x2=1'b0; //go to state_b (001)
   #10   x1=1'b0; x2=1'b0; //go to state_a (000)
   #10   x1=1'b1; x2=1'b0; //go to state_b (001)
   #10   x1=1'b1; x2=1'b1; //go to state_c (011)

   #10   x1=1'b1; x2=1'b0; //go to state_d (010)
   #10   x1=1'b0; x2=1'b0; //go to state_a (000)

   #10   x1=1'b1; x2=1'b0; //go to state_b (001)
   #10   x1=1'b1; x2=1'b1; //go to state_c (011)

   #10   x1=1'b1; x2=1'b0; //go to state_d (010)
   #10   x1=1'b0; x2=1'b0; //go to state_a (000)

   #10   x1=1'b1; x2=1'b0; //go to state_b (001)
   #10   x1=1'b1; x2=1'b1; //go to state_c (011)
   #10   x1=1'b1; x2=1'b0; //go to state_d (010)
   #10   x1=1'b1; x2=1'b1; //go to state_e (110)   //assert z1

   #10   x1=1'b1; x2=1'b0; //go to state_f (111)
   #10   x1=1'b1; x2=1'b1; //go to state_c (011)
   #10   x1=1'b1; x2=1'b0; //go to state_d (010)
   #10   x1=1'b1; x2=1'b1; //go to state_e (110)   //assert z1

   #10   x1=1'b1; x2=1'b0; //go to state_f (111)
   #10   x1=1'b0; x2=1'b0; //go to state_a (000)

   #10   $stop;
end

//instantiate the module into the test bench as a single line
asm24_bh inst1 (rst_n, x1, x2, ye, z1);

endmodule
```

Figure 4.78   (Continued)

```
x1x2 = 00, state = 000, z1 = 0
x1x2 = 10, state = 001, z1 = 0
x1x2 = 00, state = 000, z1 = 0
x1x2 = 10, state = 001, z1 = 0
x1x2 = 11, state = 011, z1 = 0
x1x2 = 10, state = 010, z1 = 0
x1x2 = 00, state = 000, z1 = 0

x1x2 = 10, state = 001, z1 = 0
x1x2 = 11, state = 011, z1 = 0
x1x2 = 10, state = 010, z1 = 0
x1x2 = 00, state = 000, z1 = 0

x1x2 = 10, state = 001, z1 = 0
x1x2 = 11, state = 011, z1 = 0
x1x2 = 10, state = 010, z1 = 0
x1x2 = 11, state = 110, z1 = 1
x1x2 = 10, state = 111, z1 = 0
x1x2 = 11, state = 011, z1 = 0
x1x2 = 10, state = 010, z1 = 0
x1x2 = 11, state = 110, z1 = 1
x1x2 = 10, state = 111, z1 = 0
x1x2 = 00, state = 000, z1 = 0
```

Figure 4.79   Outputs for the Mealy asynchronous sequential machine of Example 4.11.

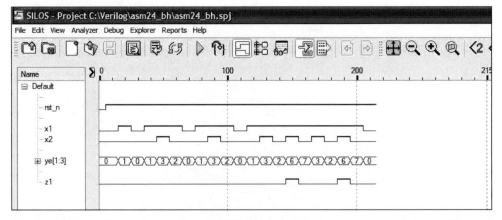

Figure 4.80   Waveforms for the Mealy asynchronous sequential machine of Example 4.11.

Example 4.12 This example repeats Example 4.10, but uses built-in primitives in the structural design. The timing diagram is reproduced in Figure 4.81. The equations for excitation variables Y_{1e} and Y_{2e} are shown in Equation 4.16 in a product-of-sums form. The equation for output z_1 is also shown in Equation 4.16. The logic diagram is shown in Figure 4.82.

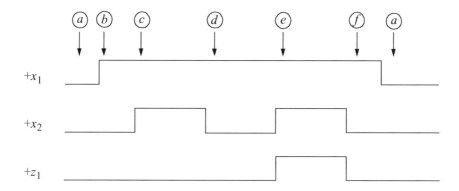

Figure 4.81 Timing diagram for the asynchronous sequential machine of Example 4.12.

$$Y_{1e} = (x_1)\,(y_{2f} + x_2')\,(y_{1f} + x_2)$$

$$Y_{2e} = (x_1)\,(y_{2f} + x_2')\,(y_{1f}' + x_2)$$

$$(4.16)$$

$$z_1 = y_{2f}'\,x_2$$

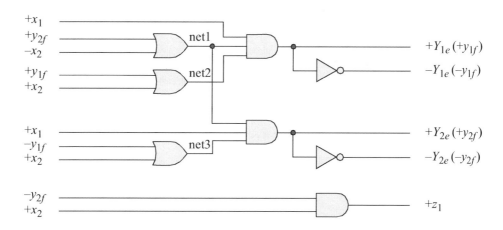

Figure 4.82 Logic diagram for Example 4.12.

The structural design module using built-in primitives is shown in Figure 4.83. The test bench module is shown in Figure 4.84, which takes the Mealy machine through a variety of paths in the primitive flow table of Figure 4.76. The outputs and the waveforms are shown in Figure 4.85 and Figure 4.86, respectively.

```
//structural asynchronous sequential machine

module asm24_struc (x1, x2, y1e, y2e, z1);

//define inputs and outputs
input x1, x2;
output y1e, y2e, z1;

//define internal nets
wire net1, net2, net3;

//design the logic for y1e
or      (net1, y2e, ~x2),
        (net2, y1e, x2);
and     (y1e, x1, net1, net2);

//design the logic for y2e
or      (net3, ~y1e, x2);
and     (y2e, net1, x1, net3);

//design the logic for output z1
and     (z1, ~y2e, x2);

endmodule
```

Figure 4.83   Structural design module for the Mealy asynchronous sequential machine of Example 4.12.

```
//test bench for the asynchronous sequential machine

module asm24_struc_tb;

//inputs are reg for test bench
//outputs are wire for test bench
reg x1, x2;
wire y1e, y2e, z1;                    //continued on next page
```

Figure 4.84   Test bench module for the Mealy asynchronous sequential machine of Example 4.12.

```
//display variables
initial
$monitor ("x1x2 = %b, state = %b, z1 = %b",
          {x1, x2}, {y1e, y2e}, z1);

//apply input vectors
initial
begin
  #0     x1 = 1'b0;
         x2 = 1'b0;

  #10    x1=1'b1; x2=1'b0; //go to state_b (001)
  #10    x1=1'b0; x2=1'b0; //go to state_a (000)
  #10    x1=1'b1; x2=1'b0; //go to state_b (001)
  #10    x1=1'b1; x2=1'b1; //go to state_c (011)

  #10    x1=1'b1; x2=1'b0; //go to state_d (010)
  #10    x1=1'b0; x2=1'b0; //go to state_a (000)

  #10    x1=1'b1; x2=1'b0; //go to state_b (001)
  #10    x1=1'b1; x2=1'b1; //go to state_c (011)

  #10    x1=1'b1; x2=1'b0; //go to state_d (010)
  #10    x1=1'b0; x2=1'b0; //go to state_a (000)

  #10    x1=1'b1; x2=1'b0; //go to state_b (001)
  #10    x1=1'b1; x2=1'b1; //go to state_c (011)
  #10    x1=1'b1; x2=1'b0; //go to state_d (010)
  #10    x1=1'b1; x2=1'b1; //go to state_e (110)   //assert z1

  #10    x1=1'b1; x2=1'b0; //go to state_f (111)
  #10    x1=1'b1; x2=1'b1; //go to state_c (011)
  #10    x1=1'b1; x2=1'b0; //go to state_d (010)
  #10    x1=1'b1; x2=1'b1; //go to state_e (110)   //assert z1

  #10    x1=1'b1; x2=1'b0;//go to state_f (111)
  #10    x1=1'b0; x2=1'b0;//go to state_a (000)

  #10    $stop;
end

//instantiate the module into the test bench as a single line
asm24_struc inst1 (x1, x2, y1e, y2e, z1);

endmodule
```

Figure 4.84   (Continued)

```
x1x2 = 00, state = 00, z1 = 0
x1x2 = 10, state = 01, z1 = 0
x1x2 = 00, state = 00, z1 = 0
x1x2 = 10, state = 01, z1 = 0
x1x2 = 11, state = 11, z1 = 0
x1x2 = 10, state = 10, z1 = 0
x1x2 = 00, state = 00, z1 = 0

x1x2 = 10, state = 01, z1 = 0
x1x2 = 11, state = 11, z1 = 0
x1x2 = 10, state = 10, z1 = 0
x1x2 = 00, state = 00, z1 = 0

x1x2 = 10, state = 01, z1 = 0
x1x2 = 11, state = 11, z1 = 0
x1x2 = 10, state = 10, z1 = 0
x1x2 = 11, state = 00, z1 = 1
x1x2 = 10, state = 01, z1 = 0
x1x2 = 11, state = 11, z1 = 0
x1x2 = 10, state = 10, z1 = 0
x1x2 = 11, state = 00, z1 = 1
x1x2 = 10, state = 01, z1 = 0
x1x2 = 00, state = 00, z1 = 0
```

Figure 4.85   Outputs for the Mealy asynchronous sequential machine of Example 4.12.

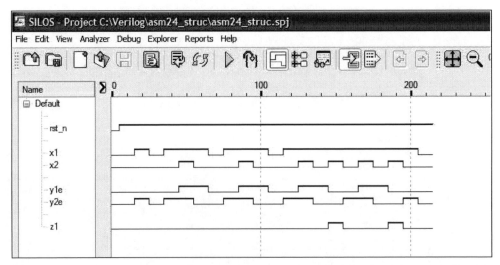

Figure 4.86   Waveforms for the Mealy asynchronous sequential machine of Example 4.12.

Example 4.13   An asynchronous sequential machine has two inputs x_1 and x_2. The machine generates two outputs z_1 and z_2 in accordance with a prescribed input sequence. Output z_1 will be asserted coincident with every second x_2 pulse, but only if x_1 is asserted for the duration of the pair of x_2 pulses. Output z_2 is asserted for every second x_2 pulse, but only if x_1 is deasserted for the duration of the pair of x_2 pulses. Input x_1 will not change state while x_2 is asserted. The outputs will never be active simultaneously, because the outputs are asserted for different values of input x_1. A representative timing diagram is shown in Figure 4.87 illustrating the two sequences that assert outputs z_1 and z_2.

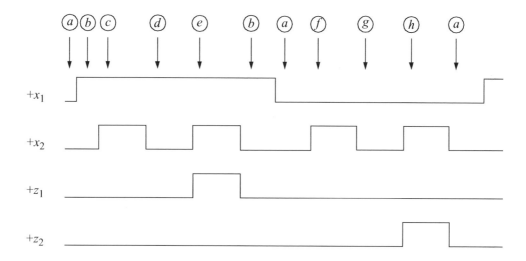

Figure 4.87   Representative timing diagram for the asynchronous sequential machine of Example 4.13.

Using traditional design techniques, the following results of the design procedure will be presented: the primitive flow table; the merged flow table; the augmented merged flow table; the combined excitation map; the individual excitation maps; the excitation equations; the output maps; and the output equations.

The primitive flow table is shown in Figure 4.88; the merged flow table in Figure 4.89; the augmented merged flow table in Figure 4.90; the combined excitation map in Figure 4.91; the individual excitation maps in Figure 4.92; the excitation equations in Equation 4.17 in a sum-of-products form; the output maps in Figure 4.93; and the output equations in Equation 4.18.

x_1x_2 00	01	11	10	z_1	z_2
ⓐ	f	–	b	0	0
a	–	c	ⓑ	0	0
–	–	ⓒ	d	0	0
a	–	e	ⓓ	0	0
–	–	ⓔ	b	1	0
g	ⓕ	–	–	0	0
ⓖ	h	–	b	0	0
a	ⓗ	–	–	0	1

Figure 4.88   Primitive flow table for the asynchronous sequential machine of Example 4.13.

	x_1x_2	00	01	11	10
1	ⓐ,ⓑ	ⓐ	f	c	ⓑ
2	ⓒ,ⓕ	g	ⓕ	ⓒ	d
3	ⓓ,ⓗ	a	ⓗ	e	ⓓ
4	ⓔ,ⓖ	ⓖ	h	ⓔ	b

Figure 4.89   Merged flow table for the asynchronous sequential machine of Example 4.13.

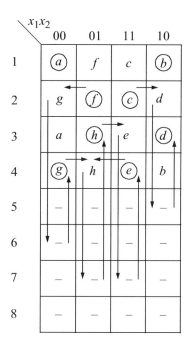

Figure 4.90   Augmented merged flow table for the asynchronous sequential machine of Example 4.13.

	$y_{1f}y_{2f}y_{3f}$	0 0	0 1	1 1	1 0
1	0 0 0	$(000)^a$	001	001	$(000)^b$
2	0 0 1	011	$(001)^f$	$(001)^c$	101
6	0 1 1	010	–	–	–
4	0 1 0	$(010)^g$	110	$(010)^e$	000
7	1 1 0	–	100	010	–
8	1 1 1	–	–	–	–
5	1 0 1	–	–	–	100
3	1 0 0	000	$(100)^h$	110	$(100)^d$

$x_1 x_2$ (column header)

$Y_{1e} Y_{2e} Y_{3e}$

Figure 4.91   Combined excitation map constructed from the augmented merged flow table of Figure 4.90 for Example 4.13.

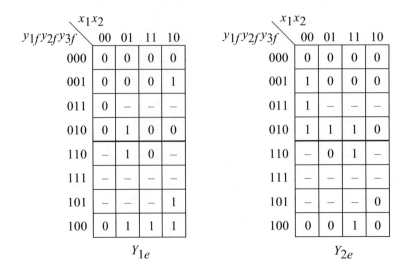

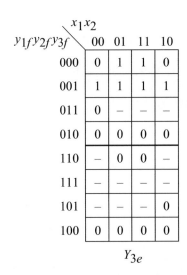

Figure 4.92   Individual excitation maps obtained from the combined excitation map of Figure 4.91 for the asynchronous sequential machine of Example 4.13.

$$Y_{1e} = y_{3f}x_1x_2' + y_{1f}y_{2f}'x_1 + y_{1f}y_{2f}'x_2 + y_{2f}x_1'x_2 + y_{1f}y_{3f}$$

$$Y_{2e} = y_{2f}x_1'x_2' + y_{3f}x_1'x_2' + y_{1f}x_1x_2 + y_{1f}'y_{2f}x_1' + y_{1f}'y_{2f}x_2 + y_{2f}x_1x_2$$

$$Y_{3e} = y_{1f}'y_{2f}'y_{3f} + y_{1f}'y_{2f}'x_2 \qquad\qquad (4.17)$$

Y_{1e} = net 1 . . . net5. Y_{2e} = net6 . . . net11. Y_{3e} = net12, net13.

$y_{1f}y_{2f}y_{3f}$ \ x_1x_2	00	01	11	10
000	0^a	0	0	0^b
001	0	0^f	0^c	0
011	0	–	–	–
010	0^g	0	1^e	0
110	–	0	–	–
111	–	–	–	–
101	–	–	–	0
100	0	0^h	–	0^d

z_1

$y_{1f}y_{2f}y_{3f}$ \ x_1x_2	00	01	11	10
000	0^a	0	0	0^b
001	0	0^f	0^c	0
011	0	–	–	–
010	0^g	–	0^e	0
110	–	–	0	–
111	–	–	–	–
101	–	–	–	0
100	0	1^h	0	0^d

z_2

Figure 4.93   Output maps for the asynchronous sequential machine of Example 4.13.

$$z_1 = y_{2f}x_1x_2$$
$$z_2 = y_{1f}x_1'x_2 \tag{4.18}$$

The dataflow design module is shown in Figure 4.94 using the continuous assignment statement, which can be applied to nets only. The continuous assignment statement uses the keyword **assign**, which has the following syntax with optional drive strength and delay.

assign [drive_strength] [delay] left-hand side target = right-hand side expression

The continuous assignment statement assigns a value to a net (**wire**) that has been previously declared — it cannot be used to assign a value to a register. Therefore, the left-hand target must be a scalar or vector net or a concatenation of scalar and vector nets. The operands on the right-hand side can be registers, nets, or function calls. The registers and nets can be declared as either scalars or vectors.

The test bench module is shown in Figure 4.95, which displays the two outputs in accordance with their respective input sequence. The outputs and waveforms are shown in Figure 4.96 and Figure 4.97, respectively.

```
//dataflow for asm

module asm_sop (rst_n, x1, x2, y1e, y2e, y3e, z1, z2);

//define inputs and outputs
input rst_n, x1, x2;
output y1e, y2e, y3e, z1, z2;

//define internal nets
wire net1, net2, net3, net4, net5, net6, net7;
wire net8, net9, net10, net11, net12, net13;

//design the logic for y1e
assign    net1 = y3e & x1 & ~x2 & rst_n,
          net2 = y1e & ~y2e & x1 & rst_n,
          net3 = y1e & ~y2e & x2 & rst_n,
          net4 = y2e & ~x1 & x2 & rst_n,
          net5 = y1e & y3e & rst_n,
          y1e = net1 | net2 | net3 | net4 | net5;

//design the logic for y2e
assign    net6 = y2e & ~x1 & ~x2 & rst_n,
          net7 = y3e & ~x1 & ~x2 & rst_n,
          net8 = y1e & x1 & x2 & rst_n,
          net9 = ~y1e & y2e & ~x1 & rst_n,
          net10 = ~y1e & y2e & x2 & rst_n,
          net11 = y2e & x1 & x2 & rst_n,
          y2e = net6 | net7 | net8 | net9 | net10 | net11;

//design the logic for y3e
assign    net12 = ~y1e & ~y2e & y3e & rst_n,
          net13 = ~y1e & ~y2e & x2 & rst_n,
          y3e = net12 | net13;

//define the logic for outputs z1 and z2
assign    z1 = y2e & x1 & x2,
          z2 = y1e & ~x1 & x2;

endmodule
```

Figure 4.94   Dataflow design module for the asynchronous sequential machine of Example 4.13 in a sum-of-products form.

```
//test bench for the asm
module asm_pos_tb;

//inputs are reg for test bench
//outputs are wire for test bench
reg rst_n, x1, x2;
wire y1e, y2e, y3e, z1, z2;

initial         //display variables
$monitor ("x1x2 = %b, state = %b, z1z2 = %b",
          {x1, x2}, {y1e, y2e, y3e}, {z1, z2});

initial          //apply input vectors
begin
   #0    rst_n = 1'b0;
         x1 = 1'b0;
         x2 = 1'b0;
   #5    rst_n = 1'b1;

   #10   x1 = 1'b1;   x2 = 1'b0;
   #10   x1 = 1'b1;   x2 = 1'b1;
   #10   x1 = 1'b1;   x2 = 1'b0;
   #10   x1 = 1'b1;   x2 = 1'b1;   //assert z1

   #10   x1 = 1'b1;   x2 = 1'b0;
   #10   x1 = 1'b0;   x2 = 1'b0;
   #10   x1 = 1'b0;   x2 = 1'b1;
   #10   x1 = 1'b0;   x2 = 1'b0;
   #10   x1 = 1'b0;   x2 = 1'b1;   //assert z2
   #10   x1 = 1'b0;   x2 = 1'b0;
   #10   x1 = 1'b1;   x2 = 1'b0;
   #10   $stop;
end

//instantiate the module into the test bench
asm_sop inst1 (
   .rst_n(rst_n),
   .x1(x1),
   .x2(x2),
   .y1e(y1e),
   .y2e(y2e),
   .y3e(y3e),
   .z1(z1),
   .z2(z2)
   );
endmodule
```

Figure 4.95   Test bench module for the asynchronous sequential machine of Example 4.13.

```
x1x2 = 00,  state = 000,  z1z2 = 00
x1x2 = 10,  state = 000,  z1z2 = 00
x1x2 = 11,  state = 001,  z1z2 = 00
x1x2 = 10,  state = 100,  z1z2 = 00
x1x2 = 11,  state = 010,  z1z2 = 10
x1x2 = 10,  state = 000,  z1z2 = 00


x1x2 = 00,  state = 000,  z1z2 = 00
x1x2 = 01,  state = 001,  z1z2 = 00
x1x2 = 00,  state = 010,  z1z2 = 00
x1x2 = 01,  state = 100,  z1z2 = 01
x1x2 = 00,  state = 000,  z1z2 = 00
x1x2 = 10,  state = 000,  z1z2 = 00
```

Figure 4.96   Outputs for the asynchronous sequential machine of Example 4.13.

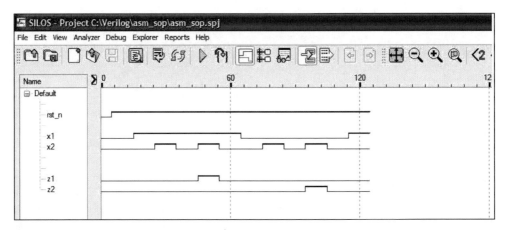

Figure 4.97   Waveforms for the asynchronous sequential machine of Example 4.13.

Example 4.14   This example repeats Example 4.13, however built-in primitives are used in the design module. Also, the excitation and output equations are in a product-of-sums form. The individual excitation maps are reproduced in Figure 4.98. The excitation equations are shown in Equation 4.19 in a product-of-sums form. The output maps are reproduced in Figure 4.99 and the output equations in Equation 4.20.

Recall that a product-of-sums is an expression in which at least one term does not contain all the variables; that is, at least one term is a proper subset of the possible variables or their complements.

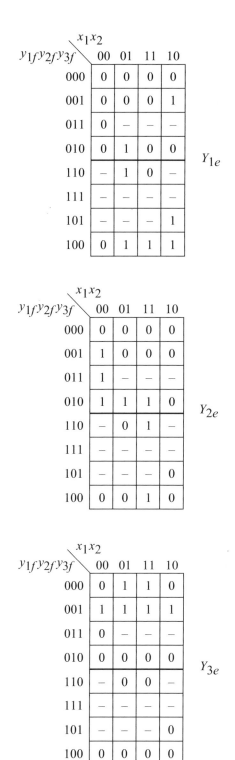

Figure 4.98   Individual excitation maps for the asynchronous sequential machine of Example 4.14.

$$Y_{1e} = (x_1 + x_2)(y_{2f}' + x_1')(y_{1f} + y_{2f} + x_2')(y_{1f} + y_{2f} + y_{3f})$$
$$(y_{1f} + x_1' + x_2')(y_{2f}' + x_2)(y_{1f} + y_{2f} + x_1) \quad (\text{net1} \ldots \text{net7})$$

$$Y_{2e} = (x_1' + x_2)(y_{1f}' + x_1)(y_{1f} + y_{2f} + y_{3f})(y_{1f} + y_{2f} + x_2')(y_{1f}' + x_2)$$
$$(y_{1f} + y_{2f} + x_1') \quad (\text{net8} \ldots \text{net13})$$

$$Y_{3e} = (y_{1f}')(y_{2f}')(y_{3f} + x_2) \quad (\text{net14}) \tag{4.19}$$

z_1 — $y_{1f}y_{2f}y_{3f}$ rows, x_1x_2 columns:

$y_{1f}y_{2f}y_{3f}$ \ x_1x_2	00	01	11	10
000	0^a	0	0	0^b
001	0	0^f	0^c	0
011	0	–	–	–
010	0^g	0	1^e	0
110	–	0	–	–
111	–	–	–	–
101	–	–	–	0
100	0	0^h	–	0^d

z_2 — $y_{1f}y_{2f}y_{3f}$ rows, x_1x_2 columns:

$y_{1f}y_{2f}y_{3f}$ \ x_1x_2	00	01	11	10
000	0^a	0	0	0^b
001	0	0^f	0^c	0
011	0	–	–	–
010	0^g	–	0^e	0
110	–	–	0	–
111	–	–	–	–
101	–	–	–	0
100	0	1^h	0	0^d

Figure 4.99   Output maps for the asynchronous sequential machine of Example 4.14.

$$z_1 = (x_1)(y_{2f})(x_2)$$
$$z_2 = (x_1')(y_{1f})(x_2) \tag{4.20}$$

The design module using built-in primitives is shown in Figure 4.100. A reset input is applied to the machine so that the outputs are initialized to a value of logic zero. The test bench module is shown in Figure 4.101, which displays the two outputs in accordance with their respective input sequence. The outputs and waveforms are shown in Figure 4.102 and Figure 4.103, respectively.

```
//built-in primitive design for pos asm

module asm_pos (rst_n, x1, x2, y1e, y2e, y3e, z1, z2);

//define inputs and outputs
input rst_n, x1, x2;
output y1e, y2e, y3e, z1, z2;


//define internal nets
wire net1, net2, net3, net4, net5, net6, net7, net8;
wire net9, net10, net11, net12, net13, net14;


//design the logic for y1e
or      (net1, x1, x2),
        (net2, ~y2e, ~x1),
        (net3, y1e, y2e, ~x2),
        (net4, y1e, y2e, y3e),
        (net5, ~y2e, x2),
        (net6, y1e, ~x1, ~x2),
        (net7, y1e, x1, y2e);
and     (y1e, rst_n, net1, net2, net3, net4, net5, net6, net7);


//design the logic for y2e
or      (net8, ~x1, x2),
        (net9, ~y1e, x1),
        (net10, y1e, y2e, y3e),
        (net11, y1e, y2e, ~x2),
        (net12, ~y1e, x2),
        (net13, y1e, y2e, ~x1);
and     (y2e, rst_n, net8, net9, net10, net11, net12, net13);


//design the logic for y3e
or      (net14, y3e, x2);
and     (y3e, rst_n, ~y1e, ~y2e, net14);


//design the logic for output z1 and output z2
and     (z1, x1, x2, y2e);
and     (z2, ~x1, x2, y1e);

endmodule
```

Figure 4.100   Design module using built-in primitives for the asynchronous sequential machine of Example 4.14.

```
//test bench for the pos asynchronous sequential machine

module asm_pos_tb;


//inputs are reg for test bench
//outputs are wire for test bench
reg rst_n, x1, x2;
wire y1e, y2e, y3e, z1, z2;


//display variables
initial
$monitor ("x1x2 = %b, state = %b, z1z2 = %b",
            {x1, x2}, {y1e, y2e, y3e}, {z1, z2});


//apply input vectors
initial
begin
    #0      rst_n = 1'b0;
            x1 = 1'b0;x2 = 1'b0;

    #5      rst_n = 1'b1;

    #10     x1 = 1'b1;   x2 = 1'b0;
    #10     x1 = 1'b1;   x2 = 1'b1;
    #10     x1 = 1'b1;   x2 = 1'b0;
    #10     x1 = 1'b1;   x2 = 1'b1;    //assert z1

    #10     x1 = 1'b1;   x2 = 1'b0;
    #10     x1 = 1'b0;   x2 = 1'b0;
    #10     x1 = 1'b0;   x2 = 1'b1;
    #10     x1 = 1'b0;   x2 = 1'b0;
    #10     x1 = 1'b0;   x2 = 1'b1;    //assert z2
    #10     x1 = 1'b0;   x2 = 1'b0;
    #10     x1 = 1'b1;   x2 = 1'b0;

    #10     $stop;

end


//instantiate the module into the test bench
asm_pos inst1 (rst_n, x1, x2, y1e, y2e, y3e, z1, z2);

endmodule
```

Figure 4.101   Test bench module for the asynchronous sequential machine of Example 4.14.

```
x1x2 = 00,  state = 000,  z1z2 = 00
x1x2 = 10,  state = 000,  z1z2 = 00
x1x2 = 11,  state = 001,  z1z2 = 00
x1x2 = 10,  state = 100,  z1z2 = 00
x1x2 = 11,  state = 010,  z1z2 = 10
x1x2 = 10,  state = 000,  z1z2 = 00


x1x2 = 00,  state = 000,  z1z2 = 00
x1x2 = 01,  state = 001,  z1z2 = 00
x1x2 = 00,  state = 010,  z1z2 = 00
x1x2 = 01,  state = 100,  z1z2 = 01
x1x2 = 00,  state = 000,  z1z2 = 00
x1x2 = 10,  state = 000,  z1z2 = 00
```

Figure 4.102   Outputs for the asynchronous sequential machine of Example 4.14.

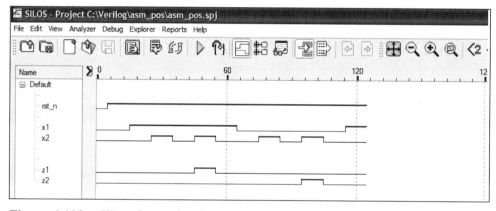

Figure 4.103   Waveforms for the asynchronous sequential machine of Example 4.14.

Example 4.15   This example repeats Example 4.13, however, the design is accomplished using behavioral modeling with the **case** statement. The waveforms are reproduced in Figure 4.104 for convenience. Since behavioral modeling is the method of implementation, only the primitive flow table is required, as shown in Figure 4.105. The behavioral design module is shown in Figure 4.106 and the test bench module is shown in Figure 4.107, which displays the two outputs in accordance with their respective input sequence. The outputs and waveforms are shown in Figure 4.108 and Figure 4.109, respectively.

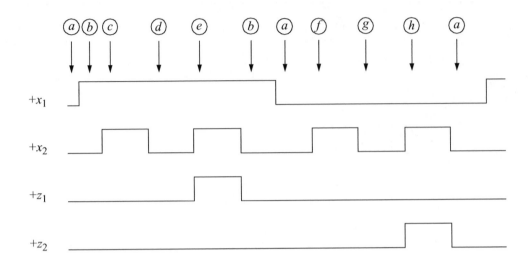

Figure 4.104   Representative timing diagram for the asynchronous sequential machine of Example 4.15.

$y_{1f}y_{2f}y_{3f}$ \ x_1x_2	00	01	11	10	z_1	z_2
$(a) = 000$	(a)	f	$-$	b	0	0
$(b) = 001$	a	$-$	c	(b)	0	0
$(c) = 011$	$-$	$-$	(c)	d	0	0
$(d) = 010$	a	$-$	e	(d)	0	0
$(e) = 110$	$-$	$-$	(e)	b	1	0
$(f) = 111$	g	(f)	$-$	$-$	0	0
$(g) = 101$	(g)	h	$-$	b	0	0
$(h) = 100$	a	(h)	$-$	$-$	0	1

Figure 4.105   Primitive flow table for the asynchronous sequential machine of Example 4.15.

```verilog
//behavioral asynchronous sequential machine

module asm30_bh (rst_n, x1, x2, ye, z1, z2);

input rst_n, x1, x2;      //define inputs and outputs
output [1:3] ye;          //inputs are wire by default
output z1, z2;

//variables are reg in always
reg [1:3] ye, next_state;
reg z1, z2;

//assign state codes; parameter defines a constant
//state names must have at least 2 characters
parameter    state_a = 3'b000,
             state_b = 3'b001,
             state_c = 3'b011,
             state_d = 3'b010,
             state_e = 3'b110,
             state_f = 3'b111,
             state_g = 3'b101,
             state_h = 3'b100;

//set next state
always @ (rst_n or x1 or x2)
begin
   if (~rst_n)
      ye <= state_a;
   else
      ye <= next_state;
end

//define outputs z1 and z2
always @ (x1 or x2 or ye)
begin
   if (ye == state_e)      //== is a logical equality
      z1 = 1'b1;
   else
      z1 = 1'b0;

   if (ye == state_h)
      z2 = 1'b1;
   else
      z2 = 1'b0;
end                        //continued on next page
```

Figure 4.106   Behavioral design module for the asynchronous sequential machine of Example 4.15.

```verilog
//determine next state
always @ (x1 or x2)
begin
   case (ye)
      state_a:
         if (x1==1'b0 & x2==1'b1)
            next_state = state_f;
         else if (x1==1'b1 & x2==1'b0)
            next_state = state_b;
         else
            next_state = state_a;

      state_b:
         if (x1==1'b0 & x2==1'b0)
            next_state = state_a;
         else if (x1==1'b1 & x2==1'b1)
            next_state = state_c;
         else
            next_state = state_b;

      state_c:
         if (x1==1'b1 & x2==1'b0)
            next_state = state_d;
         else
            next_state = state_c;

      state_d:
         if (x1==1'b0 & x2==1'b0)
            next_state = state_a;
         else if (x1==1'b1 & x2==1'b1)
            next_state = state_e;
         else
            next_state = state_d;

      state_e:
         if (x1==1'b1 & x2==1'b0)
            next_state = state_b;
         else
            next_state = state_e;

      state_f:
         if (x1==1'b0 & x2==1'b0)
            next_state = state_g;
         else
            next_state = state_f;
                                    //continued on next page
```

Figure 4.106   (Continued)

```
    state_g:
        if (x1==1'b0 & x2==1'b1)
           next_state = state_h;
        else if (x1==1'b1 & x2==1'b0)
           next_state = state_b;
        else
           next_state = state_g;

    state_h:
        if (x1==1'b0 & x2==1'b0)
           next_state = state_a;
        else
           next_state = state_h;

        default: next_state = state_a;
    endcase
end
endmodule
```

Figure 4.106   (Continued)

```
//test bench for the behavioral asm
module asm30_bh_tb;

//inputs are reg for test bench
//outputs are wire for test bench
reg rst_n, x1, x2;
wire [1:3] ye;
wire z1, z2;

//display variables
initial
$monitor ("x1x2 = %b, state = %b, z1z2 = %b",
          {x1, x2}, ye, {z1, z2});

//apply input vectors
initial
begin
    #0    rst_n = 1'b0;
          x1 = 1'b0;
          x2 = 1'b0;
    #5    rst_n = 1'b1;              //continued on next page
```

Figure 4.107   Test bench for the asynchronous sequential machine of Example 4.15.

```
    #10    x1 = 1'b1;   x2 = 1'b0;
    #10    x1 = 1'b1;   x2 = 1'b1;
    #10    x1 = 1'b1;   x2 = 1'b0;
    #10    x1 = 1'b1;   x2 = 1'b1;    //assert z1

    #10    x1 = 1'b1;   x2 = 1'b0;
    #10    x1 = 1'b0;   x2 = 1'b0;
    #10    x1 = 1'b0;   x2 = 1'b1;
    #10    x1 = 1'b0;   x2 = 1'b0;
    #10    x1 = 1'b0;   x2 = 1'b1;    //assert z2
    #10    x1 = 1'b0;   x2 = 1'b0;
    #10    x1 = 1'b1;   x2 = 1'b0;

    #10    $stop;
end

//instantiate the module into the test bench
asm30_bh inst1 (
    .rst_n(rst_n),
    .x1(x1),
    .x2(x2),
    .ye(ye),
    .z1(z1),
    .z2(z2)
    );

endmodule
```

Figure 4.107   (Continued)

```
x1x2 = 00, state = 000, z1z2 = 00
x1x2 = 10, state = 001, z1z2 = 00
x1x2 = 11, state = 011, z1z2 = 00
x1x2 = 10, state = 010, z1z2 = 00
x1x2 = 11, state = 110, z1z2 = 10
x1x2 = 10, state = 001, z1z2 = 00

x1x2 = 00, state = 000, z1z2 = 00
x1x2 = 01, state = 111, z1z2 = 00
x1x2 = 00, state = 101, z1z2 = 00
x1x2 = 01, state = 100, z1z2 = 01
x1x2 = 00, state = 000, z1z2 = 00
x1x2 = 10, state = 001, z1z2 = 00
```

Figure 4.108   Outputs for the asynchronous sequential machine of Example 4.15.

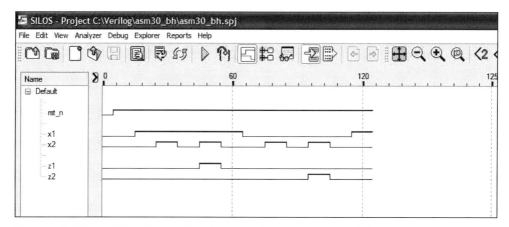

Figure 4.109 Waveforms for the asynchronous sequential machine of Example 4.15.

Example 4.16 Consider an asynchronous sequential machine with three inputs x_1, x_2, and x_3 and one output z_1. The machine operates according to the specifications defined below and the representative timing diagram of Figure 4.110.

The input signals are nonoverlapping, disjoint positive pulses of equal duration. Valid input vectors are $x_1 x_2 x_3 = 000$, 001, 010, and 100. Between each input vector in which a pulse x_i is asserted, a vector of $x_1 x_2 x_3 = 000$ is inserted, as shown in Figure 4.110.

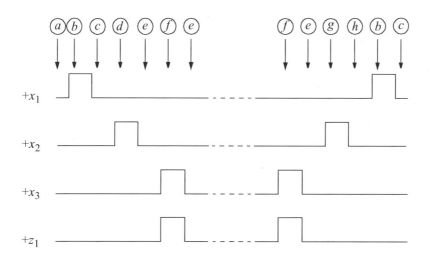

Figure 4.110 Representative timing diagram for the asynchronous sequential machine of Example 4.16.

The initial assertion of z_1 occurs coincident with the first assertion of x_3 if and only if x_3 is preceded by an input sequence of $x_1x_2 = 10, 00, 01, 00$. Thus, a valid sequence, which includes the assertion of z_1, is $x_1x_2x_3z_1 = 0000, 1000, 0000, 0100,$ $0000, 0011, 0000, 0011, \dots, 0000, 0011, 0000, 0100, 0000, 1000, 0000, \dots$, as shown in the timing diagram.

Therefore, once the initial output has been generated, every x_3 pulse will generate a z_1 pulse, provided that neither x_1 nor x_2 has been asserted during the x_3z_1 sequence. The duration of output z_1 is identical to the duration of input x_3. Input x_1 frames the $x_2x_3z_1$ sequence of pulses; input x_2 frames the x_3z_1 sequence of pulses. Simultaneous input changes will not occur.

The reduced primitive flow table is shown in Figure 4.111 in which there are no equivalent states, as obtained by analyzing the machine specifications in conjunction with the timing diagram.

$x_1x_2x_3$ 000	001	011	010	110	111	101	100	z_1
(a)	i	–	g	–	–	–	b	0
c	–	–	–	–	–	–	(b)	0
(c)	i	–	d	–	–	–	b	0
e	–	–	(d)	–	–	–	–	0
(e)	f	–	g	–	–	–	b	0
e	(f)	–	–	–	–	–	–	1
a	–	–	(g)	–	–	–	–	0
a	(i)	–	–	–	–	–	–	0

Figure 4.111   Reduced primitive flow table for the asynchronous sequential machine of Example 4.16.

The merger diagram is shown in Figure 4.112. Recall that two or more rows in a reduced primitive flow table can merge into a single row if the entries in the same column of each row satisfy one of the following requirements:

1. Identical state names
2. A state name and an unspecified entry
3. Two unspecified entries

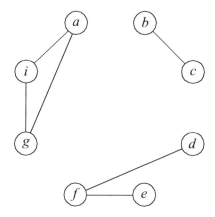

Figure 4.112 Merger diagram obtained from the reduced primitive flow table of Figure 4.111.

The merged flow table is constructed from the reduced primitive flow table using the assignments of the following partition:

$$\{ⓑ, ©\}, \{ⓓ, ①\}, \{ⓔ\}, \{ⓐ, ⑨, ①\}$$

Rows $\{ⓑ, ©\}, \{ⓓ, ①\}, \{ⓔ\}, \{ⓐ, ⑨, ①\}$ are transferred from the reduced primitive flow table to the merged flow table as shown in Figure 4.113.

		$x_1x_2x_3$ 000	001	011	010	110	111	101	100
1	ⓑ,©	©	i	–	d	–	–	–	ⓑ
2	ⓓ,①	e	①	–	ⓓ	–	–	–	–
3	ⓔ	ⓔ	f	–	g	–	–	–	b
4	ⓐ,⑨,①	ⓐ	i	–	⑨	–	–	–	b

Figure 4.113 Merged flow table for the asynchronous sequential machine of Example 4.16 obtained from the reduced primitive flow table of Figure 4.111 using the partition $\{ⓑ, ©\}, \{ⓓ, ①\}, \{ⓔ\}, \{ⓐ, ⑨, ①\}$.

The combined excitation map for Y_{1e} and Y_{2e} is shown in Figure 4.114. The individual excitation maps are shown in Figure 4.115 and the corresponding excitation equations in Equation 4.21 in both a sum-of-products form and a product-of-sums form.

Figure 4.114   Combined excitation map for the asynchronous sequential machine of Example 4.16.

Figure 4.115   Individual excitation maps for Y_{1e} and Y_{2e}.

$$Y_{1e} = y_{2f}x_2'x_3' + y_{1f}y_{2f}'x_1' + y_{1f}x_2 + y_{2f}'x_1'x_3 + y_{1f}x_1'x_3'$$

$$Y_{2e} = y_{2f}x_1'x_2' + y_{1f}'x_2 + y_{1f}'y_{2f}$$

$$(4.21)$$

$$Y_{1e} = (y_{1f} + x_2')(y_{2f} + x_1')(y_{2f}' + x_3')(y_{1f} + y_{2f} + x_2 + x_3)$$

$$Y_{2e} = (y_{2f} + x_2)(y_{1f}' + x_2')(x_1')$$

The output map for z_1 is shown in Figure 4.116 and the equation for z_1 is shown in Equation 4.22. The map is derived from the reduced primitive flow table and the merged flow table. The merged flow table shows the location of the stable states and the reduced primitive flow table specifies the value of z_1 for the corresponding stable states. The values assigned to z_1 for the intermediate transient states provide glitch-free operation for all state transition sequences.

$y_{1f}y_{2f}$ \ $x_1x_2x_3$	000	001	011	010	110	111	101	100
0 0	c 0	0	–	0	–	–	–	b 0
0 1	0	f 1	–	d 0	–	–	–	–
1 1	e 0	–	–	0	–	–	–	0
1 0	a 0	i 0	–	g 0	–	–	–	0

z_1

Figure 4.116 Output map for z_1 for the asynchronous sequential machine of Example 4.16.

$$z_1 = y_{2f}x_3 \qquad (4.22)$$

The logic diagram is shown in Figure 4.117 using the product-of-sums form of Equation 4.21. The product-of-sums form yields the fewest number of logic gates. The machine is synthesized using the logic primitives of AND, OR, and NOT. The feedback variables y_{1f} and y_{2f} become equal to the excitation variables after a delay of Δt, at which time the machine has stabilized. This is indicated by the signals $\pm y_{1f}$ and $\pm y_{2f}$ in parentheses.

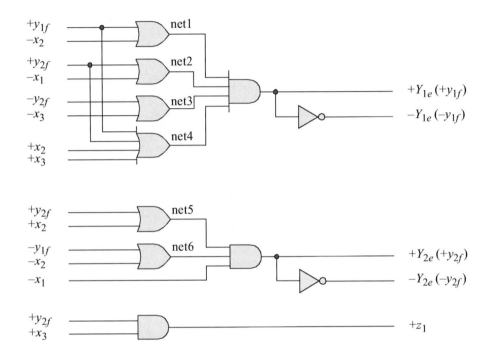

Figure 4.117   Logic diagram for the asynchronous sequential machine of Example 4.16.

The design module for the asynchronous sequential machine of Example 4.16 is shown in Figure 4.118 using built-in primitives. The test bench module is shown in Figure 4.119, which displays the timing diagram for the input variables and the output variable. The outputs and waveforms are shown in Figure 4.120 and Figure 4.121, respectively.

```
//built-in primitive design for asm

module asm_pos2 (rst_n, x1, x2, x3, y1e, y2e, z1);

//define inputs and outputs
input rst_n, x1, x2, x3;
output y1e, y2e, z1;

//define internal nets
wire net1, net2, net3, net4, net5, net6;
                                        //continued on next page
```

Figure 4.118   Design module for Example 4.16 using built-in primitives.

```
//design the logic for y1e
or      (net1, y1e, ~x2),
        (net2, y2e, ~x1),
        (net3, ~y2e, ~x3),
        (net4, y1e, y2e, x2, x3);
and     (y1e, rst_n, net1, net2, net3, net4);

//design the logic for y2e
or      (net5, y2e, x2),
        (net6, ~y1e, ~x2);
and     (y2e, rst_n, net5, net6, ~x1);

//design the logic for output z1
and     (z1, y2e, x3);

endmodule
```

Figure 4.118   (Continued)

```
//test bench for the pos asynchronous sequential machine

module asm_pos2_tb;

//inputs are reg for test bench
//outputs are wire for test bench
reg rst_n, x1, x2, x3;
wire y1e, y2e, z1;

//display variables
initial
$monitor ("x1x2x3 = %b, state = %b, z1 = %b",
          {x1, x2, x3}, {y1e, y2e}, z1);

//apply input vectors
initial
begin
   #0    rst_n = 1'b0;
         x1 = 1'b0;
         x2 = 1'b0;
         x3 = 1'b0;

   #5    rst_n = 1'b1;
                                 //continued on next page
```

Figure 4.119   Test bench module for Example 4.16 using built-in primitives.

```
    #10    x1=1'b0;    x2=1'b0;    x3=1'b0;
    #10    x1=1'b1;    x2=1'b0;    x3=1'b0;
    #10    x1=1'b0;    x2=1'b0;    x3=1'b0;
    #10    x1=1'b0;    x2=1'b1;    x3=1'b0;
    #10    x1=1'b0;    x2=1'b0;    x3=1'b0;
    #10    x1=1'b0;    x2=1'b0;    x3=1'b1;    //assert z1

    #10    x1=1'b0;    x2=1'b0;    x3=1'b0;    //state_e (11)
    #10    x1=1'b0;    x2=1'b0;    x3=1'b0;    //state_e (11)
    #10    x1=1'b0;    x2=1'b0;    x3=1'b0;    //state_e (11)
    #10    x1=1'b0;    x2=1'b0;    x3=1'b0;    //state_e (11)

    #10    x1=1'b0;    x2=1'b0;    x3=1'b1;    //assert z1
    #10    x1=1'b0;    x2=1'b0;    x3=1'b0;
    #10    x1=1'b0;    x2=1'b1;    x3=1'b0;
    #10    x1=1'b0;    x2=1'b0;    x3=1'b0;
    #10    x1=1'b1;    x2=1'b0;    x3=1'b0;
    #10    x1=1'b0;    x2=1'b0;    x3=1'b0;
    #10    $stop;
end

//instantiate the module as a single line
asm_pos2 inst1 (rst_n, x1, x2, x3, y1e, y2e, z1);

endmodule
```

Figure 4.119   (Continued)

```
x1x2x3 = 000, state = 00, z1 = 0
x1x2x3 = 100, state = 00, z1 = 0
x1x2x3 = 000, state = 00, z1 = 0
x1x2x3 = 010, state = 01, z1 = 0
x1x2x3 = 000, state = 11, z1 = 0
x1x2x3 = 001, state = 01, z1 = 1

x1x2x3 = 000, state = 11, z1 = 0

x1x2x3 = 001, state = 01, z1 = 1
x1x2x3 = 000, state = 11, z1 = 0
x1x2x3 = 010, state = 10, z1 = 0
x1x2x3 = 000, state = 10, z1 = 0
x1x2x3 = 100, state = 00, z1 = 0
x1x2x3 = 000, state = 00, z1 = 0
```

Figure 4.120   Outputs for Example 4.16.

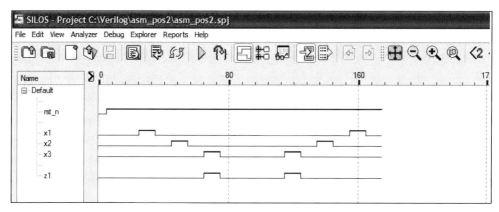

Figure 4.121   Waveforms for Example 4.16.

Example 4.17   This example repeats Example 4.16, but uses built-in primitives in a sum-of-products form. The excitation equations are reproduced in Equation 4.23 and the output equation is shown in Equation 4.24. The design module is shown in Figure 4.122. The test bench module is shown in Figure 4.123. The outputs and wave-forms are shown in Figure 4.124 and Figure 4.125, respectively.

$$Y_{1e} = y_{2f}x_2'x_3' + y_{1f}y_{2f}'x_1' + y_{1f}x_2 + y_{2f}'x_1'x_3 + y_{1f}x_1'x_3'$$

$$Y_{2e} = y_{2f}\,x_1'x_2' + y_{1f}'x_2 + y_{1f}'y_{2f} \qquad\qquad (4.23)$$

$$z_1 = y_{2f}x_3 \qquad\qquad (4.24)$$

```
//built-in primitive design for asm in a sum-of-products form

module asm_sop3_bip (rst_n, x1, x2, x3, y1e, y2e, z1);

//define inputs and outputs
input rst_n, x1, x2, x3;
output y1e, y2e, z1;

//define internal nets
wire net1, net2, net3, net4, net5, net6, net7, net8;
                                   //continued on next page
```

Figure 4.122   Design module for Example 4.17 using built-in primitives.

```
//define internal nets
wire net1, net2, net3, net4, net5, net6, net7, net8;

//design the logic for y1e
and     (net1, y2e, ~x2, ~x3, rst_n),
        (net2, y1e, ~y2e, ~x1, rst_n),
        (net3, y1e, x2, rst_n),
        (net4, ~y2e, ~x1, x3, rst_n),
        (net5, y1e, ~x1, ~x3, rst_n);
or      (y1e, net1, net2, net3, net4, net5);

//design the logic for y2e
and     (net6, y2e, ~x1, ~x2, rst_n),
        (net7, ~y1e, x2, rst_n),
        (net8, ~y1e, y2e, rst_n);
or      (y2e, net6, net7, net8);

//design the logic for output z1
and     (z1, y2e, x3);

endmodule
```

Figure 4.122   (Continued)

```
//test bench for the sop asynchronous sequential machine
module asm_sop3_bip_tb;

//inputs are reg for test bench
//outputs are wire for test bench
reg rst_n, x1, x2, x3;
wire y1e, y2e, z1;

initial     //display variables
$monitor ("x1x2x3 = %b, state = %b, z1 = %b",
            {x1, x2, x3}, {y1e, y2e}, z1);

initial     //apply input vectors
begin
    #0      rst_n = 1'b0;
            x1 = 1'b0;
            x2 = 1'b0;
            x3 = 1'b0;
    #5      rst_n = 1'b1;                    //continued on next page
```

Figure 4.123   Test bench module for Example 4.17 using built-in primitives.

```
    #10    x1 = 1'b0;   x2 = 1'b0;   x3 = 1'b0;
    #10    x1 = 1'b1;   x2 = 1'b0;   x3 = 1'b0;
    #10    x1 = 1'b0;   x2 = 1'b0;   x3 = 1'b0;
    #10    x1 = 1'b0;   x2 = 1'b1;   x3 = 1'b0;
    #10    x1 = 1'b0;   x2 = 1'b0;   x3 = 1'b0;
    #10    x1 = 1'b0;   x2 = 1'b0;   x3 = 1'b1;    //assert z1

    #10    x1 = 1'b0;   x2 = 1'b0;   x3 = 1'b0;    //state_e (11)
    #10    x1 = 1'b0;   x2 = 1'b0;   x3 = 1'b0;    //state_e (11)
    #10    x1 = 1'b0;   x2 = 1'b0;   x3 = 1'b0;    //state_e (11)
    #10    x1 = 1'b0;   x2 = 1'b0;   x3 = 1'b0;    //state_e (11)

    #10    x1 = 1'b0;   x2 = 1'b0;   x3 = 1'b1;    //assert z1
    #10    x1 = 1'b0;   x2 = 1'b0;   x3 = 1'b0;
    #10    x1 = 1'b0;   x2 = 1'b1;   x3 = 1'b0;
    #10    x1 = 1'b0;   x2 = 1'b0;   x3 = 1'b0;
    #10    x1 = 1'b1;   x2 = 1'b0;   x3 = 1'b0;
    #10    x1 = 1'b0;   x2 = 1'b0;   x3 = 1'b0;
    #10    $stop;
end

//instantiate the module into the test bench as a single line
asm_sop3_bip inst1 (rst_n, x1, x2, x3, y1e, y2e, z1);

endmodule
```

Figure 4.123   (Continued)

```
x1x2x3 = 000, state = 00, z1 = 0
x1x2x3 = 100, state = 00, z1 = 0
x1x2x3 = 000, state = 00, z1 = 0
x1x2x3 = 010, state = 01, z1 = 0
x1x2x3 = 000, state = 11, z1 = 0
x1x2x3 = 001, state = 01, z1 = 1

x1x2x3 = 000, state = 11, z1 = 0

x1x2x3 = 001, state = 01, z1 = 1
x1x2x3 = 000, state = 11, z1 = 0
x1x2x3 = 010, state = 10, z1 = 0
x1x2x3 = 000, state = 10, z1 = 0
x1x2x3 = 100, state = 00, z1 = 0
x1x2x3 = 000, state = 00, z1 = 0
```

Figure 4.124   Outputs for Example 4.17 using built-in primitives.

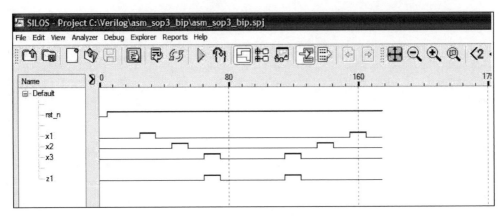

Figure 4.125   Waveforms for Example 4.17 using built-in primitives.

Example 4.18   This example repeats Example 4.16, but uses dataflow modeling in a product-of-sums form. Recall that the *continuous assignment* statement models dataflow behavior and is used to design combinational logic without using gates and interconnecting nets. The continuous assignment statement uses the keyword **assign** and provides a Boolean correspondence between the right-hand side expression and the left-hand side target.

The excitation equations are reproduced in Equation 4.25 in a sum-of-products form and the output equation is shown in Equation 4.26 for convenience. The design module is shown in Figure 4.126. The test bench module is shown in Figure 4.127. The outputs and waveforms are shown in Figure 4.128 and Figure 4.129, respectively.

$$Y_{1e} = (y_{1f} + x_2')\,(y_{2f} + x_1')\,(y_{2f}' + x_3')\,(y_{1f} + y_{2f} + x_2 + x_3)$$
$$Y_{2e} = (y_{2f} + x_2)\,(y_{1f}' + x_2')\,(x_1') \tag{4.25}$$

$$z_1 = y_{2f}x_3 \tag{4.26}$$

```
//dataflow for product-of-sums asm
module asm_df (rst_n, x1, x2, x3, y1e, y2e, z1);

//define inputs and outputs
input rst_n, x1, x2, x3;
output y1e, y2e, z1;                    //continued on next page
```

Figure 4.126   Design module using dataflow modeling for Example 4.18.

```
//define internal nets
wire net1, net2, net3, net4, net5, net6;

//design the logic for y1e
assign    net1 = (y1e | ~x2),
          net2 = (y2e | ~x1),
          net3 = (~y2e | ~x3),
          net4 = (y1e | y2e | x2 | x3),
          y1e = (net1 & net2 & net3 & net4);

//design the logic for y2e
assign    net5 = (y2e | x2),
          net6 = (~y1e | ~x2),
          y2e = (net5 & net6 & ~x1);

//design the logic for output z1
assign    z1 = (y2e & x3);

endmodule
```

Figure 4.126   (Continued)

```
//test bench for the pos asynchronous sequential machine

module asm_df_tb;

//inputs are reg for test bench
//outputs are wire for test bench
reg rst_n, x1, x2, x3;
wire y1e, y2e, z1;

//display variables
initial
$monitor ("x1x2x3 = %b, state = %b, z1 = %b",
          {x1, x2, x3}, {y1e, y2e}, z1);

initial      //apply input vectors
begin
   #0     rst_n = 1'b0;
          x1 = 1'b0;
          x2 = 1'b0;
          x3 = 1'b0;
   #5     rst_n = 1'b1;                //continued on next page
```

Figure 4.127   Test bench module for Example 4.18.

```
    #10    x1 = 1'b0;   x2 = 1'b0;   x3 = 1'b0;
    #10    x1 = 1'b1;   x2 = 1'b0;   x3 = 1'b0;
    #10    x1 = 1'b0;   x2 = 1'b0;   x3 = 1'b0;
    #10    x1 = 1'b0;   x2 = 1'b1;   x3 = 1'b0;
    #10    x1 = 1'b0;   x2 = 1'b0;   x3 = 1'b0;
    #10    x1 = 1'b0;   x2 = 1'b0;   x3 = 1'b1;   //assert z1

    #10    x1 = 1'b0;   x2 = 1'b0;   x3 = 1'b0;   //state_e (11)
    #10    x1 = 1'b0;   x2 = 1'b0;   x3 = 1'b0;   //state_e (11)
    #10    x1 = 1'b0;   x2 = 1'b0;   x3 = 1'b0;   //state_e (11)
    #10    x1 = 1'b0;   x2 = 1'b0;   x3 = 1'b0;   //state_e (11)

    #10    x1 = 1'b0;   x2 = 1'b0;   x3 = 1'b1;   //assert z1
    #10    x1 = 1'b0;   x2 = 1'b0;   x3 = 1'b0;
    #10    x1 = 1'b0;   x2 = 1'b1;   x3 = 1'b0;
    #10    x1 = 1'b0;   x2 = 1'b0;   x3 = 1'b0;
    #10    x1 = 1'b1;   x2 = 1'b0;   x3 = 1'b0;
    #10    x1 = 1'b0;   x2 = 1'b0;   x3 = 1'b0;
    #10    $stop;
end

//instantiate the module into the test bench as a single line
asm_df inst1 (rst_n, x1, x2, x3, y1e, y2e, z1);

endmodule
```

Figure 4.127   (Continued)

```
x1x2x3 = 000, state = xx, z1 = 0
x1x2x3 = 100, state = 00, z1 = 0
x1x2x3 = 000, state = 00, z1 = 0
x1x2x3 = 010, state = 01, z1 = 0
x1x2x3 = 000, state = 11, z1 = 0
x1x2x3 = 001, state = 01, z1 = 1

x1x2x3 = 000, state = 11, z1 = 0

x1x2x3 = 001, state = 01, z1 = 1
x1x2x3 = 000, state = 11, z1 = 0
x1x2x3 = 010, state = 10, z1 = 0
x1x2x3 = 000, state = 10, z1 = 0
x1x2x3 = 100, state = 00, z1 = 0
x1x2x3 = 000, state = 00, z1 = 0
```

Figure 4.128   Outputs for Example 4.18.

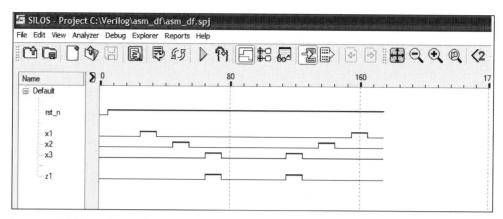

Figure 4.129   Waveforms for Example 4.18.

Example 4.19   This example repeats the previous example, but uses behavioral modeling to design the asynchronous sequential machine. The example uses the **case** statement, which is an alternative to the **if . . . else if** construct. As stated previously, the **case** statement is a multiple-way conditional branch, in which the left-hand case item is assigned the value of the right-hand expression. The **parameter** keyword declares and assigns values to the left-hand case item. For example, let the case item be state_a. Then the **case** statement shown below assigns a value of 3'b000 to state_a.

$$\textbf{parameter} \quad \text{state\_a} = 3\text{'b000};$$

The timing diagram is reproduced below in Figure 4.130 for convenience and shows the various stable states through which the machine sequences. It must be stressed that every possible input sequence should be considered to exactly replicate the operational characteristics of the machine.

The reduced primitive flow table is shown in Figure 4.131 and is the primary mechanism used in behavioral modeling, because it illustrates the various paths that the machine executes. The primitive flow table also depicts the values assigned to the various stable states; in this case, three feedback variables that are equal to the excitation variables after a delay of Δt. The primitive flow table is obtained by carefully analyzing the machine specifications in conjunction with the timing diagram. This method yields a primitive flow table in which there are no equivalent states.

The three inputs, $x_1 x_2 x_3$, are shown in the reduced primitive flow table and are assigned the following Gray code values: 000, 001, 011, 010, 110, 111, 101, 100. The Gray code allows minterm locations that are physically adjacent to also be logically adjacent. The bold center line in the primitive flow table acts as a hinge such that the columns on both sides are adjacent. For example, column $x_1 x_2 x_3 = 011$ is adjacent to column $x_1 x_2 x_3 = 111$.

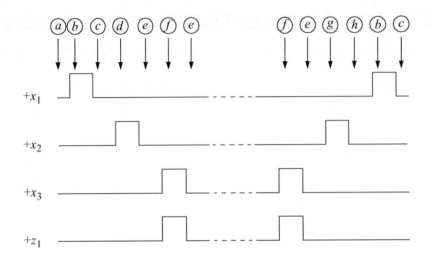

Figure 4.130   Representative timing diagram for the asynchronous sequential machine of Example 4.19.

$y_{1f}y_{2f}y_{3f}$ $\backslash$ $x_1x_2x_3$	000	001	011	010	110	111	101	100	z_1
ⓐ = 000	ⓐ	i	–	g	–	–	–	b	0
ⓑ = 001	c	–	–	–	–	–	–	ⓑ	0
ⓒ = 011	ⓒ	i	–	d	–	–	–	b	0
ⓓ = 010	e	–	–	ⓓ	–	–	–	–	0
ⓔ = 110	ⓔ	f	–	g	–	–	–	b	0
ⓕ = 111	e	ⓕ	–	–	–	–	–	–	1
ⓖ = 101	a	–	–	ⓖ	–	–	–	–	0
ⓘ = 100	a	ⓘ	–	–	–	–	–	–	0

Figure 4.131   Reduced primitive flow table for the asynchronous sequential machine of Example 4.19.

The behavioral design module is shown in Figure 4.132 using the **case** statement. The test bench module is shown in Figure 4.133. The outputs and waveforms are shown in Figure 4.134 and Figure 4.135, respectively.

```
//behavioral asynchronous sequential machine

module asm_bh (rst_n, x1, x2, x3, ye, z1);

//define inputs and outputs
input rst_n, x1, x2, x3;
output [1:3] ye;
output z1;


//do not have to declare inputs as wire
//they are wire by default


//variables are reg in always
reg [1:3] ye, next_state;
reg z1;


//assign state codes; parameter defines a constant
//state names must have at least 2 characters
parameter    state_a = 3'b000,
             state_b = 3'b001,
             state_c = 3'b011,
             state_d = 3'b010,
             state_e = 3'b110,
             state_f = 3'b111,
             state_g = 3'b101,
             state_i = 3'b100;


//set next state
always @ (rst_n or x1 or x2 or x3)
begin
   if (~rst_n)//if reset = 1.b0
      ye <= state_a;
   else
      ye <= next_state;
end
                              //continued on next page
```

Figure 4.132 Behavioral design module for Example 4.19.

```
//determine next state
always @ (x1 or x2 or x3)
begin
   case (ye)
      state_a:
         if (x1==1'b0 & x2==1'b0 & x3==1'b1)
            next_state = state_i;
         else if (x1==1'b0 & x2==1'b1 & x3==1'b0)
            next_state = state_g;
         else if (x1==1'b1 & x2==1'b0 & x3==1'b0)
            next_state = state_b;
         else
            next_state = state_a;

      state_b:
         if (x1==1'b0 & x2==1'b0 & x3==1'b0)
            next_state = state_c;
         else
            next_state = state_b;

      state_c:
         if (x1==1'b0 & x2==1'b0 & x3==1'b1)
            next_state = state_i;
         else if (x1==1'b0 & x2==1'b1 & x3==1'b0)
            next_state = state_d;
         else if (x1==1'b1 & x2==1'b0 & x3==1'b0)
            next_state = state_b;
         else
            next_state = state_c;

      state_d:
         if (x1==1'b0 & x2==1'b0 & x3==1'b0)
            next_state = state_e;
         else
            next_state = state_d;

      state_e:
         if (x1==1'b0 & x2==1'b0 & x3==1'b1)
            next_state = state_f;
         else if (x1==1'b0 & x2==1'b1 & x3==1'b0)
            next_state = state_g;
         else if (x1==1'b1 & x2==1'b0 & x3==1'b0)
            next_state = state_b;
         else
            next_state = state_e;
                                    //continued on next page
```

Figure 4.132   (Continued)

```
         state_f:
            if (x1==1'b0 & x2==1'b0 & x3==1'b0)
               next_state = state_e;
            else
               next_state = state_f;

         state_g:
            if (x1==1'b0 & x2==1'b0 & x3==1'b0)
               next_state = state_a;
            else
               next_state = state_g;

         state_i:
            if (x1==1'b0 & x2==1'b0 & x3==1'b0)
               next_state = state_a;
            else
               next_state = state_i;

         default: next_state = state_a;
      endcase
end

//define output z1
always @ (x1 or x2 or x3 or ye)
begin
   if (ye == state_f)
      z1 = 1'b1;
   else
      z1 = 1'b0;
end
endmodule
```

Figure 4.132   (Continued)

```
//test bench for the asynchronous sequential machine

module asm_bh_tb;

//inputs are reg for test bench
//outputs are wire for test bench
reg rst_n, x1, x2, x3;
wire [1:3] ye;
wire z1;                            //continued on next page
```

Figure 4.133   Test bench module for Example 4.19.

```
//display variables
initial
$monitor ("x1x2x3 = %b, state = %b, z1 = %b",
            {x1, x2, x3}, ye, z1);


//apply input vectors
initial
begin
   #0     rst_n = 1'b0;
          x1 = 1'b0;
          x2 = 1'b0;
          x3 = 1'b0;

   #5     rst_n = 1'b1;


   #10    x1 = 1'b0;  x2 = 1'b0;  x3 = 1'b0;
   #10    x1 = 1'b1;  x2 = 1'b0;  x3 = 1'b0;
   #10    x1 = 1'b0;  x2 = 1'b0;  x3 = 1'b0;
   #10    x1 = 1'b0;  x2 = 1'b1;  x3 = 1'b0;
   #10    x1 = 1'b0;  x2 = 1'b0;  x3 = 1'b0;
   #10    x1 = 1'b0;  x2 = 1'b0;  x3 = 1'b1;    //assert z1


   #10    x1 = 1'b0;  x2 = 1'b0;  x3 = 1'b0;
   #10    x1 = 1'b0;  x2 = 1'b0;  x3 = 1'b0;
   #10    x1 = 1'b0;  x2 = 1'b0;  x3 = 1'b0;
   #10    x1 = 1'b0;  x2 = 1'b0;  x3 = 1'b0;


   #10    x1 = 1'b0;  x2 = 1'b0;  x3 = 1'b1;    //assert z1
   #10    x1 = 1'b0;  x2 = 1'b0;  x3 = 1'b0;
   #10    x1 = 1'b0;  x2 = 1'b1;  x3 = 1'b0;
   #10    x1 = 1'b0;  x2 = 1'b0;  x3 = 1'b0;
   #10    x1 = 1'b1;  x2 = 1'b0;  x3 = 1'b0;
   #10    x1 = 1'b0;  x2 = 1'b0;  x3 = 1'b0;


   #10    $stop;

end

//instantiate the module into the test bench as a single line
asm_bh inst1 (rst_n, x1, x2, x3, ye, z1);

endmodule
```

Figure 4.133   (Continued)

```
x1x2x3 = 000, state = 000, z1 = 0
x1x2x3 = 100, state = 001, z1 = 0
x1x2x3 = 000, state = 011, z1 = 0
x1x2x3 = 010, state = 010, z1 = 0
x1x2x3 = 000, state = 110, z1 = 0
x1x2x3 = 001, state = 111, z1 = 1

x1x2x3 = 000, state = 110, z1 = 0

x1x2x3 = 001, state = 111, z1 = 1
x1x2x3 = 000, state = 110, z1 = 0
x1x2x3 = 010, state = 101, z1 = 0
x1x2x3 = 000, state = 000, z1 = 0
x1x2x3 = 100, state = 001, z1 = 0
x1x2x3 = 000, state = 011, z1 = 0
```

Figure 4.134   Outputs for Example 4.19.

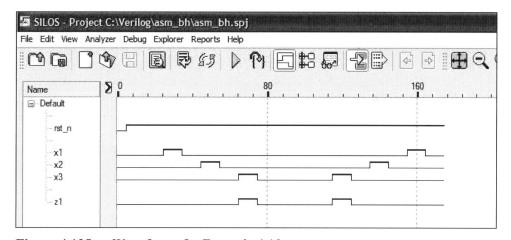

Figure 4.135   Waveforms for Example 4.19.

Example 4.20   This example repeats the previous example, but uses structural modeling in the design process by instantiating dataflow-designed logic gates. This is accomplished by adding the appropriate dataflow modules to the Project Properties screen. Each instantiated gate will be inserted into the module as a single line. The asynchronous sequential machine will be synthesized using a product-of-sums design.

Using the traditional design process, the combined excitation map is shown in Figure 4.136 and the individual excitation maps are shown in Figure 4.137. The excitation equations are listed in Equation 4.27. The output equation is listed in Equation 4.28. Both are reproduced from a previous example.

$y_{1f}y_{2f}$ \ $x_1x_2x_3$	000	001	011	010	110	111	101	100
0 0	⓪⓪ c	10	–	01	–	–	–	⓪⓪ b
0 1	11	⓪① f	–	⓪① d	–	–	–	–
1 1	①① e	01	–	10	–	–	–	10
1 0	①⓪ a	①⓪ i	–	①⓪ g	–	–	–	00

$$Y_{1e}Y_{2e}$$

Figure 4.136   Combined excitation map for the asynchronous sequential machine of Example 4.20.

$y_{1f}y_{2f}$ \ $x_1x_2x_3$	000	001	011	010	110	111	101	100
0 0	0	1	–	0	–	–	–	0
0 1	1	0	–	0	–	–	–	–
1 1	1	0	–	1	–	–	–	1
1 0	1	1	–	1	–	–	–	0

$$Y_{1e}$$

$y_{1f}y_{2f}$ \ $x_1x_2x_3$	000	001	011	010	110	111	101	100
0 0	0	0	–	1	–	–	–	0
0 1	1	1	–	1	–	–	–	–
1 1	1	1	–	0	–	–	–	0
1 0	0	0	–	0	–	–	–	0

$$Y_{2e}$$

Figure 4.137   Individual excitation maps for Y_{1e} and Y_{2e}.

$$Y_{1e} = (y_{1f} + x_2')\,(y_{2f} + x_1')\,(y_{2f}' + x_3')\,(y_{1f} + y_{2f} + x_2 + x_3)$$
$$Y_{2e} = (y_{2f} + x_2)\,(y_{1f}' + x_2')\,(x_1') \tag{4.27}$$

$$z_1 = y_{2f}\,x_3 \tag{4.28}$$

The logic diagram is shown in Figure 4.138 in a product-of-sums form. The structural design module is shown in Figure 4.139 using instantiated dataflow logic gates that are inserted as a single line. The test bench module is shown in Figure 4.140. The outputs and waveforms are shown in Figure 4.141 and Figure 4.142, respectively.

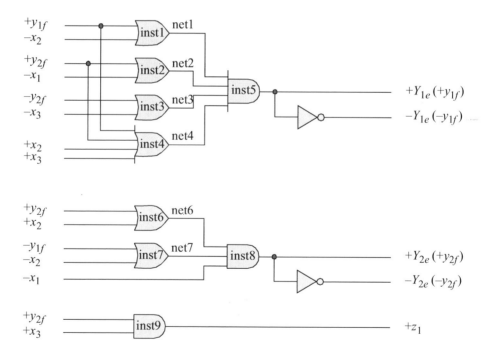

Figure 4.138   Logic diagram for the asynchronous sequential machine of Example 4.20.

```
//structural for pos asm

module asm_pos2_sngl (rst_n, x1, x2, x3, y1e, y2e, z1);

//define inputs and outputs
input rst_n, x1, x2, x3;
output y1e, y2e, z1;

//define internal nets
wire net1, net2, net3, net4, net6, net7;

//instantiate the logic for y1e
or2_df    inst1 (y1e, ~x2, net1);
or2_df    inst2 (y2e, ~x1, net2);
or2_df    inst3 (~y2e, ~x3, net3);
or4_df    inst4 (y1e, y2e, x2, x3, net4);
and4_df inst5 (net1, net2, net3, net4, y1e);

//instantiate the logic for y2e
or2_df    inst6 (y2e, x2, net6);
or2_df    inst7 (~y1e, ~x2, net7);
and3_df inst8 (net6, net7, ~x1, y2e);

//instantiate the logic for output z1
and3_df inst9 (y2e, x3, rst_n, z1);

endmodule
```

Figure 4.139   Structural design module for Example 4.20.

```
//test bench for pos asm

module asm_pos2_sngl_tb;

//inputs are reg for test bench
//outputs are wire for test bench
reg rst_n, x1, x2, x3;
wire y1e, y2e, z1;

//display variables
initial
$monitor ("x1x2x3 = %b, state = %b, z1 = %b",
          {x1, x2, x3}, {y1e, y2e}, z1);          //next page
```

Figure 4.140   Test bench module for Example 4.20.

```
//apply input vectors
initial
begin
   #0      rst_n = 1'b0;
           x1 = 1'b0;
           x2 = 1'b0;
           x3 = 1'b0;


   #5      rst_n = 1'b1;


   #10   x1 = 1'b0;   x2 = 1'b0;   x3 = 1'b0;
   #10   x1 = 1'b1;   x2 = 1'b0;   x3 = 1'b0;
   #10   x1 = 1'b0;   x2 = 1'b0;   x3 = 1'b0;
   #10   x1 = 1'b0;   x2 = 1'b1;   x3 = 1'b0;
   #10   x1 = 1'b0;   x2 = 1'b0;   x3 = 1'b0;
   #10   x1 = 1'b0;   x2 = 1'b0;   x3 = 1'b1;   //assert z1


   #10   x1 = 1'b0;   x2 = 1'b0;   x3 = 1'b0;
   #10   x1 = 1'b0;   x2 = 1'b0;   x3 = 1'b0;
   #10   x1 = 1'b0;   x2 = 1'b0;   x3 = 1'b0;
   #10   x1 = 1'b0;   x2 = 1'b0;   x3 = 1'b0;


   #10   x1 = 1'b0;   x2 = 1'b0;   x3 = 1'b1;   //assert z1
   #10   x1 = 1'b0;   x2 = 1'b0;   x3 = 1'b0;
   #10   x1 = 1'b0;   x2 = 1'b1;   x3 = 1'b0;
   #10   x1 = 1'b0;   x2 = 1'b0;   x3 = 1'b0;
   #10   x1 = 1'b1;   x2 = 1'b0;   x3 = 1'b0;
   #10   x1 = 1'b0;   x2 = 1'b0;   x3 = 1'b0;

   #10      $stop;

end

//instantiate the module into the test bench as a single line
asm_pos2_sngl inst1 (rst_n, x1, x2, x3, y1e, y2e, z1);

endmodule
```

Figure 4.140   (Continued)

```
x1x2x3 = 100, state = 00, z1 = 0
x1x2x3 = 000, state = 00, z1 = 0
x1x2x3 = 010, state = 01, z1 = 0
x1x2x3 = 010, state = 00, z1 = 0
x1x2x3 = 010, state = 01, z1 = 0
x1x2x3 = 000, state = 11, z1 = 0
x1x2x3 = 001, state = 01, z1 = 1

x1x2x3 = 000, state = 11, z1 = 0

x1x2x3 = 001, state = 01, z1 = 1
x1x2x3 = 000, state = 11, z1 = 0
x1x2x3 = 010, state = 10, z1 = 0
x1x2x3 = 000, state = 10, z1 = 0
x1x2x3 = 100, state = 00, z1 = 0
x1x2x3 = 000, state = 00, z1 = 0
```

Figure 4.141   Outputs for Example 4.20.

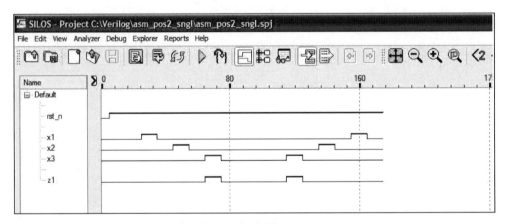

Figure 4.142   Waveforms for Example 4.20.

In conclusion, many of the examples in this chapter are repeated using different design methodologies. The different design techniques include built-in primitives, dataflow modeling, behavioral modeling, and structural modeling. Some examples use a combination of these modeling constructs. A similar test bench module can be used for each different modeling method for the same asynchronous sequential machine. Using different design methodologies illustrates alternative methods to design identical asynchronous sequential machines.

4.3   Problems

4.1   Given the merged flow table shown below, design the asynchronous sequential machine using dataflow modeling. Obtain the design module, the test bench module, the outputs, and the waveforms. The excitation equations are to be in a sum-of-products form. There is one output z_1 that is asserted for $Y_{1e}Y_{2e}x_1 = 111$.

	x_1x_2 00	01	11	10
1	ⓐ	b	ⓔ	c
2	a	ⓑ	e	ⓖ
3	ⓕ	b	ⓓ	ⓒ

4.2   Synchronize an asynchronous sequential machine, using built-in-primitives, which has two inputs x_1 and x_2 and two outputs z_1 and z_2. The two inputs may overlap, but will not change state simultaneously. Only the following sequences are valid:

$$x_1x_2 = 00 \rightarrow 10 \rightarrow 11 \rightarrow 01 \rightarrow 00$$
$$x_1x_2 = 00 \rightarrow 10 \rightarrow 11 \rightarrow 10 \rightarrow 00$$
$$x_1x_2 = 00 \rightarrow 10 \rightarrow 00$$
$$x_1x_2 = 00 \rightarrow 01 \rightarrow 00$$

Output z_1 is asserted whenever x_1 is active and x_2 is asserted or when x_2 is active and x_1 is asserted. Output z_1 will be deasserted when either x_1 or x_2 is deasserted. Output z_2 is asserted coincident with the assertion of z_1 and remains active until the deassertion of the last active input of an overlapping sequence. A representative timing diagram is shown below. Use AND gates and OR gates for the logic.

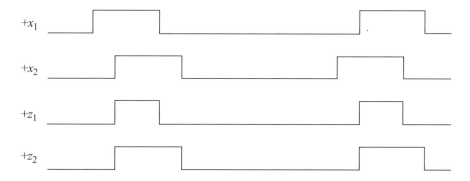

4.3   A merged flow table for an asynchronous sequential machine is shown below with the accompanying transition diagram indicating that all four rows must be adjacent, which is clearly impossible. Adjacency can be achieved by redirecting some state transitions through rows containing unspecified entries. Obtain the dataflow design module using the continuous assignment statement. Assume that there are two outputs z_1 and z_2 that satisfy the equations shown below. Then generate the test bench module. Obtain the outputs and the waveforms.

$$z_1 = Y_{1e}\, Y_{2e}{}'\, x_1 \qquad\qquad z_2 = Y_{1e}{}'\, Y_{2e}\, x_2{}'$$

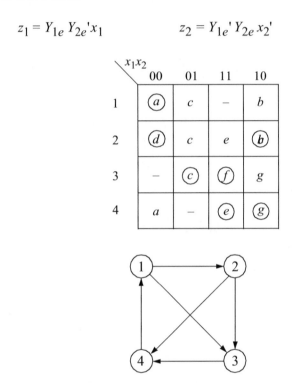

4.4   Synthesize an asynchronous sequential machine which has two inputs x_1 and x_2 and one output z_1. Output z_1 is asserted for the duration of x_2 if and only if x_1 is already asserted. Assume that the initial state of the machine is $x_1 x_2 z_1 = 000$. A representative timing diagram is shown below. Obtain the excitation equations in both a sum-of-products form and a product-of-sums form and use sum-of-products form for the design using built-in primitives.

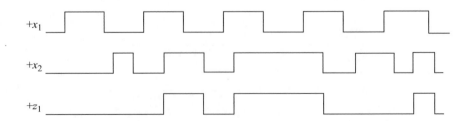

4.5   Synthesize an asynchronous sequential machine that has one input x_1 and one output z_1 which operates according to the timing diagram shown below. Assign values to the transient states in the output map such that the λ output logic will be minimized. The output response is to be as fast as possible. Use only NAND logic in a sum-of-products form.

4.6   Synthesize an asynchronous sequential machine which has one input x_1 and one output z_1. The machine operates according to the timing diagram shown below. The assertion of x_1 toggles output z_1. Assign values to the transient states in the output map such that the λ output logic will be minimized. Obtain the excitation equations in a sum-of-products form. Use NAND logic with the continuous assignment statement in the design module.

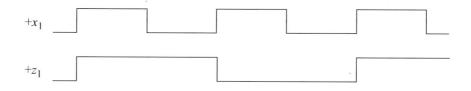

4.7   This example repeats Problem 4.6 using NOR logic. The asynchronous sequential machine has one input x_1 and one output z_1. The machine operates according to the timing diagram shown below. The assertion of x_1 toggles output z_1. Assign values to the transient states in the output map such that the λ output logic will be minimized. Obtain the excitation equations in a product-of-sums form. Use NOR logic with the continuous assignment statement.

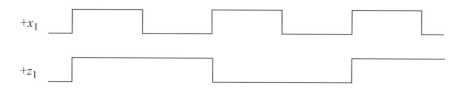

4.8   Synthesize an asynchronous sequential machine using the continuous assignment statement, which has one input x_1 and two outputs z_1 and z_2, as shown below. Output z_1 is toggled at the positive transition of x_1; output z_2 is toggled at the negative transition of x_1. Obtain the output maps for the fewest number of gates. Assume that the initial state of the machine is $x_1 z_1 z_2 = 000$.

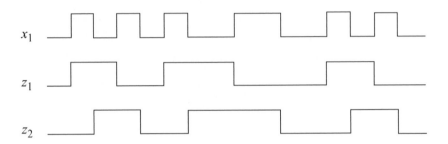

4.9   Synthesize an asynchronous sequential machine which has two inputs x_1 and x_2 and one output z_1. Output z_1 will be asserted coincident with the assertion of x_2, but only if x_1 is already asserted. The deassertion of x_2 causes the deassertion of z_1. Input x_1 will not become deasserted while x_2 is asserted. The timing diagram shown below further illustrates the operation of the machine for various states through which the machine sequences.

    Derive the primitive flow table, the merger diagram, the merged flow table, the excitation map and equation, the output map and equation, and the logic diagram. Obtain the dataflow design module, the test bench module, the outputs, and the waveforms.

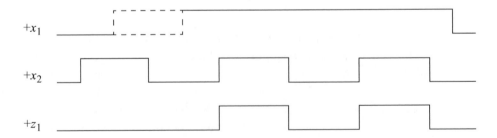

4.10   Obtain the excitation and output equations for an asynchronous sequential machine which has one input x_1 and two outputs z_1 and z_2. Output z_1 is asserted for the duration of every second x_1 pulse; output z_2 is asserted for the duration of every second z_1 pulse. The outputs are to respond as fast as possible to changes in the input vector. A representative timing diagram is shown below. Go through the design process to obtain the excitation equations and the output equations. Then obtain the design module using built-in primitives, the test bench module, the outputs, and the waveforms.

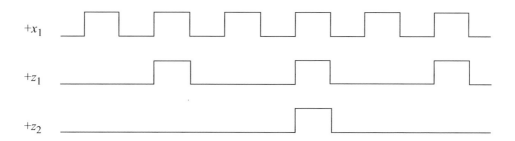

4.11   This repeats problem 4.10, but uses behavioral modeling with the **case** statement. The timing diagram is reproduced below for convenience. Obtain the behavioral design module, the test bench module, the outputs, and the waveforms.

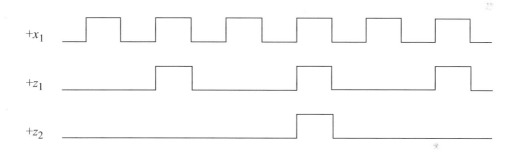

4.12   Use behavioral modeling to synthesize an asynchronous sequential machine which has two inputs x_1 and x_2 and one output z_1. Output z_1 will be asserted coincident with the assertion of the first x_2 pulse and will remain active for the duration of the first x_2 pulse. The output will be asserted only if the assertion of x_1 precedes the assertion of x_2. Input x_1 will not become deasserted while x_2 is asserted. Obtain the behavioral design module, the test bench module, the outputs, and the waveforms. A representative timing diagram is shown below.

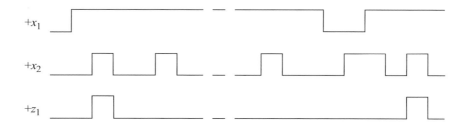

4.13   This problem repeats Problem 4.12, but uses dataflow modeling with logic gates that are instantiated and used as a single line. Synthesize an asynchronous sequential machine using dataflow modeling that has two inputs x_1 and x_2 and one output z_1. Output z_1 will be asserted coincident with the assertion of the first x_2 pulse and will remain active for the duration of the first x_2 pulse. The output will be asserted only if the assertion of x_1 precedes the assertion of x_2. Input x_1 will not become deasserted while x_2 is asserted. The λ output logic must have a minimal number of logic gates. A representative timing diagram is reproduced below for convenience.

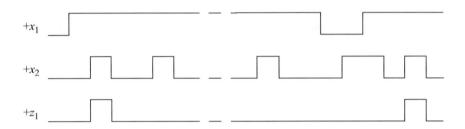

4.14   Repeat Problem 4.12 using dataflow modeling with the continuous assignment statement — which is used in dataflow modeling. The excitation equations and the output equation are reproduced below.

$$Y_{1e} = y_{1f}x_1 + y_{2f}x_2$$
$$Y_{2e} = y_{2f}x_2 + y_{1f}'x_1x_2' + y_{1f}'y_{2f}x_1$$

$$z_1 = y_{1f}y_{2f}$$

4.15   Synthesize an asynchronous sequential machine using built-in primitives, which has one input x_1 and two outputs z_1 and z_2. The machine functions as a two-output bistable multivibrator, whose operation is characterized by the timing diagram shown below. Output z_1 toggles on the positive transition of x_1 and output z_2 toggles on the negative transition of x_1. Obtain equations for z_1 and z_2 which produce the least amount of logic. Obtain the design module using built-in primitives, the test bench module, the outputs, and the waveforms.

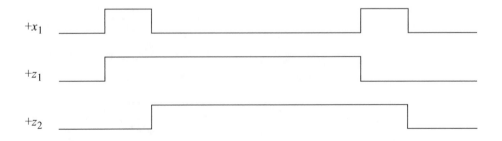

4.16   Repeat Problem 4.15 using behavioral modeling. The timing diagram is reproduced below.

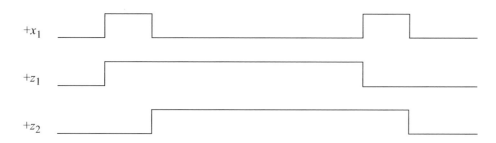

4.17   An asynchronous sequential machine has two inputs x_1 and x_2 and one output z_1. The machine operates according to the following specifications:

If $x_1 x_2 = 00$, then the state of z_1 is unchanged.
If $x_1 x_2 = 01$, then z_1 is deasserted.
If $x_1 x_2 = 10$, then z_1 is asserted.
If $x_1 x_2 = 11$, then z_1 changes state.

Design the machine using dataflow modeling. The inputs are available in both high and low assertion. Assume that the initial conditions are $x_1 x_2 z_1 = 000$. The output must change as fast as possible. There must be no output glitches. A representative timing diagram is shown below. Obtain the dataflow design module, the test bench module, the outputs, and the waveforms.

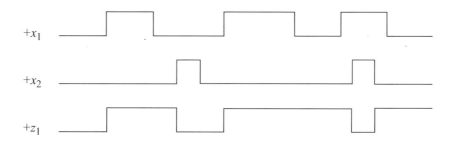

4.18   Given the merged flow table shown below, design the asynchronous sequential machine using dataflow modeling with the continuous assignment statement. Assume that output z_1 is asserted in the following stable states: ⓑ, ⓓ, ⓕ, and ⓖ. Obtain the design module and the test bench module that takes the machine through the paths to assert output z_1. The excitation equations and the output equation are to be in a sum-of-products form. Obtain the outputs and the waveforms. There should be no static-1 or static-0 hazards in the equations.

x_1x_2

	00	01	11	10
1	(a)	c	(b)	d
2	e	(f)	b	(d)
3	a	(c)	b	(g)
4	(e)	f	$-$	g

4.19   An asynchronous sequential machine has two inputs x_1 and x_2 and one output z_1. Input x_1 will always be asserted whenever x_2 is asserted; that is, there will never be a situation where x_1 is deasserted and x_2 is asserted. Output z_1 is asserted coincident with every third x_2 pulse and remains active for the duration of x_2. A representative timing diagram is shown below. Synthesize the asynchronous sequential machine using behavioral modeling in the design module. Obtain the test bench module to show two assertions of output z_1. Obtain the outputs and the waveforms.

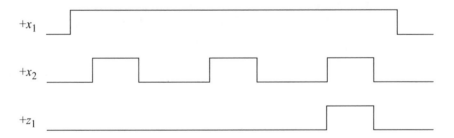

4.20   Repeat Problem 4.19 using dataflow modeling by instantiating logic gates that were designed using dataflow modeling and using them as a single line. Obtain the excitation and output equations in a sum-of-products form and use them in the dataflow design. Obtain the test bench module to show two assertions of output z_1. Obtain the outputs and the waveforms. The timing diagram is duplicated below.

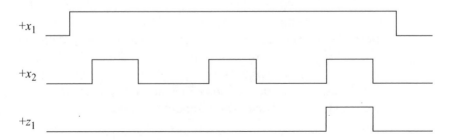

5.1   *Introduction*
5.2   *Synthesis Examples*
5.3   *Problems*

5

Synthesis of Pulse-Mode Asynchronous Sequential Machines Using Verilog HDL

This chapter implements pulse-mode asynchronous sequential machine designs using Verilog HDL. The designs will be accomplished by utilizing one or more of the following modeling methods for each design: built-in primitive gates, dataflow modeling, behavioral modeling, structural modeling. These four modeling methods are briefly summarized in Section 5.1 for convenience.

5.1   Introduction

This section briefly describes the four modeling methods of the Verilog hardware description language that will be used to design the pulse-mode asynchronous sequential machines. Several different types of pulse-mode asynchronous sequential machines will be designed using Verilog HDL.

5.1.1   Built-In Primitive Gates

These gates describe a net and have one or more scalar inputs, but only one scalar output. The multiple-input gates are **and**, **nand**, **or**, **nor**, **xor**, and **xnor**. The output signal is listed first, followed by the inputs in any order. Two or more instances of the same type of gate can be specified in the same construct, as shown below. Note that only the last instantiation has a semicolon terminating the line. All previous lines are terminated by a comma.

$$\textbf{gate\_type} \;\; \text{inst1 (output\_1, input\_11, input\_12, } \ldots \text{, input\_1}n),$$
$$\text{inst2 (output\_2, input\_21, input\_22, } \ldots \text{, input\_2}n),$$

$$.$$
$$.$$

$$\text{inst}m \text{ (output\_}m\text{, input\_}m1\text{, input\_}m2, \ldots \text{, input\_}mn);$$

5.1.2   Dataflow Modeling

This method is at a higher level of abstraction than gate-level modeling using built-in primitives. The *continuous assignment* statement models dataflow behavior and is used to design combinational logic without using gates and interconnecting nets. Continuous assignment statements provide a Boolean correspondence between the right-hand side expression and the left-hand side target. The continuous assignment statement uses the keyword **assign** and has the following syntax with optional drive strength and delay:

assign [drive_strength] [delay] left-hand side target = right-hand side expression

The **assign** statement continuously monitors the right-hand side expression. If a variable changes value, then the expression is evaluated and the result is assigned to the target after any specified delay. If no delay is specified, then the default delay is zero. The continuous assignment statement can be considered to be a form of behavioral modeling, because the behavior of the circuit is specified, not the implementation.

5.1.3   Behavioral Modeling

This method describes the behavior of a digital system and is not concerned with the direct implementation of logic gates but more on the architecture of the system. This is an algorithmic approach to hardware implementation and represents a higher level of abstraction than the previous modeling methods.

A Verilog module that is designed using behavioral modeling contains no internal structural details, it simply defines the behavior of the hardware in an abstract, algorithmic description. Verilog contains two structured procedure statements or behaviors: **initial** and **always**.

Initial statement   An **initial** statement executes only once beginning at time zero, then suspends execution. An **initial** statement provides a method to initialize and monitor variables before the variables are used in a module; it can also be used to generate waveforms. For a given time unit, all statements within the **initial** block execute sequentially. The syntax for an **initial** statement is shown below.

> **initial** [optional timing control] procedural statement or
> block of procedural statements

Always statement The **always** statement executes the behavioral statements within the **always** block repeatedly in a looping manner and begins execution at time zero. Execution of the statements continues indefinitely until the simulation is terminated. The syntax for the **always** statement is shown below.

> **always** [optional timing control] procedural statement or
> block of procedural statements

Conditional statements Conditional statements alter the flow within a behavior based upon certain conditions. The choice among alternative statements depends on the Boolean value of an expression. The alternative statements can be a single statement or a block of statements delimited by the keywords **begin** . . . **end**. The keywords **if** and **else** are used in conditional statements as shown below. The **case** statement — defined in Section 4.1.3 — is an alternative to conditional statements.

```
//no else statement
if (expression) statement1;          //if expression is true, then statement1 is executed.
//one else statement                 //choice of two statements. Only one is executed.
if (expression) statement1;          //if expression is true, then statement1 is executed.
else statement2;                     //if expression is false, then statement2 is executed.


//nested if-else if                  //choice of multiple statements. Only one is execut-
                                     ed.
if (expression1) statement1;         //if expression1 is true, then statement1 is executed.
else if (expression2) statement2;    //if expression2 is true, then statement2 is executed.
else if (expression3) statement3;    //if expression3 is true, then statement3 is executed.
else default statement;
```

While loop The **while** loop executes a procedural statement or a block of procedural statements as long as a Boolean expression returns a value of true. When the procedural statements are executed, the Boolean expression is reevaluated. If the evaluation of the expression is false, then the **while** loop is terminated and control is passed to the next statement in the module. If the expression is false before the loop is initially entered, then the **while** loop is never executed. The syntax for a **while** statement is as follows:

> **while** (expression)
> procedural statement or block of procedural statements

5.1.4   Structural Modeling

Structural modeling consists of instantiating one or more of the following design objects: built-in primitives, user-defined primitives (UDPs), design modules.

Instantiation means to use one or more lower-level modules — including built-in primitives — that are interconnected in the construction of a higher-level structural module. A module can be a logic gate, an adder, a multiplexer, a counter, or some other logical function. The objects that are instantiated are called *instances*. Structural modeling is described by the interconnection of these lower-level logic primitives or modules.

5.2   Synthesis Examples

The examples which follow illustrate the synthesis procedure for pulse-mode asynchronous sequential machines using a timing diagram and/or a verbal specification. In order to prevent possible race conditions and associated timing problems when two or more inputs change value simultaneously, it will be assumed that only one input variable will change state at a time. This is referred to as a *fundamental-mode model*, further defined with the following characteristics:

1. Only one input will change at a time.
2. No other input will change until the machine has sequenced to a stable state.

Reliability of pulse-mode machines can be increased by inserting delay circuits of an appropriate duration in the output networks of the storage elements or by delaying the clock input to the storage elements. The aggregate delay of the storage elements and the delay circuit must be of sufficient duration so that the input pulse will be deasserted before the storage element output signals arrive at the δ next-state logic.

The techniques that are commonly used to insert delays in the storage element outputs are: An even number of inverters are connected in series with each latch output; a linear delay circuit is connected in series with each latch output; an edge-triggered D flip-flop is connected in series with each latch output; or a T flip-flop with a delay circuit from the T input is connected to the clock input. The D flip-flops are set to the state of the latches, but are triggered on the trailing edge of the input pulses. Thus, the flip-flop outputs — and therefore the state of the machine as represented by the SR latch outputs — are received at the δ next-state logic only when the active input pulse has been deasserted. The SR latches and the flip-flops constitute a master-slave relationship.

The synthesis procedures will be accomplished using two different types of storage elements: SR latches configured as T flip-flops, and SR latches with D flip-flops arranged in a master-slave configuration.

Example 5.1 A Moore pulse-mode asynchronous sequential machine will be designed that has two inputs x_1 and x_2 and one output z_1. The storage elements consist of SR latches and D flip-flops. The machine operates according to the representative timing diagram shown in Figure 5.1. The corresponding state diagram is shown in Figure 5.2.

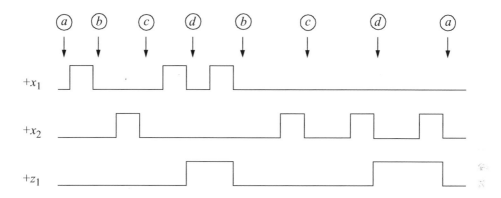

Figure 5.1 Representative timing diagram for the Moore pulse-mode sequential machine of Example 5.1.

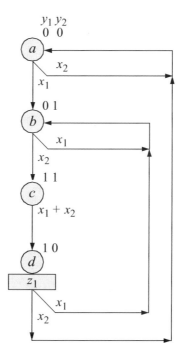

Figure 5.2 State diagram for the pulse-mode sequential machine of Example 5.1.

Table 5.1 presents the same information as the state diagram, but in a tabular representation. Although there are two input variables, only one input can be asserted at a time; therefore, only two combinations are listed for x_1 and x_2: $x_1 x_2 = 10$ and 01. The remaining two combinations, $x_1 x_2 = 00$ and 11, are not used in pulse-mode synthesis. If $x_1 x_2 = 00$, then the machine does not change state; if $x_1 x_2 = 11$, then this represents an invalid combination, since input pulses cannot occur simultaneously.

Table 5.1 Next-State Table for the Moore Pulse-Mode Machine of Example 5.1

State name	Present state $y_1 y_2$	Inputs $x_1 x_2$	Next state $y_1 y_2$	Output z_1
a	0 0	1 0	0 1	0
	0 0	0 1	0 0	0
b	0 1	1 0	0 1	0
	0 1	0 1	1 1	0
c	1 1	1 0	1 0	0
	1 1	0 1	1 0	0
d	1 0	1 0	0 1	1
	1 0	0 1	0 0	1

Each latch requires two input maps, one map for x_1 and one map for x_2, as shown in the input maps of Figure 5.3. The maps are arranged such that the maps corresponding to each latch are in the same row, and each column of maps corresponds to a separate input. The map entries are defined as follows: S and s indicate that the latch will be set or remain set, respectively; R and r indicate that the latch will be reset or remain reset, respectively.

The map entries correlate directly to the entries in the next-state table. For example, in state $y_1 y_2 = 10$, latch Ly_1 will be reset if x_1 is pulsed, as indicated by the letter R in minterm location 2 of the map in row Ly_1, column x_1. Also, in state $y_1 y_2 = 10$, latch Ly_2 will be set if x_1 is pulsed, as indicated by the letter S in minterm location 2 of the map in row Ly_2, column x_1.

Now consider the effect when input x_2 is asserted in state $y_1 y_2 = 10$. When x_2 is activated, a reset pulse is applied to latch Ly_1, as shown in the next-state table. Since latch Ly_1 was set, the letter R is entered in minterm location 2 of the map for latch Ly_1 in row Ly_1, column x_2. The assertion of x_2 will also generate a reset pulse to latch Ly_2. However, since latch Ly_2 is already reset, an x_2 pulse causes latch Ly_2 to remain in a reset state, as specified by the letter r in minterm location 2 of the map for latch Ly_2 in row Ly_2, column x_2.

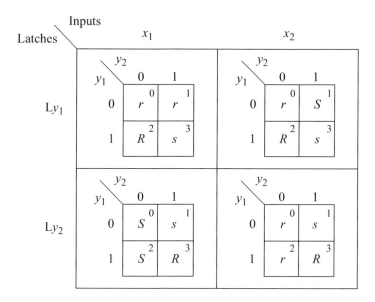

Figure 5.3   Input maps for the Moore pulse-mode sequential machine of Example 5.1.

The input equations are obtained directly from the input maps and are shown in Equation 5.1. Since the input equations refer to the latches, the equations are labelled as follows: SLy_1, RLy_1, SLy_2, and RLy_2, indicating the set (S) and reset (R) of the latches (L), respectively. The D flip-flops are labelled simply y_1 and y_2. Latch Ly_1 will be set if flip-flop y_2 is set and x_2 is pulsed.   Latch Ly_1 will be reset if y_2 is reset and either x_1 or x_2 is pulsed.   Thus, the set and reset conditions for latch Ly_1 are $SLy_1 = y_2 x_2$ and $RLy_1 = y_2'(x_1 + x_2)$, respectively.

Similarly, latch Ly_2 will be set if flip-flop y_2 is reset and x_1 is pulsed.   Latch y_2 will be reset if flip-flops y_1 and y_2 are both set and x_1 is pulsed or if flip-flop y_1 is set and x_2 is pulsed.   Thus, the set and reset conditions for latch Ly_2 are $SLy_2 = y_2' x_1$ and $RLy_2 = y_1 y_2 x_1 + y_1 x_2$, respectively.

$$SLy_1 = y_2 x_2$$

$$RLy_1 = y_2'(x_1 + x_2)$$

$$SLy_2 = y_2' x_1$$

$$RLy_2 = y_1 y_2 x_1 + y_1 x_2 \qquad (5.1)$$

Since the pulse-mode machine of Example 5.1 is a Moore machine, the output is a function of the present state only. Thus, output z_1 is asserted if flip-flops y_1 and y_2 are set and reset, respectively, yielding the equation of Equation 5.2.

$$z_1 = y_1 y_2'$$

(5.2)

The logic diagram for the pulse-mode asynchronous sequential machine of Example 5.1 is shown in Figure 5.4 using SR latches and D flip-flops in a master-slave configuration together with the net names. There is an implied reset for the SR latches and the D flip-flops. The machine is designed from the input and output equations of Equation 5.1 and Equation 5.2, respectively. The design module using built-in primitives and D flip-flops — that were designed using behavioral modeling — is shown in Figure 5.5. The test bench module is shown in Figure 5.6. The outputs and waveforms are shown in Figure 5.7 and Figure 5.8, respectively.

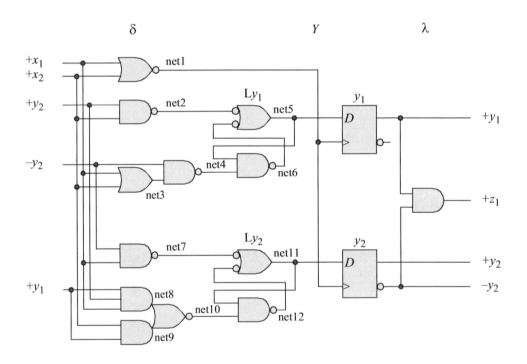

Figure 5.4   Logic diagram for the Moore pulse-mode sequential machine of Example 5.1.

```verilog
//moore pulse-mode asm using bip with D flip-flops
module pm_asm_moore (rst_n, x1, x2, y1, y2, z1);

input rst_n, x1, x2;      //define inputs and outputs
output y1, y2, z1;

wire net1, net2, net3, net4, net5,   //define internal nets
      net6, net7, net8, net9, net10, net11, net12;

//design the D flip-flop clock
nor    (net1, x1, x2);

//design the logic for latch Ly1 and D flip-flop y1
nand   (net2, y2, x2);
or     (net3, x1, x2);
nand   (net4, ~y2, net3);
nand   (net5, net2, net6),
       (net6, net5, net4, rst_n);

//instantiate the D flip-flop for y1
d_ff_bh inst1 (
   .rst_n(rst_n),
   .clk(net1),
   .d(net5),
   .q(y1)
   );

//design the logic for latch Ly2 and D flip-flop y2
nand   (net7, ~y2, x1);
and    (net8, y1, y2, x1),
       (net9, x2, y1);
nor    (net10, net8, net9);
nand   (net11, net7, net12),
       (net12, net11, net10, rst_n);

//instantiate the D flip-flop for y2
d_ff_bh inst2 (
   .rst_n(rst_n),
   .clk(net1),
   .d(net11),
   .q(y2)
   );

//design the logic for output z1
and    (z1, y1, ~y2);
endmodule
```

Figure 5.5   Design module using built-in primitives and *D* flip-flops for Example 5.1.

```
//test bench for moore pulse-mode asm

module pm_asm_moore_tb;

//inputs are reg for test bench
//outputs are wire for test bench
reg rst_n, x1, x2;
wire y1, y2, z1;

initial      //display variables
$monitor ("x1x2 = %b, state = %b, z1 = %b",
          {x1, x2}, {y1, y2}, z1);

initial      //apply input sequence
begin
   #0    rst_n = 1'b0;
         x1 = 1'b0;    x2 = 1'b0;
   #5    rst_n = 1'b1;
//-------------------------------------------------
   #10   x1 = 1'b0;   x2 = 1'b0;
   #10   x2 = 1'b1;   x2 = 1'b0;

   #10   x1 = 1'b1;   x2 = 1'b0;
   #10   x1 = 1'b0;   x2 = 1'b0;
   #10   x1 = 1'b0;   x2 = 1'b1;
   #10   x1 = 1'b0;   x2 = 1'b0;
   #10   x1 = 1'b1;   x2 = 1'b0;
   #10   x1 = 1'b0;   x2 = 1'b0;    //assert z1
   #10   x1 = 1'b1;   x2 = 1'b0;    //assert z1
   #10   x1 = 1'b0;   x2 = 1'b0;
//-------------------------------------------------
   #20   x1 = 1'b0;   x2 = 1'b1;
   #10   x1 = 1'b0;   x2 = 1'b0;
   #10   x1 = 1'b0;   x2 = 1'b1;
   #10   x1 = 1'b0;   x2 = 1'b0;    //assert z1
   #20   x1 = 1'b0;   x2 = 1'b1;    //assert z1
   #10   x1 = 1'b0;   x2 = 1'b0;
//-------------------------------------------------
   #10   x1 = 1'b0;   x2 = 1'b0;

   $stop;
end

//instantiate the module into the test bench as a single line
pm_asm_moore inst1 (rst_n, x1, x2, y1, y2, z1);

endmodule
```

Figure 5.6   Test bench module for Example 5.1.

```
x1x2 = 00, state = 00, z1 = 0
x1x2 = 00, state = 00, z1 = 0
x1x2 = 10, state = 00, z1 = 0
x1x2 = 00, state = 01, z1 = 0
x1x2 = 01, state = 01, z1 = 0
x1x2 = 00, state = 11, z1 = 0
x1x2 = 10, state = 11, z1 = 0
x1x2 = 00, state = 10, z1 = 1
x1x2 = 10, state = 10, z1 = 1
x1x2 = 00, state = 01, z1 = 0

x1x2 = 01, state = 01, z1 = 0
x1x2 = 00, state = 11, z1 = 0
x1x2 = 01, state = 11, z1 = 0
x1x2 = 00, state = 10, z1 = 1
x1x2 = 01, state = 10, z1 = 1
x1x2 = 00, state = 00, z1 = 0
```

Figure 5.7   Outputs for Example 5.1.

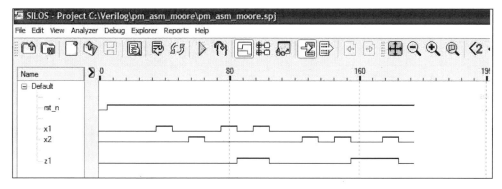

Figure 5.8   Waveforms for Example 5.1.

Example 5.2   A T flip-flop will be designed in this example. A T flip-flop has an input T and two outputs: y_1 and y_1'. If the flip-flop is reset, then an active pulse on the T input will toggle the flip-flop to the set state; if the flip-flop is set, then a pulse on the T input will toggle the flip-flop to the reset state. The T flip-flop characteristics are shown in Table 5.2.

The T flip-flop utilized in this example incorporates a D flip-flop, an exclusive-OR circuit, and a delay circuit as a **buf** built-in primitive, as shown in Figure 5.9. The T input connects to the clock input of the D flip-flop through a delay circuit, which allows the clock input to be delayed until the signal on the D input has stabilized. When T has a value of 0, the next state is the same as the present state; when T has a value of 1, the next state is the complement of the present state.

Table 5.2 *T* Flip-Flop Characteristics

Present state $Y_{j(t)}$	Input T	Next state $Y_{k(t+1)}$	State transition sequence
0	0	0	$0 \rightarrow 0$
1	0	1	$1 \rightarrow 1$
0	1	1	$0 \rightarrow 1$
1	1	0	$1 \rightarrow 0$

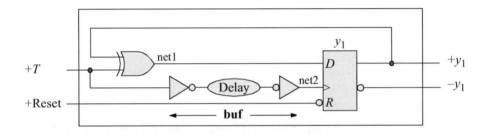

Figure 5.9   Alternative *T* flip-flop configuration.

The design module for the *T* flip-flop is shown in Figure 5.10. The test bench module is shown in Figure 5.11. The outputs and waveforms are shown in Figure 5.12 and Figure 5.13, respectively.

```
//T flip-flop design using a D flip-flop and an xor

module t_ff_da (rst_n, t, y1);

input rst_n, t;        //define inputs and output
output y1;

wire net1, net2;        //net2 is the T input delayed

//define the logic for the T flip-flop
xor      (net1, t, y1);     //flip-flop D input
buf      (net2, t);         //flip-flop clk input delayed
                                     //continued on next page
```

Figure 5.10   Design module for a *T* flip-flop.

```
//instantiate the D flip-flop
d_ff_bh inst1 (
    .rst_n(rst_n),
    .clk(net2),
    .d(net1),
    .q(y1)
    );

endmodule
```

Figure 5.10   (Continued)

```
//test bench for the T-flop-flop
module t_ff_da_tb;

//inputs are reg for test bench
//outputs are wire for test bench
reg rst_n, t;
wire y1;

initial      //display variables
$monitor ($time, "ns, t = %b, y1 = %b", t, y1);

//define input sequence
initial
begin
    #0     rst_n = 1'b0;
           t = 1'b0;
    #5     rst_n = 1'b1;
//-----------------------------
    #20    t = 1'b1;
    #10    t = 1'b0;
    #30    t = 1'b1;
    #10    t = 1'b0;
    #20    t = 1'b1;
    #10    t = 1'b0;
    #10    t = 1'b1;
    #10    $stop;
end

//instantiate the module into the test bench as a single line
t_ff_da inst1 (rst_n, t, y1);

endmodule
```

Figure 5.11   Test bench for the *T* flip-flop.

```
0ns,    t = 0,  y1 = 0
25ns,   t = 1,  y1 = 1
35ns,   t = 0,  y1 = 1
65ns,   t = 1,  y1 = 0
75ns,   t = 0,  y1 = 0
95ns,   t = 1,  y1 = 1
105ns,  t = 0,  y1 = 1
115ns,  t = 1,  y1 = 0
```

Figure 5.12   Outputs for the T flip-flop.

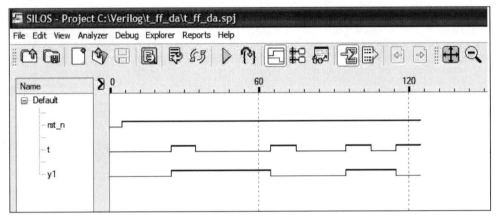

Figure 5.13   Waveforms for the T flip-flop.

Example 5.3   A Moore pulse-mode asynchronous sequential machine will be designed that has two inputs x_1 and x_2 and one output z_1. This example repeats Example 5.1 that used built-in primitives and storage elements which consisted of SR latches and D flip-flops. However, the machine in this example will be designed using logic gates that were designed using dataflow modeling and with D flip-flops that were designed using behavioral modeling.

The output of each latch connects to the D input of the associated flip-flop forming a master-slave relationship. Since the D flip-flops are clocked on the trailing edge of the positive input pulses, state changes are not fed back to the δ next-state logic until the active input has been deasserted. Clocking the flip-flops on the negative edge of the positive input pulses delays the next state from affecting the input logic while an input pulse is still active. Thus, the machine operates in a deterministic manner.

The machine operates according to the representative timing diagram shown in Figure 5.14, which is reproduced from Example 5.1 for convenience. The corresponding state diagram is also reproduced for convenience and is shown in Figure 5.15.

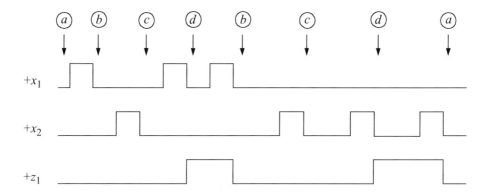

Figure 5.14   Representative timing diagram for the Moore pulse-mode sequential machine of Example 5.3.

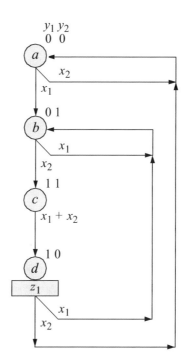

Figure 5.15   State diagram for the Moore pulse-mode sequential machine of Example 5.3.

The input equations and output equation are shown in Equation 5.3 and Equation 5.4, respectively. The logic diagram is shown in Figure 5.16 using NAND, NOR, AND, and OR instantiated modules. The design module is shown in Figure 5.17. The test bench module is shown in Figure 5.18. The outputs and waveforms are shown in Figure 5.19 and Figure 5.20, respectively.

$$SLy_1 = y_2x_2$$
$$RLy_1 = y_2'(x_1 + x_2)$$

$$SLy_2 = y_2'x_1$$
$$RLy_2 = y_1y_2x_1 + y_1x_2 \qquad (5.3)$$

$$z_1 = y_1y_2' \qquad (5.4)$$

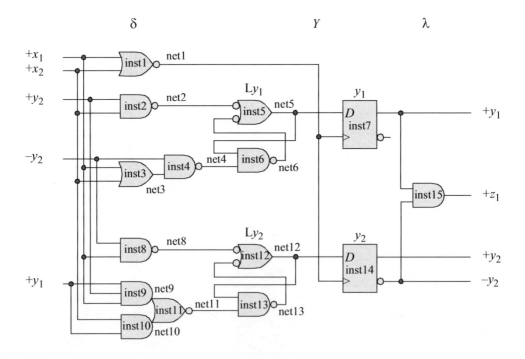

Figure 5.16   Logic diagram for the Moore pulse-mode sequential machine of Example 5.3.

```
//moore structural pulse-mode asm using instantiation

module pm_asm_moore2 (rst_n, x1, x2, y1, y2, z1);

//define inputs and outputs
input rst_n, x1, x2;
output y1, y2, z1;

//define internal nets
wire net1, net2, net3, net4, net5, net6, net7;
wire net8, net9, net10, net11, net12, net13;

//define the D flip-flop clock
nor2_df  inst1 (x1, x2, net1);

//------------------------------------------------------
//design the logic for latch Ly1 and D flip-flop y1
nand2_df inst2 (y2, x2, net2);
or2_df   inst3 (x1, x2, net3);
nand2_df inst4 (~y2, net3, net4);

//latch Ly1
nand2_df inst5 (net2, net6, net5);
nand3_df inst6 (net5, net4, rst_n, net6);

//instantiate the D flip-flop for y1
d_ff_bh inst7 (
   .rst_n(rst_n),
   .clk(net1),
   .d(net5),
   .q(y1)
   );


//------------------------------------------------------
//design the logic for latch Ly2 and D flip-flop y2
nand2_df inst8 (~y2, x1, net8);
and3_df  inst9 (y1, y2, x1, net9);
and2_df  inst10 (x2, y1, net10);
nor2_df  inst11 (net9, net10, net11);

//latch Ly2
nand2_df inst12 (net8, net13, net12);
nand3_df inst13 (net12, net11, rst_n, net13);

                               //continued on next page
```

Figure 5.17   Design module for the Moore pulse-mode asynchronous sequential machine of Example 5.3.

```
//instantiate the D flip-flop for y2
d_ff_bh inst14 (
   .rst_n(rst_n),
   .clk(net1),
   .d(net12),
   .q(y2)
   );

//design the logic for output z1
and2_df  inst15 (y1, ~y2, z1);

endmodule
```

Figure 5.17   (Continued)

```
//test bench for moore pulse-mode asm

module pm_asm_moore2_tb;

//inputs are reg for test bench
//outputs are wire for test bench
reg rst_n, x1, x2;
wire y1, y2, z1;

//display variables
initial
$monitor ("x1x2 = %b, state = %b, z1 = %b",
            {x1, x2}, {y1, y2}, z1);

//apply input sequence
initial
begin
   #0     rst_n = 1'b0;
          x1 = 1'b0;
          x2 = 1'b0;

   #5     rst_n = 1'b1;

//----------------------------------------------------
   #10    x1 = 1'b0;   x2 = 1'b0;
   #10    x1 = 1'b1;   x2 = 1'b0;

                              //continued on next page
```

Figure 5.18   Test bench module for the Moore pulse-mode asynchronous sequential machine of Example 5.3.

```
    #10    x1 = 1'b1;   x2 = 1'b0;
    #10    x1 = 1'b0;   x2 = 1'b0;
    #10    x1 = 1'b0;   x2 = 1'b1;
    #10    x1 = 1'b0;   x2 = 1'b0;
    #10    x1 = 1'b1;   x2 = 1'b0;
    #10    x1 = 1'b0;   x2 = 1'b0;   //assert z1
    #10    x1 = 1'b1;   x2 = 1'b0;   //assert z1
    #10    x1 = 1'b0;   x2 = 1'b0;
//-----------------------------------------------
    #20    x1 = 1'b0;   x2 = 1'b1;
    #10    x1 = 1'b0;   x2 = 1'b0;
    #10    x1 = 1'b0;   x2 = 1'b1;
    #10    x1 = 1'b0;   x2 = 1'b0;   //assert z1
    #20    x1 = 1'b0;   x2 = 1'b1;   //assert z1
    #10    x1 = 1'b0;   x2 = 1'b0;
//-----------------------------------------------
    #10    x1 = 1'b0;   x2 = 1'b0;
//-----------------------------------------------
   $stop;
end

//instantiate the module into the test bench as a single line
pm_asm_moore2 inst1 (rst_n, x1, x2, y1, y2, z1);
endmodule
```

Figure 5.18   (Continued)

```
x1x2 = 00, state = 00, z1 = 0
x1x2 = 10, state = 00, z1 = 0
x1x2 = 00, state = 01, z1 = 0
x1x2 = 01, state = 01, z1 = 0
x1x2 = 00, state = 11, z1 = 0
x1x2 = 10, state = 11, z1 = 0
x1x2 = 00, state = 10, z1 = 1
x1x2 = 10, state = 10, z1 = 1
x1x2 = 00, state = 01, z1 = 0

x1x2 = 01, state = 01, z1 = 0
x1x2 = 00, state = 11, z1 = 0
x1x2 = 01, state = 11, z1 = 0
x1x2 = 00, state = 10, z1 = 1
x1x2 = 01, state = 10, z1 = 1
x1x2 = 00, state = 00, z1 = 0
```

Figure 5.19   Outputs for the pulse-mode machine of Example 5.3.

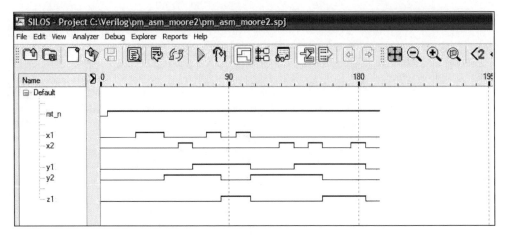

Figure 5.20   Waveforms for the pulse-mode machine of Example 5.3.

Example 5.4   A Moore pulse-mode asynchronous sequential machine will be designed that has two inputs x_1 and x_2 and one output z_1. This example repeats Example 5.1 in which the storage elements consisted of *SR* latches and *D* flip-flops. However, the machine in this example will be designed using built-in primitives and *T* flip-flops. The timing diagram is reproduced in Figure 5.21 and the state diagram is reproduced in Figure 5.22.

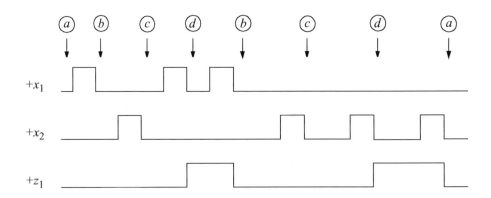

Figure 5.21   Timing diagram for Example 5.4.

The input maps, obtained from the state diagram, are shown in Figure 5.23 using the entry *T* in appropriate minterm locations. The *T* entry indicates that the state of the machine will change state; that is, it will toggle from 0 to 1 or toggle from 1 to 0. The entries that must be considered are the *T* entries, since these are the only entries for a *T* flip-flop that result in a change of state for y_1 and y_2. The *s* and *r* entries cannot combine with the T in the minimization process, since these entries maintain a constant flip-flop state, whereas a T will change the state of the corresponding flip-flop.

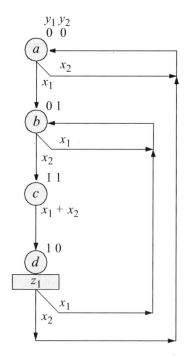

Figure 5.22   State diagram for Example 5.4.

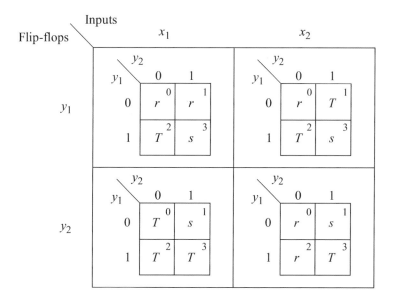

Figure 5.23   Input maps for the Moore pulse-mode machine of Example 5.4.

The input equations to toggle the T flip-flops — obtained from the input maps — are shown in Equation 5.5. The output equation for z_1 is obtained from the state diagram and is shown in Equation 5.6. The logic diagram is designed from the input and output equations and is shown in Figure 5.24.

$$Ty_1 = y_1y_2'x_1 + y_1'y_2x_2 + y_1y_2'x_2$$
$$Ty_2 = y_2'x_1 + y_1x_1 + y_1y_2x_2 \qquad\qquad (5.5)$$

$$z_1 = y_1y_2' \qquad\qquad (5.6)$$

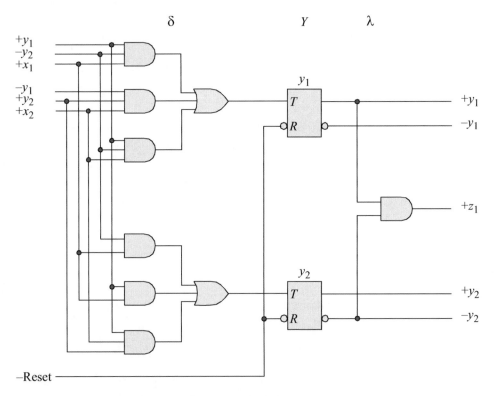

Figure 5.24   Logic diagram for Example 5.4.

The design module is shown in Figure 5.25 using built-in primitives and instantiated T flip-flops that were also designed with built-in primitives. The outputs of the T flip-flops are delayed by 11 time units, using a **buf** built-in primitive before being fed back to the δ next-state logic. This allows the inputs to be deasserted before the state of the machine is received at the input logic, which is a characteristic of a *fundamental-mode model*. The test bench module is shown in Figure 5.26. The outputs and waveforms are shown in Figure 5.27 and Figure 5.28, respectively.

```verilog
//moore pulse-mode asm using built-in primitives and T ff

module pm_asm_moore3 (rst_n, x1, x2, y1, y2, z1);

//define inputs and outputs
input rst_n, x1, x2;
output y1, y2, z1;

//define internal nets
wire net1, net2, net3, net4, net5, net6, net7, net8;
wire nety1, nety2;

//----------------------------------------
//define the logic for flip-flop y1
and    (net1, y1, ~y2, x1),
       (net2, ~y1, y2, x2),
       (net3, y1, ~y2, x2);
or     (net4, net1, net2, net3);

//instantiate the T flip-flop as a single line
t_ff_da inst1 (rst_n, net4, nety1);     //(rst_n, t, y1)

buf    #11 (y1, nety1);

//----------------------------------------
//define the logic for flip-flop y2
and    (net5, ~y2, x1),
       (net6, y1, x1),
       (net7, y1, y2, x2);
or     (net8, net5, net6, net7);

//instantiate the T flip-flop as a single line
t_ff_da inst2 (rst_n, net8, nety2);     //(rst_n, t, y2)

buf    #11 (y2, nety2);

//----------------------------------------
//define the logic for output z1
and    (z1, y1, ~y2);

endmodule
```

Figure 5.25   Design module for the Moore pulse-mode asynchronous sequential machine of Example 5.4.

```
//test bench for moore pulse-mode asm
module pm_asm_moore3_tb;

reg rst_n, x1, x2;    //inputs are reg for test bench
wire y1, y2, z1;      //outputs are wire for test bench

initial                //display variables
$monitor ("x1x2 = %b, state = %b, z1 = %b",
          {x1, x2}, {y1, y2}, z1);

initial                    //apply input sequence
begin
    #0    rst_n = 1'b0;
          x1 = 1'b0;
          x2 = 1'b0;

    #5    rst_n = 1'b1;
//-------------------------------------------------
    #10   x1 = 1'b0;   x2 = 1'b0;
    #10   x1 = 1'b1;   x2 = 1'b0;

    #10   x1 = 1'b0;   x2 = 1'b0;
    #10   x1 = 1'b0;   x2 = 1'b1;
    #10   x1 = 1'b0;   x2 = 1'b0;
    #10   x1 = 1'b1;   x2 = 1'b0;
    #10   x1 = 1'b0;   x2 = 1'b0;    //assert z1
    #10   x1 = 1'b1;   x2 = 1'b0;    //assert z1
    #10   x1 = 1'b0;   x2 = 1'b0;
//-------------------------------------------------
    #10   x1 = 1'b0;   x2 = 1'b0;
    #10   x1 = 1'b0;   x2 = 1'b1;
    #10   x1 = 1'b0;   x2 = 1'b0;
    #10   x1 = 1'b0;   x2 = 1'b1;
    #10   x1 = 1'b0;   x2 = 1'b0;    //assert z1
    #20   x1 = 1'b0;   x2 = 1'b1;    //assert z1
    #10   x1 = 1'b0;   x2 = 1'b0;
//-------------------------------------------------
    #10   x1 = 1'b0;   x2 = 1'b0;
//-------------------------------------------------
    $stop;
end

//instantiate the module into the test bench as a single line
pm_asm_moore3 inst1 (rst_n, x1, x2, y1, y2, z1);

endmodule
```

Figure 5.26   Test bench module for the Moore pulse-mode asynchronous sequential machine of Example 5.4.

```
x1x2 = 00, state = 00, z1 = 0
x1x2 = 10, state = 00, z1 = 0
x1x2 = 00, state = 00, z1 = 0
x1x2 = 00, state = 01, z1 = 0
x1x2 = 01, state = 01, z1 = 0
x1x2 = 00, state = 01, z1 = 0
x1x2 = 00, state = 11, z1 = 0
x1x2 = 10, state = 11, z1 = 0
x1x2 = 00, state = 11, z1 = 0

x1x2 = 00, state = 10, z1 = 1
x1x2 = 10, state = 10, z1 = 1
x1x2 = 00, state = 10, z1 = 1

x1x2 = 00, state = 01, z1 = 0
x1x2 = 01, state = 01, z1 = 0
x1x2 = 00, state = 01, z1 = 0
x1x2 = 00, state = 11, z1 = 0
x1x2 = 01, state = 11, z1 = 0
x1x2 = 00, state = 11, z1 = 0

x1x2 = 00, state = 10, z1 = 1
x1x2 = 01, state = 10, z1 = 1
x1x2 = 00, state = 10, z1 = 1

x1x2 = 00, state = 00, z1 = 0
```

Figure 5.27   Outputs for the Moore pulse-mode asynchronous sequential machine of Example 5.4.

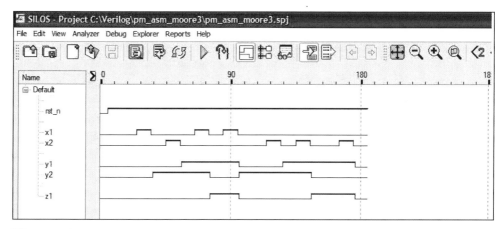

Figure 5.28   Waveforms for the Moore pulse-mode asynchronous sequential machine of Example 5.4.

Example 5.5   Using built-in primitives and D flip-flops, a Mealy machine will be synthesized that operates according to the following specifications: The machine has two inputs x_1 and x_2 and one output z_1. The inputs are pulses and will never be active concurrently. Output z_1 is also a pulse and is asserted coincident with x_2 whenever x_2 immediately follows exactly two x_1 pulses, as shown in the timing diagram of Figure 5.29. No output will be generated for three or more consecutive x_1 pulses. For this occurrence, the machine will be reinitialized by the next x_2 pulse. Assume that timing restrictions for pulse width and duty cycle have been satisfied.

The operation of the machine is graphically depicted by the state diagram of Figure 5.30. Since the inputs will not be asserted simultaneously, the state transition sequence depends on the occurrence of only a single pulse, either x_1 or x_2.

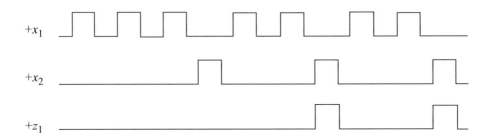

Figure 5.29   Representative timing diagram for the Mealy pulse-mode machine of Example 5.5.

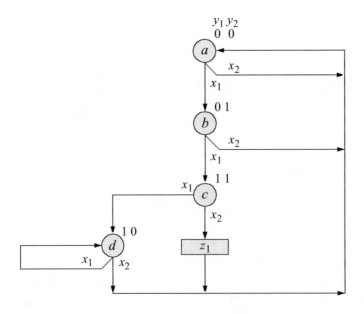

Figure 5.30   State diagram for the Mealy pulse-mode machine of Example 5.5.

The input maps are shown in Figure 5.31, as obtained from the state diagram. The input equations are shown in Equation 5.7. The output map is shown in Figure 5.32 and the output equation is shown in Equation 5.8. The logic diagram, derived from the input and output equations, is shown in Figure 5.33.

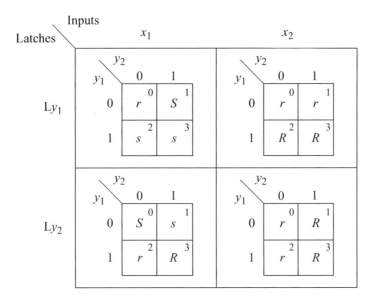

Figure 5.31   Input maps for the Mealy pulse-mode machine of Figure 5.30.

$$SLy_1 = y_2x_1$$
$$RLy_1 = x_2$$

$$SLy_2 = y_1'x_1$$
$$RLy_2 = y_1x_1 + x_2 \tag{5.7}$$

Figure 5.32   Output map for z_1 for the Mealy pulse-mode machine.

$$z_1 = y_1y_2x_2 \tag{5.8}$$

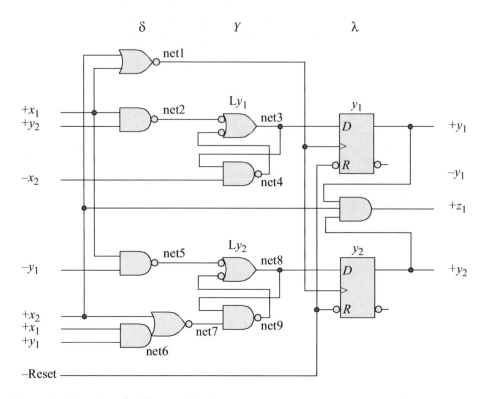

Figure 5.33   Logic diagram for the Mealy pulse-mode machine of Example 5.5.

The design module using built-in primitives and instantiated D flip-flops that were designed using behavioral modeling is shown in Figure 5.34. The test bench module is shown in Figure 5.35, which takes the machine through a sequence of inputs to assert output z_1. The outputs and waveforms are shown in Figure 5.36 and Figure 5.37, respectively.

```
//mealy pulse-mode asm using built-in primitives
module pm_asm13 (rst_n, x1, x2, y1, y2, z1);

//define inputs and outputs
input rst_n, x1, x2;
output y1, y2, z1;

//define internal nets
wire net1, net2, net3, net4, net5, net6, net7, net8, net9;
                                        //continued on next page
```

Figure 5.34   Design module for the Mealy pulse-mode machine of Example 5.5.

```
//design the D flip-flop clock
nor    (net1, x1, x2);

//--------------------------------------
//design the logic for latch Ly1
nand   (net2, x1, y2),
       (net3, net2, net4),
       (net4, net3, ~x2, rst_n);

//instantiate the D flip-flop for y1 as a single line
d_ff_bh inst1 (rst_n, net1, net3, y1);  //rst_n, clk, d, q)

//--------------------------------------
//design the logic for latch Ly2
nand(net5, x1, ~y1);
and    (net6, x1, y1);
nor    (net7, x2, net6);
nand   (net8, net5, net9),
       (net9, net8, net7, rst_n);

//instantiate the D flip-flop for y2 as a single line
d_ff_bh inst2 (rst_n, net1, net8, y2);  //rst_n, clk, d, q)

//--------------------------------------
//design the logic for output z1
and    (z1, y1, y2, x2);

endmodule
```

Figure 5.34   (Continued)

```
//test bench for mealy pulse-mode asm
module pm_asm13_tb;

//inputs are reg for test bench
//outputs are wire for test bench
reg rst_n, x1, x2;
wire y1, y2, z1;

//display variables
initial
$monitor ("x1x2 = %b, state = %b, z1 = %b",
          {x1, x2}, {y1, y2}, z1);
                                    //continued on next page
```

Figure 5.35   Test bench for the Mealy pulse-mode machine of Example 5.5.

```
//apply input sequence
initial
begin
   #0     rst_n = 1'b0;        //reset to state_a (00)
          x1 = 1'b0;
          x2 = 1'b0;

   #5     rst_n = 1'b1;

//--------------------------------------------------
   #10    x1 = 1'b0;   x2 = 1'b0;   //a
   #10    x1 = 1'b1;   x2 = 1'b0;   //b

   #10    x1 = 1'b0;   x2 = 1'b0;
   #10    x1 = 1'b1;   x2 = 1'b0;   //c

   #10    x1 = 1'b0;   x2 = 1'b0;
   #10    x1 = 1'b0;   x2 = 1'b1;   //a

   #10    x1 = 1'b0;   x2 = 1'b0;
   #10    x1 = 1'b0;   x2 = 1'b1;   //z1, back to a

//--------------------------------------------------
   #10    x1 = 1'b0;   x2 = 1'b0;   //a
   #10    x1 = 1'b1;   x2 = 1'b0;   //b

   #10    x1 = 1'b0;   x2 = 1'b0;
   #10    x1 = 1'b1;   x2 = 1'b0;   //c

   #10    x1 = 1'b0;   x2 = 1'b0;
   #10    x1 = 1'b1;   x2 = 1'b0;   //d

   #10    x1 = 1'b0;   x2 = 1'b0;
   #10    x1 = 1'b0;   x2 = 1'b1;   //a

//--------------------------------------------------
   #10    x1 = 1'b0;   x2 = 1'b0;
   #10    x1 = 1'b1;   x2 = 1'b0;   //b

   #10    x1 = 1'b0;   x2 = 1'b0;
   #10    x1 = 1'b1;   x2 = 1'b0;   //c

   #10    x1 = 1'b0;   x2 = 1'b0;
   #10    x1 = 1'b0;   x2 = 1'b1;   //z1, back to a
                                    //continued on next page
```

Figure 5.35   (Continued)

```
   #10   x1 = 1'b0;   x2 = 1'b0;
   #10   x1 = 1'b0;   x2 = 1'b1;

   #10   x1 = 1'b0;   x2 = 1'b0;

//-----------------------------------------------
   $stop;
end

//instantiate the module into the test bench as a single line
pm_asm13 inst1 (rst_n, x1, x2, y1, y2, z1);

endmodule
```

Figure 5.35   (Continued)

```
x1x2 = 00, state = 00, z1 = 0
x1x2 = 10, state = 00, z1 = 0
x1x2 = 00, state = 01, z1 = 0
x1x2 = 10, state = 01, z1 = 0
x1x2 = 00, state = 11, z1 = 0
x1x2 = 01, state = 11, z1 = 1
x1x2 = 00, state = 00, z1 = 0

x1x2 = 01, state = 00, z1 = 0
x1x2 = 00, state = 00, z1 = 0
x1x2 = 10, state = 00, z1 = 0
x1x2 = 00, state = 01, z1 = 0
x1x2 = 10, state = 01, z1 = 0
x1x2 = 00, state = 11, z1 = 0
x1x2 = 10, state = 11, z1 = 0
x1x2 = 00, state = 10, z1 = 0
x1x2 = 01, state = 10, z1 = 0
x1x2 = 00, state = 00, z1 = 0
x1x2 = 10, state = 00, z1 = 0
x1x2 = 00, state = 01, z1 = 0
x1x2 = 10, state = 01, z1 = 0
x1x2 = 00, state = 11, z1 = 0
x1x2 = 01, state = 11, z1 = 1
x1x2 = 00, state = 00, z1 = 0
x1x2 = 01, state = 00, z1 = 0
```

Figure 5.36   Outputs for the Mealy pulse-mode machine of Example 5.5.

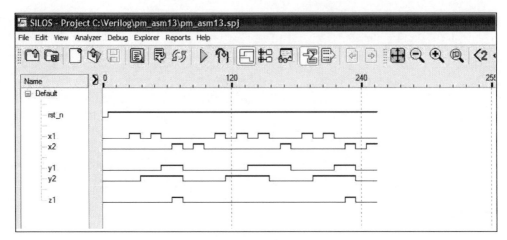

Figure 5.37   Waveforms for the Mealy pulse-mode machine of Example 5.5.

Example 5.6   In this example, built-in primitives will be utilized for the δ next-state logic and T flip-flops will be used as the storage elements for a Moore machine which operates according to the following specifications: A pulse on input x_2 will assert output z_1. Output z_1 will be deasserted by the second x_1 pulse in a sequence of consecutive x_1 pulses, but only if the two x_1 pulses are preceded by an x_2 pulse, as shown in the timing diagram of Figure 5.38.

Figure 5.39 depicts the state diagram for this machine. Since this is a Moore machine, the choice of state codes is critical to ensure that output z_1 will not glitch. The machine is initially reset to state a ($y_1y_2 = 00$). Pulses on input x_1 maintain the machine in state a, whereas an x_2 pulse sequences the machine to state b, where output z_1 is asserted as a level. The continued assertion of x_2 pulses in state b causes the machine to remain in state b, according to the machine specifications. The next x_1 pulse sequences the machine to state c, where z_1 remains asserted. It is only after the second x_1 pulse that the machine proceeds to state a, where z_1 is deasserted and the process repeats. The input maps are illustrated in Figure 5.40.

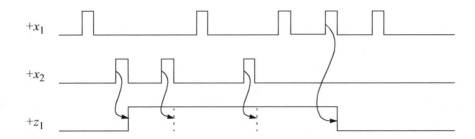

Figure 5.38   Representative timing diagram for the Moore pulse-mode machine of Example 5.6.

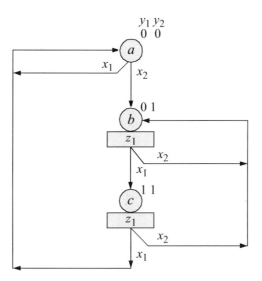

Figure 5.39   State diagram for the Moore pulse-mode machine of Example 5.6. There is one unused state: $y_1 y_2 = 10$.

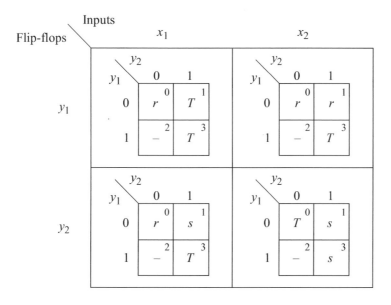

Figure 5.40   Input maps for the Moore pulse-mode machine of Figure 5.39 for Example 5.6.

The maps representing flip-flop y_1 are in the row corresponding to y_1; the maps representing flip-flop y_2 are in row y_2. The input variables x_1 and x_2 denote the column headings. The following different types of map entries are shown in Figure 5.40 and are obtained directly from the state diagram using the attributes of the T flip-flop:

> T indicates that the flip-flop toggles from 0 to 1 or from 1 to 0.
> s indicates that the flip-flop remains set.
> r indicates that the flip-flop remains reset.

To derive the input equations, refer to the input maps of Figure 5.40. The entries that must be considered are the T entries, since these are the only entries for a T flip-flop that result in a change of state for y_1 and y_2. The unused state $y_1 y_2 = 10$ can be used for minimization. The s and r entries cannot combine with the T in the minimization process, since these entries maintain a constant flip-flop state, whereas a T will change the state of the corresponding flip-flop.

The equations for toggling flip-flops y_1 and y_2 consist of the terms shown in Equation 5.9. Refer to the input map in row y_1, column x_1, state $y_1 y_2 = 01$. Flip-flop y_1 will toggle from 0 to 1 if input x_1 is pulsed and the machine is in state $y_1 y_2 = 01$ or from 1 to 0 if x_1 is pulsed in state $y_1 y_2 = 11$. The T entries in minterm locations 1 and 3 can combine, resulting in the term $y_2 x_1$. The other occurrence where y_1 will toggle (from 1 to 0) is: if x_2 is pulsed in state $y_1 y_2 = 11$. These conditions, when combined with the unused state $y_1 y_2 = 10$, yield two toggle terms, $y_2 x_1$ and $y_1 x_2$. Thus, the complete toggle equation for flip-flop y_1 is $Ty_1 = y_2 x_1 + y_1 x_2$. The toggle equation for flip-flop y_2 is obtained in a similar manner.

$$Ty_1 = y_2 x_1 + y_1 x_2$$

$$Ty_2 = y_1 x_1 + y_2' x_2 \qquad (5.9)$$

Output z_1 will be asserted whenever flip-flop y_2 is set. The equation for z_1 is shown in Equation 5.10. The logic diagram is shown in Figure 5.41. The storage elements are T flip-flops of the type shown in Figure 5.9. Correct operation of the machine can be verified by applying an appropriate sequence of x_1 and x_2 pulses and observing that the machine functions in accordance with the machine specifications as depicted by the state diagram of Figure 5.39.

$$z_1 = y_2 \qquad (5.10)$$

The design module is shown in Figure 5.42. The test bench module is shown in Figure 5.43. The outputs and waveforms are shown in Figure 5.44 and Figure 5.45, respectively.

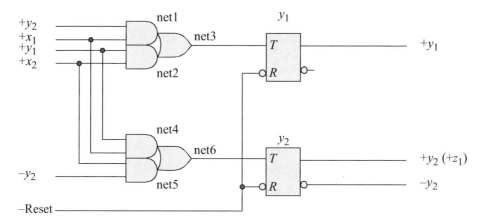

Figure 5.41   Logic diagram for the pulse-mode Moore machine of Figure 5.39 of Example 5.6 using positive-input T flip-flops for the storage elements.

```
//moore pulse-mode asm using built-in primitives

module pm_asm14 (rst_n, x1, x2, y1, y2, z1);

//define inputs and outputs
input rst_n, x1, x2;
output y1, y2, z1;


//define internal nets
wire net1, net2, net3, net4, net5, net6, nety1, nety2;


//-------------------------------------
//design the logic for flip-flop y1
and     (net1, x1, y2),
        (net2, x2, y1);
or      (net3, net1, net2);

//instantiate the T flip-flop
t_ff_da inst1 (rst_n, net3, nety1);    //rst_n, t, y1

buf     #4 (y1, nety1);
                                //continued on next page
```

Figure 5.42   Design module using built-in primitives and T flip-flops for the Moore pulse-mode machine of Example 5.6.

```
//-------------------------------------
//design the logic for flip-flop y2
and     (net4, x1, y1),
        (net5, x2, ~y2);
or      (net6, net4, net5);

//instantiate the T flip-flop
t_ff_da inst2 (rst_n, net6, nety2);    //rst_n, t, y1

buf    #4 (y2, nety2);

//-------------------------------------
//design the logic for output z1
assign z1 = y2;

endmodule
```

Figure 5.42   (Continued)

```
//test bench for the moore pulse=-mode asm

module pm_asm14_tb;

//inputs are reg for test bench
//outputs are wire for test bench
reg rst_n, x1, x2;
wire y1, y2, z1;

//display variables
initial
$monitor ("x1x2 = %b, state = %b, z1 = %b",
          {x1, x2}, {y1, y2}, z1);

//apply input sequence
initial
begin
   #0     rst_n = 1'b0;
          x1 = 1'b0;
          x2 = 1'b0;

   #5     rst_n = 1'b1;
                                //continued on next page
```

Figure 5.43   Test bench module for the Moore pulse-mode machine of Example 5.6

```
//-------------------------------------
   #5    x1 = 1'b1;
   #3    x1 = 1'b0;
   #7    x2 = 1'b1;      //20
   #3    x2 = 1'b0;
   #7    x2 = 1'b1;      //30
   #3    x2 = 1'b0;

   #7    x1 = 1'b1;      //40
   #3    x1 = 1'b0;
   #7    x2 = 1'b1;      //50
   #3    x2 = 1'b0;
   #7    x1 = 1'b1;      //60
   #3    x1 = 1'b0;

   #7    x1 = 1'b1;      //70
   #3    x1 = 1'b0;
   #7    x1 = 1'b1;      //80
   #3    x1 = 1'b0;

   #10   $stop;
end

//-------------------------------------
//instantiate the module into the test bench as a single line
pm_asm14 inst1 (rst_n, x1, x2, y1, y2, z1);

endmodule
```

Figure 5.43   (Continued)

```
x1x2 = 00, state = 00, z1 = 0    x1x2 = 01, state = 11, z1 = 1
x1x2 = 10, state = 00, z1 = 0    x1x2 = 01, state = 01, z1 = 1
x1x2 = 00, state = 00, z1 = 0    x1x2 = 00, state = 01, z1 = 1
x1x2 = 01, state = 00, z1 = 0    x1x2 = 10, state = 01, z1 = 1
x1x2 = 01, state = 01, z1 = 1    x1x2 = 10, state = 11, z1 = 1
x1x2 = 00, state = 01, z1 = 1    x1x2 = 00, state = 11, z1 = 1
x1x2 = 01, state = 01, z1 = 1    x1x2 = 10, state = 11, z1 = 1
x1x2 = 00, state = 01, z1 = 1    x1x2 = 10, state = 00, z1 = 0
x1x2 = 10, state = 01, z1 = 1    x1x2 = 00, state = 00, z1 = 0
x1x2 = 10, state = 11, z1 = 1    x1x2 = 10, state = 00, z1 = 0
x1x2 = 00, state = 11, z1 = 1    x1x2 = 00, state = 00, z1 = 0
```

Figure 5.44   Outputs for the pulse-mode machine of Example 5.6.

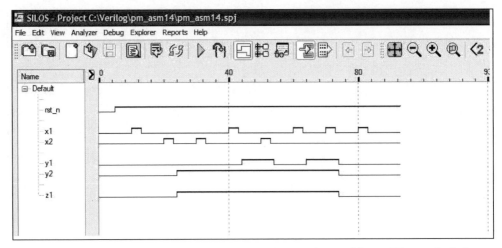

Figure 5.45   Waveforms for the Moore pulse-mode machine of Example 5.6.

Example 5.7   A Mealy machine will be synthesized which has three pulse input variables x_1, x_2, and x_3 and one output z_1 that is asserted coincident with x_3 whenever the sequence $x_1 x_2 x_3 = 100, 010, 001$ occurs. The storage elements will consist of SR latches and positive-edge-triggered D flip-flops.

A representative timing diagram displaying valid input sequences and corresponding outputs is shown in Figure 5.46. The state diagram is shown in Figure 5.47. State code assignment is arbitrary, since input pulses trigger all state transitions and the machine does not begin to sequence to the next state until the input pulse, which initiated the transition, has been deasserted. Thus, output z_1 will not glitch.

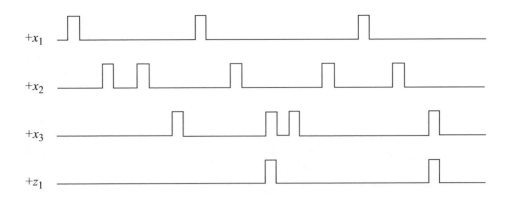

Figure 5.46   Representative timing diagram for the pulse-mode Mealy machine of Example 5.7.

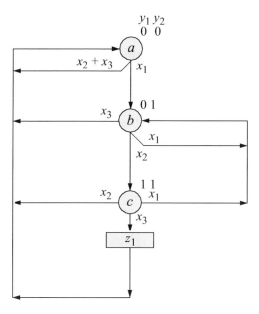

Figure 5.47 State diagram for the Mealy pulse-mode machine of Example 5.7.

A tabular representation of the state diagram is shown in Table 5.3. Since only one input variable can be active at a time, only three combinations are listed: $x_1 x_2 x_3 = 100, 010$, and 001. With the exception of $x_1 x_2 x_3 = 000$, all other combinations of the inputs are invalid.

**Table 5.3 Next-State Table for the Mealy
Pulse-Mode Machine of Figure 5.47**

State name	Present state $y_1\ y_2$	Inputs $x_1\ x_2\ x_3$	Next state $y_1\ y_2$	Output z_1
a	0 0	1 0 0	0 1	0
	0 0	0 1 0	0 0	0
	0 0	0 0 1	0 0	0
b	0 1	1 0 0	0 1	0
	0 1	0 1 0	1 1	0
	0 1	0 0 1	0 0	0
c	1 1	1 0 0	0 1	0
	1 1	0 1 0	0 0	0
	1 1	0 0 1	0 0	1

The input maps are shown in Figure 5.48. Each latch requires three input maps, one each for $x_1, x_2,$ and x_3. As in previous examples, the maps are arranged such that the maps corresponding to each latch are in the same row, and each column of maps corresponds to a unique input. The map entries are defined as follows:

S indicates that the latch will be set.
s indicates that the latch will remain set.
R indicates that the latch will be reset.
r indicates that the latch will remain reset.

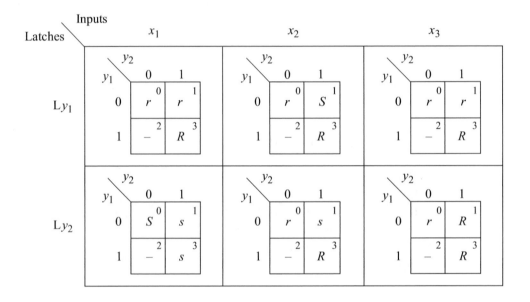

Figure 5.48   Input maps for the Mealy pulse-mode machine of Figure 5.47.

The map entries are obtained as in previous examples. Refer to the state diagram and minterm location 0 of the map in row Ly_1, column x1. In state a ($y_1 y_2 = 00$), if x_1 is pulsed, then the machine sequences to state b ($y_1 y_2 = 01$) where flip-flop y_1 remains reset. Thus, the letter r is inserted in minterm location 0. In the same map, minterm location 3 contains the entry R. That is, in state c ($y_1 y_2 = 11$), flip-flop y_1 is reset if x_1 is pulsed. In a similar manner, the remaining input maps are derived.

When obtaining the equations for the latches from the input maps, only the upper-case letters must be considered. The lowercase letters and the unused states are used only if they contribute to a minimized equation. The set and reset input equations are listed in Equation 5.11, where SLy_1, RLy_1 and SLy_2, RLy_2 are the set and reset equations for latches Ly_1 and Ly_2, respectively. Note that all equations contain an input variable x_i, since the machine is triggered by input pulses.

$$SL_{y_1} = y_1' y_2 x_2$$
$$RL_{y_1} = x_1 + y_1 x_2 + x_3$$

$$SL_{y_2} = x_1$$
$$RL_{y_2} = y_1 x_2 + x_3 \tag{5.11}$$

The output map for z_1 is shown in Figure 5.49. Since z_1 is asserted coincident with x_3, input x_3 is used as a map-entered variable in state c ($y_1 y_2 = 11$) and combines with the unused state to yield Equation 5.12, which specifies a Mealy pulse-mode asynchronous sequential machine.

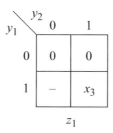

Figure 5.49   Output map for z_1 for the pulse-mode machine of Figure 5.47.

$$z_1 = y_1 x_3 \tag{5.12}$$

The logic diagram for this Mealy machine is shown in Figure 5.50, using SR latches and D flip-flops in a master-slave relationship with an implied reset. The logic is synthesized from the input and output equations of Equation 5.11 and Equation 5.12, respectively. Each of the three mutually exclusive input pulses is inverted through the three-input NOR gate.

When the pulses are active, a low voltage level is applied to the clock inputs of D flip-flops y_1 and y_2. The active level of the pulses also sets or resets latches L_{y_1} and L_{y_2}, depending on the present state and the present input. The latches stabilize to their respective next states while the input pulse is still active and provide the next-state values to the D inputs of flip-flops y_1 and y_2. When the input pulse is deasserted, a positive transition is applied to the clock inputs of flip-flops y_1 and y_2, which then sequence the machine to the next state.

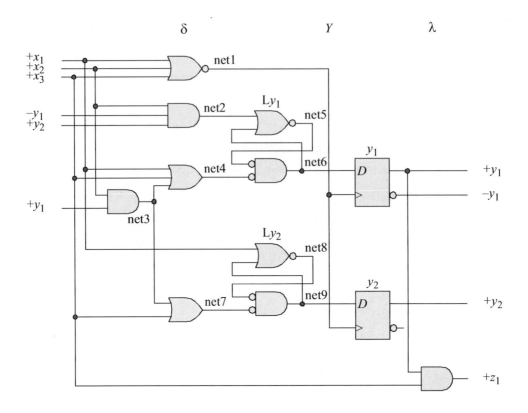

Figure 5.50   Logic diagram for the Mealy pulse-mode machine of Figure 5.47, using *SR* latches and *D* flip-flops in a master-slave configuration.

The design module is shown in Figure 5.51 using built-in primitives and *D* flip-flops that were designed using behavioral modeling. The test bench module is shown in Figure 5.52. The outputs and waveforms are shown in Figure 5.53 and Figure 5.54, respectively.

```
//mealy pulse-mode asm using built-in primitives
module pm_asm16 (rst_n, rst, x1, x2, x3, y1, y2, z1);

//define inputs and outputs
input rst_n, rst, x1, x2, x3;
output y1, y2, z1;

//define internal nets
wire net1, net2, net3, net4, net5, net6, net7, net8, net9;
                            //continued on next page
```

Figure 5.51   Design module for the Mealy pulse-mode machine of Example 5.7.

```verilog
//design the D flip-flop clock
nor    (net1, x1, x2, x3);


//------------------------------------
//design the logic for latch y1
and    (net2, x2, ~y1, y2),
       (net3, x2, y1);

or     (net4, x1, x3, net3);

nor    (net5, net2, net6),
       (net6, net5, net4, rst);

//instantiate the D flip-flop for y1
d_ff_bh inst1 (
   .rst_n(rst_n),
   .clk(net1),
   .d(net6),
   .q(y1)
   );


//------------------------------------
//design the logic for latch y2
or     (net7, net3, x3);

nor    (net8, x1, net9),
       (net9, net8, net7, rst);

//instantiate the D flip-flop for y2
d_ff_bh inst2 (
   .rst_n(rst_n),
   .clk(net1),
   .d(net9),
   .q(y2)
   );




//------------------------------------
//design the logic for output z1
and    (z1, y1, x3);

endmodule
```

Figure 5.51   (Continued)

```
//test bench for mealy pulse-mode asm

module pm_asm16_tb;

//inputs are reg for test bench
//outputs are wire for test bench
reg rst_n, rst, x1, x2, x3;
wire y1, y2, z1;

//display variables
initial
$monitor ("x1x2x3 = %b, state = %b, z1 = %b",
          {x1, x2, x3}, {y1, y2}, z1);

//define input sequence
initial
begin
   #0     rst_n = 1'b0;
          rst = 1'b1;
          x1 = 1'b0;
          x2 = 1'b0;
          x3 = 1'b0;

   #5     rst_n = 1'b1;
          rst = 1'b0;

//--------------------------------------------------------------
   #5     x1 = 1'b1;   x2 = 1'b0;   x3 = 1'b0;   //10
   #5     x1 = 1'b0;   x2 = 1'b0;   x3 = 1'b0;   //15

   #10    x1 = 1'b0;   x2 = 1'b1;   x3 = 1'b0;   //25
   #5     x1 = 1'b0;   x2 = 1'b0;   x3 = 1'b0;   //30

   #10    x1 = 1'b0;   x2 = 1'b1;   x3 = 1'b0;   //40
   #5     x1 = 1'b0;   x2 = 1'b0;   x3 = 1'b0;   //45

   #10    x1 = 1'b0;   x2 = 1'b0;   x3 = 1'b1;   //55
   #5     x1 = 1'b0;   x2 = 1'b0;   x3 = 1'b0;   //60

   #10    x1 = 1'b1;   x2 = 1'b0;   x3 = 1'b0;   //70
   #5     x1 = 1'b0;   x2 = 1'b0;   x3 = 1'b0;   //75

   #10    x1 = 1'b0;   x2 = 1'b1;   x3 = 1'b0;   //85
   #5     x1 = 1'b0;   x2 = 1'b0;   x3 = 1'b0;   //90
                                    //continued on next page
```

Figure 5.52   Test bench module for the Mealy pulse-mode machine of Example 5.7.

```
    #10    x1 = 1'b0;  x2 = 1'b0;  x3 = 1'b1;   //100, assert z1
    #5     x1 = 1'b0;  x2 = 1'b0;  x3 = 1'b0;   //105

    #10    x1 = 1'b0;  x2 = 1'b0;  x3 = 1'b1;   //115
    #5     x1 = 1'b0;  x2 = 1'b0;  x3 = 1'b0;   //120

    #10    x1 = 1'b0;  x2 = 1'b1;  x3 = 1'b0;   //130
    #5     x1 = 1'b0;  x2 = 1'b0;  x3 = 1'b0;   //135

    #10    x1 = 1'b1;  x2 = 1'b0;  x3 = 1'b0;   //145
    #5     x1 = 1'b0;  x2 = 1'b0;  x3 = 1'b0;   //150

    #10    x1 = 1'b0;  x2 = 1'b1;  x3 = 1'b0;   //160
    #5     x1 = 1'b0;  x2 = 1'b0;  x3 = 1'b0;   //165

    #10    x1 = 1'b0;  x2 = 1'b0;  x3 = 1'b1;   //175, assert z1
    #5     x1 = 1'b0;  x2 = 1'b0;  x3 = 1'b0;   //180

    #10    $stop;
end

//instantiate the module into the test bench as a single line
pm_asm16 inst1 (rst_n, rst, x1, x2, x3, y1, y2, z1);

endmodule
```

Figure 5.52   (Continued)

```
x1x2x3 = 000, state = 00, z1 = 0   x1x2x3 = 000, state = 00, z1 = 0
x1x2x3 = 100, state = 00, z1 = 0   x1x2x3 = 001, state = 00, z1 = 0
x1x2x3 = 000, state = 01, z1 = 0   x1x2x3 = 000, state = 00, z1 = 0
x1x2x3 = 010, state = 01, z1 = 0   x1x2x3 = 010, state = 00, z1 = 0
x1x2x3 = 000, state = 11, z1 = 0   x1x2x3 = 000, state = 00, z1 = 0
x1x2x3 = 010, state = 11, z1 = 0   x1x2x3 = 100, state = 00, z1 = 0
x1x2x3 = 000, state = 00, z1 = 0   x1x2x3 = 000, state = 01, z1 = 0
x1x2x3 = 001, state = 00, z1 = 0   x1x2x3 = 010, state = 01, z1 = 0
x1x2x3 = 000, state = 00, z1 = 0   x1x2x3 = 000, state = 11, z1 = 0
x1x2x3 = 100, state = 00, z1 = 0   x1x2x3 = 001, state = 11, z1 = 1
x1x2x3 = 000, state = 01, z1 = 0
x1x2x3 = 010, state = 01, z1 = 0   x1x2x3 = 000, state = 00, z1 = 0
x1x2x3 = 000, state = 11, z1 = 0
x1x2x3 = 001, state = 11, z1 = 1
```

Figure 5.53   Outputs for the Mealy pulse-mode machine of Example 5.7.

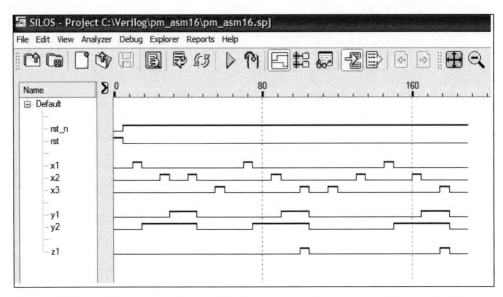

Figure 5.54   Waveforms for the Mealy pulse-mode machine of Example 5.7.

Example 5.8   A Mealy machine will be synthesized which has three pulse input variables x_1, x_2, and x_3 and one output z_1 that is asserted coincident with x_3 whenever the sequence $x_1 x_2 x_3 = 100, 010, 001$ occurs. The storage elements will consist of T flip-flops. Recall that a T flip-flop has a T input, a Reset input, and two outputs y_1 and y_1'. If the flip-flop is reset, then an active pulse on the T input will toggle the flip-flop to the set state; if the flip-flop is set, then a pulse on the T input will toggle the flip-flop to the reset state.

The T flip-flop utilized in this example incorporates a D flip-flop and an exclusive-OR circuit. The T input connects to the clock input of the D flip-flop through a delay circuit, which allows the clock input to be delayed until the signal on the D input has stabilized. When T has a value of 0, the next state is the same as the present state; when T has a value of 1, the next state is the complement of the present state.

A representative timing diagram displaying valid input sequences and corresponding outputs is shown in Figure 5.55. The state diagram is shown in Figure 5.56.

The input maps, obtained from the state diagram, are shown in Figure 5.57 using the entry T in appropriate minterm locations. The T entry indicates that the state of the flip-flop will change state; that is, it will toggle from 0 to 1 or toggle from 1 to 0. The entries that must be considered are the T entries, since these are the only entries for a T flip-flop that result in a change of state for y_1 and y_2. The s and r entries cannot combine with the T in the minimization process, since these entries maintain a constant flip-flop state, whereas a T will change the state of the corresponding flip-flop.

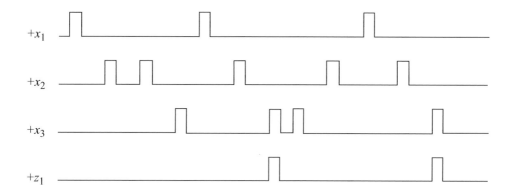

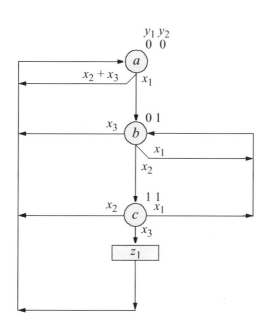

Figure 5.55 Representative timing diagram for the pulse-mode Mealy machine of Example 5.8.

Figure 5.56 State diagram for the pulse-mode Mealy machine of Example 5.8.

The input equations obtained from the input maps are shown in Equation 5.13. The "don't care" entries are used to minimize the input equations. The output equation is shown in Equation 5.14 and is obtained directly from the state diagram in state c.

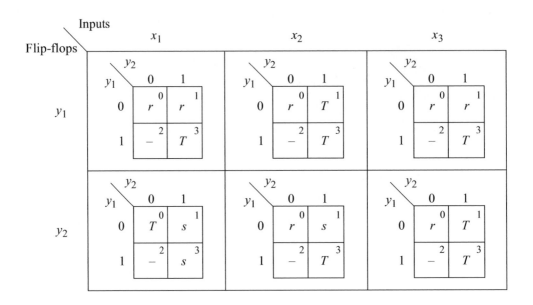

Figure 5.57   Input maps for the pulse-mode Mealy machine of Example 5.8.

$$Ty_1 = y_1x_1 + y_2x_2 + y_1x_3 \tag{5.13}$$

$$Ty_2 = y_2'x_1 + y_1x_2 + y_2x_3$$

$$z_1 = y_1x_3 \tag{5.14}$$

The logic diagram is shown in Figure 5.58 and displays the net names for all of the internal nets. The design module is shown in Figure 5.59 using T flip-flops and built-in primitives for **and**, **or**, and **buf**. Recall that built-in primitive gates are used to describe a net and have one or more scalar inputs, but only one scalar output. The output signal is listed first, followed by the inputs in any order. The gates represent combinational logic functions and can be instantiated into a module, as follows, where the instance name is optional:

gate_type inst1 (output, input_1, input_2, . . . , input_n);

The test bench module is shown in Figure 5.60 and takes the machine through various states represented in the timing diagram of Figure 5.55. The outputs and waveforms are shown in Figure 5.61 and Figure 5.62, respectively.

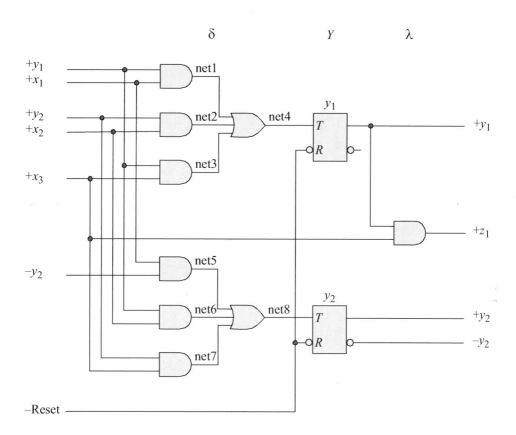

Figure 5.58   Logic diagram for the pulse-mode Mealy machine of Example 5.8.

```
//mealy pulse-mode asm using built-in primitives and T ff

module pm_asm16_tff (rst_n, x1, x2, x3, y1, y2, z1);

//define inputs and outputs
input rst_n, x1, x2, x3;
output y1, y2, z1;

//define internal nets
wire net1, net2, net3, net4, net5, net6, net7, net8;
wire nety1, nety2;

                              //continued on next page
```

Figure 5.59   Design module for the pulse-mode Mealy machine of Example 5.8.

```
//-----------------------------------------------
//design the logic for T flip-flop y1
and     (net1, y1, x1),
        (net2, y2, x2),
        (net3, y1, x3);
or      (net4, net1, net2, net3);

//instantiate the T flip-flop as a single line
t_ff_da inst1 (rst_n, net4, nety1);    //rst_n, t, y1

buf     #11 (y1, nety1);

//-----------------------------------------------
//design the logic for T flip-flop y2
and     (net5, ~y2, x1),
        (net6, y1, x2),
        (net7, y2, x3);
or      (net8, net5, net6, net7);

//instantiate the T flip-flop as a single line
t_ff_da inst2 (rst_n, net8, nety2);    //rst_n, t, y1

buf     #11 (y2, nety2);

//-----------------------------------------------
//design the logic for output z1
and     (z1, y1, x3);

endmodule
```

Figure 5.59   (Continued)

```
//test bench for mealy pulse-mode asm
module pm_asm16_tff_tb;

//inputs are reg for test bench
//outputs are wire for test bench
reg rst_n, x1, x2, x3;
wire y1, y2, z1;

initial      //display variables
$monitor ("x1x2x3 = %b, state = %b, z1 = %b",
          {x1, x2, x3}, {y1, y2}, z1);   //continued next pg
```

Figure 5.60   Test bench module for the pulse-mode Mealy machine of Example 5.8.

```
initial      //apply input sequence
begin
   #0     rst_n = 1'b0;
          x1 = 1'b0;   x2 = 1'b0;   x3 = 1'b0;
   #5     rst_n = 1'b1;
//----------------------------------------------------------
   #5     x1 = 1'b1;   x2 = 1'b0;   x3 = 1'b0;   //10
   #5     x1 = 1'b0;   x2 = 1'b0;   x3 = 1'b0;   //15

   #10    x1 = 1'b0;   x2 = 1'b1;   x3 = 1'b0;   //25
   #5     x1 = 1'b0;   x2 = 1'b0;   x3 = 1'b0;   //30

   #10    x1 = 1'b0;   x2 = 1'b1;   x3 = 1'b0;   //40
   #5     x1 = 1'b0;   x2 = 1'b0;   x3 = 1'b0;   //45

   #10    x1 = 1'b0;   x2 = 1'b0;   x3 = 1'b1;   //55
   #5     x1 = 1'b0;   x2 = 1'b0;   x3 = 1'b0;   //60

   #10    x1 = 1'b1;   x2 = 1'b0;   x3 = 1'b0;   //70
   #5     x1 = 1'b0;   x2 = 1'b0;   x3 = 1'b0;   //75

   #10    x1 = 1'b0;   x2 = 1'b1;   x3 = 1'b0;   //85
   #5     x1 = 1'b0;   x2 = 1'b0;   x3 = 1'b0;   //90

   #10    x1 = 1'b0;   x2 = 1'b0;   x3 = 1'b1;   //100, assert z1
   #5     x1 = 1'b0;   x2 = 1'b0;   x3 = 1'b0;   //105

   #10    x1 = 1'b0;   x2 = 1'b0;   x3 = 1'b1;   //115
   #5     x1 = 1'b0;   x2 = 1'b0;   x3 = 1'b0;   //120

   #10    x1 = 1'b0;   x2 = 1'b1;   x3 = 1'b0;   //130
   #5     x1 = 1'b0;   x2 = 1'b0;   x3 = 1'b0;   //135

   #10    x1 = 1'b1;   x2 = 1'b0;   x3 = 1'b0;   //145
   #5     x1 = 1'b0;   x2 = 1'b0;   x3 = 1'b0;   //150

   #10    x1 = 1'b0;   x2 = 1'b1;   x3 = 1'b0;   //160
   #5     x1 = 1'b0;   x2 = 1'b0;   x3 = 1'b0;   //165

   #10    x1 = 1'b0;   x2 = 1'b0;   x3 = 1'b1;   //175, assert z1
   #5     x1 = 1'b0;   x2 = 1'b0;   x3 = 1'b0;   //180
   #10    $stop;
end

//instantiate the module into the test bench as a single line
pm_asm16_tff inst1 (rst_n, x1, x2, x3, y1, y2, z1);
endmodule
```

Figure 5.60   (Continued)

```
x1x2x3 = 000, state = 00, z1 = 0
x1x2x3 = 100, state = 00, z1 = 0
x1x2x3 = 000, state = 00, z1 = 0
x1x2x3 = 000, state = 01, z1 = 0
x1x2x3 = 010, state = 01, z1 = 0
x1x2x3 = 000, state = 01, z1 = 0

x1x2x3 = 000, state = 11, z1 = 0
x1x2x3 = 010, state = 11, z1 = 0
x1x2x3 = 000, state = 11, z1 = 0
x1x2x3 = 000, state = 00, z1 = 0
x1x2x3 = 001, state = 00, z1 = 0
x1x2x3 = 000, state = 00, z1 = 0

x1x2x3 = 100, state = 00, z1 = 0
x1x2x3 = 000, state = 00, z1 = 0
x1x2x3 = 000, state = 01, z1 = 0
x1x2x3 = 010, state = 01, z1 = 0
x1x2x3 = 000, state = 01, z1 = 0
x1x2x3 = 000, state = 11, z1 = 0

x1x2x3 = 001, state = 11, z1 = 1

x1x2x3 = 000, state = 11, z1 = 0
x1x2x3 = 000, state = 00, z1 = 0
x1x2x3 = 001, state = 00, z1 = 0
x1x2x3 = 000, state = 00, z1 = 0
x1x2x3 = 010, state = 00, z1 = 0
x1x2x3 = 000, state = 00, z1 = 0

x1x2x3 = 100, state = 00, z1 = 0
x1x2x3 = 000, state = 00, z1 = 0
x1x2x3 = 000, state = 01, z1 = 0
x1x2x3 = 010, state = 01, z1 = 0
x1x2x3 = 000, state = 01, z1 = 0
x1x2x3 = 000, state = 11, z1 = 0

x1x2x3 = 001, state = 11, z1 = 1

x1x2x3 = 000, state = 11, z1 = 0
x1x2x3 = 000, state = 00, z1 = 0
```

Figure 5.61   Outputs for the pulse-mode Mealy machine of Example 5.8.

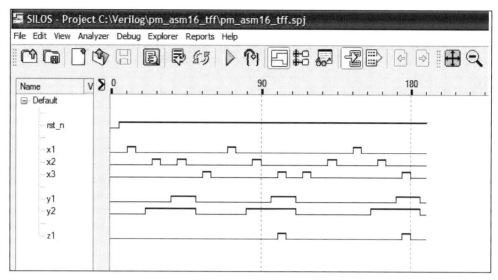

Figure 5.62   Waveforms for the pulse-mode Mealy machine of Example 5.8.

Example 5.9   A Mealy pulse-mode asynchronous sequential machine will be synthesized which has two inputs x_1 and x_2 and one output z_1. Output z_1 is asserted coincident with every second x_2 pulse, if and only if the pair of x_2 pulses is immediately preceded by an x_1 pulse. **and**, **nand** and **nor** built-in primitives will be utilized, plus inverters as required. The storage elements will consist of *SR* latches and *D* flip-flops in a master-slave configuration. Output z_1 is asserted as a logic 1 level.

A representative timing diagram is illustrated in Figure 5.63 and the state diagram is shown in Figure 5.64. The input maps are shown in Figure 5.65 and the input equations are shown in Equation 5.15. The output equation is shown in Equation 5.16. The logic diagram is displayed in Figure 5.66.

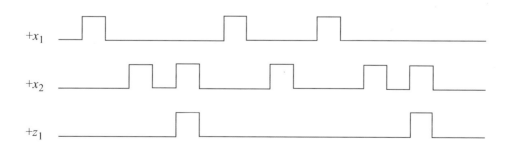

Figure 5.63   Representative timing diagram for the Mealy pulse-mode machine of Example 5.9.

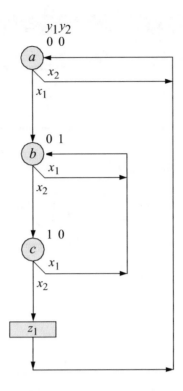

Figure 5.64   State diagram for the Mealy pulse-mode machine of Example 5.9.

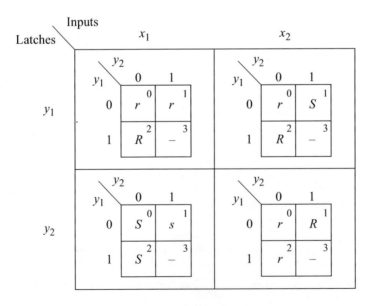

Figure 5.65   Input maps for the Mealy pulse-mode machine of Example 5.9.

$$Sy_1 = y_2x_2$$

$$Ry_1 = x_1 + y_1x_2$$

$$Sy_2 = x_1$$

$$Ry_2 = x_2 \tag{5.15}$$

$$z_1 = y_1x_2 \tag{5.16}$$

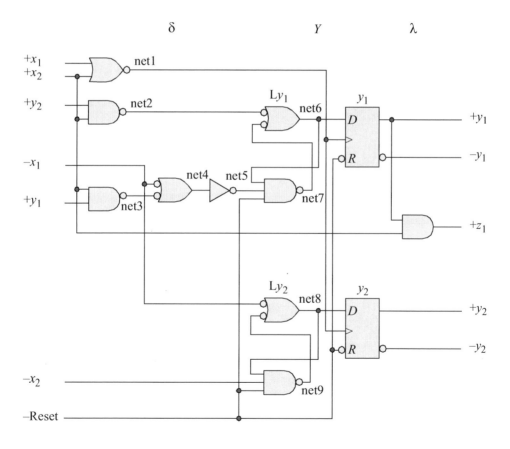

Figure 5.66   Logic diagram for the Mealy pulse-mode machine of Example 5.9.

The design module is shown in Figure 5.67 using built-in primitives and D flip-flops that were designed using behavioral modeling. The D flip-flops are instantiated as a single line. The test bench module is shown in Figure 5.68. The outputs and waveforms and shown in Figure 5.69 and Figure 5.70, respectively.

```
//dataflow mealy pulse-mode asm using built-in primitives

module pm_asm12 (rst_n, x1, x2, y1, y2, z1);

//define inputs and outputs
input rst_n, x1, x2;
output y1, y2, z1;

//define internal nets
wire net1, net2, net3, net4, net5, net6, net7, net8, net9;

//design the D flip-flop clock
nor     (net1, x1, x2);


//-----------------------------------
//design the logic for latch Ly1
nand    (net2, x2, y2),
        (net3, x2, y1),
        (net4, ~x1, net3);
not     (net5, net4);
nand    (net6, net2, net7),
        (net7, net6, net5, rst_n);

//instantiate the D flip-flop for y1
d_ff_bh inst1 (rst_n, net1, net6, y1);   //rst_n, clk, d, q


//-----------------------------------
//design the logic for latch Ly2
nand    (net8, ~x1, net9),
        (net9, net8, ~x2, rst_n);

//instantiate the D flip-flop for y2
d_ff_bh inst2 (rst_n, net1, net8, y2);   //rst_n, clk, d, q

//-----------------------------------
//design the logic for output z1
and     (z1, y1, x2);

endmodule
```

Figure 5.67   Design module for the Mealy pulse-mode machine of Example 5.9.

```
//test bench for pulse-mode
//asynchronous sequential machine

module pm_asm12_tb;

//inputs are reg for test bench
//outputs are wire for test bench
reg rst_n, x1, x2;
wire y1, y2, z1;

//display variables
initial
$monitor ("x1x2 = %b, state = %b, z1 = %b",
          {x1, x2}, {y1, y2}, z1);

//define input sequence
initial
begin
   #0     rst_n = 1'b0;   //reset to state_a (00)
          x1 = 1'b0;
          x2 = 1'b0;

   #5     rst_n = 1'b1;

   #10    x1 = 1'b0; x2 = 1'b1;
   #10    x1 = 1'b0; x2 = 1'b0; //remain in state_a (00)

   #10    x1 = 1'b1; x2 = 1'b0;
   #10    x1 = 1'b0; x2 = 1'b0; //go to state_b (01)

   #10    x1 = 1'b1; x2 = 1'b0;
   #10    x1 = 1'b0; x2 = 1'b0; //remain in state_b (01)

   #10    x1 = 1'b0; x2 = 1'b1;
   #10    x1 = 1'b0; x2 = 1'b0; //go to state_c (10)

   #10    x1 = 1'b0; x2 = 1'b1;
   #10    x1 = 1'b0; x2 = 1'b0; //go to state_a (00); assert z1

   #10    x1 = 1'b1; x2 = 1'b0;
   #10    x1 = 1'b0; x2 = 1'b0; //go to state_b (01)

   #10    x1 = 1'b0; x2 = 1'b1;
   #10    x1 = 1'b0; x2 = 1'b0; //go to state_c (10)

                         //continued on next page
```

Figure 5.68   Test bench module for the Mealy pulse-mode machine of Example 5.9.

```
    #10     x1 = 1'b0; x2 = 1'b1;
    #10     x1 = 1'b0; x2 = 1'b0; //go to state_a (00); assert z1

    #10     x1 = 1'b0; x2 = 1'b1;
    #10     x1 = 1'b0; x2 = 1'b0; //remain in state_a (00)

    #10     $stop;

end

//instantiate the module into the test bench as a single line
pm_asm12 inst1 (rst_n, x1, x2, y1, y2, z1);

endmodule
```

Figure 5.68   (Continued)

```
x1x2 = 00, state = 00, z1 = 0
x1x2 = 01, state = 00, z1 = 0
x1x2 = 00, state = 00, z1 = 0
x1x2 = 10, state = 00, z1 = 0
x1x2 = 00, state = 01, z1 = 0
x1x2 = 10, state = 01, z1 = 0
x1x2 = 00, state = 01, z1 = 0
x1x2 = 01, state = 01, z1 = 0
x1x2 = 00, state = 10, z1 = 0
x1x2 = 01, state = 10, z1 = 1
x1x2 = 00, state = 00, z1 = 0

x1x2 = 10, state = 00, z1 = 0
x1x2 = 00, state = 01, z1 = 0
x1x2 = 01, state = 01, z1 = 0
x1x2 = 00, state = 10, z1 = 0
x1x2 = 01, state = 10, z1 = 1
x1x2 = 00, state = 00, z1 = 0

x1x2 = 01, state = 00, z1 = 0
x1x2 = 00, state = 00, z1 = 0
```

Figure 5.69   Outputs for the Mealy pulse-mode machine of Example 5.9.

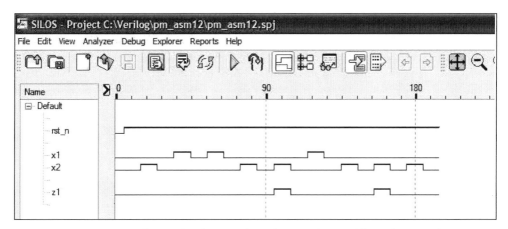

Figure 5.70   Waveforms for the Mealy pulse-mode machine of Example 5.9.

Example 5.10   A Mealy pulse-mode asynchronous sequential machine will be synthesized which has two inputs x_1 and x_2 and one output z_1. Output z_1 is asserted coincident with every second x_2 pulse, if and only if the pair of x_2 pulses is immediately preceded by an x_1 pulse. This example repeats Example 5.9, however **nand** built-in primitives will be utilized together with storage elements that consist of T flip-flops. Output z_1 is asserted as a logic 1 level.

A represented timing diagram is illustrated in Figure 5.71 and the state diagram is shown in Figure 5.72. The input maps are shown in Figure 5.73 and the input equations are shown in Equation 5.17. The output equation is shown in Equation 5.18. The logic diagram is displayed in Figure 5.74.

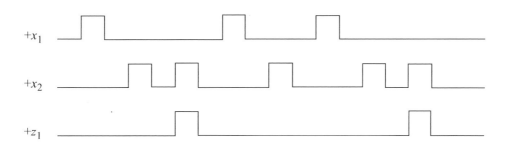

Figure 5.71   Representative timing diagram for the Mealy pulse-mode machine of Example 5.10.

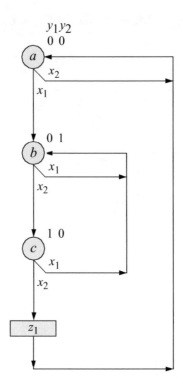

Figure 5.72   State diagram for the Mealy pulse-mode machine of Example 5.10.

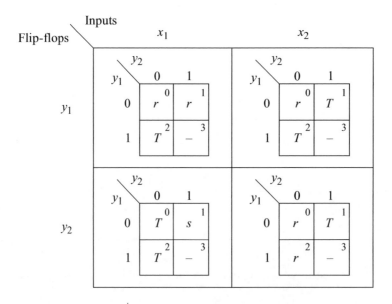

Figure 5.73   Input maps for the pulse-mode Mealy machine of Figure 5.72 for Example 5.10.

$$Ty_1 = y_1x_1 + y_1x_2 + y_2x_2$$

$$Ty_2 = y_2'x_1 + y_2x_2 \qquad\qquad (5.17)$$

$$z_1 = y_1x_2 \qquad\qquad (5.18)$$

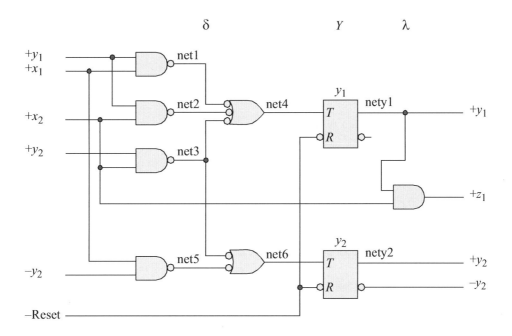

Figure 5.74   Logic diagram for the pulse-mode Mealy machine of Figure 5.72 of Example 5.10 using positive-input T flip-flops for the storage elements.

The design module for the pulse-mode asynchronous sequential machine of Example 5.10 is shown in Figure 5.75 using **nand** and **buf** built-in primitives and T flip-flops. The T flip-flops were designed using built-in primitives and D flip-flops that were designed using behavioral modeling. The T flip-flops are instantiated into the design module as a single line. The test bench module is shown in Figure 5.76, which takes the machine through various input sequences to illustrate the representative timing diagram. The outputs and waveforms are shown in Figure 5.77 and Figure 5.78, respectively.

```
//mealy pulse-mode asm using T flip-flops and bip
module pm_asm15 (rst_n, x1, x2, y1, y2, z1);

input rst_n, x1, x2;      //define inputs and outputs
output y1, y2, z1;

//define internal nets
wire net1, net2, net3, net4, net5, net6, nety1, nety2;

//-------------------------------------------
//design the logic for flip-flop y1
nand    (net1, x1, y1),
        (net2, x2, y1),
        (net3, x2, y2),
        (net4, net1, net2, net3);

//instantiate the T flip-flop
t_ff_da inst1 (rst_n, net4, y1);    //rst_n, t, y1

buf    #11 (y1, nety1);

//-------------------------------------------
//design the logic for flip-flop y2
nand    (net5, x1, ~y2),
        (net6, net3, net5);

//instantiate the T flip-flop
t_ff_da inst2 (rst_n, net6, y2);    //rst_n, t, y1

buf    #11 (y2, nety2);

//-------------------------------------------
//design the logic for output z1
assign z1 = y1 & x2;
```

Figure 5.75   Design module for the pulse-mode Mealy machine of Example 5.10.

```
//test bench for the mealy pulse-mode asm
module pm_asm15_tb;

//inputs are reg for test bench
//outputs are wire for test bench
reg rst_n, x1, x2;
wire y1, y2, z1;                        //continued on next page
```

Figure 5.76   Test bench for the pulse-mode Mealy machine of Example 5.10.

```
//display variables
initial
$monitor ("x1x2 = %b, state = %b, z1 = %b",
          {x1, x2}, {y1, y2}, z1);

//define input sequence
initial
begin
   #0    rst_n = 1'b0;          //reset to state_a (00)
         x1 = 1'b0; x2 = 1'b0;

   #5    rst_n = 1'b1;

   #10   x1 = 1'b0; x2 = 1'b1;
   #10   x1 = 1'b0; x2 = 1'b0; //remain in state_a (00)

   #10   x1 = 1'b1; x2 = 1'b0;
   #10   x1 = 1'b0; x2 = 1'b0; //go to state_b (01)

   #10   x1 = 1'b1; x2 = 1'b0;
   #10   x1 = 1'b0; x2 = 1'b0; //remain in state_b (01)

   #10   x1 = 1'b0; x2 = 1'b1;
   #10   x1 = 1'b0; x2 = 1'b0; //go to state_c (10)

   #10   x1 = 1'b0; x2 = 1'b1;
   #10   x1 = 1'b0; x2 = 1'b0; //go to state_a (00); assert z1

   #10   x1 = 1'b1; x2 = 1'b0;
   #10   x1 = 1'b0; x2 = 1'b0; //go to state_b (01)

   #10   x1 = 1'b0; x2 = 1'b1;
   #10   x1 = 1'b0; x2 = 1'b0; //go to state_c (10)

   #10   x1 = 1'b0; x2 = 1'b1;
   #10   x1 = 1'b0; x2 = 1'b0; //go to state_a (00); assert z1

   #10   x1 = 1'b0; x2 = 1'b1;
   #10   x1 = 1'b0; x2 = 1'b0; //remain in state_a (00)

   #10   $stop;
end

//instantiate the module into the test bench as a single line
pm_asm15 inst1 (rst_n, x1, x2, y1, y2, z1);

endmodule
```

Figure 5.76   (Continued)

```
x1x2 = 00, state = 00, z1 = 0
x1x2 = 01, state = 00, z1 = 0
x1x2 = 00, state = 00, z1 = 0
x1x2 = 10, state = 00, z1 = 0
x1x2 = 00, state = 00, z1 = 0
x1x2 = 00, state = 01, z1 = 0
x1x2 = 10, state = 01, z1 = 0
x1x2 = 00, state = 01, z1 = 0
x1x2 = 01, state = 01, z1 = 0
x1x2 = 00, state = 01, z1 = 0
x1x2 = 00, state = 10, z1 = 0
x1x2 = 01, state = 10, z1 = 1
x1x2 = 00, state = 10, z1 = 0

x1x2 = 00, state = 00, z1 = 0
x1x2 = 10, state = 00, z1 = 0
x1x2 = 00, state = 00, z1 = 0
x1x2 = 00, state = 01, z1 = 0
x1x2 = 01, state = 01, z1 = 0
x1x2 = 00, state = 01, z1 = 0
x1x2 = 00, state = 10, z1 = 0
x1x2 = 01, state = 10, z1 = 1
x1x2 = 00, state = 10, z1 = 0

x1x2 = 00, state = 00, z1 = 0
x1x2 = 01, state = 00, z1 = 0
x1x2 = 00, state = 00, z1 = 0
```

Figure 5.77   Outputs for the pulse-mode machine of Example 5.10.

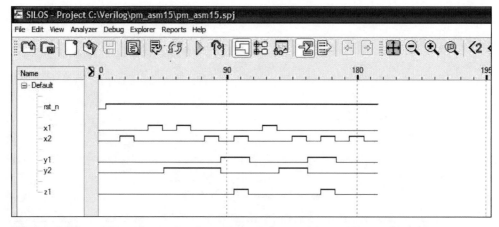

Figure 5.78   Waveforms for the pulse-mode machine of Example 5.10.

Example 5.11   A state diagram is shown in Figure 5.79 for a Moore pulse-mode asynchronous sequential machine with three inputs x_1, x_2, and x_3, and two outputs z_1 and z_2. Dataflow modules for the logic gates will be instantiated into the structural module. Also, D flip-flops are used in the implementation. The input maps and equations are shown in Figure 5.80 and Equation 5.19, respectively. The output equations are shown in Equation 5.20 and the logic diagram is shown in Figure 5.81. The design module is shown in Figure 5.82 and the test bench module that takes the machine through various sequences to assert output z_1 and output z_2 is shown in Figure 5.83. The outputs and waveforms are shown in Figure 5.84 and Figure 5.85, respectively.

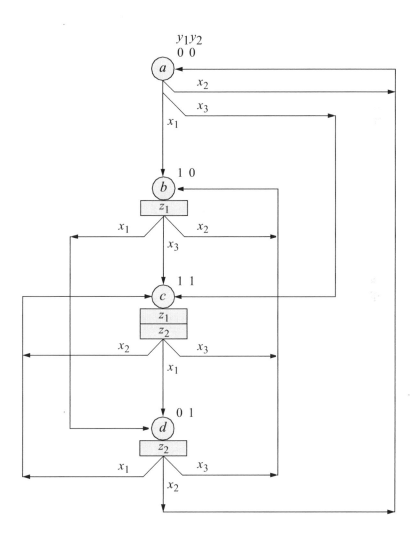

Figure 5.79   State diagram for the Moore pulse-mode machine of Example 5.11.

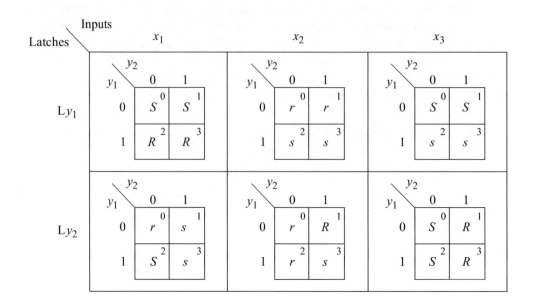

Figure 5.80   Input maps for the Moore pulse-mode machine of Example 5.11.

$$SLy_1 = y_1'x_1 + x_3$$

$$RLy_1 = y_1x_1$$

$$SLy_2 = y_1x_1 + y_2'x_3$$

$$RLy_2 = y_1'x_2 + y_2x_3 \qquad (5.19)$$

$$z_1 = y_1$$

$$z_2 = y_2 \qquad (5.20)$$

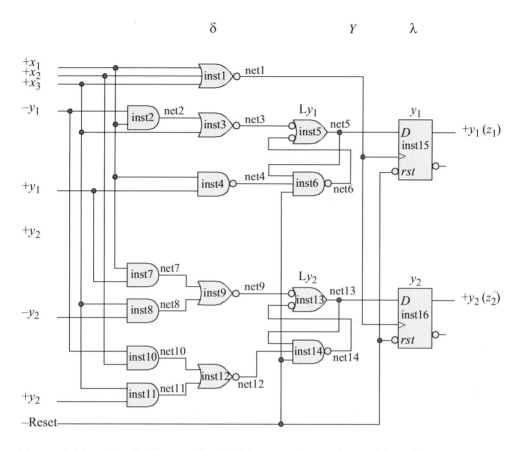

Figure 5.81   Logic diagram for the Moore pulse-mode machine of Example 5.11.

```verilog
//moore pulse-mode asm using instantiated dataflow
//modules and D flip-flops

module pm_asm_moore9 (rst_n, x1, x2, x3, y1, y2, z1, z2);

//design inputs and outputs
input rst_n, x1, x2, x3;
output y1, y2, z1, z2;

//define internal nets
wire net1, net2, net3, net4, net5, net6, net7, net8;
wire net9, net10, net11, net12, net13, net14;
                                //continued on next page
```

Figure 5.82   Design module for the Moore pulse-mode machine of Example 5.11.

```
//------------------------------------------------
//define the D flip-flop clock
nor3_df  inst1 (x1, x2, x3, net1);


//------------------------------------------------
//design the logic for latch Ly1 and D flip-flop y1
and2_df  inst2 (~y1, x1, net2);
nor2_df  inst3 (net2, x3, net3);
nand2_df inst4 (x1, y1, net4);


//latch Ly1
nand2_df inst5 (net3, net6, net5);
nand3_df inst6 (net5, net4, rst_n, net6);


//instantiate the D flip-flop as a single line
d_ff_bh inst15 (rst_n, net1, net5, y1);   //rst_n, clk, d, q


//------------------------------------------------
//design the logic for latch Ly2 and D flip-flop y2
and2_df  inst7 (x1, y1, net7);
and2_df  inst8 (x3, ~y2, net8);
nor2_df  inst9 (net7, net8, net9);
and2_df  inst10 (~y1, x2, net10);
and2_df  inst11 (x3, y2, net11);
nor2_df  inst12 (net10, net11, net12);


//latch Ly2
nand2_df inst13 (net9, net14, net13);
nand3_df inst14 (net13, net12, rst_n, net14);


//instantiate the D flip-flop as a single line
d_ff_bh inst16 (rst_n, net1, net13, y2);   //rst_n, clk, d, q


//design the logic for outputs z1 and z2
assign    z1 = y1;
assign    z2 = y2;

endmodule
```

Figure 5.82   (Continued)

```
//test bench for moore pulse-mode asm
module pm_asm_moore9_tb;

reg rst_n, x1, x2, x3;   //inputs are reg for test bench
wire y1, y2, z1, z2;     //outputs are wire for test bench

initial                  //display inputs and outputs
$monitor ("x1x2x3 = %b, state = %b, z1z2 = %b",
          {x1, x2, x3}, {y1, y2}, {z1, z2});

initial                  //define input sequence
begin
    #0    rst_n = 1'b0;  x1 = 1'b0;  x2 = 1'b0;  x3 = 1'b0;
    #5    rst_n = 1'b1;

    #10   x1=1'b0;    x2=1'b0;    x3=1'b0;

    #10   x1=1'b1;    x2=1'b0;    x3=1'b0;    //→b, assert z1
    #10   x1=1'b0;    x2=1'b0;    x3=1'b0;

    #10   x1=1'b1;    x2=1'b0;    x3=1'b0;    //→d, assert z2
    #10   x1=1'b0;    x2=1'b0;    x3=1'b0;

    #10   x1=1'b0;    x2=1'b0;    x3=1'b1;    //→b, assert z1
    #10   x1=1'b0;    x2=1'b0;    x3=1'b0;

    #10   x1=1'b0;    x2=1'b0;    x3=1'b1;    //→c, assert z1,z2
    #10   x1=1'b0;    x2=1'b0;    x3=1'b0;

    #10   x1=1'b1;    x2=1'b0;    x3=1'b0;    //→d, assert z2
    #10   x1=1'b0;    x2=1'b0;    x3=1'b0;

    #10   x1=1'b1;    x2=1'b0;    x3=1'b0;    //→c, assert z1,z2
    #10   x1=1'b0;    x2=1'b0;    x3=1'b0;

    #10   x1=1'b1;    x2=1'b0;    x3=1'b0;    //→d, assert z2
    #10   x1=1'b0;    x2=1'b0;    x3=1'b0;

    #10   x1=1'b0;    x2=1'b1;    x3=1'b0;    //→a
    #10   x1=1'b0;    x2=1'b0;    x3=1'b0;
    #20   $stop;
end

//instantiate the module into the test bench
pm_asm_moore9 inst1 (rst_n, x1, x2, x3, y1, y2, z1, z2);
endmodule
```

Figure 5.83   Test bench module for the Moore pulse-mode machine of Example 5.11.

```
x1x2x3 = 000, state = 00, z1z2 = 00
x1x2x3 = 100, state = 00, z1z2 = 00
x1x2x3 = 000, state = 10, z1z2 = 10
x1x2x3 = 100, state = 10, z1z2 = 10

x1x2x3 = 000, state = 01, z1z2 = 01
x1x2x3 = 001, state = 01, z1z2 = 01
x1x2x3 = 000, state = 10, z1z2 = 10
x1x2x3 = 001, state = 10, z1z2 = 10

x1x2x3 = 000, state = 11, z1z2 = 11
x1x2x3 = 100, state = 11, z1z2 = 11
x1x2x3 = 000, state = 01, z1z2 = 01
x1x2x3 = 100, state = 01, z1z2 = 01

x1x2x3 = 000, state = 11, z1z2 = 11
x1x2x3 = 100, state = 11, z1z2 = 11
x1x2x3 = 000, state = 01, z1z2 = 01
x1x2x3 = 010, state = 01, z1z2 = 01

x1x2x3 = 000, state = 00, z1z2 = 00
```

Figure 5.84   Outputs for the Moore pulse-mode machine of Example 5.11.

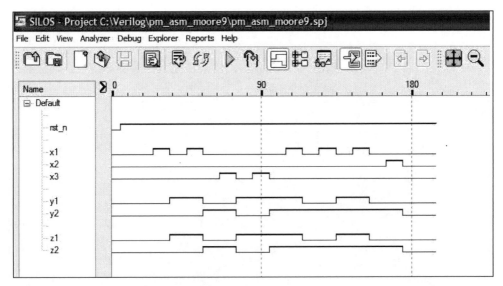

Figure 5.85   Waveforms for the Moore pulse-mode machine of Example 5.11.

5.3   Problems

5.1   Design a Moore pulse-mode asynchronous sequential machine which has two inputs x_1 and x_2 and one output z_1. The deassertion of every second consecutive x_1 pulse will assert output z_1 as a level. The output will remain set for all following contiguous x_1 pulses. The output will be deasserted at the trailing edge of the second of two consecutive x_2 pulses. A state diagram is shown below that represents the complete sequencing for this Moore machine.

Obtain the design module using the continuous assignment statement, *SR* latches, and *D* flip-flops. The *D* flip-flops are to be designed using behavioral modeling. Obtain the test bench module that takes the machine through the various sequences required to generate the output assertion and deassertion as specified above. Obtain the outputs and the waveforms.

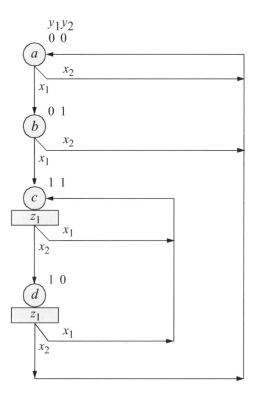

5.2   Design a Moore pulse-mode asynchronous sequential machine which has two inputs x_1 and x_2 and one output z_1. The deassertion of every second consecutive x_1 pulse will assert output z_1 as a level. The output will remain set for all following contiguous x_1 pulses. The output will be deasserted at the trailing edge of the second of two consecutive x_2 pulses. Built-in primitives are to be utilized together with T flip-flops. The state diagram is shown below. Obtain the design module and the test bench module that takes the machine through the various sequences required to generate the output assertion and deassertion as specified above. Obtain the outputs and the waveforms.

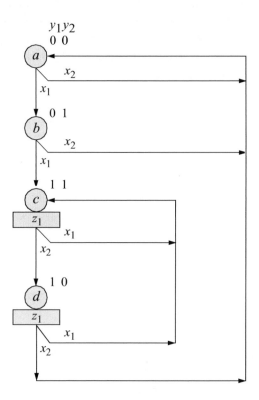

5.3   Design a Moore pulse-mode asynchronous sequential machine which has two inputs x_1 and x_2 and one output z_1. The deassertion of every second consecutive x_1 pulse will assert output z_1 as a level. The output will remain set for all following contiguous x_1 pulses. The output will be deasserted at the trailing edge of the second of two consecutive x_2 pulses. Use **nand** built-in primitives together with SR latches and D flip-flops. The state diagram is shown below and represents the complete sequencing for this Moore machine.

Obtain the design module using D flip-flops that were designed using behavioral modeling. Obtain the test bench module that takes the machine through the various sequences required to generate the output assertion and deassertion as specified above. Obtain the outputs and the waveforms.

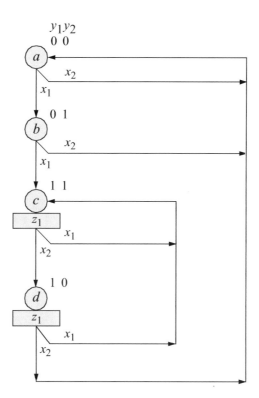

5.4 Design a Moore pulse-mode asynchronous sequential machine which has two inputs x_1 and x_2 and one output z_1. The deassertion of every second consecutive x_1 pulse will assert output z_1 as a level. The output will remain set for all following contiguous x_1 pulses. The output will be deasserted at the trailing edge of the second of two consecutive x_2 pulses. Use **nor** built-in primitives together with SR latches, D flip-flops, and inverters where required. The state diagram is shown below and represents the complete sequencing for this Moore machine. Obtain the design module using D flip-flops that are designed using behavioral modeling. Obtain the test bench module that takes the machine through the various sequences required to generate the output assertion and deassertion as specified above. Obtain the outputs and the waveforms.

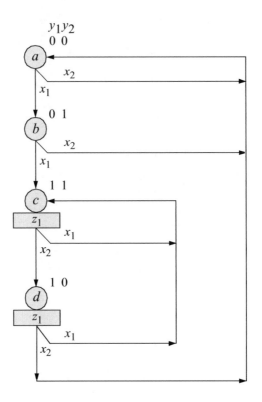

5.5   Synthesize a Mealy pulse-mode asynchronous sequential machine which has two inputs x_1 and x_2 and one output z_1. Output z_1 is asserted coincident with every second x_2 pulse, if and only if the pair of x_2 pulses is immediately preceded by an x_1 pulse. Use logic gates that were designed using dataflow modeling. The storage elements will consist of SR latches and D flip-flops in a master-slave configuration. Output z_1 is asserted as a logic 1 level. A representative timing diagram and the state diagram are shown below. Obtain the design module, the test bench module, the outputs, and the waveforms.

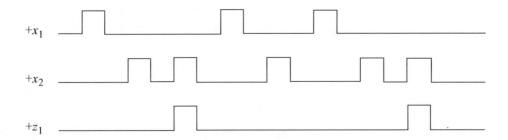

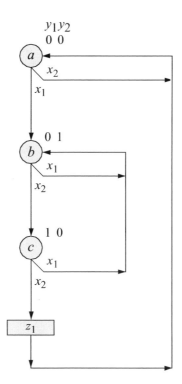

5.6   Design a Mealy pulse-mode asynchronous sequential machine which has two inputs x_1 and x_2 and one output z_1. For a Mealy machine, the outputs are a function of the present states and the present inputs. Output z_1 is asserted co-incident with the x_2 pulse if the x_2 pulse is immediately preceded by a pair of x_1 pulses. Use SR latches and D flip-flops in a master-slave configuration. Use NAND logic for all gates and latches except for output z_1, which will be an AND gate.

Derive the state diagram, input maps, output map, and logic diagram. Then use the continuous assignment statement to design the module. Obtain the test bench, outputs, and waveforms.

5.7   Given the state diagram shown below for a Moore pulse-mode asynchronous sequential machine, synthesize the machine using SR latches and D flip-flops in a master-slave configuration. Obtain the input maps and equations, the output equations, and the logic diagram using **and**, **nand**, and **nor** built-in primitives for the δ next-state logic and latches. Then develop the design module using built-in primitives and the test bench module that takes the machine through various sequences to assert output z_1 and output z_2. Obtain the outputs and the waveforms.

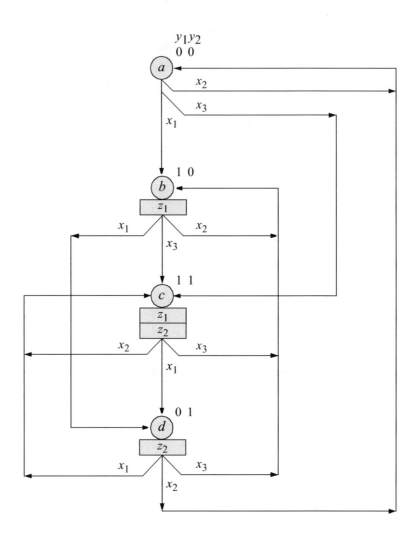

5.8   A state diagram is shown below for a Moore pulse-mode asynchronous sequential machine with three inputs x_1, x_2, and x_3, and two outputs z_1 and z_2. Use T flip-flops in the implementation. Obtain the input maps and equations, the output equations, and the logic diagram using AND and OR gates for the δ next-state logic. Obtain the design module using **and** and **or** built-in primitives and T flip-flops. Then develop the test bench module that takes the machine through various sequences to assert output z_1 and output z_2. Obtain the outputs, and the waveforms.

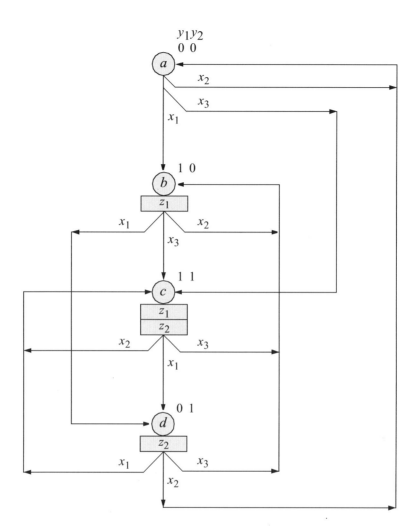

5.9   Design a Mealy pulse-mode asynchronous sequential machine which has
two inputs x_1 and x_2 and one output z_1. Output z_1 is asserted coincident with
every second x_2 pulse, if and only if the pair of x_2 pulses is immediately preceded by an x_1 pulse. Use the continuous assignment statement of dataflow
modeling in the implementation. The storage elements will consist of SR
latches and D flip-flops in a master-slave configuration. The design will be
implemented primarily with NOR logic for the SR latches and the logic primitives. A representative timing diagram is shown below.

Generate a state diagram that depicts all possible state transition
sequences that conform to the functional specifications. Obtain the input
maps and equations, the output equation, and the logic diagram. Obtain the
design module, the test bench module, the outputs, and the waveforms.

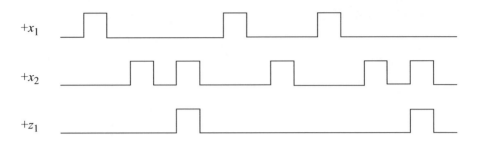

5.10   Design a Moore pulse-mode asynchronous sequential machine which has three inputs x_1, x_2, and x_3 and one output z_1. Output z_1 will be asserted coincident with the assertion of the x_3 pulse if and only if the x_3 pulse was preceded by an x_1 pulse followed by an x_2 pulse. That is, the input vector must be $x_1x_2x_3$ = 100, 000, 010, 000, 001 to assert z_1. Output z_1 will be deasserted at the next x_1 pulse or x_2 pulse.

A representative timing diagram is shown below. Obtain the state diagram, the input maps and equations, the output equation, and the logic diagram using NOR gates for the SR latches and D flip-flops that were designed using behavioral modeling. Obtain the design module using the continuous assignment statement, the test bench module, the outputs, and the waveforms.

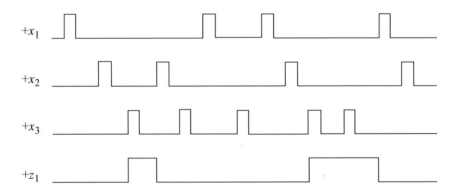

5.11   Design a Moore pulse-mode asynchronous sequential machine which has three inputs x_1, x_2, and x_3 and one output z_1. Output z_1 will be asserted coincident with the assertion of the x_3 pulse if and only if the x_3 pulse was preceded by an x_1 pulse followed by an x_2 pulse. That is, the input vector must be $x_1x_2x_3$ = 100, 000, 010, 000, 001 to assert z_1. Output z_1 will be deasserted at the next x_1 pulse or x_2 pulse. Use structural modeling that instantiates dataflow modules for the δ next-state logic and the SR latches. Instantiate the dataflow modules as single lines. Instantiate the D flip-flops as single lines that were designed using behavioral modeling. The timing diagram is shown below. Obtain the design module, the test bench module, the outputs, and the waveforms.

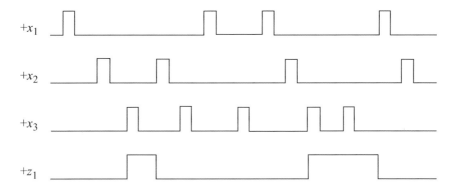

5.12 Design a Mealy pulse-mode asynchronous sequential machine that has three input variables x_1, x_2, and x_3 and one output z_1 that is asserted coincident with x_3 whenever the sequence $x_1 x_2 x_3 = 100, 010, 001$ occurs. The storage elements consist of *SR* latches and positive-edge-triggered *D* flip-flops in a master-slave configuration.

A representative timing diagram displaying valid input sequences and corresponding outputs is shown below. Obtain the state diagram that represents the functional operation of the machine. State code assignment is arbitrary for the state diagram, since input pulses trigger all state transitions and the machine does not begin to sequence to the next state until the input pulse, which initiated the transition, has been deasserted. Thus, output z_1 will not glitch. Obtain the input maps, the input equations, the output equation, and the logic diagram. Then obtain the structural design module using dataflow modeling for the logic primitives, which are instantiated as a single line. Use NOR logic for the *SR* latches. Use behavioral modeling for the *D* flip-flop. Obtain the test bench, the outputs, and the waveforms.

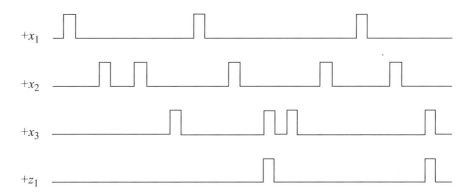

5.13   Given the state diagram shown below for a Moore pulse-mode asynchronous sequential machine, implement the machine using NAND gates for the *SR* latches and *D* flip-flops as the storage elements in a master-slave configuration. Use any type of gates for the logic primitives.

   Derive the input maps, the input equations, the output equations, and the the logic diagram. Generate the design module using the continuous assignment statement construct for the logic primitives and latches. Instantiate positive-edge-triggered *D* flip-flops that were designed using behavioral modeling. Obtain the test bench module, the outputs, and the waveforms.

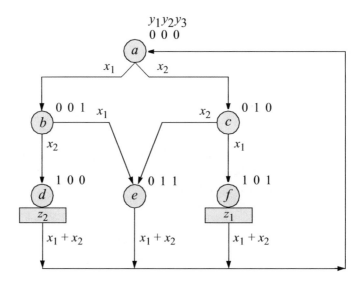

5.14   Given the state diagram shown below for a Moore pulse-mode asynchronous sequential machine, implement the machine using built-in primitives gates for the δ next-state logic and the *SR* latches. This problem uses built-in primitives and *D* flip-flops — instantiated as a single line — as the storage elements in a master-slave configuration with the latches.

   Derive the input maps, the input equations, the output equations, and the the logic diagram. Obtain the design module and instantiate positive-edge-triggered *D* flip-flops that were designed using behavioral modeling. Obtain the test bench module, the outputs, and the waveforms.

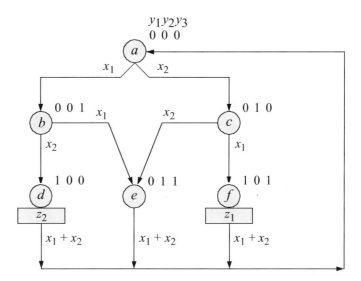

5.15   Given the state diagram shown below for a Moore pulse-mode asynchronous
        sequential machine, implement the machine using built-in primitives gates
        for the δ next-state logic and T flip-flops — instantiated as a single line — as
        the storage elements.

        Derive the input maps, the input equations, the output equations, and the
        the logic diagram. Obtain the design module, the test bench module, the out-
        puts, and the waveforms.

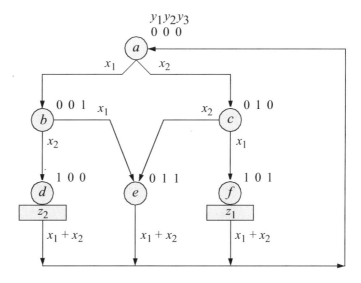

5.16   Design a Mealy pulse-mode asynchronous sequential machine which has two inputs x_1 and x_2 and one output z_1. For a Mealy machine, the outputs are a function of the present states and the present inputs. Output z_1 is asserted coincident with the x_2 pulse if the x_2 pulse is immediately preceded by a pair of x_1 pulses.

    Use SR latches and instantiate D flip-flops as a single line. The D flip-flops are designed using behavioral modeling. Use NAND dataflow modules logic for all gates and latches except for output z_1, which will be an AND gate. Derive the state diagram, the input maps and equations, the output equation, and the logic diagram. Then obtain the design module using logic gates that were designed using dataflow modeling and instantiated as a single line. Obtain the test bench, the outputs, and the waveforms.

5.17   Design a Mealy pulse-mode asynchronous sequential machine which has two inputs x_1 and x_2 and one output z_1. Output z_1 is asserted coincident with the x_2 pulse if the x_2 pulse is immediately preceded by a pair of x_1 pulses.

    Use SR latches and instantiate D flip-flops as a single line. Use **nand** built-in primitives for all logic gates and latches except for output z_1, which will be an **and** built-in primitive. Derive the state diagram, the input maps and equations, the output equation, and the logic diagram. Then obtain the design module, the test bench module, the outputs, and the waveforms.

5.18   Design a Mealy pulse-mode asynchronous sequential machine which has two inputs x_1 and x_2 and one output z_1. Output z_1 is asserted coincident with the x_2 pulse if the x_2 pulse is immediately preceded by a pair of x_1 pulses.

    Use SR latches and instantiate D flip-flops as a single line. Use NAND, NOR, and AND logic with the continuous assignment statement for all logic gates and latches. Derive the state diagram, the input maps and equations, the output equation, and the logic diagram. Then obtain the design module, the test bench module, the outputs, and the waveforms.

5.19   Design a Mealy pulse-mode asynchronous sequential machine which has two inputs x_1 and x_2 and one output z_1. Output z_1 is asserted coincident with the x_2 pulse if the x_2 pulse is immediately preceded by a pair of x_1 pulses.

    Use SR latches and D flip-flops. Instantiate the D flip-flops as a single line. Use instantiated dataflow modules for the AND and OR gates for the δ next-state logic and NOR logic for the latches. The **not** built-in primitive can also be utilized in the design. Instantiate the dataflow modules for all logic gates as a single line. Derive the state diagram, the input maps and equations, the output equation, and the logic diagram. Then obtain the design module, the test bench module, the outputs, and the waveforms.

Appendix A

Event Queue

Event management in Verilog hardware description language (HDL) is controlled by an event queue. Verilog modules generate events in the test bench, which provide stimulus to the module under test. These events can then produce new events by the modules under test. Since the Verilog HDL Language Reference Manual (LRM) does not specify a method of handling events, the simulator must provide a way to arrange and schedule these events in order to accurately model delays and obtain the correct order of execution. The manner of implementing the event queue is vendor-dependent.

Time in the event queue advances when every event that is scheduled in that time step is executed. Simulation is finished when all event queues are empty. An event at time t may schedule another event at time t or at time $t + n$.

A.1 Event Handling for Dataflow Assignments

Dataflow constructs consist of continuous assignments using the **assign** statement. The assignment occurs whenever simulation causes a change to the right-hand side expression. Unlike procedural assignments, continuous assignments are order independent — they can be placed anywhere in the module.

Consider the logic diagram shown in Figure A.1 which is represented by the two dataflow modules of Figure A.2 and Figure A.3. The test bench for both modules is shown in Figure A.4. The only difference between the two dataflow modules is the reversal of the two **assign** statements. The order in which the two statements execute is not defined by the Verilog HDL LRM; therefore, the order of execution is indeterminate.

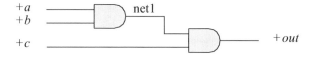

Figure A.1   Logic diagram to demonstrate event handling.

```
module dataflow (a, b, c, out);    module dataflow (a, b, c, out)

input a, b, c;                     input a, b, c;
output out;                        output out;

wire a, b, c;                      wire a, b, c;
wire out;                          wire out;

//define internal net             //define internal net
wire net1;                        wire net1;

assign net1 = a & b;              assign out = net1 & c;
assign out = net1 & c;            assign net1 = a & b;

endmodule                         endmodule
```

Figure A.2   Dataflow module 1.   **Figure A.3**   Dataflow module 2.

```
module dataflow_tb;               end
                                  //instantiate the module
reg test_a, test_b, test_c;       dataflow inst1
wire test_out;                        .a(test_a),
                                      .b(test_b),
initial                               .c(test_c),
begin                                 .out(test_out)
    test_a = 1'b1;                    );
    test_b = 1'b0;
    test_c = 1'b0;                endmodule

    #10    test_b = 1'b1;
           test_c = 1'b1;
    #10    $stop;
```

Figure A.4   Test bench for Figure A.2 and Figure A.3.

Assume that the simulator executes the assignment order shown in Figure A.2 first. When the simulator reaches time unit #10 in the test bench, it will evaluate the right-hand side of *test_b = 1'b1;* and place its value in the event queue for an immediate scheduled assignment. Since this is a blocking statement, the next statement will not execute until the assignment has been made. Figure A.5 represents the event queue after the evaluation. The input signal *b* will assume the value of *test_b* through instantiation.

Event queue					
Scheduled event 5	Scheduled event 4	Scheduled event 3	Scheduled event 2	Scheduled event 1	Time unit
				test_b ← 1'b1 b ← 1'b1	$t = \#10$
←				Order of execution	

Figure A.5   Event queue after execution of *test_b = 1'b1;*.

After the assignment has been made, the simulator will execute the *test_c = 1'b1;* statement by evaluating the right-hand side, and then placing its value in the event queue for immediate assignment. The new event queue is shown in Figure A.6. The entry that is not shaded represents an executed assignment.

Event queue					
Scheduled event 5	Scheduled event 4	Scheduled event 3	Scheduled event 2	Scheduled event 1	Time unit
			test_c ← 1'b1 c ← 1'b1	test_b ← 1'b1 b ← 1'b1	$t = \#10$
←				Order of execution	

Figure A.6   Event queue after execution of *test_c = 1'b1;*.

When the two assignments have been made, time unit #10 will have ended in the test bench, which is the top-level module in the hierarchy. The simulator will then enter the instantiated dataflow module during this same time unit and determine that events have occurred on input signals *b* and *c* and execute the two continuous assignments. At this point, inputs *a*, *b*, and *c* will be at a logic 1 level. However, *net1* will still contain a logic 0 level as a result of the first three assignments that executed at time #0 in the test bench. Thus, the statement *assign out = net1 & c;* will evaluate to a logic 0, which will be placed in the event queue and immediately assigned to *out*, as shown in Figure A.7.

Event queue					
Scheduled event 5	Scheduled event 4	Scheduled event 3	Scheduled event 2	Scheduled event 1	Time unit
		out ← 1'b0 test_out ← 1'b0	test_c ← 1'b1 c ← 1'b1	test_b ← 1'b1 b ← 1'b1	t = #10
				Order of execution	

Figure A.7   Event queue after execution of *assign out = net1 & c;*.

The simulator will then execute the *assign net1 = a & b;* statement in which the right-hand side evaluates to a logic 1 level. This will be placed on the queue and immediately assigned to *net1* as shown in Figure A.8.

Event queue					
Scheduled event 5	Scheduled event 4	Scheduled event 3	Scheduled event 2	Scheduled event 1	Time unit
	net1 ← 1'b1	out ← 1'b0 test_out ← 1'b0	test_c ← 1'b1 c ← 1'b1	test_b ← 1'b1 b ← 1'b1	t = #10
				Order of execution	

Figure A.8   Event queue after execution of *assign net1 = a & b;*.

When the assignment has been made to *net1*, the simulator will recognize this as an event on *net1*, which will cause all statements that use *net1* to be reevaluated. The only statement to be reevaluated is *assign out = net1 & c;*. Since both *net1* and *c* equal a logic 1 level, the right-hand side will evaluate to a logic 1, resulting in the event queue shown in Figure A.9.

Event queue					
Scheduled event 5	Scheduled event 4	Scheduled event 3	Scheduled event 2	Scheduled event 1	Time unit
out ← 1'b1 test_out ← 1'b1	net1 ← 1'b1	out ← 1'b0 test_out ← 1'b0	test_c ← 1'b1 c ← 1'b1	test_b ← 1'b1 b ← 1'b1	$t = \#10$
←───────────────────────────────────				Order of execution	

Figure A.9   Event queue after execution of *assign out = net1 & c;*.

The test bench signal *test_out* must now be updated because it is connected to *out* through instantiation. Because the signal *out* is not associated with any other statements within the module, the output from the module will now reflect the correct output. Since all statements within the dataflow module have been processed, the simulator will exit the module and return to the test bench. All events have now been processed; therefore, time unit #10 is complete and the simulator will advance the simulation time.

Since the order of executing the **assign** statements is irrelevant, processing of the dataflow events will now begin with the *assign net1 = a & b;* statement to show that the result is the same. The event queue is shown in Figure A.10.

Event queue					
Scheduled event 5	Scheduled event 4	Scheduled event 3	Scheduled event 2	Scheduled event 1	Time unit
		net1 ← 1'b1	test_c ← 1'b1 c ← 1'b1	test_b ← 1'b1 b ← 1'b1	$t = \#10$
←───────────────────────────────────				Order of execution	

Figure A.10   Event queue beginning with the statement *assign net1 = a & b;*.

Once the assignment to *net1* has been made, the simulator recognizes this as a new event on *net1*. The existing event on input *c* requires the evaluation of statement *assign out = net1 & c;*. The right-hand side of the statement will evaluate to a logic 1, and will be placed on the event queue for immediate assignment, as shown in Figure A.11.

Event queue					
Sched-uled event 5	Scheduled event 4	Scheduled event 3	Scheduled event 2	Scheduled event 1	Time unit
	out ← 1'b1 test_out ← 1'b1	net1 ← 1'b1	test_c ← 1'b1 c ← 1'b1	test_b ← 1'b1 b ← 1'b1	$t = \#10$
←				Order of execution	

Figure A.11   Event queue after execution of *assign out = net1 & c;*.

A.2 Event Handling for Blocking Assignments

The blocking assignment operator is the equal (=) symbol. A blocking assignment evaluates the right-hand side arguments and completes the assignment to the left-hand side before executing the next statement; that is, the assignment *blocks* other assignments until the current assignment has been executed.

Example A.1   Consider the code segment shown in Figure A.12 using blocking assignments in conjunction with the event queue of Figure A.13. There are no interstatement delays and no intrastatement delays associated with this code segment. In the first blocking assignment, the right-hand side is evaluated and the assignment is scheduled in the event queue. Program flow is blocked until the assignment is executed. This is true for all blocking assignment statements in this code segment. The assignments all occur in the same simulation time step t.

```
always @ (x2 or x3 or x5 or x7)
begin
   x1 = x2 | x3;
   x4 = x5;
   x6 = x7;
end
```

Figure A.12   Code segment with blocking assignments.

Event queue					
Scheduled event 5	Scheduled event 4	Scheduled event 3	Scheduled event 2	Scheduled event 1	Time unit
		x6 ← x7 (*t*)	x4 ← x5 (*t*)	x1 ← x2 | x3 (*t*)	*t*
			←	Order of execution	

Figure A.13   Event queue for Figure A.12.

Example A.2   The code segment shown in Figure A.14 contains an interstatement delay. Both the evaluation and the assignment are delayed by two time units. When the delay has taken place, the right-hand side is evaluated and the assignment is scheduled in the event queue as shown in Figure A.15. The program flow is blocked until the assignment is executed.

```
always @ (x2)
begin
   #2 x1 = x2;
end
```

Figure A.14   Blocking statement with interstatement delay.

Event queue					
Scheduled event 5	Scheduled event 4	Scheduled event 3	Scheduled event 2	Scheduled event 1	Time unit
					t
				x1 ← x2 (*t* + 2)	*t* + 2
			←	Order of execution	

Figure A.15   Event queue for Figure A.14.

Example A.3   The code segment of Figure A.16 shows three statements with interstatement delays of $t + 2$ time units. The first statement does not execute until simulation time $t + 2$ as shown in Figure A.17. The right-hand side $(x_2 \mid x_3)$ is evaluated at the current simulation time which is $t + 2$ time units, and then assigned to the left-hand side. At $t + 2$, x_1 receives the output of $x_2 \mid x_3$.

```
always @ (x2 or x3 or x5 or x7)
begin
    #2 x1 = x2 | x3;
    #2 x4 = x5;
    #2 x6 = x7;
end
```

Figure A.16 Code segment for delayed blocking assignment with interstatement delays.

		Event queue			
Scheduled event 5	Scheduled event 4	Scheduled event 3	Scheduled event 2	Scheduled event 1	Time unit
					t
				$x1 \leftarrow x2 \mid x3 \ (t + 2)$	$t + 2$
				$x4 \leftarrow x5 \ (t + 4)$	$t + 4$
				$x6 \leftarrow x7 \ (t + 6)$	$t + 6$
				Order of execution	

Figure A.17 Event queue for Figure A.16.

Example A.4 The code segment in Figure A.18 shows three statements using blocking assignments with intrastatement delays. Evaluation of $x_3 = \#2 \ x_4$ and $x_5 = \#2 \ x_6$ is blocked until x_2 has been assigned to x_1, which occurs at $t + 2$ time units. When the second statement is reached, it is scheduled in the event queue at time $t + 2$, but the assignment to x_3 will not occur until $t + 4$ time units. The evaluation in the third statement is blocked until the assignment is made to x_3. Figure A.19 shows the event queue.

```
always @ (x2 or x4 or x6)
begin
    x1 = #2 x2;      //first statement
    x3 = #2 x4;      //second statement
    x5 = #2 x6;      //third statement
end
```

Figure A.18 Code segment using blocking assignments with interstatement delays.

Event queue					
Scheduled event 5	Scheduled event 4	Scheduled event 3	Scheduled event 2	Scheduled event 1	Time unit
					t
				$x1 \leftarrow x2\,(t)$	$t + 2$
				$x3 \leftarrow x4\,(t + 2)$	$t + 4$
				$x5 \leftarrow x6\,(t + 4)$	$t + 6$
←				Order of execution	

Figure A.19   Event queue for the code segment of Figure A.18.

A.3 Event Handling for Nonblocking Assignments

Whereas blocking assignments block the sequential execution of an **always** block until the simulator performs the assignment, nonblocking statements evaluate each statement in succession and place the result in the event queue. Assignment occurs when all of the **always** blocks in the module have been processed for the current time unit. The assignment may cause new events that require further processing by the simulator for the current time unit.

Example A.5   For nonblocking statements, the right-hand side is evaluated and the assignment is scheduled at the end of the queue. The program flow continues and the assignment occurs at the end of the time step. This is shown in the code segment of Figure A.20 and the event queue of Figure A.21.

```
always @ (posedge clk)
begin
    x1 <= x2;
end
```

Figure A.20   Code segment for a nonblocking assignment.

Event queue					
Scheduled event 5	Scheduled event 4	Scheduled event 3	Scheduled event 2	Scheduled event 1	Time unit
x1 ← x2 (*t*)					*t*
			Order of execution		

Figure A.21   Event queue for Figure A.20.

Example A.6   The code segment of Figure A.22 shows a nonblocking statement with an interstatement delay. The evaluation is delayed by the timing control, and then the right-hand side expression is evaluated and assignment is scheduled at the end of the queue. Program flow continues and assignment is made at the end of the current time step as shown in the event queue of Figure A.23.

```
always @ (posedge clk)
begin
    #2 x1 <= x2;
end
```

Figure A.22   Nonblocking assignment with interstatement delay.

Event queue					
Scheduled event 5	Scheduled event 4	Scheduled event 3	Scheduled event 2	Scheduled event 1	Time unit
					t
x1 ← x2 (*t* + 2)					*t* + 2
			Order of execution		

Figure A.23   Event queue for Figure A.22.

Example A.7   The code segment of Figure A.24 shows a nonblocking statement with an intrastatement delay. The right-hand side expression is evaluated and assignment is delayed by the timing control and is scheduled at the end of the queue. Program flow continues and assignment is made at the end of the current time step as shown in the event queue of Figure A.25.

```
always @ (posedge clk)
begin
   x1 <= #2 x2;
end
```

Figure A.24   Nonblocking assignment with intrastatement delay.

Event queue					
Scheduled event 5	Scheduled event 4	Scheduled event 3	Scheduled event 2	Scheduled event 1	Time unit
					t
$x1 \leftarrow x2\ (t)$					$t+2$
				Order of execution	

Figure A.25   Event queue for Figure A.24.

Example A.8   The code segment of Figure A.26 shows nonblocking statements with intrastatement delays. The right-hand side expressions are evaluated and assignment is delayed by the timing control and is scheduled at the end of the queue. Program flow continues and assignment is made at the end of the current time step as shown in the event queue of Figure A.27.

```
always @ (posedge clk)
begin
   x1 <= #2 x2;
   x3 <= #2 x4;
   x5 <= #2 x6;
end
```

Figure A.26   Nonblocking assignments with intrastatement delays.

Event queue					
Scheduled event 5	Scheduled event 4	Scheduled event 3	Scheduled event 2	Scheduled event 1	Time unit
					t
x5 ← x6 (t)	x3 ← x4 (t)	x1 ← x2 (t)			$t+2$
◄───────────────────────────				Order of execution	

Figure A.27   Event queue for Figure A.26.

Example A.9   Figure A.28 shows a code segment using nonblocking assignment with an intrastatement delay. The right-hand expression is evaluated at the current time. The assignment is scheduled, but delayed by the timing control #2. This method allows for propagation delay through a logic element; for example, a D flip-flop. The event queue is shown in Figure A.29.

```
always @ (posedge clk)
begin
    q <= #2 d;
end
```

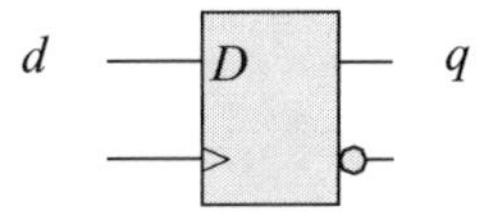

Figure A.28   Code segment using intrastatement delay with blocking assignment.

Event queue					
Scheduled event 5	Scheduled event 4	Scheduled event 3	Scheduled event 2	Scheduled event 1	Time unit
					t
				q ← d (t)	$t+2$
◄───────────────────────────				Order of execution	

Figure A.29   Event queue for the code segment of Figure A.28.

A.4   Event Handling for Mixed Blocking and Nonblocking Assignments

All nonblocking assignments are placed at the end of the queue while all blocking assignments are placed at the beginning of the queue in their respective order of evaluation. Thus, for any given simulation time t, all blocking statements are evaluated and assigned first, then all nonblocking statements are evaluated.

This is the reason why combinational logic requires the use of blocking assignments while sequential logic, such as flip-flops, requires the use of nonblocking assignments.   In this way, Verilog events can model real hardware in which combinational logic at the input to a flip-flop can stabilize before the clock sets the flip-flop to the state of the input logic. Therefore, blocking assignments are placed at the top of the queue to allow the input data to be stable, whereas nonblocking assignments are placed at the bottom of the queue to be executed after the input data has stabilized.

The logic diagram of Figure A.30 illustrates this concept for two multiplexers connected to the D inputs of their respective flip-flops. The multiplexers represent combinational logic; the D flip-flops represent sequential logic. The behavioral module is shown in Figure A.31 and the event queue is shown in Figure A.32.

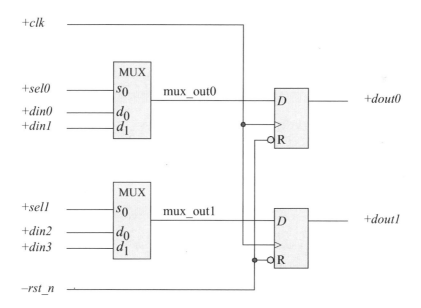

Figure A.30   Combinational logic connected to sequential logic to illustrate the use of blocking and nonblocking assignments.

Because multiplexers are combinational logic, the outputs *mux_out0* and *mux_out1* are placed at the beginning of the queue, as shown in Figure A.32. Nets *mux_out0* and *mux_out1* are in separate **always** blocks; therefore, the order in which they are placed in the queue is arbitrary and can differ with each simulator. The result, however, is the same. If *mux_out0* and *mux_out1* were placed in the same **always** block, then the order in which they are placed in the queue must be the same order as they appear in the **always** block.

Because *dout0* and *dout1* are sequential, they are placed at the end of the queue. Since they appear in separate **always** blocks, the order of their placement in the queue is irrelevant. Once the values of *mux_out0* and *mux_out1* are assigned in the queue, their values will then be used in the assignment of *dout0* and *dout1*; that is, the state of *mux_out0* and *mux_out1* will be set into the *D* flip-flops at the next positive clock transition and assigned to *dout0* and *dout1*.

```
//behavioral module with combinational and sequential logic
//to illustrate their placement in the event queue

module mux_plus_flop (clk, rst_n,
      din0, din1, sel0, dout0,
      din2, din3, sel1, dout1);

input clk, rst_n;
input din0, din1, sel0;
input din2, din3, sel1;
output dout0, dout1;

reg mux_out0, mux_out1;
reg dout0, dout1;

//combinational logic for multiplexers
always @ (din0 or din1 or sel0)
begin
   if (sel0)
      mux_out0 = din1;
   else
      mux_out0 = din0;
end

always @ (din2 or din3 or sel1)
begin
   if (sel1)
      mux_out1 = din3;
   else
      mux_out1 = din2;
end
//continued on next page
```

Figure A.31   Mixed blocking and nonblocking assignments that represent combinational and sequential logic.

```
//sequential logic for D flip-flops
always @ (posedge clk or negedge rst_n)
begin
   if (~rst_n)
      dout0 <= 1'b0;
   else
      dout0 <= mux_out0;
end

always @ (posedge clk or negedge rst_n)
begin
   if (~rst_n)
      dout1 <= 1'b0;
   else
      dout1 <= mux_out1;
end

endmodule
```

Figure A.31   (Continued)

Event queue					
Scheduled event 4	Scheduled event 3	N/A	Scheduled event 2	Scheduled event 1	Time unit
dout1 ← mux_out1 (t)	dout0 ← mux_out0 (t)		mux_out1 ← din3 (t)	mux_out0 ← din1 (t)	t
			←	Order of execution	

Figure A.32   Event queue for Figure A.31.

Appendix B

Verilog Project Procedure

- **Create a folder** (Do only once)

   Local disk (C:) > New Folder <Verilog> > Enter > Exit local disk C.

- **Create a project** (Do for each project)

   Bring up Silos Simulation Environment.
      File > Close Project. Minimize Silos.
   Local disk (C:) > Verilog > File > New Folder <new folder name> Enter.
   Exit Local disk (C:). Maximize Silos.
   File > New Project.
   Create New Project. Save In: Verilog folder.
      Click new folder name. Open.
   Create New Project. Filename: Give project name — usually same name
      as the folder name. Save
   Project Properties > Cancel.

- **File > New**

   .
   . Design module code goes here
   .

- **File > Save As > File name: <filename.v> > Save**

- **Compile code**

   Edit > Project Properties > Add. Select one or more files to add.
      Click on the file > Open.
   Project Properties. The selected files are shown > OK.
   Load/Reload Input Files. This compiles the code.
   Check screen output for errors. "Simulation stopped at the end of time 0"
      indicates no compilation errors.

- **Test bench**
    File > New

        .
        .   Test bench module code goes here
        .

- **File > Save As > File name: < filename.v> > Save.**

- **Compile test bench**
    Edit > Project Properties > Add. Select one or more files to add.
        Click on the file > Open
    Project Properties. The selected files are shown > OK.
    Load/Reload Input Files. This compiles the code.
    Check screen output for errors. "Simulation stopped at end of time 0"
        indicates no compilation errors.

- **Binary Output and Waveforms**
    For binary output: click on the GO icon.
    For waveforms: click on the Analyzer icon.
            Click on the Explorer icon. The signals are listed in Silos Explorer.
            Click on the desired signal names.
            Right click. Add Signals to Analyzer.
            Waveforms are displayed.
            Exit Silos Explorer.

- **Change Time Scale**
    With the waveforms displayed, click on Analyzer > X-Axis > Timescale
        Enter Time / div > OK

- **Exit the project**
    Close the waveforms, module, and test bench.
    File > Close Project.

Appendix C

Answers to Select Problems

Chapter 1   Introduction to Verilog HDL

1.2   Design a circuit using built-in primitive NAND gates that satisfies the following specifications: $3 < N \le 8$ and $10 \le N < 15$. Obtain the Karnaugh map, the equation in a sum-of-products form, and the logic diagram using NAND gates. Then obtain the design module using built-in primitives, the test bench module, the outputs, and the waveforms.

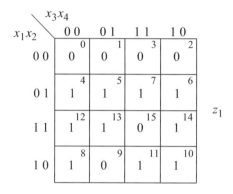

$$z_1 = x_1'x_2 + x_2x_3' + x_1x_4' + x_1x_2'x_3$$

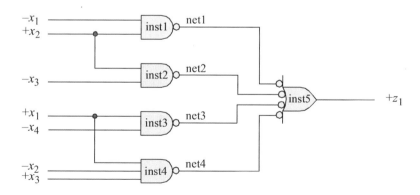

```
//built-in primitive number range
module num_range3 (x1, x2, x3, x4, z1);
input x1, x2, x3, x4;
output z1;

nand   inst1 (net1, ~x1, x2),
       inst2 (net2, x2, ~x3),
       inst3 (net3, x1, ~x4),
       inst4 (net4, x1, ~x2, x3);

nand   inst5 (z1, net1, net2, net3, net4);
endmodule
```

```
//test bench for number range module
module num_range3_tb;

//inputs are reg for test bench
reg x1, x2, x3, x4;

//outputs are wire for test bench
wire z1;

//apply input vectors
initial
begin: apply_stimulus
   reg [4:0] invect;
   for (invect = 0; invect < 16; invect = invect + 1)
      begin
         {x1, x2, x3, x4} = invect [4:0];
         #10 $display ("x1 x2 x3 x4 = %b, z1 = %b",
                        {x1, x2, x3, x4}, z1);
      end
end

//instantiate the module into the test bench
num_range3 inst1 (
   .x1(x1),
   .x2(x2),
   .x3(x3),
   .x4(x4),
   .z1(z1)
   );
endmodule
```

```
x1 x2 x3 x4 = 0000,  z1 = 0
x1 x2 x3 x4 = 0001,  z1 = 0
x1 x2 x3 x4 = 0010,  z1 = 0
x1 x2 x3 x4 = 0011,  z1 = 0
x1 x2 x3 x4 = 0100,  z1 = 1
x1 x2 x3 x4 = 0101,  z1 = 1
x1 x2 x3 x4 = 0110,  z1 = 1
x1 x2 x3 x4 = 0111,  z1 = 1
x1 x2 x3 x4 = 1000,  z1 = 1
x1 x2 x3 x4 = 1001,  z1 = 0
x1 x2 x3 x4 = 1010,  z1 = 1
x1 x2 x3 x4 = 1011,  z1 = 1
x1 x2 x3 x4 = 1100,  z1 = 1
x1 x2 x3 x4 = 1101,  z1 = 1
x1 x2 x3 x4 = 1110,  z1 = 1
x1 x2 x3 x4 = 1111,  z1 = 0
```

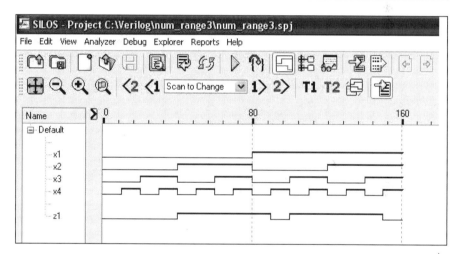

1.7   Given the Karnaugh map shown below, obtain the equation for output z_1 in a sum-of-products notation and the corresponding logic diagram using AND and OR gates. Then use dataflow modeling for the design module and generate a test bench. Obtain the outputs and the waveforms.

x_1x_2 \\ x_3x_4	0 0	0 1	1 1	1 0
0 0	1 ⁰	1 ¹	0 ³	0 ²
0 1	0 ⁴	1 ⁵	1 ⁷	1 ⁶
1 1	0 ¹²	0 ¹³	1 ¹⁵	1 ¹⁴
1 0	1 ⁸	0 ⁹	0 ¹¹	0 ¹⁰

z_1

$$z_1 = x_2'x_3'x_4' + x_1'x_3'x_4 + x_2x_3$$

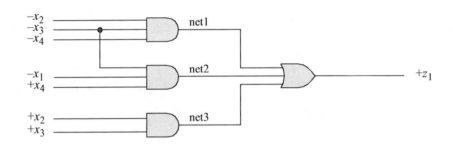

```
//dataflow for a sum-of-products equation
module sop_eqn_df2 (x1, x2, x3, x4, z1);

//define inputs and output
input x1, x2, x3, x4;
output z1;

//define internal nets
wire net1, net2, net3;

//design logic
assign    net1 = ~x2 & ~x3 & ~x4,
          net2 = ~x1 & ~x3 & x4,
          net3 = x2 & x3;

assign    z1 = net1 | net2 | net3;

endmodule
```

```
//test bench for the dataflow sop
module sop_eqn_df2_tb;

reg x1, x2, x3, x4;
wire z1;

//apply input vectors and display variables
initial
begin: apply_stimulus
   reg [4:0] invect;
   for (invect=0; invect<16; invect=invect+1)
      begin
         {x1, x2, x3, x4} = invect [4:0];
         #10 $display ("x1 x2 x3 x4 = %b, z1 = %b",
                       {x1, x2, x3, x4}, z1);
      end
end

//instantiate the module into the test bench
sop_eqn_df2 inst1 (
   .x1(x1),
   .x2(x2),
   .x3(x3),
   .x4(x4),
   .z1(z1)
   );

endmodule
```

```
x1 x2 x3 x4 = 0000, z1 = 1
x1 x2 x3 x4 = 0001, z1 = 1
x1 x2 x3 x4 = 0010, z1 = 0
x1 x2 x3 x4 = 0011, z1 = 0
x1 x2 x3 x4 = 0100, z1 = 0
x1 x2 x3 x4 = 0101, z1 = 1
x1 x2 x3 x4 = 0110, z1 = 1
x1 x2 x3 x4 = 0111, z1 = 1
x1 x2 x3 x4 = 1000, z1 = 1
x1 x2 x3 x4 = 1001, z1 = 0
x1 x2 x3 x4 = 1010, z1 = 0
x1 x2 x3 x4 = 1011, z1 = 0
x1 x2 x3 x4 = 1100, z1 = 0
x1 x2 x3 x4 = 1101, z1 = 0
x1 x2 x3 x4 = 1110, z1 = 1
x1 x2 x3 x4 = 1111, z1 = 1
```

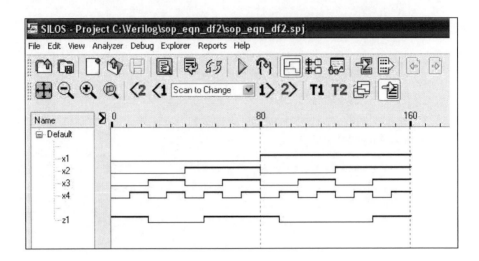

1.11   Use the three bitwise operators of AND (&), OR (|), and exclusive-OR (^) to implement the logical operations shown below. Obtain the dataflow design module, the test bench module for eight variations of the three 4-bit operands a, b, and c, the outputs, and the waveforms.

$$z_1 = (a \, \& \, b) \,|\, c$$
$$z_2 = (a \wedge b) \, \& \, c$$
$$z_3 = (a \,|\, c) \wedge b$$

```
//dataflow bitwise operators
module bitwise3 (a, b, c, z1, z2, z3);

input  [3:0] a, b, c;
output [3:0] z1, z2, z3;

assign    z1 = (a & b) | c,
          z2 = (a ^ b) & c,
          z3 = (a | c) ^ b;
endmodule
```

```
//test bench for bitwise operators
module bitwise3_tb;

reg [3:0] a, b, c;
wire [3:0] z1, z2, z3;

//display variables
initial
$monitor ("a=%b, b=%b, c=%b, z1=%b, z2=%b, z3=%b",
          a, b, c, z1, z2, z3);

//apply input vectors
initial
begin
   #0    a = 4'b0001;b = 4'b0001;c = 4'b0001;
   #10   a = 4'b0011;b = 4'b0011;c = 4'b0011;
   #10   a = 4'b1111;b = 4'b0000;c = 4'b1000;
   #10   a = 4'b0000;b = 4'b1000;c = 4'b0000;

   #10   a = 4'b0100;b = 4'b0110;c = 4'b0111;
   #10   a = 4'b0111;b = 4'b0000;c = 4'b1000;
   #10   a = 4'b0000;b = 4'b0000;c = 4'b0000;
   #10   a = 4'b1111;b = 4'b1111;c = 4'b1111;

   #10   $stop;
end

//instantiate the module into the test bench
bitwise3 inst1 (
   .a(a),
   .b(b),
   .c(c),
   .z1(z1),
   .z2(z2),
   .z3(z3)
   );
endmodule
```

```
z1 = (a & B) | c
z2 = (a ^ b) & c
z3 = (a | c) ^ b

a=0001, b=0001, c=0001, z1=0001, z2=0000, z3=0000
a=0011, b=0011, c=0011, z1=0011, z2=0000, z3=0000
a=1111, b=0000, c=1000, z1=1000, z2=1000, z3=1111
a=0000, b=1000, c=0000, z1=0000, z2=0000, z3=1000

a=0100, b=0110, c=0111, z1=0111, z2=0010, z3=0001
a=0111, b=0000, c=1000, z1=1000, z2=0000, z3=1111
a=0000, b=0000, c=0000, z1=0000, z2=0000, z3=0000
a=1111, b=1111, c=1111, z1=1111, z2=0000, z3=0000
```

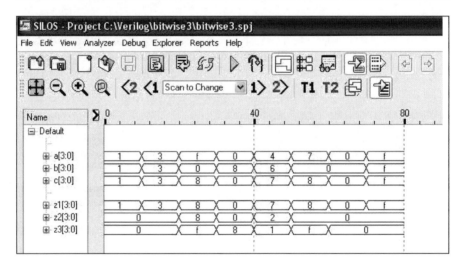

1.19   Use behavioral modeling to design a full adder. A full adder has three scalar inputs a, b, and cin; there are two scalar outputs sum and $cout$. Obtain the design module, the test bench module for all combinations of the inputs, the outputs, and the waveforms. The equations for sum and $cout$ are shown below.

$$sum = a'b'cin + a'b\,cin' + ab'cin' + ab\,cin$$

$$= a \oplus b \oplus cin$$

$$cout = a'b\,cin + ab'cin + ab\,cin' + ab\,cin$$

$$= ab + a\,cin + b\,cin$$

```verilog
//behavioral full adder
module full_adder_bh (a, b, cin, sum, cout);

input a, b, cin;
output sum, cout;

wire a, b, cin;
reg sum, cout;

always @ (a or b or cin)
begin
   sum = a ^ b ^ cin;
   cout = (a & b) | (a & cin) | (b & cin);
end

endmodule
```

```verilog
//test bench for behavioral full adder
module full_adder_bh_tb;

reg a, b, cin;
wire sum, cout;

//apply input vectors and display variables
initial
begin: apply_stimulus
   reg [3:0] invect;
   for (invect = 0; invect < 8; invect = invect + 1)
      begin
         {a, b, cin} = invect [3:0];
         #10 $display ("a b cin = %b, cout = %b,
                        sum = %b", {a, b, cin}, cout, sum);
      end
end

//instantiate the module into the test bench
full_adder_bh inst1 (
   .a(a),
   .b(b),
   .cin(cin),
   .sum(sum),
   .cout(cout)
   );

endmodule
```

```
a b cin = 000, cout = 0, sum = 0
a b cin = 001, cout = 0, sum = 1
a b cin = 010, cout = 0, sum = 1
a b cin = 011, cout = 1, sum = 0

a b cin = 100, cout = 0, sum = 1
a b cin = 101, cout = 1, sum = 0
a b cin = 110, cout = 1, sum = 0
a b cin = 111, cout = 1, sum = 1
```

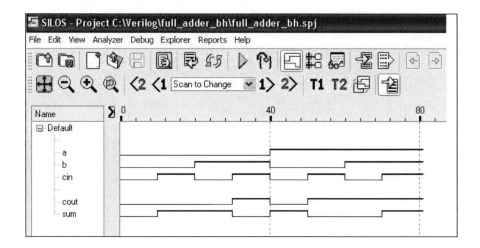

1.22   Design a 4:1 multiplexer using a combination of behavioral modeling and dataflow modeling. The multiplexer has four data inputs, which are specified as a 4-bit vector $d[3:0]$, two select inputs, specified as a 2-bit vector $s[1:0]$, one scalar enable input *enbl*, and one scalar output z_1. Obtain the design module and the test bench module containing eight combinations of the data inputs. Obtain the outputs and the waveforms.

```
//behavioral/dataflow 4:1 multiplexer
module mux4_bh_df (d, s, enbl, z1);

input [3:0] d;
input [1:0] s;
input enbl;
output z1;

wire [3:0] d;
wire [1:0] s;
wire enbl;
wire net0, net1, net2, net3;
reg z1;

assign    net0 = (enbl & ~s[1] & ~s[0] & d[0]),
          net1 = (enbl & ~s[1] & s[0] & d[1]),
          net2 = (enbl & s[1] & ~s[0] & d[2]),
          net3 = (enbl & s[1] & s[0] & d[3]);

always @ (net0 or net1 or net2 or net3)
begin
   z1 = (net0 || net1 || net2 || net3);
end

endmodule
```

```verilog
//test bench for 4:1 multiplexer
module mux4_bh_df_tb;

reg [3:0] d;
reg [1:0] s;
reg enbl;
wire z1;

//display variables
initial
$monitor ("select = %b, data = %b, z1 = %b", s, d, z1);

//apply input vectors
initial
begin
   #0    s = 2'b00;   d = 4'b0001;   enbl = 1'b1;
   #10   s = 2'b00;   d = 4'b0100;   enbl = 1'b1;

   #10   s = 2'b01;   d = 4'b1010;   enbl = 1'b1;
   #10   s = 2'b01;   d = 4'b1100;   enbl = 1'b1;

   #10   s = 2'b10;   d = 4'b1100;   enbl = 1'b1;
   #10   s = 2'b10;   d = 4'b1000;   enbl = 1'b1;

   #10   s = 2'b11;   d = 4'b1100;   enbl = 1'b1;
   #10   s = 2'b11;   d = 4'b0111;   enbl = 1'b1;

   #10   $stop;
end

//instantiate the module into the test bench
mux4_bh_df inst1 (
   .d(d),
   .s(s),
   .enbl(enbl),
   .z1(z1)
   );
endmodule
```

```
select = 00, data = 0001, z1 = 1
select = 00, data = 0100, z1 = 0
select = 01, data = 1010, z1 = 1
select = 01, data = 1100, z1 = 0
select = 10, data = 1100, z1 = 1
select = 10, data = 1000, z1 = 0
select = 11, data = 1100, z1 = 1
select = 11, data = 0111, z1 = 0
```

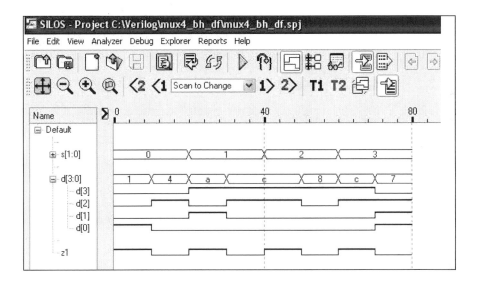

1.25   Use behavioral modeling with the **case** statement to design a 6-function logic
        unit for the following six functions: add, subtract, multiply, AND, OR, and
        exclusive-OR. The operands are 4-bit vectors: $a[3:0]$ and $b[3:0]$. Obtain the
        design module and the test bench module for four variations of the operands
        for each function. Obtain the outputs and waveforms.

```
//demonstrate arithmetic operations
module arith_log_ops (a, b, opcode, rslt);

input [3:0] a, b;
input [2:0] opcode;
output [7:0] rslt;

reg [7:0] rslt;

parameter    addop = 3'b000,
             subop = 3'b001,
             mulop = 3'b010,
             andop = 3'b011,
             orop  = 3'b100,
             xorop = 3'b101;

always @ (a or b or opcode)
begin
   case (opcode)
      addop: rslt = a + b;
      subop: rslt = a - b;
      mulop: rslt = a * b;
      andop: rslt = a & b;
      orop:  rslt = a | b;
      xorop: rslt = a ^ b;
   default: rslt = 8'bxxxxxxxx;
   endcase
end

endmodule
```

```
//arithmetic operations test bench
module arith_log_ops_tb;

reg [3:0] a, b;
reg [2:0] opcode;
wire [7:0] rslt ;

initial
$monitor ("a = %b, b = %b, opcode = %b, rslt = %b",
          a , b, opcode, rslt);

                              //continued on next page
```

```
initial
begin
   #0    a = 4'b0011;b = 4'b0111;opcode = 3'b000;//add
   #10   a = 4'b0111;b = 4'b0001;opcode = 3'b000;
   #10   a = 4'b0011;b = 4'b0110;opcode = 3'b000;
   #10   a = 4'b1011;b = 4'b0100;opcode = 3'b000;

   #10   a = 4'b1111;b = 4'b1111;opcode = 3'b001;//sub
   #10   a = 4'b1000;b = 4'b0101;opcode = 3'b001;
   #10   a = 4'b1110;b = 4'b0111;opcode = 3'b001;
   #10   a = 4'b1111;b = 4'b1110;opcode = 3'b001;

   #10   a = 4'b1110;b = 4'b1110;opcode = 3'b010;//mul
   #10   a = 4'b1111;b = 4'b1111;opcode = 3'b010;
   #10   a = 4'b0110;b = 4'b0110;opcode = 3'b010;
   #10   a = 4'b0111;b = 4'b0111;opcode = 3'b010;

   #10   a = 4'b1000;b = 4'b0010;opcode = 3'b011;//and
   #10   a = 4'b1111;b = 4'b0110;opcode = 3'b011;
   #10   a = 4'b0110;b = 4'b0010;opcode = 3'b011;
   #10   a = 4'b0111;b = 4'b0011;opcode = 3'b011;

   #10   a = 4'b0111;b = 4'b0011;opcode = 3'b100;//or
   #10   a = 4'b0100;b = 4'b0011;opcode = 3'b100;
   #10   a = 4'b1101;b = 4'b0111;opcode = 3'b100;
   #10   a = 4'b0110;b = 4'b1100;opcode = 3'b100;

   #10   a = 4'b0110;b = 4'b1100;opcode = 3'b101;//xor
   #10   a = 4'b0100;b = 4'b1100;opcode = 3'b101;
   #10   a = 4'b1110;b = 4'b0100;opcode = 3'b101;
   #10   a = 4'b0111;b = 4'b0000;opcode = 3'b101;

   #10   $stop;
end

//instantiate the module into the test bench
arith_log_ops inst1 (
   .a(a),
   .b(b),
   .opcode(opcode),
   .rslt(rslt)
   );

endmodule
```

```
a = 0011, b = 0111, opcode = 000, rslt = 00001010//add
a = 0111, b = 0001, opcode = 000, rslt = 00001000
a = 0011, b = 0110, opcode = 000, rslt = 00001001
a = 1011, b = 0100, opcode = 000, rslt = 00001111
a = 1111, b = 1111, opcode = 001, rslt = 00000000//sub
a = 1000, b = 0101, opcode = 001, rslt = 00000011
a = 1110, b = 0111, opcode = 001, rslt = 00000111
a = 1111, b = 1110, opcode = 001, rslt = 00000001
a = 1110, b = 1110, opcode = 010, rslt = 11000100//mul
a = 1111, b = 1111, opcode = 010, rslt = 11100001
a = 0110, b = 0110, opcode = 010, rslt = 00100100
a = 0111, b = 0111, opcode = 010, rslt = 00110001
a = 1000, b = 0010, opcode = 011, rslt = 00000000//and
a = 1111, b = 0110, opcode = 011, rslt = 00000110
a = 0110, b = 0010, opcode = 011, rslt = 00000010
a = 0111, b = 0011, opcode = 011, rslt = 00000011
a = 0111, b = 0011, opcode = 100, rslt = 00000111//or
a = 0100, b = 0011, opcode = 100, rslt = 00000111
a = 1101, b = 0111, opcode = 100, rslt = 00001111
a = 0110, b = 1100, opcode = 100, rslt = 00001110
a = 0110, b = 1100, opcode = 101, rslt = 00001010//xor
a = 0100, b = 1100, opcode = 101, rslt = 00001000
a = 1110, b = 0100, opcode = 101, rslt = 00001010
a = 0111, b = 0000, opcode = 101, rslt = 00000111
```

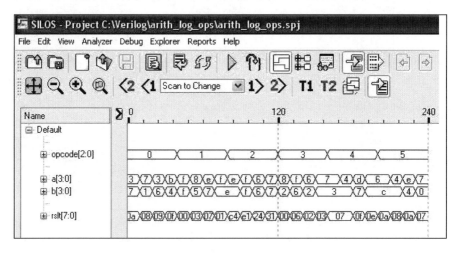

1.27   Design a structural 4-bit, *[3:0]*, binary-to-excess-3 code converter by instantiating behavioral full adders into the design. The excess-3 code will contain five bits to include the carry out of the high-order bit position of *adder[3]*. For example, `binary = 1111, excess3 = 10010`. Obtain the design module and the test bench module for all 16 combinations of the binary inputs. Obtain the outputs and the waveforms.

```verilog
//structural binary-to-excess3 conversion
//by instantiating full adders
module binary_excess3_struc (bin, ex3);

input [3:0] bin;
output [4:0] ex3;

//define internal nets
wire cout0, cout1, cout2;

//instantiate the full adder for binary bit[0]
full_adder_bh inst1 (
   .a(bin[0]),      //binary input[0], low order
   .b(1'b1),        //adder b input
   .cin(1'b0),
   .sum(ex3[0]),    //excess3[0], low order
   .cout(cout0)
   );

//instantiate the full adder for binary bit[1]
full_adder_bh inst2 (
   .a(bin[1]),
   .b(1'b1),
   .cin(cout0),
   .sum(ex3[1]),
   .cout(cout1)
   );

//instantiate the full adder for binary bit[2]
full_adder_bh inst3 (
   .a(bin[2]),
   .b(1'b0),
   .cin(cout1),
   .sum(ex3[2]),
   .cout(cout2)
   );

//instantiate the full adder for binary bit[3]
full_adder_bh inst4 (
   .a(bin[3]),
   .b(1'b0),
   .cin(cout2),
   .sum(ex3[3]),
   .cout(ex3[4])
   );

endmodule
```

```
//test bench for binary to excess3 conversion
module binary_excess3_struc_tb;

reg [3:0] bin;
wire [4:0] ex3;

//apply input vectors
initial
begin: apply_stimulus
   reg [4:0] invect;
   for (invect = 0; invect < 16; invect = invect + 1)
      begin
         bin = invect [4:0];
         #10 $display ("binary = %b, excess3 = %b",
                        bin, ex3);
      end
end

//instantiate the module into the test bench
binary_excess3_struc inst1 (
   .bin(bin),
   .ex3(ex3)
   );

endmodule
```

```
binary = 0000, excess3 = 00011
binary = 0001, excess3 = 00100
binary = 0010, excess3 = 00101
binary = 0011, excess3 = 00110

binary = 0100, excess3 = 00111
binary = 0101, excess3 = 01000
binary = 0110, excess3 = 01001
binary = 0111, excess3 = 01010

binary = 1000, excess3 = 01011
binary = 1001, excess3 = 01100
binary = 1010, excess3 = 01101
binary = 1011, excess3 = 01110

binary = 1100, excess3 = 01111
binary = 1101, excess3 = 10000
binary = 1110, excess3 = 10001
binary = 1111, excess3 = 10010
```

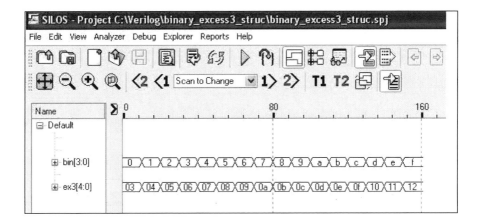

1.29   Design a logic circuit that will generate a high logic level on output z_1 if a 4-bit binary number $x[3:0]$ has a value less than or equal to five or greater than nine. Obtain the structural design module and the test bench module for all 16 combinations of the inputs. Obtain the outputs and the waveforms.

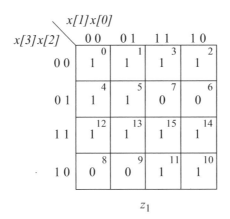

$$z_1 = x[3]'\,x[2]' + x[3]\,x[2] + x[2]\,x[1]' + x[2]'\,x[1]$$
$$= (x[3] \oplus x[2])' + (x[2] \oplus x[1])$$

```verilog
//structural for a number in the range >=5 N > 9
module num_range5 (x, z1);

input [3:0] x;
output z1;

//define internal nets
wire net1, net2;

//instantiate the logic gates
xnor2_df inst1 (
    .x1(x[3]),
    .x2(x[2]),
    .z1(net1)
    );

xor2_df inst2 (
    .x1(x[2]),
    .x2(x[1]),
    .z1(net2)
    );

or2_df inst3 (
    .x1(net1),
    .x2(net2),
    .z1(z1)
    );

endmodule
```

```
//test bench for num_range5 module
module num_range5_tb;

reg [3:0] x;
wire z1;

//apply input vectors and display variables
initial
begin: apply_stimulus
   reg [4:0] invect;
   for (invect = 0; invect < 16; invect = invect + 1)
      begin
         x = invect [4:0];
         #10 $display ("x = %b, z1=%b", x, z1);
      end
end

//instantiate the module into the test bench
num_range5 inst1 (
   .x(x),
   .z1(z1)
   );

endmodule
```

```
x = 0000, z1=1
x = 0001, z1=1
x = 0010, z1=1
x = 0011, z1=1

x = 0100, z1=1
x = 0101, z1=1
x = 0110, z1=0
x = 0111, z1=0

x = 1000, z1=0
x = 1001, z1=0
x = 1010, z1=1
x = 1011, z1=1

x = 1100, z1=1
x = 1101, z1=1
x = 1110, z1=1
x = 1111, z1=1
```

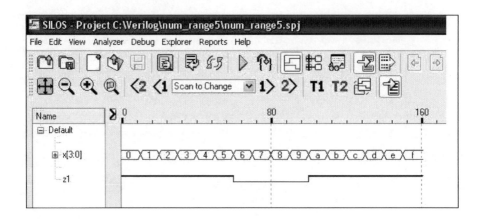

1.31   Given the logic diagram shown below, obtain the minimum product-of-sums equation, then design a structural module using NOR gates to implement the equation. Then design the test bench using all 16 combinations of the four input variables. Verify the results by displaying the outputs and the waveforms.

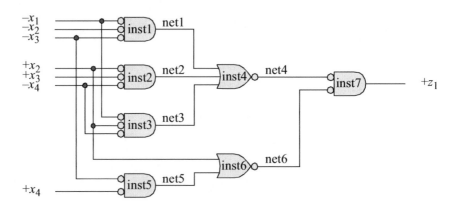

$$z_1 = (x_1 x_2 x_3 + x_2' x_3' x_4 + x_1 x_2' x_4)(x_2 + x_3 x_4')$$

$$= (x_1 x_2 x_3 x_2 + x_1 x_2 x_3 x_3 x_4') + (x_2' x_3' x_4 x_2 + x_2' x_3' x_4 x_3 x_4') +$$
$$(x_1 x_2' x_4 x_2 + x_1 x_2' x_4 x_3 x_4')$$

$$= x_1 x_2 x_3 + x_1 x_2 x_3 x_4'$$

$$= x_1 x_2 x_3 (1 + x_4')$$

$$= x_1 x_2 x_3$$

```verilog
//structural logic equation as a product of sums
//using only NOR logic
module log_eqn_pos_nor (x1, x2, x3, x4, z1);

input x1, x2, x3, x4;
output z1;

//define internal nets
wire net1, net2, net3, net4, net5, net6;

//instantiate the logic gates
nor3_df inst1 (
    .x1(~x1),
    .x2(~x2),
    .x3(~x3),
    .z1(net1)
    );

nor3_df inst2 (
    .x1(x2),
    .x2(x3),
    .x3(~x4),
    .z1(net2)
    );

nor3_df inst3 (
    .x1(~x1),
    .x2(x2),
    .x3(~x4),
    .z1(net3)
    );

nor3_df inst4 (
    .x1(net1),
    .x2(net2),
    .x3(net3),
    .z1(net4)
    );

nor2_df inst5 (
    .x1(~x3),
    .x2(x4),
    .z1(net5)
    );
```

//continued on next page

```
nor2_df inst6 (
   .x1(x2),
   .x2(net5),
   .z1(net6)
   );

nor2_df inst7 (
   .x1(net4),
   .x2(net6),
   .z1(z1)
   );

endmodule
```

```
//test bench for the product of sums using NOR gates
module log_eqn_pos_nor_tb;

reg x1, x2, x3, x4;
wire z1;

//apply input vectors and display variables
initial
begin: apply_stimulus
   reg [4:0] invect;
   for (invect = 0; invect < 16; invect = invect + 1)
      begin
         {x1, x2, x3, x4} = invect [4:0];
         #10 $display ("x1 x2 x3 x4 = %b, z1 = %b",
                        {x1, x2, x3, x4}, z1);
      end
end

//instantiate the module into the test bench
log_eqn_pos_nor inst1 (
   .x1(x1),
   .x2(x2),
   .x3(x3),
   .x4(x4),
   .z1(z1)
   );

endmodule
```

```
x1 x2 x3 x4 = 0000, z1 = 0
x1 x2 x3 x4 = 0001, z1 = 0
x1 x2 x3 x4 = 0010, z1 = 0
x1 x2 x3 x4 = 0011, z1 = 0

x1 x2 x3 x4 = 0100, z1 = 0
x1 x2 x3 x4 = 0101, z1 = 0
x1 x2 x3 x4 = 0110, z1 = 0
x1 x2 x3 x4 = 0111, z1 = 0

x1 x2 x3 x4 = 1000, z1 = 0
x1 x2 x3 x4 = 1001, z1 = 0
x1 x2 x3 x4 = 1010, z1 = 0
x1 x2 x3 x4 = 1011, z1 = 0

x1 x2 x3 x4 = 1100, z1 = 0
x1 x2 x3 x4 = 1101, z1 = 0
x1 x2 x3 x4 = 1110, z1 = 1
x1 x2 x3 x4 = 1111, z1 = 1
```

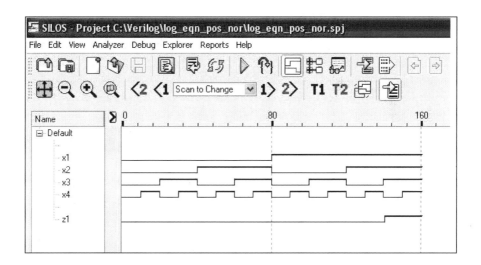

Chapter 2   Synthesis of Synchronous Sequential Machines 1 Using Verilog HDL

2.2   Determine the counting sequence for the counter shown below by designing the counter using structural modeling with built-in primitives and D flip-flops that were designed using behavioral modeling. The D flip-flops have an implied reset input. Obtain the structural design module, the test bench module, the outputs, and the waveforms. The counter is reset initially; that is, $y_1 y_2 = 00$, where y_2 is the low-order flip-flop.

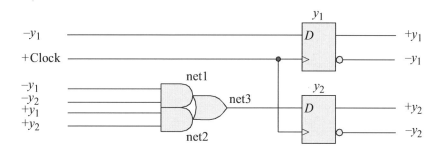

```
//structural counter using built-in primitives

module ctr_struc2 (rst_n, clk, y);

input rst_n, clk;
output [1:2] y;

//define internal nets
wire net1, net2, net3;

//instantiate the logic for flip-flop y[1]
d_ff_bh inst4 (
  .rst_n(rst_n),
  .clk(clk),
  .d(~y[1]),
  .q(y[1])
  );

//instantiate the logic for flip-flop y[2]
and2_df inst1 (
  .x1(~y[1]),
  .x2(~y[2]),
  .z1(net1)
  );
                              //continued on next page
```

743

```
and2_df inst2 (
  .x1(y[1]),
  .x2(y[2]),
  .z1(net2)
  );

nor2_df inst3 (
  .x1(net1),
  .x2(net2),
  .z1(net3)
  );

d_ff_bh inst5 (
  .rst_n(rst_n),
  .clk(clk),
  .d(net3),
  .q(y[2])
  );

endmodule
```

```
//test bench for ctr_struc2

module ctr_struc2_tb;

reg rst_n, clk;
wire [1:2] y;

//display outputs
initial
$monitor ("count = %b", y);

//define reset
initial
begin
  #0  rst_n = 1'b0;
  #5  rst_n = 1'b1;
end

//define clock
initial
begin
  clk = 1'b0;
  forever
    #10 clk = ~clk;
end
                              //continued on next page
```

```
//define length of simulation
initial
begin
  #60 $finish;
end

//instantiate the module into the test bench
ctr_struc2 inst1 (
  .rst_n(rst_n),
  .clk(clk),
  .y(y)
  );

endmodule
```

```
count = 00
count = 10
count = 01
count = 11
count = 00
```

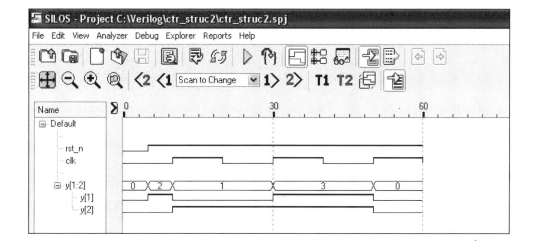

2.5   Design a modulo-11 counter with no self-starting state using structural modeling with built-in primitives and D flip-flops that were designed using behavioral modeling. Obtain the design module, the test bench module, the outputs, and the waveforms.

y_1	y_2	y_3	y_4
0	0	0	0
0	0	0	1
0	0	1	0
0	0	1	1
0	1	0	0
0	1	0	1
0	1	1	0
0	1	1	1
1	0	0	0
1	0	0	1
1	0	1	0
0	0	0	0

$y_3 y_4$

$y_1 y_2$	0 0	0 1	1 1	1 0
0 0	0 ^{0}	0 ^{1}	0 ^{3}	0 ^{2}
0 1	0 ^{4}	0 ^{5}	1 ^{7}	0 ^{6}
1 1	– ^{12}	– ^{13}	– ^{15}	– ^{14}
1 0	1 ^{8}	1 ^{9}	– ^{11}	0 ^{10}

y_1

$y_3 y_4$

$y_1 y_2$	0 0	0 1	1 1	1 0
0 0	0 ^{0}	0 ^{1}	1 ^{3}	0 ^{2}
0 1	1 ^{4}	1 ^{5}	0 ^{7}	1 ^{6}
1 1	– ^{12}	– ^{13}	– ^{15}	– ^{14}
1 0	0 ^{8}	0 ^{9}	– ^{11}	0 ^{10}

y_2

$Dy_1 = y_1 y_3' + y_2 y_3 y_4$

$Dy_2 = y_2 y_3' + y_2' y_3 y_4 + y_2 y_4'$

y_3y_4

y_1y_2	0 0	0 1	1 1	1 0
0 0	0 [0]	1 [1]	0 [3]	1 [2]
0 1	0 [4]	1 [5]	0 [7]	1 [6]
1 1	– [12]	– [13]	– [15]	– [14]
1 0	0 [8]	1 [9]	– [11]	0 [10]

y_3

y_3y_4

y_1y_2	0 0	0 1	1 1	1 0
0 0	1 [0]	0 [1]	0 [3]	1 [2]
0 1	1 [4]	0 [5]	0 [7]	1 [6]
1 1	– [12]	– [13]	– [15]	– [14]
1 0	1 [8]	0 [9]	– [11]	0 [10]

y_4

$$Dy_3 = y_3'y_4 + y_1'y_3y_4'$$

$$Dy_4 = y_3'y_4' + y_1'y_4'$$
$$= y_4'(y_1' + y_3')$$

```
//structural modulo-11 counter

module ctr_mod11_struc_d (rst_n, clk, y);

input rst_n, clk;
output [1:4] y;

//define internal nets
wire net1, net2, ner3, net4, net5, net6, net7, net8, net9,
     net10, net11, net12;

//-------------------------------------------
//instantiate the logic for flip-flop y[1]
and (net1, y[1], ~y[3]);
and (net2, y[2], y[3], y[4]);
or  (net3, net1, net2);

d_ff_bh inst1 (
 .rst_n(rst_n),
 .clk(clk),
 ..d(net3),
 .q(y[1])
 );

                              //continued on next page
```

```
//---------------------------------------------
//instantiate the logic for flip-flop y[2]
and  (net4, y[2], ~y[3]);
and  (net5, ~y[2], y[3], y[4]);
and  (net6, y[2], ~y[4]);
or   (net7, net4, net5, net6);

d_ff_bh inst2 (
  .rst_n(rst_n),
  .clk(clk),
  .d(net7),
  .q(y[2])
  );

//---------------------------------------------
//instantiate the logic for flip-flop y[3]
and  (net8, ~y[3], y[4]);
and  (net9, ~y[1], y[3], ~y[4]);
or   (net10, net8, net9);

d_ff_bh inst3 (
  .rst_n(rst_n),
  .clk(clk),
  .d(net10),
  .q(y[3])
  );

//---------------------------------------------
//instantiate the logic for flip-flop y[4]
or   (net11, ~y[1], ~y[3]);
and  (net12, ~y[4], net11);

d_ff_bh inst4 (
  .rst_n(rst_n),
  .clk(clk),
  .d(net12),
  .q(y[4])
  );

endmodule
```

```verilog
//test bench for modulo-11 counter
module ctr_mod11_struc_d_tb;

//define inputs and outputs
reg rst_n, clk;
wire [1:4] y;

//display outputs
initial
$monitor ("count = %b", y);

//define reset
initial
begin
  #0 rst_n = 1'b0;
  #5 rst_n = 1'b1;
end

//define clock
initial
begin
  clk = 1'b0;
  forever
    #10 clk = ~clk;
end

//define length of simulation
initial
begin
  #200 $finish;
end

//instantiate the module into the test bench
ctr_mod11_struc_d inst1 (
  .rst_n(rst_n),
  .clk(clk),
  .y(y)
  );

endmodule
```

```
count = 0000        count = 0110
count = 0001        count = 0111
count = 0010        count = 1000
count = 0011        count = 1001
count = 0100        count = 1010
count = 0101        count = 0000
```

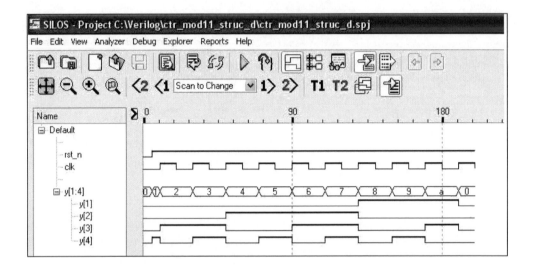

2.9   Obtain the input equations for flip-flops y_1 and y_4 only, for a BCD counter which counts in the sequence shown below. The equations are to be in minimum form. Use JK flip-flops. There is no self-starting state.

$y_1y_2y_3y_4 = 0000, 0001, 0010, 0011, 0100, 0101, 0110, 0111, 1000, 1001, 0000, \cdots .$

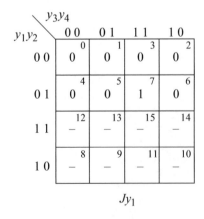

$Jy_1 = y_2y_3y_4$

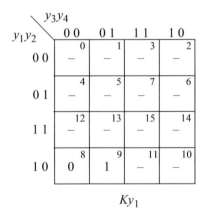

$Ky_1 = y_4$

Karnaugh map Jy_4:

y_1y_2 \ y_3y_4	0 0	0 1	1 1	1 0
0 0	1 (0)	— (1)	— (3)	1 (2)
0 1	1 (4)	— (5)	— (7)	1 (6)
1 1	— (12)	— (13)	— (15)	— (14)
1 0	1 (8)	— (9)	— (11)	— (10)

$$Jy_4$$
$$Jy_4 = 1$$

Karnaugh map Ky_4:

y_1y_2 \ y_3y_4	0 0	0 1	1 1	1 0
0 0	— (0)	1 (1)	1 (3)	— (2)
0 1	— (4)	1 (5)	1 (7)	— (6)
1 1	— (12)	— (13)	— (15)	— (14)
1 0	— (8)	1 (9)	— (11)	— (10)

$$Ky_4$$
$$Ky_4 = 1$$

2 .13   Generate a reduced state diagram for a Moore machine which generates an output z_1 whenever a serial, 4-bit binary word on an input line x_1 is greater than or equal to six. The first bit received in each word is the high-order bit. There is no space between words. Output z_1 is asserted during the fourth bit of a word. Then implement the state diagram in behavioral modeling. Assert output z_1 at time t_2 and deassert z_1 at time t_3. Obtain the design module, the test bench module, the outputs, and the waveforms.

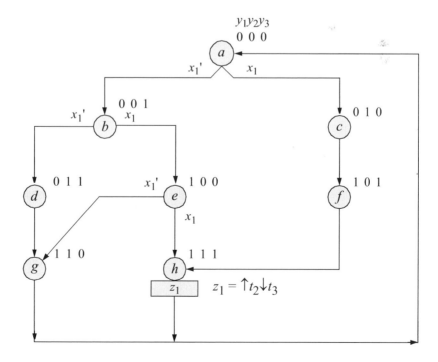

```verilog
//behavioral moore ssm to detect greater/equal to 6
module moore_ssm_ge6 (rst_n, clk, x1, y, z1);

//define inputs and outputs
input rst_n, clk, x1;
output [1:3] y;
output z1;

//variables are reg in always
reg [1:3] y, next_state;
reg z1;

//assign state codes
//parameter defines a constant
parameter    state_a = 3'b000,
             state_b = 3'b001,
             state_c = 3'b010,
             state_d = 3'b011,
             state_e = 3'b100,
             state_f = 3'b101,
             state_g = 3'b110,
             state_h = 3'b111;

//set next state
always @ (posedge clk)
begin
   if (~rst_n)              //if rst_n = 0 (~rst_n is true)
     y <= state_a;          //go to state_a (000)
   else
     y <= next_state;       //else go to next_state
end

//determine output
always @ (y or clk)
begin
   if (y == state_h)
     begin
       if (~clk)
         z1 = 1'b1;
       else
         z1 = 1'b0;
     end

   else
     z1 = 1'b0;
end

                                    //continued on next page
```

```verilog
//determine next state
always @ (x1 or y)
begin
   case (y)//case is a multiway conditional branch
      state_a://if y = state_a, do if ... else
         if (~x1)
            next_state = state_b;
         else
            next_state = state_c;

      state_b:
         if (~x1)
            next_state = state_d;
         else
            next_state = state_e;

      state_c: next_state = state_f;

      state_d: next_state = state_g;

      state_e:
         if (~x1)
            next_state = state_g;
         else
            next_state = state_h;

      state_f: next_state = state_h;

      state_g: next_state = state_a;

      state_h: next_state = state_a;

      default: next_state = state_a;
   endcase
end

endmodule
```

```
//test bench for moore_ssm_ge6

module moore_ssm_ge6_tb;

reg rst_n, clk, x1;          //inputs are reg for test bench
wire [1:3] y;                //outputs are wire for test bench
wire z1;

//display variables
initial
$monitor ("x1 = %b, state = %b, z1 = %b", x1, y, z1);

//define clock
initial
begin
   clk = 1'b0;
   forever
      #10 clk = ~clk;
end

//define input sequence
initial
begin
   #0   rst_n = 1'b0;        //reset to state_a (000)
   x1 = 1'b0;

   #5   rst_n = 1'b1;        //deassert reset
//--------------------------------------------------------
               @ (posedge clk)    //go to state_a (000)
   x1 = 1'b0; @ (posedge clk)    //go to state_b (001)
   x1 = 1'b0; @ (posedge clk)    //go to state_d (011)
   x1 = $random; @ (posedge clk) //go to state_g (110)
   x1 = $random; @ (posedge clk) //go to state_a (000)

                            //continued on next page
```

```
//---------------------------------------------------------
   x1 = 1'b0; @ (posedge clk)      //go to state_b (001)
   x1 = 1'b1; @ (posedge clk)      //go to state_e (100)
   x1 = 1'b1; @ (posedge clk)      //go to state_h (111)
                                   //assert z1
   x1 = $random; @ (posedge clk) //go to state_a (000)

//---------------------------------------------------------
   x1 = 1'b0; @ (posedge clk)      //go to state_b (001)
   x1 = 1'b1; @ (posedge clk)      //go to state_e (100)
   x1 = 1'b0; @ (posedge clk)      //go to state_g (110)
   x1 = $random; @ (posedge clk) //go to state_a (000)

//---------------------------------------------------------
   x1 = 1'b1; @ (posedge clk)      //go to state_c (010)
   x1 = $random; @ (posedge clk) //go to state_f (101)
   x1 = $random; @ (posedge clk) //go to state_h (111)
                                   //assert z1
   x1 = $random; @ (posedge clk) //go to state_a (000)

//---------------------------------------------------------
   #20   $stop;

end

//---------------------------------------------------------
//instantiate the module into the test bench
moore_ssm_ge6 inst1 (
 .rst_n(rst_n),
 .clk(clk),
 .x1(x1),
 .y(y),
 .z1(z1)
 );

endmodule
```

```
x1 = 0, state = xxx, z1 = 0
x1 = 0, state = 000, z1 = 0
x1 = 0, state = 001, z1 = 0
x1 = 0, state = 011, z1 = 0
x1 = 1, state = 110, z1 = 0

x1 = 0, state = 000, z1 = 0
x1 = 1, state = 001, z1 = 0
x1 = 1, state = 100, z1 = 0
x1 = 1, state = 111, z1 = 0
x1 = 1, state = 111, z1 = 1

x1 = 0, state = 000, z1 = 0
x1 = 1, state = 001, z1 = 0
x1 = 0, state = 100, z1 = 0
x1 = 1, state = 110, z1 = 0

x1 = 1, state = 000, z1 = 0
x1 = 1, state = 010, z1 = 0
x1 = 1, state = 101, z1 = 0
x1 = 1, state = 111, z1 = 0
x1 = 1, state = 111, z1 = 1

x1 = 1, state = 000, z1 = 0
```

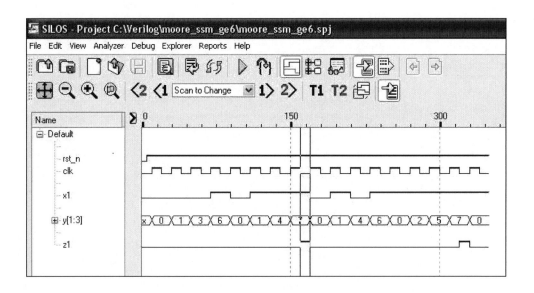

2.19   Generate a reduced state diagram for a Mealy machine which detects a 4-bit word of 1001 on a serial input line x_1. If a correct sequence is detected, then a conditional output z_1 is generated. There is no spacing between words. There is also no overlapping of words. Assert z_1 from time t_2 to time t_3. Obtain the behavioral design module, the test bench module, the outputs, and the waveforms.

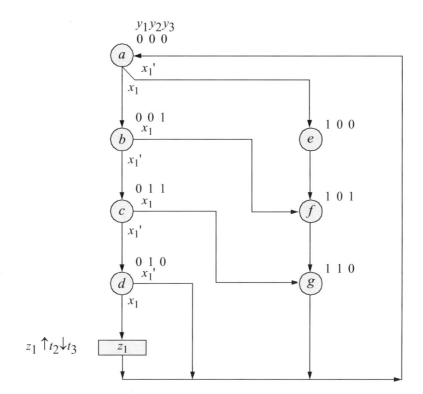

```verilog
//behavioral detect 1001
module detect_1001_bh (rst_n, clk, x1, y, z1);

//define inputs and output
input rst_n, clk, x1;
output [1:3] y;
output z1;

reg [1:3] y, next_state;     //variables are reg in always
wire z1;

//assign state codes, parameter defines a constant
parameter  state_a = 3'b000,
           state_b = 3'b001,
           state_c = 3'b011,
           state_d = 3'b010,
           state_e = 3'b100,
           state_f = 3'b101,
           state_g = 3'b110;

//set next state
always @ (posedge clk)
begin
   if (~rst_n)          //reset = 1'b0
      y <= state_a;     //y <= state_a (000)
   else
      y = next_state;
end

//determine output
assign z1 = ((~y[1]) && (y[2]) && (~y[3]) && x1 && ~clk);

//determine next state
always @ (y or x1)
begin
   case (y)
      state_a:
         if (x1)
            next_state = state_b;
         else
            next_state = state_e;

      state_b:
         if (~x1)
            next_state = state_c;
         else
            next_state = state_f;
                              //continued on next page
```

```
      state_c:
         if (~x1)
            next_state = state_d;
         else
            next_state = state_g;

      state_d:
         if (x1)
            next_state = state_a;
         else
            next_state = state_a;

      state_e: next_state = state_f;

      state_f: next_state = state_g;

      state_g: next_state = state_a;

      default: next_state =  state_a;

   endcase
end
endmodule
```

```
//test bench to detect 1001

module detect_1001_bh_tb;

reg rst_n, clk, x1;      //inputs are reg for test bench
wire [1:3] y;            //outputs are wire for test bench
wire z1;


//display variables
initial
$monitor ("x1 = %b,  state = %b,  z1 = %b",  x1,  y,  z1);


//define clock
initial
begin
   clk = 1'b0;
   forever
      #10 clk = ~clk;
end

                                    //continued on next page
```

```
//define input sequence
initial
begin
   #0  rst_n = 1'b0;                    //reset to state_a (000)
       x1 = 1'b0;
   #5  rst_n = 1'b1;                    //deassert reset
//-----------------------------------------------------------
                @ (posedge clk)     //go to state_a (000)
   x1 = 1'b1; @ (posedge clk)       //go to state_b (001)
   x1 = 1'b0; @ (posedge clk)       //go to state_c (011)
   x1 = 1'b0; @ (posedge clk)       //go to state_d (010)
   x1 = 1'b1; @ (posedge clk)       //go to state_a (000)
                                    //assert z1 at t2
//-----------------------------------------------------------
   x1 = 1'b0; @ (posedge clk)       //go to state_e (100)
   x1 = $random; @ (posedge clk)    //go to state_f (101)
   x1 = $random; @ (posedge clk)    //go to state_g (110)
   x1 = $random; @ (posedge clk)    //go to state_a (000)

   #20  $stop;
end

//instantiate the module into the test bench
detect_1001_bh inst1 (
  .rst_n(rst_n),
  .clk(clk),
  .x1(x1),
  .y(y),
  .z1(z1)
  );

endmodule
```

```
x1 = 0,  state = xxx,  z1 = 0
x1 = 1,  state = 000,  z1 = 0
x1 = 0,  state = 001,  z1 = 0
x1 = 0,  state = 011,  z1 = 0
x1 = 1,  state = 010,  z1 = 0
x1 = 1,  state = 010,  z1 = 1

x1 = 0,  state = 000,  z1 = 0
x1 = 0,  state = 100,  z1 = 0
x1 = 1,  state = 101,  z1 = 0
x1 = 1,  state = 110,  z1 = 0
x1 = 1,  state = 000,  z1 = 0
```

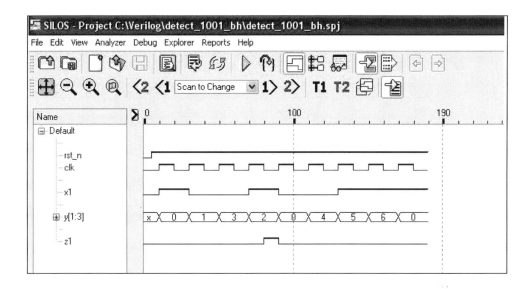

2.23   Select state codes for states d and e for the Moore machine shown below so that there will be no output glitches. Consider all state transitions. Then design a structural module for the Moore machine using built-in primitives and D flip-flops. Obtain the test bench module, the outputs, and the waveforms.

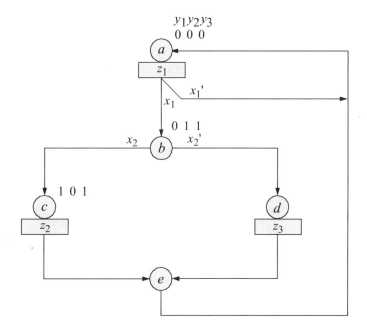

State d: $y_1 y_2 y_3 = 110$. State e: $y_1 y_2 y_3 = 100$.
Unused states: $y_1 y_2 y_3 = 001, 010, 111$.

Dy_1

y_1 \ y_2y_3	0 0	0 1	1 1	1 0
0	0 (0)	− (1)	1 (3)	− (2)
1	0 (4)	1 (5)	− (7)	1 (6)

$$Dy_1 = y_3 + y_2$$

Dy_2

y_1 \ y_2y_3	0 0	0 1	1 1	1 0
0	x_1 (0)	− (1)	x_2' (3)	− (2)
1	0 (4)	0 (5)	− (7)	0 (6)

$$Dy_2 = y_1'y_2'x_1 + y_2y_3x_2'$$

Dy_3

y_1 \ y_2y_3	0 0	0 1	1 1	1 0
0	x_1 (0)	− (1)	x_2 (3)	− (2)
1	0 (4)	0 (5)	− (7)	0 (6)

$$Dy_3 = y_1'y_2'x_1 + y_2y_3x_2$$

```verilog
//structural moore with no glitches
module moore_no_glitch_d (rst_n, clk, x1, x2, y, z1, z2, z3);

//define inputs and outputs
input rst_n, clk, x1, x2;
output [1:3] y;
output z1, z2, z3;

//define internal nets
wire net1, net2, net3, net4, net5, net6, net7;

//-----------------------------------------------
//instantiate the logic for flip-flop y[1]
or (net1, y[3], y[2]);

d_ff_bh inst1 (
  .rst_n(rst_n),
  .clk(clk),
  .d(net1),
  .q(y[1])
  );
```
//continued on next page

```
//-------------------------------------------------
//instantiate the logic for flip-flop y[2]
and  (net2, ~y[1], ~y[2], x1);
and  (net3, y[2], y[3], ~x2);
or   (net4, net2, net3);

d_ff_bh inst2 (
  .rst_n(rst_n),
  .clk(clk),
  .d(net4),
  .q(y[2])
  );

//-------------------------------------------------
//instantiate the logic for flip-flop y[3]
and  (net5, ~y[1], ~y[2], x1);
and  (net6, y[2], y[3], x2);
or   (net7, net5, net6);

d_ff_bh inst3 (
  .rst_n(rst_n),
  .clk(clk),
  .d(net7),
  .q(y[3])
  );

//define the outputs
assign  z1 = ~y[1] && ~y[2] && ~y[3];
assign  z2 = y[1] && ~y[2] && y[3];
assign  z3 = y[1] && y[2] && ~y[3];

endmodule
```

```
//test bench for moore no glitch machine

module moore_no_glitch_d_tb;

reg rst_n, clk, x1, x2;    //inputs are reg for test bench
wire [1:3] y;              //outputs are wire for test bench
wire z1, z2, z3;

//display variables
initial
$monitor ("x1 x1 = %b, state = %b, z1 = %b, z2 = %b, z3 = %b",
          {x1, x2}, y, z1, z2, z3);
                                    //continued on next page
```

```verilog
//define clock
initial
begin
   clk = 1'b0;
   forever
      #10 clk = ~clk;
end

//define input sequence
initial
begin
   #0 rst_n = 1'b0;              //reset to state_a (00)
      x1 = 1'b0;
      x2 = 1'b0;

   #5 rst_n = 1'b1;             //deassert reset
//--------------------------------------------------------
   x1 = 1'b0; x2 = $random;@ (posedge clk)    //assert z1
   x1 = 1'b1; x2 = $random;@ (posedge clk)    //go to state_b
   x1 = $random; x2 = 1'b1; @ (posedge clk)   //go to state_c
                                              //assert z2
   x1 = $random; x2 = $random; @ (posedge clk) //go to state_e
   x1 = $random; x2 = $random; @ (posedge clk) //go to state_a
                                              //assert z1
//--------------------------------------------------------
   x1 = 1'b1; x2 = $random; @ (posedge clk)   //go to state_b
   x1 = $random; x2 = 1'b0; @ (posedge clk)   //go to state_d
                                              //assert z3
   x1 = $random; x2 = $random; @ (posedge clk) //go to state_a
//--------------------------------------------------------

   #10 $stop;
end

//instantiate the module into the test bench
moore_no_glitch_d inst1 (
  .rst_n(rst_n),
  .clk(clk),
  .x1(x1),
  .x2(x2),
  .y(y),
  .z1(z1),
  .z2(z2),
  .z3(z3)
  );

endmodule
```

```
x1 x2 = 00, state = 000, z1 = 1, z2 = 0, z3 = 0
x1 x2 = 11, state = 000, z1 = 1, z2 = 0, z3 = 0

x1 x2 = 11, state = 011, z1 = 0, z2 = 0, z3 = 0
x1 x2 = 11, state = 101, z1 = 0, z2 = 1, z3 = 0

x1 x2 = 11, state = 100, z1 = 0, z2 = 0, z3 = 0
x1 x2 = 10, state = 000, z1 = 1, z2 = 0, z3 = 0
x1 x2 = 10, state = 011, z1 = 0, z2 = 0, z3 = 0
x1 x2 = 10, state = 110, z1 = 0, z2 = 0, z3 = 1

x1 x2 = 10, state = 100, z1 = 0, z2 = 0, z3 = 0
```

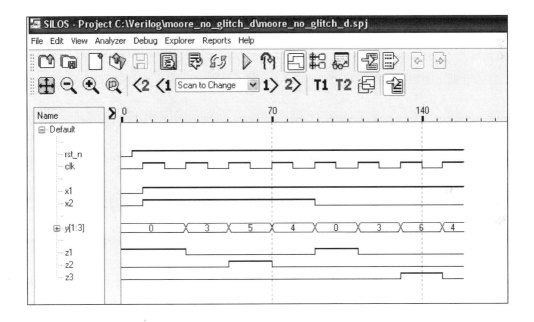

Chapter 3   Synthesis of Synchronous Sequential Machines 2 Using Verilog HDL

3.3   The state diagram for a Moore synchronous sequential machine is shown below. Implement the machine using linear-select multiplexers for the δ next-state logic, D flip-flops for the storage elements, and any additional logic for the λ output logic. Use x_1 and x_2 as map-entered variables. Obtain the structural design module, the test bench module, the outputs, and the waveforms.

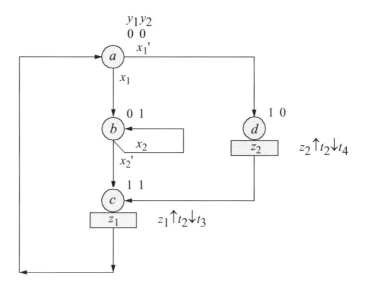

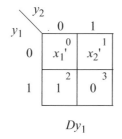

Dy_1

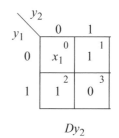

Dy_2

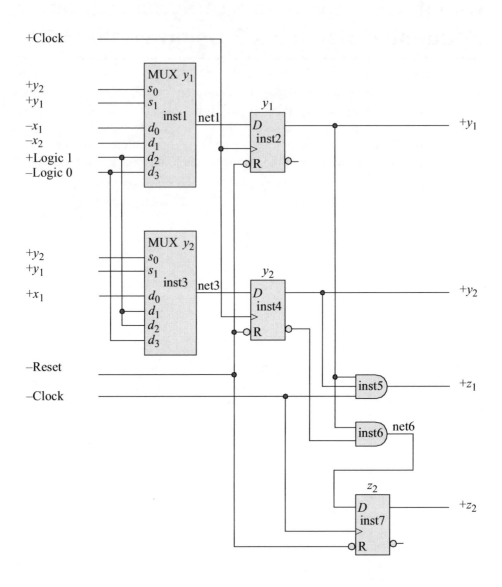

```verilog
//structural moore using multiplexers
module moore_mux_linear (rst_n, clk, x1, x2, y, z1, z2);

input rst_n, clk, x1, x2;       //define inputs and outputs
output [1:2] y;
output z1, z2;

wire net1, net3;                //define internal nets
//-----------------------------------------
//instantiate the logic for flip-flop y[1]
mux_4to1_struc inst1 (
   .d({1'b0, 1'b1, ~x2, ~x1}),
   .s({y[1], y[2]}),
   .z1(net1)
   );

d_ff_bh insr2 (
   .rst_n(rst_n),
   .clk(clk),
   .d(net1),
   .q(y[1])
   );
//-----------------------------------------
//instantiate the logic for flip-flop y[2]
mux_4to1_struc inst3 (
   .d({1'b0, 1'b1, 1'b1, x1}),
   .s({y[1], y[2]}),
   .z1(net3)
   );

d_ff_bh insr4 (
   .rst_n(rst_n),
   .clk(clk),
   .d(net3),
   .q(y[2])
   );
//-----------------------------------------
//instantiate the logic for output z1
and inst5 (z1, y[1], y[2], ~clk);
//-----------------------------------------
//instantiate the logic for output z2
and inst6 (net6, y[1], ~y[2]);
d_ff_bh insr7 (
   .rst_n(rst_n),
   .clk(~clk),
   .d(net6),
   .q(z2)
   );
endmodule
```

```verilog
//test bench for the moore machine using multiplexers
module moore_mux_linear_tb;

reg rst_n, clk, x1, x2;     //inputs are reg for test bench
wire [1:2] y;               //outputs are wire for test bench
wire z1, z2;

initial                     //display variables
$monitor ("x1 x2 = %b, state = %b, z1 z2 = %b",
           {x1, x2}, y, {z1, z2});

initial                     //define clock
begin
   clk = 1'b0;
   forever
      #10 clk = ~clk;
end

//define input sequence
initial
begin
   #0 rst_n = 1'b0;         //reset to state_a (00)
      x1 = 1'b1; x2 = 1'b0;

   #5 rst_n = 1'b1;

//-----------------------------------------------------------
   x1 = 1'b1;   x2 = 1'b1;@ (posedge clk)//go to state_b (01)
   x1 = 1'b1;   x2 = 1'b1;@ (posedge clk)//go to state_b (01)
   x1 = 1'b1;   x2 = 1'b0;@ (posedge clk)//go to state_c (11)
                                         //assert z1, t2 -- t3
   x1 = 1'b0;   x2 = 1'b0;@ (posedge clk)//go to state_a (00)

//-----------------------------------------------------------
   x1 = 1'b0;   x2 = 1'b0;@ (posedge clk)//go to state_d (10)
                                         //assert z2, t2 -- t4
   x1 = 1'b0;   x2 = 1'b0;@ (posedge clk)//go to state_c (11)
                                         //assert z1, t2 -- t3
   x1 = 1'b0;   x2 = 1'b0;@ (posedge clk)//go to state_a (00)

//-----------------------------------------------------------
   #11    $stop;
end

//-----------------------------------------------------------
//instantiate the module into the test bench
moore_mux_linear inst1 (rst_n, clk, x1, x2, y, z1, z2);

endmodule
```

```
x1 x2 = 10, state = 00, z1 z2 = 00
x1 x2 = 11, state = 01, z1 z2 = 00
x1 x2 = 10, state = 01, z1 z2 = 00
x1 x2 = 00, state = 11, z1 z2 = 00
x1 x2 = 00, state = 11, z1 z2 = 10
x1 x2 = 00, state = 00, z1 z2 = 00
x1 x2 = 00, state = 10, z1 z2 = 00
x1 x2 = 00, state = 10, z1 z2 = 01
x1 x2 = 00, state = 11, z1 z2 = 01
x1 x2 = 00, state = 11, z1 z2 = 10
x1 x2 = 00, state = 00, z1 z2 = 00
```

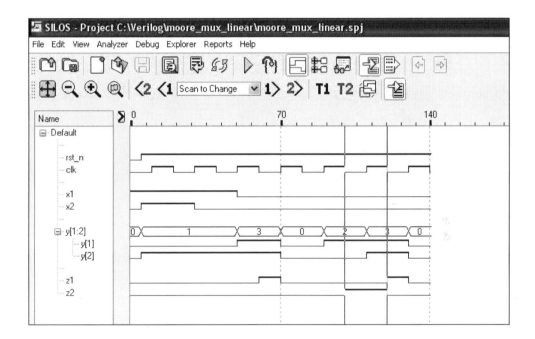

3.7   Given the state diagram shown below for a Moore machine, implement the design using nonlinear-select multiplexers for the δ next-state logic, D flip-flops for the storage elements, and continuous assignment statements for the λ output logic. Outputs z_1, z_2, and z_3 are asserted at time t_2 and deasserted at t_3. Obtain the structural design module, the test bench module, the outputs, and the waveforms.

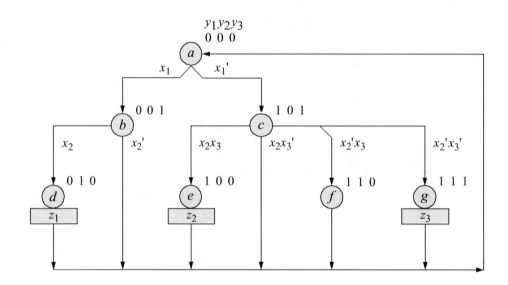

y_1 \ y_2y_3	0 0	0 1	1 1	1 0
0	x_1' 0	0 1	– 3	0 2
1	0 4	$x_1' + x_3$ 5	0 7	0 6

Dy_1

y_2 \ y_1y_3	0 0	0 1	1 1	1 0
0	x_1' 0	0 1	$x_1' + x_3$ 5	0 4
1	0 2	– 3	0 7	0 6
	d_0	d_1	d_3	d_2

Dy_1

Permuted map for Dy_1

y_1 \ y_2y_3	0 0	0 1	1 1	1 0
0	0 [0]	x_2 [1]	$-$ [3]	0 [2]
1	0 [4]	x_2' [5]	0 [7]	0 [6]

Dy_2

y_2 \ y_1y_3	0 0	0 1	1 1	1 0
0	0 [0]	x_2 [1]	x_2' [5]	0 [4]
1	0 [2]	$-$ [3]	0 [7]	0 [6]
	d_0	d_1	d_3	d_2

Dy_2 Permuted map for Dy_2

y_1 \ y_2y_3	0 0	0 1	1 1	1 0
0	1 [0]	0 [1]	$-$ [3]	0 [2]
1	0 [4]	$x_2'x_3'$ [5]	0 [7]	0 [6]

Dy_3

y_2 \ y_1y_3	0 0	0 1	1 1	1 0
0	1 [0]	0 [1]	$x_2'x_3'$ [5]	0 [4]
1	0 [2]	$-$ [3]	0 [7]	0 [6]
	d_0	d_1	d_3	d_2

Dy_3 Permuted map for Dy_3

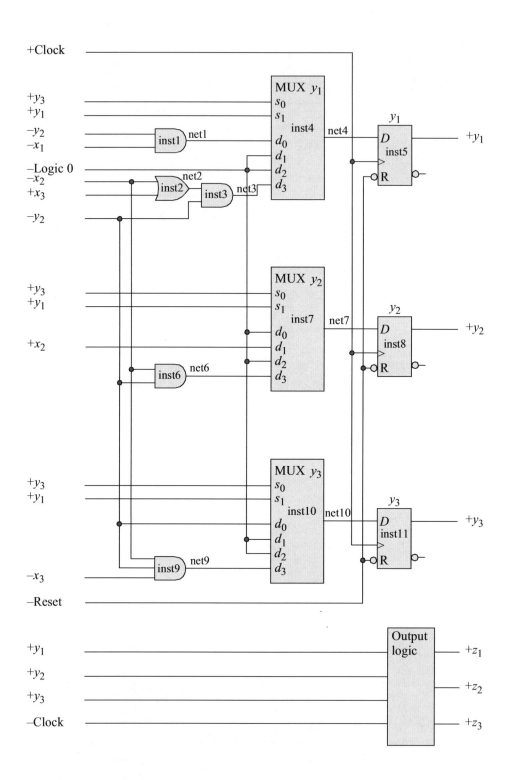

```
//structural moore nonlinear multiplexer with three outputs

module mux_nonlinear10 (rst_n, clk, x, y, z1, z2, z3);

//define inputs and outputs
input rst_n, clk;
input [1:3] x;
output [1:3] y;
output z1, z2, z3;

//define internal nets
wire net1, net2, net3, net4, net6, net7, net9, net10;

//-------------------------------------------
//instantiate the logic for flip-flop y[1]
and inst1 (net1, ~y[2], ~x[1]);
or  inst2 (net2, ~x[2], x[3]);
and inst3 (net3, net2, ~y[2]);

mux_4to1_struc inst4 (
    .d({net3, 1'b0, 1'b0, net1}),
    .s({y[1], y[3]}),
    .z1(net4)
    );

d_ff_bh inst5 (
    .rst_n(rst_n),
    .clk(clk),
    .d(net4),
    .q(y[1])
    );

//-------------------------------------------
//instantiate the logic for flip-flop y[2
and inst6 (net6, ~y[2], ~x[2]);

mux_4to1_struc inst7 (
    .d({net6, 1'b0, x[2], 1'b0}),
    .s({y[1], y[3]}),
    .z1(net7)
    );

d_ff_bh inst8 (
    .rst_n(rst_n),
    .clk(clk),
    .d(net7),
    .q(y[2])
    );
```
                                    //continued on next page

```
//-------------------------------------------------
//instantiate the logic for flip-flop y[3]
and inst9 (net9, ~y[2], ~x[2], ~x[3]);

mux_4to1_struc inst10 (
    .d({net9, 1'b0, 1'b0, ~y[2]}),
    .s({y[1], y[3]}),
    .z1(net10)
    );

d_ff_bh inst11 (
    .rst_n(rst_n),
    .clk(clk),
    .d(net10),
    .q(y[3])
    );

//-------------------------------------------------
//instantiate the logic for the outputs
assign z1 = (~y[1] & y[2] & ~y[3] & ~ clk);
assign z2 = (y[1] & ~y[2] & ~y[3] & ~clk);
assign z3 = (y[1] & y[2] & y[3] & ~clk);

endmodule
```

```
//test bench for moore nonlineat multiplexer with 3 outputs
module mux_nonlinear10_tb;

//inputs are reg for test bench
//outputs are wire for test bench
reg rst_n, clk;
reg [1:3] x;
wire [1:3] y;
wire z1, z2, z3;

//display variables
initial
$monitor ("inputs = %b, state = %b, z1 z2 z3 = %b",
            x, y, {z1, z2, z3});

//define clock
initial
begin
    clk = 1'b0;
    forever
        #10 clk = ~clk;
end                                  //continued on next page
```

```
//define input sequence
initial
begin
   #0  rst_n = 1'b0;
       x = 3'b100;

   #5  rst_n = 1'b1;

//----------------------------------------------------------

   x = 3'b010;  @ (posedge clk)   //go to state_d (010)
                                  //assert z1
   x = 3'b000;  @ (posedge clk)   //go to state_a (000)

//----------------------------------------------------------

   x = 3'b011;  @ (posedge clk)   //go to state_c (101)
   x = 3'b011;  @ (posedge clk)   //go to state_e (100)
                                  //assert z2
   x = 3'b000;  @ (posedge clk)   //go to state_a (000)

//----------------------------------------------------------

   x = 3'b011;  @ (posedge clk)   //go to state_c (101)
   x = 3'b100;  @ (posedge clk)   //go to state_g (111)
                                  //assert z3
   x = 3'b000;  @ (posedge clk)   //go to state_a (000)

//----------------------------------------------------------

   x = 3'b011;  @ (posedge clk)   //go to state_c (101)
   x = 3'b101;  @ (posedge clk)   //go to state_f (110)
   x = 3'b000;  @ (posedge clk)   //go to state_a (000)

//----------------------------------------------------------
   #10    $stop;

end

//----------------------------------------------------------
//instantiate the module into the test bench
mux_nonlinear10 inst1 (rst_n, clk, x, y, z1, z2, z3);

endmodule
```

```
inputs = 100, state = 000, z1 z2 z3 = 000
inputs = 010, state = 001, z1 z2 z3 = 000
inputs = 000, state = 010, z1 z2 z3 = 000
inputs = 000, state = 010, z1 z2 z3 = 100


inputs = 011, state = 000, z1 z2 z3 = 000
inputs = 011, state = 101, z1 z2 z3 = 000
inputs = 000, state = 100, z1 z2 z3 = 000
inputs = 000, state = 100, z1 z2 z3 = 010


inputs = 011, state = 000, z1 z2 z3 = 000
inputs = 100, state = 101, z1 z2 z3 = 000
inputs = 000, state = 111, z1 z2 z3 = 000
inputs = 000, state = 111, z1 z2 z3 = 001


inputs = 011, state = 000, z1 z2 z3 = 000
inputs = 101, state = 101, z1 z2 z3 = 000
inputs = 000, state = 110, z1 z2 z3 = 000
inputs = 000, state = 000, z1 z2 z3 = 000
```

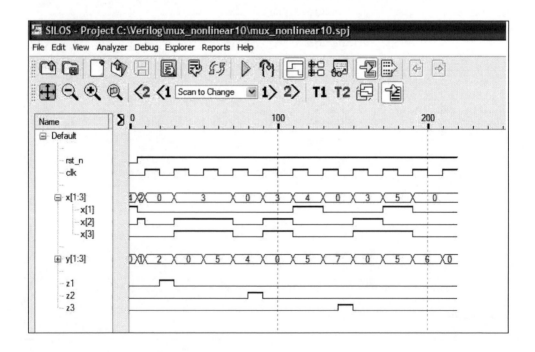

3.11   Design a 4-bit Johnson counter using D flip-flops. The counter counts in the following sequence: 0000, 1000, 1100, 1110, 1111, 0111, 0011, 0001, 0000. Obtain the structural design module, the test bench module, the outputs, and the waveforms.

y_1y_2 \ y_3y_4	0 0	0 1	1 1	1 0
0 0	1 (0)	0 (1)	0 (3)	– (2)
0 1	– (4)	– (5)	0 (7)	– (6)
1 1	1 (12)	– (13)	0 (15)	1 (14)
1 0	1 (8)	– (9)	– (11)	– (10)

Dy_1

$$Dy_1 = y_4{'}$$

y_1y_2 \ y_3y_4	0 0	0 1	1 1	1 0
0 0	0 (0)	0 (1)	0 (3)	– (2)
0 1	– (4)	– (5)	0 (7)	– (6)
1 1	1 (12)	– (13)	1 (15)	1 (14)
1 0	1 (8)	– (9)	– (11)	– (10)

Dy_2

$$Dy_2 = y_1$$

y_1y_2 \ y_3y_4	0 0	0 1	1 1	1 0
0 0	0 (0)	0 (1)	0 (3)	– (2)
0 1	– (4)	– (5)	1 (7)	– (6)
1 1	1 (12)	– (13)	1 (15)	1 (14)
1 0	0 (8)	– (9)	– (11)	– (10)

Dy_3

$$Dy_3 = y_2$$

y_1y_2 \ y_3y_4	0 0	0 1	1 1	1 0
0 0	0 (0)	0 (1)	1 (3)	– (2)
0 1	– (4)	– (5)	1 (7)	– (6)
1 1	0 (12)	– (13)	1 (15)	1 (14)
1 0	0 (8)	– (9)	– (11)	– (10)

Dy_4

$$Dy_4 = y_3$$

```
//structural for 4-bit johnson counter
module ctr_johnson4_dff (set_n, rst_n, clk, y);

//define inputs and outputs
input set_n, rst_n, clk;
output [1:4] y;

//instantiate the logic for y[1], y[2], y[3], and y[4]
d_ff_bh inst1 (
    .rst_n(rst_n),
    .clk(clk),
    .d(~y[4]),
    .q(y[1])
    );

d_ff_bh inst2 (
    .rst_n(rst_n),
    .clk(clk),
    .d(y[1]),
    .q(y[2])
    );

d_ff_bh inst3 (
    .rst_n(rst_n),
    .clk(clk),
    .d(y[2]),
    .q(y[3])
    );

d_ff_bh inst4 (
    .rst_n(rst_n),
    .clk(clk),
    .d(y[3]),
    .q(y[4])
    );

endmodule
```

```verilog
//test bench for the 4-bit johnson counter
module ctr_johnson4_dff_tb;

reg set_n, rst_n, clk;       //inputs are reg for test bench
wire [1:4] y;                //outputs are wire for test bench

initial                      //display count
$monitor ("count = %b", y);

//initialize the counter
initial
begin
   #0 rst_n = 1'b0;
   #5 rst_n = 1'b1;
end

//define clock
initial
begin
   clk = 1'b0;
   forever
      #10 clk = ~clk;
end

//define length of simulation
initial
begin
#140 $finish;
end

//instantiate the module into the test bench
ctr_johnson4_dff inst1 (
   .set_n(set_n),
   .rst_n(rst_n),
   .clk(clk),
   .y(y)
   );
endmodule
```

```
count = 0000
count = 1000
count = 1100
count = 1110
count = 1111
count = 0111
count = 0011
count = 0001
count = 0000
```

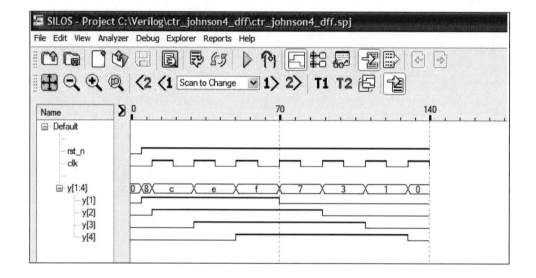

3.14   Design a parity-checked Mealy synchronous sequential machine that generates an output z_1 whenever the sequence 1001 is detected on a serial data input line x_1. Overlapping sequences are valid. Output z_1 is asserted at time t_2 and deasserted at t_3. The parity flip-flop maintains odd parity over the state flip-flops and the parity flip-flop itself. Use built-in primitives for the δ next-state logic and the output logic. This problem repeats Problem 3.13, but uses JK flip-flops as the storage elements. Obtain the structural design module, the test bench module, the outputs, and the waveforms.

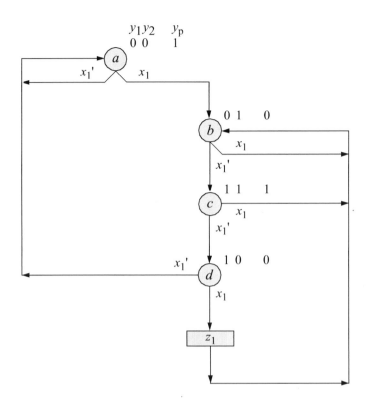

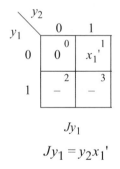

$$Jy_1$$

$$Jy_1 = y_2 x_1'$$

$$Ky_1$$

$$Ky_1 = x_1 + y_2'$$

$$Jy_2$$

$$Jy_2 = x_1$$

$$Ky_2$$

$$Ky_2 = y_1 x_1'$$

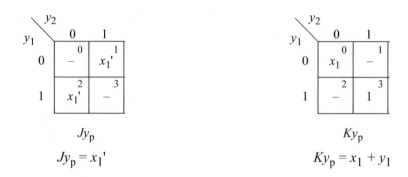

$$Jy_p$$

$$Jy_p = x_1'$$

$$Ky_p$$

$$Ky_p = x_1 + y_1$$

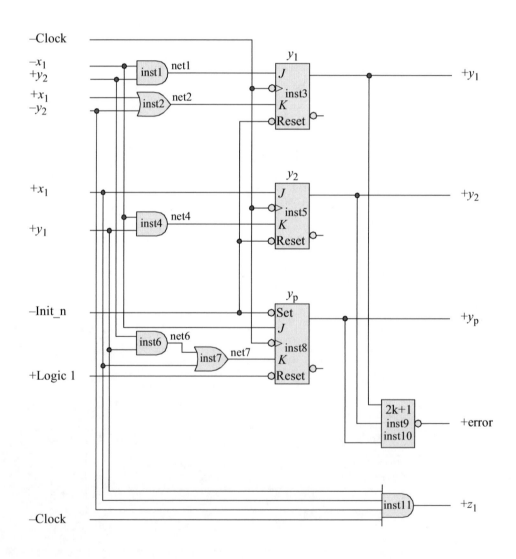

```
//structural parity-checked mealy machine using JK
//flip-flops to detect overlapping sequences of 1001
module mealy_par_chk_jk (init_n, clk, x1, y, yp, z1, err);

input init_n, clk, x1;            //define inputs and outputs
output [1:2] y;
output yp, z1, err;

wire net1, net2, net4, net6, net7;  //define internal nets
//-------------------------------------------------
//instantiate the logic for flip-flop y[1]
and inst1 (net1, ~x1, y[2]);
or  inst2 (net2, x1, ~y[2]);

jkff_neg_clk inst3 (
   .set_n(1'b1),
   .rst_n(init_n),
   .clk(clk),
   .j(net1),
   .k(net2),
   .q(y[1])
   );
//-------------------------------------------------
//instantiate the logic for flip-flop y[2]
and inst4 (net4, ~x1, y[1]);

jkff_neg_clk inst5 (
   .set_n(1'b1),
   .rst_n(init_n),
   .clk(clk),
   .j(x1),
   .k(net4),
   .q(y[2])
   );
//-------------------------------------------------
//instantiate the logic for flip-flop yp
and inst6 (net6, y[1], y[2]);
or  inst7 (net7, net6, x1);

jkff_neg_clk inst8 (
   .set_n(init_n),
   .rst_n(1'b1),
   .clk(clk),
   .j(~x1),
   .k(net7),
   .q(yp)
   );

                                  //continued on next page
```

```
//-------------------------------------------------
//instantiate the logic for the parity error
xor   inst9 (net9, y[1], y[2]);
xnor inst10 (err, net9, yp);

//-------------------------------------------------
//instantiate the logic for output z1
and #1 inst11 (z1, y[1], ~y[2], x1, ~clk);

endmodule
```

```
//test bench for the parity-checked mealy machine
//using JK flip-flops

module mealy_par_chk_jk_tb;

//inputs are reg for test bench
//outputs are wire for test bench
reg   init_n, clk, x1;
wire [1:2] y;
wire yp, z1, err;

//display variables
initial
$monitor ("x1 = %b, state = %b, yp = %b, output = %b,
           parity_err = %b", x1, y, yp, z1, err);

//define clock
initial
begin
   clk = 1'b0;
   forever
      #10 clk = ~clk;
end

//define input sequence
initial
begin
   #0 init_n = 1'b0;        //reset to state_a (00)
      x1 = 1'b0;

   #5 init_n = 1'b1;        //deassert reset

                              //continued on next page
```

```
//--------------------------------------------------------------
   x1 = 1'b0;   @ (negedge clk)    //go to state_a (00)
   x1 = 1'b1;   @ (negedge clk)    //go to state_b (01)
   x1 = 1'b0;   @ (negedge clk)    //go to state_c (11)
   x1 = 1'b0;   @ (negedge clk)    //go to state_d (10)
   x1 = 1'b1;   @ (negedge clk)    //go to state_b (01)
                                   //assert z1 t2 -- t3
//--------------------------------------------------------------
   x1 = 1'b0;   @ (negedge clk)    //go to state_c (11)
   x1 = 1'b0;   @ (negedge clk)    //go to state_d (10)
   x1 = 1'b1;   @ (negedge clk)    //go to state_b (01)
                                   //assert z1 t2 -- t3
//--------------------------------------------------------------
   x1 = 1'b0;   @ (negedge clk)    //go to state_c (11)
   x1 = 1'b0;   @ (negedge clk)    //go to state_d (10)
   x1 = 1'b0;   @ (negedge clk)    //go to state_a (00)
//--------------------------------------------------------------
   x1 = 1'b1;   @ (negedge clk)    //go to state_b (01)
   x1 = 1'b0;   @ (negedge clk)    //go to state_c (11)
   x1 = 1'b1;   @ (negedge clk)    //go to state_b (01)
//--------------------------------------------------------------
   x1 = 1'b0;   @ (negedge clk)    //go to state_c (11)
   x1 = 1'b0;   @ (negedge clk)    //go to state_d (10)
   x1 = 1'b1;   @ (negedge clk)    //go to state_b (01)
                                   //assert z1 t2 -- t3

//--------------------------------------------------------------
   #10    $stop;
end

//--------------------------------------------------------------
//instantiate the module into the test bench
mealy_par_chk_jk inst1 (
   .init_n(init_n),
   .clk(clk),
   .x1(x1),
   .y(y),
   .yp(yp),
   .z1(z1),
   .err(err)
   );

endmodule
```

```
x1 = 0, state = 00, yp = 1, output = 0, parity_err = 0
x1 = 1, state = 00, yp = 1, output = 0, parity_err = 0
x1 = 0, state = 01, yp = 0, output = 0, parity_err = 0
x1 = 0, state = 11, yp = 1, output = 0, parity_err = 0
x1 = 1, state = 10, yp = 0, output = 0, parity_err = 0
x1 = 1, state = 10, yp = 0, output = 1, parity_err = 0
x1 = 1, state = 10, yp = 0, output = 0, parity_err = 0
x1 = 0, state = 01, yp = 0, output = 0, parity_err = 0
x1 = 0, state = 11, yp = 1, output = 0, parity_err = 0
x1 = 1, state = 10, yp = 0, output = 0, parity_err = 0
x1 = 1, state = 10, yp = 0, output = 1, parity_err = 0
x1 = 1, state = 10, yp = 0, output = 0, parity_err = 0
x1 = 0, state = 01, yp = 0, output = 0, parity_err = 0
x1 = 0, state = 11, yp = 1, output = 0, parity_err = 0
x1 = 0, state = 10, yp = 0, output = 0, parity_err = 0
x1 = 1, state = 00, yp = 1, output = 0, parity_err = 0
x1 = 0, state = 01, yp = 0, output = 0, parity_err = 0
x1 = 1, state = 11, yp = 1, output = 0, parity_err = 0
x1 = 0, state = 01, yp = 0, output = 0, parity_err = 0
x1 = 0, state = 11, yp = 1, output = 0, parity_err = 0
x1 = 1, state = 10, yp = 0, output = 0, parity_err = 0
x1 = 1, state = 10, yp = 0, output = 1, parity_err = 0
x1 = 1, state = 10, yp = 0, output = 0, parity_err = 0
x1 = 1, state = 01, yp = 0, output = 0, parity_err = 0
```

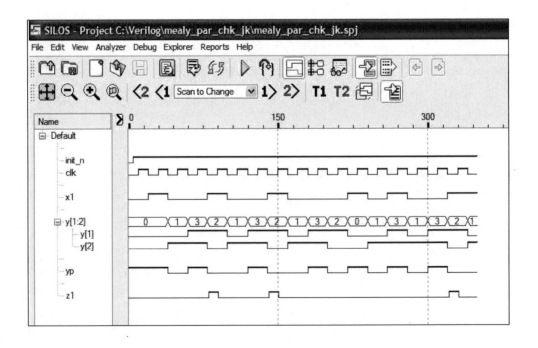

Chapter 4   Synthesis of Asynchronous Sequential Machines Using Verilog HDL

4.2   Synchronize an asynchronous sequential machine, using built-in-primitives, which has two inputs x_1 and x_2 and two outputs z_1 and z_2. The two inputs may overlap, but will not change state simultaneously. Only the following sequences are valid:

$$x_1 x_2 = 00 \rightarrow 10 \rightarrow 11 \rightarrow 01 \rightarrow 00$$
$$x_1 x_2 = 00 \rightarrow 10 \rightarrow 11 \rightarrow 10 \rightarrow 00$$
$$x_1 x_2 = 00 \rightarrow 10 \rightarrow 00$$
$$x_1 x_2 = 00 \rightarrow 01 \rightarrow 00$$

Output z_1 is asserted whenever x_1 is active and x_2 is asserted or when x_2 is active and x_1 is asserted. Output z_1 will be deasserted when either x_1 or x_2 is deasserted. Output z_2 is asserted coincident with the assertion of z_1 and remains active until the deassertion of the last active input of an overlapping sequence. A representative timing diagram is shown below. Use AND gates and OR gates for the logic.

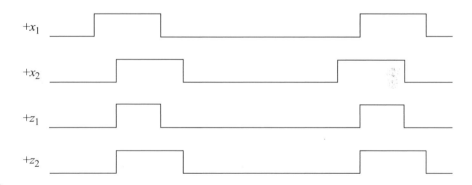

x_1x_2 00	01	11	10	z_1	z_2
ⓐ	e	–	b	0	0
a	–	c	ⓑ	0	0
–	d	ⓒ	–	1	1
a	ⓓ	–	–	0	1
a	ⓔ	f	–	0	0
–	–	ⓕ	g	1	1
a	–	–	ⓖ	0	1

Primitive flow table

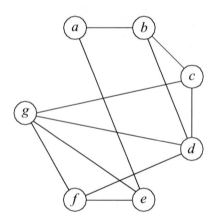

Merger diagram

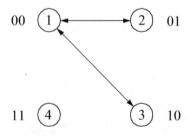

Transition diagram

$y_{1f} y_{2f}$ \ $x_1 x_2$	0 0	0 1	1 1	1 0
1 0 0	(00) a	10	–	01
2 0 1	00	(01) d	(01) c	(01) b
4 1 1	–	–	–	–
3 1 0	00	(10) e	(10) f	(10) g

$$Y_{1e}\, Y_{2e}$$

Combined excitation map

$y_{1f} y_{2f}$ \ $x_1 x_2$	0 0	0 1	1 1	1 0
0 0	0	1	–	0
0 1	0	0	0	0
1 1	–	–	–	–
1 0	0	1	1	1

$$Y_{1e}$$

$y_{1f} y_{2f}$ \ $x_1 x_2$	0 0	0 1	1 1	1 0
0 0	0	0	–	1
0 1	0	1	1	1
1 1	–	–	–	–
1 0	0	0	0	0

$$Y_{2e}$$

Individual excitation maps

$$Y_{1e} = y_{1f} x_1 + y_{2f}' x_2 \qquad\qquad Y_{2e} = y_{2f} x_2 + y_{1f}' x_1$$

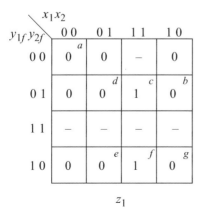

$y_{1f} y_{2f}$ \ $x_1 x_2$	0 0	0 1	1 1	1 0
0 0	0 a	0	–	0
0 1	0	0 d	1 c	0 b
1 1	–	–	–	–
1 0	0	0 e	1 f	0 g

$$z_1$$

$$z_1 = x_1 x_2$$

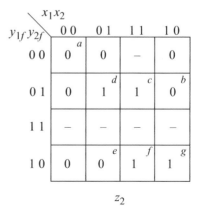

$y_{1f} y_{2f}$ \ $x_1 x_2$	0 0	0 1	1 1	1 0
0 0	0 a	0	–	0
0 1	0	1 d	1 c	0 b
1 1	–	–	–	–
1 0	0	0 e	1 f	1 g

$$z_2$$

$$z_2 = y_{2f} x_2 + y_{1f} x_1$$

```
//asm using built-in-primitives

module asm_bip2 (rst_n, x1, x2, y1e, y2e, z1, z2);


//define inputs and outputs
input rst_n, x1, x2;
output y1e, y2e, z1, z2;


//define internal nets
wire net1, net2, net3, net4;


//design the logic for y1e
and    (net1, y1e, x1, rst_n),
       (net2, ~y2e, x2, rst_n);
or     (y1e, net1, net2);


//design the logic for y2e
and    (net3, y2e, x2, rst_n),
       (net4, ~y1e, x1, rst_n);
or     (y2e, net3, net4);


//design the logic for outputs z1 and z2
and    (z1, x1, x2);
or     (z2, net1, net3);

endmodule
```

```verilog
//test bench for the asm using built-in-primitives
module asm_bip2_tb;

//inputs are reg for test bench
//outputs are wire for test bench
reg rst_n, x1, x2;
wire y1e, y2e, z1, z2;

//display variables
initial
$monitor ("x1x2 = %b, state = %b, z1z2 = %b",
          {x1, x2}, {y1e, y2e}, {z1, z2});

//apply input vectors
initial
begin
   #0    rst_n = 1'b0;
         x1 = 1'b0;
         x2 = 1'b0;

   #5    rst_n = 1'b1;

   #10   x1 = 1'b0;   x2 = 1'b0;
   #10   x1 = 1'b1;   x2 = 1'b0;
   #10   x1 = 1'b1;   x2 = 1'b1;        //z1 = 1, z2 = 1
   #10   x1 = 1'b0;   x2 = 1'b1;        //z1 = 0, z2 = 1
   #10   x1 = 1'b0;   x2 = 1'b0;        //z1 = 0, z2 = 0
   #10   x1 = 1'b0;   x2 = 1'b0;

   #10   x1 = 1'b0;   x2 = 1'b1;        //z1 = 0, z2 = 0
   #10   x1 = 1'b1;   x2 = 1'b1;        //z1 = 1, z2 = 1
   #10   x1 = 1'b1;   x2 = 1'b0;        //z1 = 0, z2 = 1
   #10   x1 = 1'b0;   x2 = 1'b0;        //z1 = 0, z2 = 0

   #10   $stop;
end

//instantiate the module into the test bench
asm_bip2 inst1 (
   .rst_n(rst_n),
   .x1(x1),
   .x2(x2),
   .y1e(y1e),
   .y2e(y2e),
   .z1(z1),
   .z2(z2)
   );

endmodule
```

```
x1x2 = 00, state = 00, z1z2 = 00
x1x2 = 10, state = 01, z1z2 = 00
x1x2 = 11, state = 01, z1z2 = 11
x1x2 = 01, state = 01, z1z2 = 01
x1x2 = 00, state = 00, z1z2 = 00

x1x2 = 01, state = 10, z1z2 = 00
x1x2 = 11, state = 10, z1z2 = 11
x1x2 = 10, state = 10, z1z2 = 01
x1x2 = 00, state = 00, z1z2 = 00
```

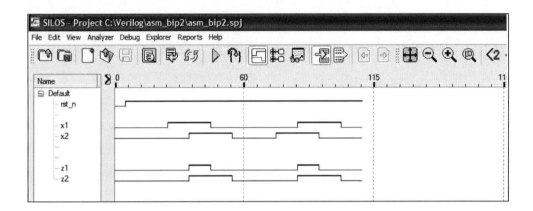

4.6   Synthesize an asynchronous sequential machine which has one input x_1 and one output z_1. The machine operates according to the timing diagram shown below. The assertion of x_1 toggles output z_1. Assign values to the transient states in the output map such that the λ output logic will be minimized. Obtain the excitation equations in a sum-of-products form. Use NAND logic with the continuous assignment statement in the design module.

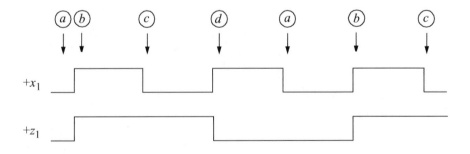

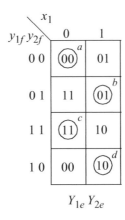

Primitive flow table

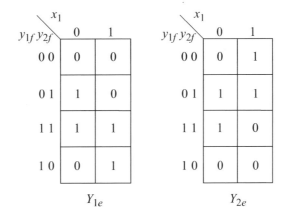

Combined excitation map

Individual excitation maps

$$Y_{1e} = y_{1f}x_1 + y_{2f}x_1' + y_{1f}y_{2f} \qquad Y_{2e} = y_{1f}'x_1 + y_{2f}x_1' + y_{1f}'y_{2f}$$

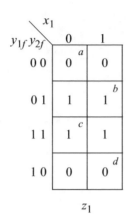

Output map

$$z_1 = y_{2f}$$

```
//dataflow nand for asynchronous sequential machine

module asm_nand (rst_n, x1, y1e, y2e, z1);

//define inputs and outputs
input rst_n, x1;
output y1e, y2e, z1;

//define internal nets
wire net1, net2, net3, net4, net5, net6;

//design the logic for y1e
assign    net1 = ~(y2e & ~x1 & rst_n),
          net2 = ~(y1e & x1 & rst_n),
          net3 = ~(y1e & y2e & rst_n),
          y1e = ~(net1 & net2 & net3);

//design the logic for y2e
assign    net4 = ~(y2e & ~x1 & rst_n),
          net5 = ~(~y1e & x1 & rst_n),
            net6 = ~(~y1e & y2e & rst_n),
      y2e = ~(net1 & net4 & net5);

//define the logic for output z1
assign    z1 = y2e;

endmodule
```

```
//test bench for the nand asm

module asm_nand_tb;

//inputs are reg for test bench
//outputs are wire for test bench
reg rst_n, x1;
wire y1e, y2e, z1;

//display variables
initial
$monitor ("x1 = %b, state = %b, z1 = %b", x1, {y1e, y2e}, z1);

//apply input vectors
initial
begin
   #0      rst_n = 1'b0;
           x1 = 1'b0;

   #5      rst_n = 1'b1;

   #10     x1 = 1'b0;
   #10     x1 = 1'b1;   //assert z1
   #10     x1 = 1'b0;   //assert z1
   #10     x1 = 1'b1;
   #10     x1 = 1'b0;
   #10     x1 = 1'b1;   //assert z1
   #10     x1 = 1'b0;   //assert z1
   #10     x1 = 1'b1;

   #10     $stop;
end

//instantiate the module into the test bench as a single line
asm_nand inst1 (
   .rst_n(rst_n),
   .x1(x1),
   .y1e(y1e),
   .y2e(y2e),
   .z1(z1)
   );

endmodule
```

```
x1 = 0, state = 00, z1 = 0
x1 = 1, state = 01, z1 = 1
x1 = 0, state = 11, z1 = 1
x1 = 1, state = 10, z1 = 0

x1 = 0, state = 00, z1 = 0
x1 = 1, state = 01, z1 = 1
x1 = 0, state = 11, z1 = 1
x1 = 1, state = 10, z1 = 0
```

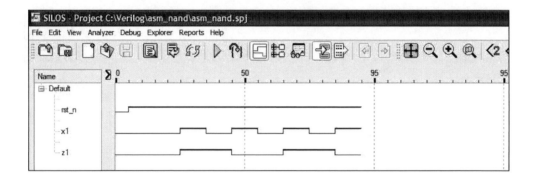

4.10   Obtain the excitation and output equations for an asynchronous sequential machine which has one input x_1 and two outputs z_1 and z_2. Output z_1 is asserted for the duration of every second x_1 pulse; output z_2 is asserted for the duration of every second z_1 pulse. The outputs are to respond as fast as possible to changes in the input vector. A representative timing diagram is shown below. Go through the design process to obtain the excitation equations and the output equations. Then obtain the design module using built-in primitives, the test bench module, the outputs, and the waveforms.

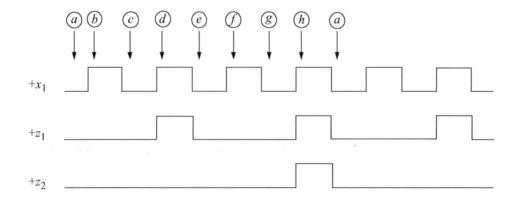

x_1	0	1	z_1	z_2
	(a)	b	0	0
	c	(b)	0	0
	(c)	d	0	0
	e	(d)	1	0
	(e)	f	0	0
	g	(f)	0	0
	(g)	h	0	0
	a	(h)	1	1

Primitive flow table

x_1 $y_{1f}y_{2f}y_{3f}$	0	1
0 0 0	(000) a	001
0 0 1	011	(001) b
0 1 1	(011) c	010
0 1 0	110	(010) d
1 1 0	(110) e	111
1 1 1	101	(111) f
1 0 1	(101) g	100
1 0 0	000	(100) h

$Y_{1e}Y_{2e}Y_{3e}$

Combined excitation map

x_1 $y_{1f}y_{2f}y_3$	0	1
000	0	0
001	0	0
011	0	0
010	1	0
110	1	1
111	1	1
101	1	1
100	0	1

Y_{1e}

x_1 $y_{1f}y_{2f}y_3$	0	1
000	0	0
001	1	0
011	1	1
010	1	1
110	1	1
111	0	1
101	0	0
100	0	0

Y_{2e}

x_1 $y_{1f}y_{2f}y_3$	0	1
000	0	1
001	1	1
011	1	0
010	0	0
110	0	1
111	1	1
101	1	0
100	0	0

Y_{3e}

Individual excitation maps

$y_{1f}y_{2f}y_3$ \ x_1	0	1
000	0^a	0
001	0	0^b
011	0^c	1
010	0	1^d
110	0^e	0
111	0	0^f
101	0^g	1
100	0	1^h

z_1

$y_{1f}y_{2f}y_3$ \ x_1	0	1
000	0^a	0
001	0	0^b
011	0^c	0
010	0	0^d
110	0^e	0
111	0	0^f
101	0^g	1
100	0	1^h

z_2

Output maps

net1	net2	net3	net4

$$Y_{1e} = y_{1f}x_1 + y_{1f}y_{3f} + y_{2f}y_{3f}'x_1' + y_{1f}y_{2f}$$

net5	net6	net7	net8

$$Y_{2e} = y_{2f}y_{3f}' + y_{2f}x_1 + y_{1f}'y_{3f}x_1' + y_{1f}'y_{2f}$$

net9	net10	net11	net12	net13

$$Y_{3e} = y_{3f}x_1' + y_{1f}'y_{2f}'x_1 + y_{1f}y_{2f}x_1 + y_{1f}'y_{2f}'y_{3f} + y_{1f}y_{2f}y_{3f}$$

net14	net15

$$z_1 = y_{1f}'y_{2f}x_1 + y_{1f}y_{2f}'x_1$$

$$z_2 = y_{1f}y_{2f}'x_1$$

```verilog
//asm using built-in primitives

module asm13_bip (rst_n, x1, y1e, y2e, y3e, z1, z2);

//define inputs and outputs
input rst_n, x1;
output y1e, y2e, y3e, z1, z2;

//define internal nets
wire net1, net2, net3, net4, net5, net6, net7, net8;
wire net9, net10, net11, net12, net13, net14, net15;

//design the logic for y1e
and     (net1, y1e, x1, rst_n),
        (net2, y1e, y3e, rst_n),
        (net3, y2e, ~y3e, ~x1, rst_n),
        (net4, y1e, y2e, rst_n);

or      (y1e, net1, net2, net3, net4);

//design the logic for y2e
and     (net5, y2e, ~y3e, rst_n),
        (net6, y2e, x1, rst_n),
        (net7, ~y1e, y3e, ~x1, rst_n),
        (net8, ~y1e, y2e, rst_n);

or      (y2e, net5, net6, net7, net8);

//design the logic for y3e
and     (net9, y3e, ~x1, rst_n),
        (net10, ~y1e, ~y2e, x1, rst_n),
        (net11, y1e, y2e, x1, rst_n),
        (net12, ~y1e, ~y2e, y3e, rst_n),
        (net13, y1e, y2e, y3e, rst_n);

or      (y3e, net9, net10, net11, net12, net13);

//design the logic for outputs z1 and z2
and     (net14, ~y1e, y2e, x1),
        (net15, y1e, ~y2e, x1);

or      (z1, net14, net15);

and     (z2, y1e, ~y2e, x1);

endmodule
```

```
//test bench for the bip asm

module asm13_bip_tb;

//inputs are reg for test bench
//outputs are wire for test bench
reg rst_n, x1;
wire y1e, y2e, y3e, z1, z2;

//display variables
initial
$monitor ("x1 = %b, state = %b, z1z2 = %b",
          x1, {y1e, y2e, y3e}, {z1, z2});

//apply input vectors
initial
begin
   #0     rst_n = 1'b0;
          x1 = 1'b0;
   #5     rst_n = 1'b1;

   #10    x1 = 1'b0;   //state_a
   #10    x1 = 1'b1;   //state_b
   #10    x1 = 1'b0;   //state_c
   #10    x1 = 1'b1;   //state_d, assert z1

   #10    x1 = 1'b0;   //state_e
   #10    x1 = 1'b1;   //state_f
   #10    x1 = 1'b0;   //state_g
   #10    x1 = 1'b1;   //state_h, assert z1 and z2

   #10    x1 = 1'b0;   //state_a
   #10    x1 = 1'b1;   //state_b
   #10    x1 = 1'b0;   //state_c
   #10    x1 = 1'b1;   //state_d, assert z1

   #10    x1 = 1'b0;   //state_e
   #10    x1 = 1'b1;   //state_f
   #10    x1 = 1'b0;   //state_g
   #10    x1 = 1'b1;   //state_h, assert z1 and z2

   #10    x1 = 1'b0;   //state_a
   #10    $stop;
end

//instantiate the module into the test bench as a single line
asm13_bip inst1 (rst_n, x1, y1e, y2e, y3e, z1, z2);

endmodule
```

```
x1 = 0, state = 000, z1z2 = 00
x1 = 1, state = 001, z1z2 = 00
x1 = 0, state = 011, z1z2 = 00
x1 = 1, state = 010, z1z2 = 10

x1 = 0, state = 110, z1z2 = 00
x1 = 1, state = 111, z1z2 = 00
x1 = 0, state = 101, z1z2 = 00
x1 = 1, state = 100, z1z2 = 11

x1 = 0, state = 000, z1z2 = 00
x1 = 1, state = 001, z1z2 = 00
x1 = 0, state = 011, z1z2 = 00
x1 = 1, state = 010, z1z2 = 10

x1 = 0, state = 110, z1z2 = 00
x1 = 1, state = 111, z1z2 = 00
x1 = 0, state = 101, z1z2 = 00
x1 = 1, state = 100, z1z2 = 11

x1 = 0, state = 000, z1z2 = 00
```

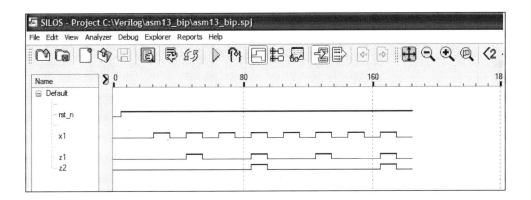

4.15   Synthesize an asynchronous sequential machine using built-in primitives, which has one input x_1 and two outputs z_1 and z_2. The machine functions as a two-output bistable multivibrator, whose operation is characterized by the timing diagram shown below. Output z_1 toggles on the positive transition of x_1 and output z_2 toggles on the negative transition of x_1. Obtain equations for z_1 and z_2 which produce the least amount of logic. Obtain the design module using built-in primitives, the test bench module, the outputs, and the waveforms.

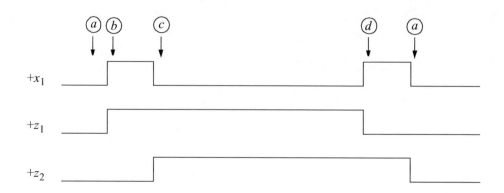

	x_1 0	1	z_1	z_2
	ⓐ	b	0	0
	c	ⓑ	1	0
	ⓒ	d	1	1
	a	ⓓ	0	1

Primitive flow table. There are no equivalent states, because each state has different outputs.

Merger diagram in which no rows can merge because both columns for x_1 have at least one different state in each selected row.

x_1

$y_{1f}y_{2f}$	0	1
0 0	(00) a	01
0 1	11	(01) b
1 1	(11) c	10
1 0	00	(10) d

$Y_{1e}Y_{2e}$

x_1

$y_{1f}y_{2f}$	0	1
0 0	0	0
0 1	1	0
1 1	1	1
1 0	0	1

Y_{1e}

x_1

$y_{1f}y_{2f}$	0	1
0 0	0	1
0 1	1	1
1 1	1	0
1 0	0	0

Y_{2e}

Combined excitation map and individual excitation maps.

$$Y_{1e} = x_1'y_{2f} + x_1y_{1f} + y_{1f}y_{2f}$$
$$Y_{2e} = x_1'y_{2f} + x_1y_{1f}' + y_{1f}'y_{2f}$$

x_1

$y_{1f}y_{2f}$	0	1
0 0	00	-0 a
0 1	$1-$	10 b
1 1	11	-1 c
1 0	$0-$	01 d

z_1z_2

x_1

$y_{1f}y_{2f}$	0	1
0 0	0	$-$
0 1	1	1
1 1	1	$-$
1 0	0	0

z_1

x_1

$y_{1f}y_{2f}$	0	1
0 0	0	0
0 1	$-$	0
1 1	1	1
1 0	$-$	1

z_2

Output maps

$$z_1 = y_{2f}$$
$$z_2 = y_{1f}$$

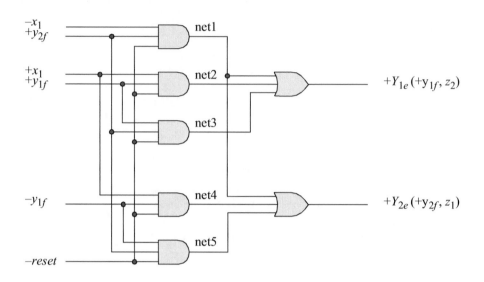

```
//asm using built-in primitives

module asm_bip4 (rst_n, x1, y1e, y2e, z1, z2);

//define the inputs and  outputs
input rst_n, x1;
output y1e, y2e, z1, z2;

//define internal nets
wire net1, net2, net3, net4, net5;

//define the logic for y1e
and     (net1, ~x1, y2e, rst_n),
        (net2, x1, y1e, rst_n),
        (net3, y1e, y2e, rst_n);

or      (y1e, net1, net2, net3);

//define the logic for y2e
and     (net4, x1, ~y1e, rst_n),
        (net5, ~y1e, y2e, rst_n);

or      (y2e, net1, net4, net5);

//define the logic for outputs z1 and z2
or      (z1, y2e),
        (z2, y1e);

endmodule
```

```verilog
//test bench for the asm using built-in primitives

module asm_bip4_tb;

//inputs are reg for test bench
//outputs are wire for test bench
reg rst_n, x1;
wire y1e, y2e, z1, z2;

//display variables
initial
$monitor ("x1 = %b, state = %b, z1z2 = %b",
          x1, {y1e, y2e}, {z1, z2});

//apply input vectors
initial
begin
   #0    rst_n = 1'b0;
         x1 = 1'b0;

   #5    rst_n = 1'b1;

   #10   x1 = 1'b0;
   #10   x1 = 1'b1;   //state_b, assert z1
   #10   x1 = 1'b0;   //state_c, assert z2

   #30   x1 = 1'b0;

   #10   x1 = 1'b1;   //state_d, deassert z1
   #10   x1 = 1'b0;   //state_a, deassert z2

   #10   x1 = 1'b0;

   #10   $stop;
end

//instantiate the module into the test bench as a single line
asm_bip4 inst1 (rst_n, x1, y1e, y2e, z1, z2);

endmodule
```

```
x1 = 0, state = 00, z1z2 = 00
x1 = 1, state = 01, z1z2 = 10
x1 = 0, state = 11, z1z2 = 11
x1 = 1, state = 10, z1z2 = 01
x1 = 0, state = 00, z1z2 = 00
```

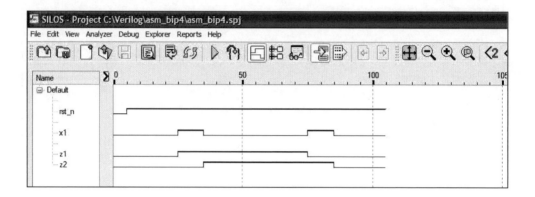

4.18   Given the merged flow table shown below, design the asynchronous sequential machine using dataflow modeling with the continuous assignment statement. Assume that output z_1 is asserted in the following stable states: ⓑ,
ⓓ, ⓕ, and ⓖ. Obtain the design module and the test bench module that
takes the machine through the paths to assert output z_1. The excitation equations and the output equation are to be in a sum-of-products form. Obtain the
outputs and the waveforms. There should be no static-1 or static-0 hazards in
the equations.

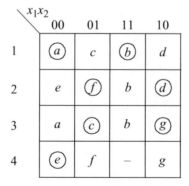

Merged flow table

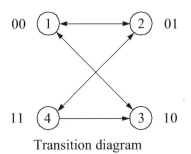

Transition diagram

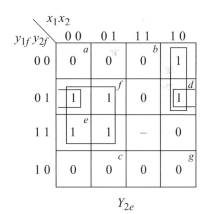

Combined excitation map

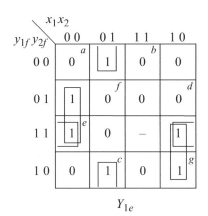

Y_{1e} Y_{2e}

Individual excitation maps

	net1	net2	net3	net4

$$Y_{1e} = y_{2f}x_1'x_2' + y_{2f}'x_1'x_2 + y_{1f}x_1x_2' + y_{1f}y_{2f}x_2'$$

	net5	net6	net7

$$Y_{2e} = y_{2f}x_1' + y_{1f}'x_1x_2' + y_{1f}'y_{2f}x_2'$$

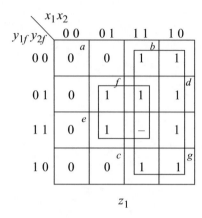

Output map

$$z_1 = x_1 + y_{2f}x_2$$

```
//dataflow asynchronous sequential machine

module asm21a (rst_n, x1, x2, y1e, y2e, z1);

input rst_n, x1, x2;
output y1e, y2e, z1;

//define internal nets
wire net1, net2, net3, net4, net5, net6, net7;

//design the logic for y1e
assign    net1 = y2e & ~x1 & ~x2 & rst_n,
          net2 = ~y2e & ~x1 & x2 & rst_n,
          net3 = y1e & x1 & ~x2 & rst_n,
          net4 = y1e & y2e & ~x2 & rst_n,
          y1e = net1 | net2 | net3 | net4;

//design the logic for y2e
assign    net5 = y2e & ~x1 & rst_n,
          net6 = ~y1e & x1 & ~x2 & rst_n,
          net7 = ~y1e & y2e & ~x2 & rst_n,
          y2e = net5 | net6 | net7;

//design for output z1
assign    z1 = x1 | (y2e & x2);

endmodule
```

```verilog
//test bench the asynchronous sequential machine
module asm21a_tb;

//inputs are reg for test bench
//outputs are wire for test bench
reg rst_n, x1, x2;
wire y1e, y2e, z1;

//display variables
initial
$monitor ("x1x2 = %b, state = %b, z1 = %b",
          {x1, x2}, {y1e, y2e}, z1);


//apply input vectors
initial
begin
   #0    rst_n = 1'b0;  x1 = 1'b0;  x2 = 1'b0;
   #5    rst_n = 1'b1;

   #10   x1 = 1'b0;  x2 = 1'b0;  //state_a
   #10   x1 = 1'b1;  x2 = 1'b0;  //state_d, assert z1

   #10   x1 = 1'b0;  x2 = 1'b0;  //state_e
   #10   x1 = 1'b1;  x2 = 1'b0;  //state_g, assert z1

   #10   x1 = 1'b1;  x2 = 1'b1;  //state_b, assert z1

   #10   x1 = 1'b0;  x2 = 1'b1;  //state_c
   #10   x1 = 1'b1;  x2 = 1'b0;  //state_d
   #10   x1 = 1'b0;  x2 = 1'b0;  //state_e
   #10   x1 = 1'b0;  x2 = 1'b1;  //state_f, assert z1

   #10   x1 = 1'b0;  x2 = 1'b0;  //state_e

   #10   x1 = 1'b0;  x2 = 1'b0;
   #10   $stop;
end

//instantiate the module into the test bench
asm21a inst1 (
   .rst_n(rst_n),
   .x1(x1),
   .x2(x2),
   .y1e(y1e),
   .y2e(y2e),
   .z1(z1)
   );

endmodule
```

```
x1x2 = 00, state = 00, z1 = 0
x1x2 = 10, state = 01, z1 = 1

x1x2 = 00, state = 11, z1 = 0
x1x2 = 10, state = 10, z1 = 1
x1x2 = 11, state = 00, z1 = 1

x1x2 = 01, state = 10, z1 = 0
x1x2 = 10, state = 10, z1 = 1
x1x2 = 00, state = 00, z1 = 0
x1x2 = 01, state = 10, z1 = 0
x1x2 = 00, state = 00, z1 = 0
```

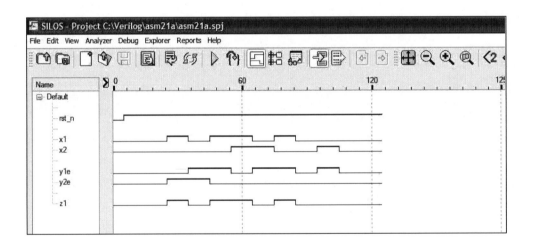

Chapter 5   Synthesis of Pulse-Mode Asynchronous Sequential Machines Using Verilog HDL

5.3   Design a Moore pulse-mode asynchronous sequential machine which has two inputs x_1 and x_2 and one output z_1. The deassertion of every second consecutive x_1 pulse will assert output z_1 as a level. The output will remain set for all following contiguous x_1 pulses. The output will be deasserted at the trailing edge of the second of two consecutive x_2 pulses. Use NAND built-in primitives together with SR latches and D flip-flops. The state diagram is shown below and represents the complete sequencing for this Moore machine.

Obtain the design module using D flip-flops that were designed using behavioral modeling. Obtain the test bench module that takes the machine through the various sequences required to generate the output assertion and deassertion as specified above. Obtain the outputs and the waveforms.

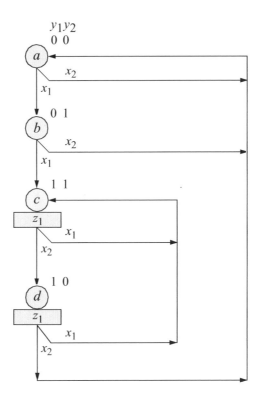

The input maps and input equations are shown below together with the equation for output z_1.

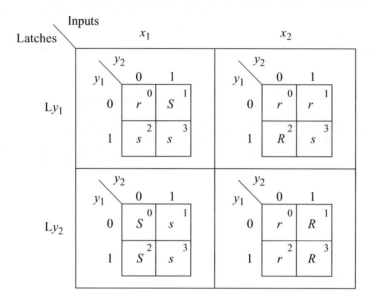

$$SLy_1 = y_2 x_1$$

$$RLy_1 = y_2' x_2$$

$$SLy_2 = x_1$$

$$RLy_2 = x_2$$

$$z_1 = y_1$$

```
//moore pulse-mode asm using built-in primitives and D ff

module pm_asm_moore6 (rst_n, x1, x2, y1, y2, z1);

//define inputs and outputs
input rst_n, x1, x2;
output y1, y2, z1;

//define internal nets
wire net1, net2, net3, net4, net5, net6, net7;

                                //continued on next page
```

```verilog
//------------------------------------------------
//design the D flip-flop clock
nor    (net1, x1, x2);

//------------------------------------------------
//design the logic for latch Ly1
nand   (net2, x1, y2),
       (net3, x2, ~y2);

nand   (net4, net2, net5),
       (net5, net4, net3, rst_n);

//instantiate the D flip-flop for y1 as a single line
d_ff_bh inst1 (rst_n, net1, net4, y1);   //rst_n, clk, d, q

//------------------------------------------------
//design the logic for latch Ly2
nand   (net6, ~x1, net7),
       (net7, net6, ~x2, rst_n);

//instantiate the D flip-flop for y2 as a single line
d_ff_bh inst2 (rst_n, net1, net6, y2);   //rst_n, clk, d, q

//------------------------------------------------
//design the logic for output z1
assign   z1 = y1;

endmodule
```

```verilog
//test bench for moore pulse-mode asm
module pm_asm_moore6_tb;

//inputs are reg for test bench
//outputs are wire for test bench
reg rst_n, x1, x2;
wire y1, y2, z1;

//display inputs and outputs --------------------------
initial
$monitor ("x1x2 = %b, state = %b, z1 = %b",
          {x1, x2}, {y1, y2}, z1);

//define input sequence ------------------------------
initial
begin
    #0    rst_n = 1'b0;    //reset to state_a(00); no output
          x1 = 1'b0;   x2 = 1'b0;
    #5    rst_n = 1'b1;            //continued on next page
```

```
   #10    x1 = 1'b1;   //→ state_b
   #10    x1 = 1'b0;   //no output

   #10    x1 = 1'b1;   //→ state_c
   #10    x1 = 1'b0;   //assert output z1

   #10    x1 = 1'b1;   //remain in state_c
   #10    x1 = 1'b0;   //output z1 remains asserted

   #20    x2 = 1'b1;   //→ state_d
   #10    x2 = 1'b0;   //output z1 remains asserted

   #10    x1 = 1'b1;   //→ state_c(11)
   #10    x1 = 1'b0;   //output z1 remains asserted

   #10    x2 = 1'b1;   //→ state_d(10)
   #10    x2 = 1'b0;   //output z1 remains asserted

   #10    x2 = 1'b1;   //→ state_a
   #10    x2 = 1'b0;   //deassert output z1

   #30    $stop;
end

//instantiate the module into the test bench -------------
pm_asm_moore6 inst1 (rst_n, x1, x2, y1, y2, z1);

endmodule
```

```
x1x2 = 00, state = 00, z1 = 0
x1x2 = 10, state = 00, z1 = 0
x1x2 = 00, state = 01, z1 = 0
x1x2 = 10, state = 01, z1 = 0

x1x2 = 00, state = 11, z1 = 1
x1x2 = 10, state = 11, z1 = 1
x1x2 = 00, state = 11, z1 = 1
x1x2 = 01, state = 11, z1 = 1
x1x2 = 00, state = 10, z1 = 1
x1x2 = 10, state = 10, z1 = 1
x1x2 = 00, state = 11, z1 = 1
x1x2 = 01, state = 11, z1 = 1
x1x2 = 00, state = 10, z1 = 1
x1x2 = 01, state = 10, z1 = 1

x1x2 = 00, state = 00, z1 = 0
```

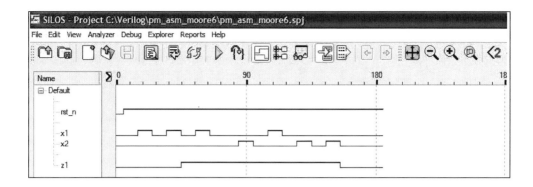

5.8   A state diagram is shown below for a Moore pulse-mode asynchronous sequential machine with three inputs x_1, x_2, and x_3, and two outputs z_1 and z_2. T flip-flops are used in the implementation. Obtain the input maps and equations, the output equations, and the logic diagram using AND and OR gates for the δ next-state logic. Obtain the design module using AND and OR built-in primitives and T flip-flops. Then develop the test bench module that takes the machine through various sequences to assert output z_1 and output z_2. Obtain the outputs and the waveforms.

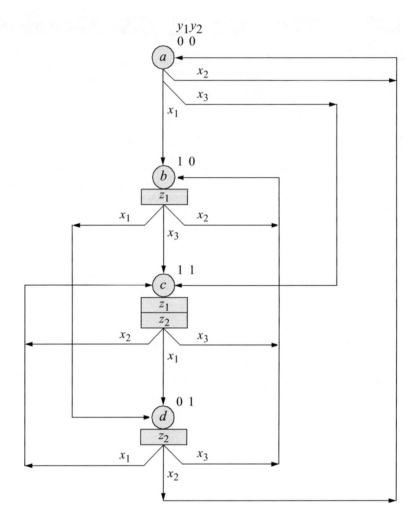

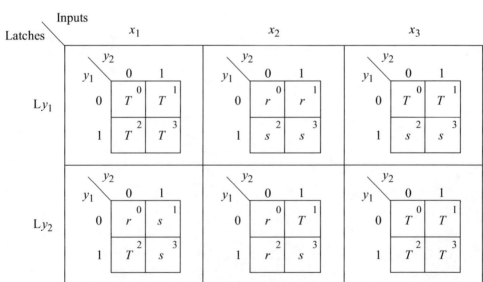

$$Ty_1 = x_1 + y_1'x_3$$

$$Ty_2 = y_1y_2'x_1 + y_1'y_2x_2 + x_3$$

$$z_1 = y_1$$

$$z_2 = y_2$$

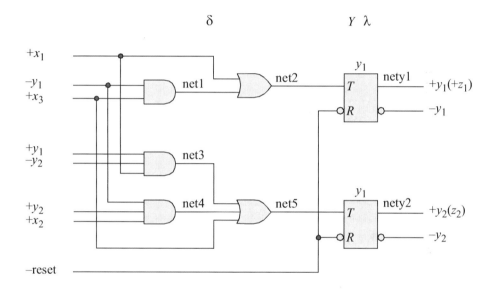

```
//moore pulse-mode asm using built-in primitives and T ff

module pm_asm_moore8 (rst_n, x1, x2, x3, y1, y2, z1, z2);

//define inputs and outputs
input rst_n, x1, x2, x3;
output y1, y2, z1, z2;

//define internal nets
wire net1, net2, net3, net4, net5, nety1, nety2;

//-----------------------------------------------
                              //continued on next page
```

```
//-------------------------------------------------
//define the logic for flip-flop y1
and     (net1, ~y1, x3);
or      (net2, x1, net1);

//instantiate the T flip-flop as a single line
t_ff_da inst1 (rst_n, net2, nety1);    //rst_n, t, y1

buf     #11 (y1, nety1);

//-------------------------------------------------
//define the logic for flip-flop y2
and     (net3, y1, ~y2, x1),
        (net4, ~y1, y2, x2);
or      (net5, net3, net4, x3);

//instantiate the T flip-flop as a single line
t_ff_da inst2 (rst_n, net5, nety2);    //rst_n, t, y1

buf     #11 (y2, nety2);

//-------------------------------------------------
//define the logic for outputs z1 and z2
assign   z1 = y1;
assign   z2 = y2;

endmodule
```

```
//test bench for moore pulse-mode asm

module pm_asm_moore8_tb;

//inputs are reg for test bench
//outputs are wire for test bench
reg rst_n, x1, x2, x3;
wire y1, y2, z1, z2;

//display inputs and outputs
initial
$monitor ("x1x2x3 = %b, state = %b, z1z2 = %b",
          {x1, x2, x3}, {y1, y2}, {z1, z2});

                                //continued on next page
```

```
//define input sequence
initial
begin
   #0     rst_n = 1'b0;
          x1 = 1'b0;
          x2 = 1'b0;
          x3 = 1'b0;

   #5     rst_n = 1'b1;

   #10    x1=1'b0;    x2=1'b0;    x3=1'b0;

   #10    x1=1'b1;    x2=1'b0;    x3=1'b0;    //→ b, assert z1
   #10    x1=1'b0;    x2=1'b0;    x3=1'b0;

   #10    x1=1'b1;    x2=1'b0;    x3=1'b0;    //→ d, assert z2
   #10    x1=1'b0;    x2=1'b0;    x3=1'b0;

   #10    x1=1'b0;    x2=1'b0;    x3=1'b1;    //→ b, assert z1
   #10    x1=1'b0;    x2=1'b0;    x3=1'b0;

   #10    x1=1'b0;    x2=1'b0;    x3=1'b1;    //→ c, assert z1,z2
   #10    x1=1'b0;    x2=1'b0;    x3=1'b0;

   #10    x1=1'b1;    x2=1'b0;    x3=1'b0;    //→ d, assert z2
   #10    x1=1'b0;    x2=1'b0;    x3=1'b0;

   #10    x1=1'b1;    x2=1'b0;    x3=1'b0;    //→ c, assert z1,z2
   #10    x1=1'b0;    x2=1'b0;    x3=1'b0;

   #10    x1=1'b1;    x2=1'b0;    x3=1'b0;    //→ d, assert z2
   #10    x1=1'b0;    x2=1'b0;    x3=1'b0;

   #10    x1=1'b0;    x2=1'b1;    x3=1'b0;    //→ a
   #10    x1=1'b0;    x2=1'b0;    x3=1'b0;

   #20    $stop;

end

//instantiate the module into the test bench
pm_asm_moore8 inst1 (rst_n, x1, x2, x3, y1, y2, z1, z2);

endmodule
```

```
x1x2x3 = 000, state = 00, z1z2 = 00
x1x2x3 = 100, state = 00, z1z2 = 00
x1x2x3 = 000, state = 00, z1z2 = 00
x1x2x3 = 000, state = 10, z1z2 = 10

x1x2x3 = 100, state = 10, z1z2 = 10
x1x2x3 = 000, state = 10, z1z2 = 10
x1x2x3 = 000, state = 01, z1z2 = 01
x1x2x3 = 001, state = 01, z1z2 = 01

x1x2x3 = 000, state = 01, z1z2 = 01
x1x2x3 = 000, state = 10, z1z2 = 10
x1x2x3 = 001, state = 10, z1z2 = 10
x1x2x3 = 000, state = 10, z1z2 = 10

x1x2x3 = 000, state = 11, z1z2 = 11
x1x2x3 = 100, state = 11, z1z2 = 11
x1x2x3 = 000, state = 11, z1z2 = 11
x1x2x3 = 000, state = 01, z1z2 = 01

x1x2x3 = 100, state = 01, z1z2 = 01
x1x2x3 = 000, state = 01, z1z2 = 01
x1x2x3 = 000, state = 11, z1z2 = 11
x1x2x3 = 100, state = 11, z1z2 = 11

x1x2x3 = 000, state = 11, z1z2 = 11
x1x2x3 = 000, state = 01, z1z2 = 01
x1x2x3 = 010, state = 01, z1z2 = 01
x1x2x3 = 000, state = 01, z1z2 = 01

x1x2x3 = 000, state = 00, z1z2 = 00
```

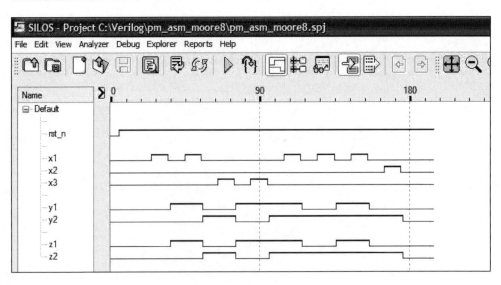

5.12   Design a Mealy pulse-mode asynchronous sequential machine that has three input variables x_1, x_2, and x_3 and one output z_1 that is asserted coincident with x_3 whenever the sequence $x_1 x_2 x_3 = 100, 010, 001$ occurs. The storage elements consist of *SR* latches and positive-edge-triggered *D* flip-flops in a master-slave configuration.

   A representative timing diagram displaying valid input sequences and corresponding outputs is shown below. Obtain the state diagram that represents the functional operation of the machine. State code assignment is arbitrary for the state diagram, since input pulses trigger all state transitions and the machine does not begin to sequence to the next state until the input pulse, which initiated the transition, has been deasserted. Thus, output z_1 will not glitch. Obtain the input maps, the input equations, the output equation, and the logic diagram. Then obtain the structural design module using dataflow modeling for the logic primitives, which are instantiated as a single line. Use NOR logic for the *SR* latches. Use behavioral modeling for the *D* flip-flop. Obtain the test bench, the outputs, and the waveforms.

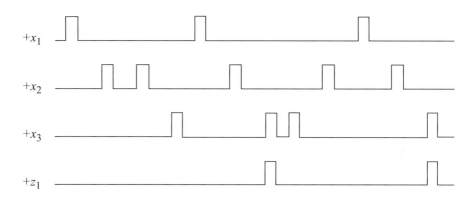

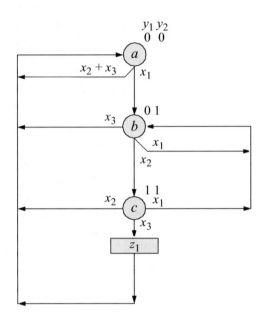

Inputs

Latches

Ly_1

x_1

y_1 \ y_2	0	1
0	r	r
1	$-$	R

x_2

y_1 \ y_2	0	1
0	r	S
1	$-$	R

x_3

y_1 \ y_2	0	1
0	r	r
1	$-$	R

Ly_2

x_1

y_1 \ y_2	0	1
0	S	s
1	$-$	s

x_2

y_1 \ y_2	0	1
0	r	s
1	$-$	R

x_3

y_1 \ y_2	0	1
0	r	R
1	$-$	R

$$SLy_1 = y_1'y_2x_2 \qquad RLy_1 = x_1 + y_1x_2 + x_3$$

$$SLy_2 = x_1 \qquad RLy_2 = y_1x_2 + x_3$$

$$z_1 = y_1x_3$$

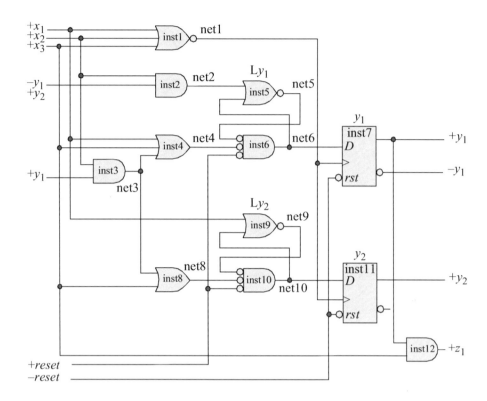

```
//structural module for a Mealy pulse-mode asm

module pm_asm6a (rst_n, rst, x1, x2, x3, y1, y2, z1);

//define inputs and outputs
input rst_n, rst, x1, x2, x3;
output y1, y2, z1;

//define internal nets
wire net1, net2, net3, net4, net5, net6, net8, net9, net10;

//----------------------------------------------------
//design for clock
nor3_df  inst1 (x1, x2, x3, net1);

//----------------------------------------------------
//design for latch Ly1
and3_df  inst2 (x2, ~y1, y2, net2);

and2_df  inst3 (x2, y1, net3);
                                  //continued on next page
```

```
or3_df    inst4 (x1, x3, net3, net4);

nor2_df   inst5 (net2, net6, net5);

nor3_df   inst6 (net5, net4, rst, net6);


//design for D flip-flop y1
d_ff_bh   inst7 (rst_n, net1, net6, y1);    //rst_n, clk, d, q

//-------------------------------------------------
//design for latch Ly2
or2_df    inst8 (net3, x3, net8);

nor2_df   inst9 (x1, net10, net9);

nor3_df   inst10 (net9, net8, rst, net10);


//design for D flip-flop y2
d_ff_bh   inst11 (rst_n, net1, net10, y2);   //rst_n, clk, d, q

//-------------------------------------------------
//design for z1
and2_df   inst12 (y1, x3, z1);

endmodule
```

```
//test bench for Mealy pulse-mode machine
module pm_asm6a_tb;

//inputs are reg for test bench
//outputs are wire for test bench
reg rst_n, rst, set_n, x1, x2, x3;
wire y1, y2, z1;

//display variables
initial
$monitor ("x1=%b, x2=%b, x3=%b, state=%b, z1=%b",
          x1, x2, x3, {y1, y2}, z1);

//apply stimulus
initial
begin
   #0    rst = 1'b1;         //reset latches
         rst_n = 1'b0;       //reset flip-flops to state_a
                                   //continued on next page
```

```
            x1 = 1'b0;   x2 = 1'b0;   x3 = 1'b0;

   #5     rst = 1'b0;         //remove reset from latches
          rst_n = 1'b1;       //remove reset from flip-flops

   #10    x1 = 1'b0;   x2 = 1'b1;   x3 = 1'b0;
   #10    x1 = 1'b0;   x2 = 1'b0;   x3 = 1'b0;   //→ state_a

   #10    x1 = 1'b0;   x2 = 1'b0;   x3 = 1'b1;
   #10    x1 = 1'b0;   x2 = 1'b0;   x3 = 1'b0;   //→ state_a

   #10    x1 = 1'b1;   x2 = 1'b0;   x3 = 1'b0;
   #10    x1 = 1'b0;   x2 = 1'b0;   x3 = 1'b0;   //→ state_b

   #10    x1 = 1'b0;   x2 = 1'b0;   x3 = 1'b1;
   #10    x1 = 1'b0;   x2 = 1'b0;   x3 = 1'b0;   //→ state_a

   #10    x1 = 1'b1;   x2 = 1'b0;   x3 = 1'b0;
   #10    x1 = 1'b0;   x2 = 1'b0;   x3 = 1'b0;   //→ state_b

   #10    x1 = 1'b0;   x2 = 1'b1;   x3 = 1'b0;
   #10    x1 = 1'b0;   x2 = 1'b0;   x3 = 1'b0;   //→ state_c

   #10    x1 = 1'b1;   x2 = 1'b0;   x3 = 1'b0;
   #10    x1 = 1'b0;   x2 = 1'b0;   x3 = 1'b0;   //→ state_b

   #10    x1 = 1'b0;   x2 = 1'b1;   x3 = 1'b0;
   #10    x1 = 1'b0;   x2 = 1'b0;   x3 = 1'b0;   //→ state_c
   #10    x1 = 1'b0;   x2 = 1'b1;   x3 = 1'b0;
   #10    x1 = 1'b0;   x2 = 1'b0;   x3 = 1'b0;   //→ state_a

   #10    x1 = 1'b1;   x2 = 1'b0;   x3 = 1'b0;
   #10    x1 = 1'b0;   x2 = 1'b0;   x3 = 1'b0;   //→ state_b

   #10    x1 = 1'b0;   x2 = 1'b1;   x3 = 1'b0;
   #10    x1 = 1'b0;   x2 = 1'b0;   x3 = 1'b0;   //→ state_c

   #10    x1 = 1'b0;   x2 = 1'b0;   x3 = 1'b1;
   #10    x1 = 1'b0;   x2 = 1'b0;   x3 = 1'b0;   //z1; → state_a

   #10    x1 = 1'b0;   x2 = 1'b1;   x3 = 1'b0;
   #10    x1 = 1'b0;   x2 = 1'b0;   x3 = 1'b0;   //→ state_a
   #10    $stop;
end

//instantiate the module into the test bench
pm_asm6a inst1 (rst_n, rst, x1, x2, x3, y1, y2, z1);

endmodule
```

```
x1=0,  x2=0,  x3=0,  state=00,  z1=0
x1=0,  x2=1,  x3=0,  state=00,  z1=0
x1=0,  x2=0,  x3=0,  state=00,  z1=0
x1=0,  x2=0,  x3=1,  state=00,  z1=0
x1=0,  x2=0,  x3=0,  state=00,  z1=0
x1=1,  x2=0,  x3=0,  state=00,  z1=0
x1=0,  x2=0,  x3=0,  state=01,  z1=0
x1=0,  x2=0,  x3=1,  state=01,  z1=0
x1=0,  x2=0,  x3=0,  state=00,  z1=0
x1=1,  x2=0,  x3=0,  state=00,  z1=0
x1=0,  x2=0,  x3=0,  state=01,  z1=0
x1=0,  x2=1,  x3=0,  state=01,  z1=0
x1=0,  x2=0,  x3=0,  state=11,  z1=0
x1=1,  x2=0,  x3=0,  state=11,  z1=0
x1=0,  x2=0,  x3=0,  state=01,  z1=0
x1=0,  x2=1,  x3=0,  state=01,  z1=0
x1=0,  x2=0,  x3=0,  state=11,  z1=0
x1=0,  x2=1,  x3=0,  state=11,  z1=0
x1=0,  x2=0,  x3=0,  state=00,  z1=0
x1=1,  x2=0,  x3=0,  state=00,  z1=0
x1=0,  x2=0,  x3=0,  state=01,  z1=0
x1=0,  x2=1,  x3=0,  state=01,  z1=0
x1=0,  x2=0,  x3=0,  state=11,  z1=0
x1=0,  x2=0,  x3=1,  state=11,  z1=1
x1=0,  x2=0,  x3=0,  state=00,  z1=0
x1=0,  x2=1,  x3=0,  state=00,  z1=0
x1=0,  x2=0,  x3=0,  state=00,  z1=0
```

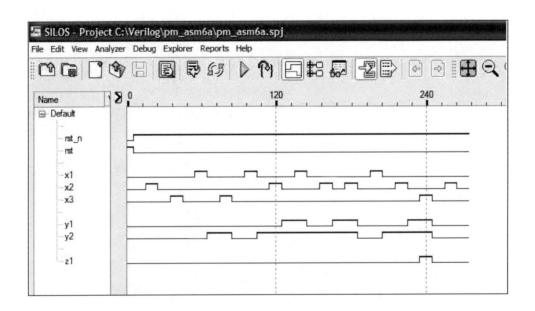

5.15   Given the state diagram shown below for a Moore pulse-mode asynchronous sequential machine, implement the machine using built-in primitives gates for the δ next-state logic and T flip-flops — instantiated as a single line — as the storage elements.

Derive the input maps, the input equations, the output equations, and the the logic diagram. Obtain the design module, the test bench module, the outputs, and the waveforms.

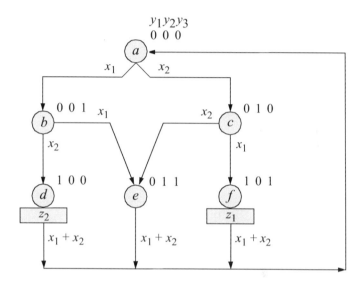

Inputs x_1 x_2

Flip-flops

y_1

y_1 \ y_2y_3	00	01	11	10
0	r (0)	r (1)	r (3)	T (2)
1	T (4)	T (5)	$-$ (7)	$-$ (6)

y_1 \ y_2y_3	00	01	11	10
0	r (0)	T (1)	r (3)	r (2)
1	T (4)	T (5)	$-$ (7)	$-$ (6)

y_2

y_1 \ y_2y_3	00	01	11	10
0	r (0)	T (1)	T (3)	T (2)
1	r (4)	r (5)	$-$ (7)	$-$ (6)

y_1 \ y_2y_3	00	01	11	10
0	T (0)	r (1)	T (3)	s (2)
1	r (4)	r (5)	$-$ (7)	$-$ (6)

y_3

y_1 \ y_2y_3	00	01	11	10
0	T (0)	s (1)	T (3)	T (2)
1	r (4)	T (5)	$-$ (7)	$-$ (6)

y_1 \ y_2y_3	00	01	11	10
0	r (0)	T (1)	T (3)	T (2)
1	r (4)	T (5)	$-$ (7)	$-$ (6)

$$Ty_1 = y_2y_3'x_1 + y_1x_1 + y_2'y_3x_2 + y_1x_2$$

$$Ty_2 = y_2x_1 + y_1'y_3x_1 + y_1'y_2'y_3'x_2 + y_2y_3x_2$$

$$Ty_3 = y_1'y_3'x_1 + y_2y_3x_1 + y_1y_3x_1 + y_2x_2 + y_3x_2$$

$$z_1 = y_1y_2'y_3$$

$$z_2 = y_1y_2'y_3'$$

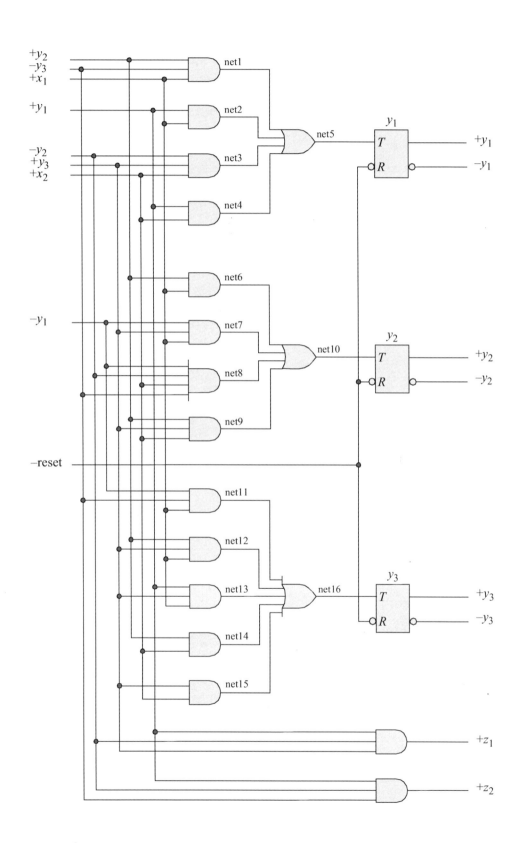

```
//moore pulse-mode asm using built-in primitives and T ff

module pm_asm10c (rst_n, x1, x2, y1, y2, y3, z1, z2);

//define inputs and outputs
input rst_n, x1,x2;
output y1, y2, y3, z1, z2;

//define internal nets
wire   net1, net2, net3, net4, net5, net6, net7,
       net8, net9, net10, net11, net12, net13,
       net14, net15, net16, nety1, nety2, nety3;

//---------------------------------------------------
//design the logic for flip-flop y1
and    (net1, y2, ~y3, x1),
       (net2, y1, x1),
       (net3, ~y1, y3, x2),
       (net4, y1, x2);

or     net5, net1, net2, net3, net4);

//instantiate the T flip-flop for y1
t_ff_da inst1 (rst_n, net5, nety1);   //rst_n, t, y1

buf    #6 (y1, nety1);

//---------------------------------------------------
//design the logic for flip-flop y2
and    (net6, y2, x1),
       (net7, ~y1, y3, x1),
       (net8, ~y1, ~y2, ~y3, x2),
       (net9, y2, y3, x2);

or     (net10, net6, net7, net8, net9);

//instantiate the T flip-flop for y2
t_ff_da inst2 (rst_n, net10, nety2);   //rst_n, t, y1

buf    #6 (y2, nety2);

                          //continued on next page
```

```
//-----------------------------------------------------
//design the logic for flip-flop y3
and     (net11, ~y1, ~y3, x1),
        (net12, y2, y3, x1),
        (net13, y1, y3, x1),
        (net14, y2, x2),
        (net15, y3, x2);

or      (net16, net11, net12, net13, net14, net15);

//instantiate the T flip-flop for y3
t_ff_da inst3 (rst_n, net16, nety3);    //rst_n, t, y1

buf    #6 (y3, nety3);

//design the logic for outputs z1 and z2
and     (z1, y1, ~y2, y3),
        (z2, y1, ~y2, ~y3);

endmodule
```

```
//test bench for moore pulse-mode asm

module pm_asm10c_tb;

//inputs are reg for test bench
//outputs are wire for test bench
reg rst_n, x1, x2;
wire y1, y2, y3, z1, z2;

//display inputs and outputs
initial
$monitor ("x1x2 = %b, y1y2y3 = %b, z1z2 = %b",
          {x1, x2}, {y1, y2, y3}, {z1, z2});

//define input sequence
initial
begin
   #0     rst_n = 1'b0;   //reset to state_a
          x1 = 1'b0;
          x2 = 1'b0;

   #5     rst_n = 1'b1;

                              //continued on next page
```

```
   #10    x1 = 1'b1;   x2 = 1'b0;
   #10    x1 = 1'b0;   x2 = 1'b0;

   #10    x1 = 1'b1;   x2 = 1'b0;
   #10    x1 = 1'b0;   x2 = 1'b0;

   #10    x1 = 1'b1;   x2 = 1'b0;
   #10    x1 = 1'b0;   x2 = 1'b0;

   #10    x1 = 1'b0;   x2 = 1'b1;
   #10    x1 = 1'b0;   x2 = 1'b0;

   #10    x1 = 1'b0;   x2 = 1'b1;
   #10    x1 = 1'b0;   x2 = 1'b0;

   #10    x1 = 1'b0;   x2 = 1'b1;
   #10    x1 = 1'b0;   x2 = 1'b0;

   #10    x1 = 1'b1;   x2 = 1'b0;
   #10    x1 = 1'b0;   x2 = 1'b0;

   #10    x1 = 1'b0;   x2 = 1'b1;
   #10    x1 = 1'b0;   x2 = 1'b0;

   #10    x1 = 1'b1;   x2 = 1'b0;
   #10    x1 = 1'b0;   x2 = 1'b0;

      $stop;
end

//instantiate the module into the test bench
pm_asm10c inst1 (rst_n, x1, x2, y1, y2, y3, z1, z2);

endmodule
```

```
x1x2 = 00,  y1y2y3 = 000,  z1z2 = 00
x1x2 = 10,  y1y2y3 = 000,  z1z2 = 00
x1x2 = 10,  y1y2y3 = 001,  z1z2 = 00
x1x2 = 00,  y1y2y3 = 001,  z1z2 = 00
x1x2 = 00,  y1y2y3 = 011,  z1z2 = 00
x1x2 = 10,  y1y2y3 = 011,  z1z2 = 00
x1x2 = 10,  y1y2y3 = 000,  z1z2 = 00
x1x2 = 00,  y1y2y3 = 000,  z1z2 = 00
x1x2 = 10,  y1y2y3 = 000,  z1z2 = 00
x1x2 = 10,  y1y2y3 = 001,  z1z2 = 00
x1x2 = 00,  y1y2y3 = 001,  z1z2 = 00
x1x2 = 00,  y1y2y3 = 011,  z1z2 = 00
x1x2 = 01,  y1y2y3 = 011,  z1z2 = 00

x1x2 = 01,  y1y2y3 = 100,  z1z2 = 01
x1x2 = 00,  y1y2y3 = 100,  z1z2 = 01
x1x2 = 01,  y1y2y3 = 100,  z1z2 = 01

x1x2 = 01,  y1y2y3 = 000,  z1z2 = 00
x1x2 = 00,  y1y2y3 = 000,  z1z2 = 00
x1x2 = 00,  y1y2y3 = 010,  z1z2 = 00
x1x2 = 01,  y1y2y3 = 010,  z1z2 = 00
x1x2 = 01,  y1y2y3 = 011,  z1z2 = 00
x1x2 = 00,  y1y2y3 = 011,  z1z2 = 00

x1x2 = 00,  y1y2y3 = 101,  z1z2 = 10
x1x2 = 10,  y1y2y3 = 101,  z1z2 = 10

x1x2 = 10,  y1y2y3 = 000,  z1z2 = 00
x1x2 = 00,  y1y2y3 = 000,  z1z2 = 00
x1x2 = 01,  y1y2y3 = 000,  z1z2 = 00
x1x2 = 01,  y1y2y3 = 010,  z1z2 = 00
x1x2 = 00,  y1y2y3 = 010,  z1z2 = 00
x1x2 = 00,  y1y2y3 = 011,  z1z2 = 00
x1x2 = 10,  y1y2y3 = 011,  z1z2 = 00
x1x2 = 10,  y1y2y3 = 000,  z1z2 = 00
```

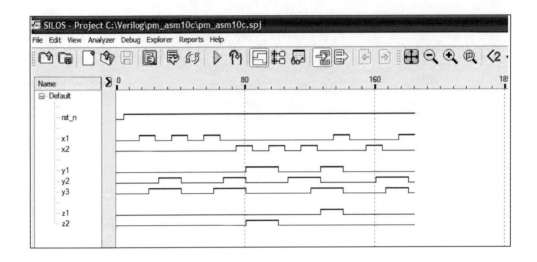

5.18   Design a Mealy pulse-mode asynchronous sequential machine which has two inputs x_1 and x_2 and one output z_1. Output z_1 is asserted coincident with the x_2 pulse if the x_2 pulse is immediately preceded by a pair of x_1 pulses.

Use SR latches and instantiate D flip-flops as a single line. Use NAND, NOR, and AND logic with the continuous assignment statement for all logic gates and latches. Derive the state diagram, the input maps and equations, the output equation, and the logic diagram. Then obtain the design module, the test bench module, the outputs, and the waveforms.

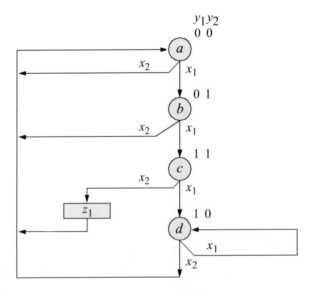

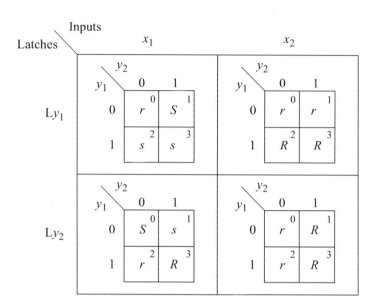

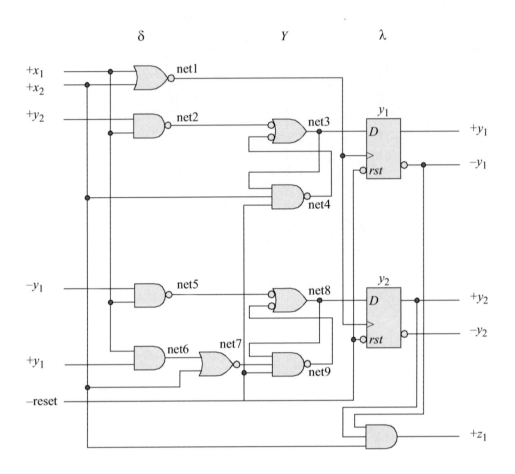

```
//mealy pulse-mode asm using continuous assignment and D ff

module pm_asm_mealy6 (rst_n, x1, x2, y1, y2, z1);

//define inputs and outputs
input rst_n, x1, x2;
output y1, y2, z1;

//define internal nets
wire net1, net2, net3, net4, net5, net6, net7, net8, net9;

//----------------------------------------------------
//design the D flip-flop clock
assign   net1 = ~(x1 | x2);

                                  //continued on next page
```

```verilog
//-------------------------------------------------------
//design the logic for latch Ly1
assign   net2 = ~(x1 & y2),
         net3 = ~(net2 & net4),
         net4 = ~(net3 & x2 & rst_n);

//instantiate the D flip-flop for y1 as a single line
d_ff_bh inst1 (rst_n, net1, net3, y1);   //rst_n, clk, d, q

//-------------------------------------------------------
//design the logic for latch Ly2
assign   net5 = ~(~y1 & x1),
         net6 = (x1 & y1),
         net7 = ~(net6 | x2),
         net8 = ~(net5 & net9),
         net9 = ~(net8 & net7 & rst_n);

//instantiate the D flip-flop for y2 as a single line
d_ff_bh inst2 (rst_n, net1, net8, y2);   //rst_n, clk, d, q

//-------------------------------------------------------
//design the logic for output z1
assign   z1 = y1 & y2 & x2;

endmodule
```

```verilog
//test bench for pulse-mode asynchronous sequential machine
module pm_asm_mealy6_tb;

//inputs are reg for test bench
//outputs are wire for test bench
reg rst_n, x1, x2;
wire y1, y2, z1;

//display inputs and outputs ---------------------------
initial
$monitor ("x1x2 = %b, y1y2 = %b, z1 = %b",
          {x1, x2}, {y1, y2}, z1);

//define input sequence ---------------------------------
initial
begin
   #0 rst_n = 1'b0;      //reset to state_a; no output
      x1 = 1'b0;
      x2 = 1'b0;
   #5 rst_n = 1'b1;                    //continued on next page
```

```
   #10    x1 = 1'b1;   //→ state_b
   #10    x1 = 1'b0;

   #10    x1 = 1'b1;   //→ state_c
   #10    x1 = 1'b0;

   #10    x2 = 1'b1;   //set z1; → state_a
   #10    x2 = 1'b0;

   #10    x1 = 1'b1;   //→ state_b
   #10    x1 = 1'b0;

   #10    x1 = 1'b1;   //→ state_c
   #10    x1 = 1'b0;

   #10    x1 = 1'b1;   //→ state_d
   #10    x1 = 1'b0;

   #20    x2 = 1'b1;   //→ state_a
   #10    x2 = 1'b0;

   #10    x1 = 1'b1;   //→ state_b
   #10    x1 = 1'b0;

   #10    x1 = 1'b1;   //→ state_c
   #10    x1 = 1'b0;


   #10    x2 = 1'b1;   //set z1; → state_a
   #10    x2 = 1'b0;

   #30    $stop;
end

//instantiate the module into the test bench -------------
pm_asm_mealy6 inst1 (rst_n, x1, x2, y1, y2, z1);

endmodule
```

```
x1x2 = 00, y1y2 = 00, z1 = 0
x1x2 = 10, y1y2 = 00, z1 = 0
x1x2 = 00, y1y2 = 01, z1 = 0
x1x2 = 10, y1y2 = 01, z1 = 0
x1x2 = 00, y1y2 = 11, z1 = 0

x1x2 = 01, y1y2 = 11, z1 = 1

x1x2 = 00, y1y2 = 00, z1 = 0
x1x2 = 10, y1y2 = 00, z1 = 0
x1x2 = 00, y1y2 = 01, z1 = 0
x1x2 = 10, y1y2 = 01, z1 = 0
x1x2 = 00, y1y2 = 11, z1 = 0
x1x2 = 10, y1y2 = 11, z1 = 0
x1x2 = 00, y1y2 = 10, z1 = 0
x1x2 = 01, y1y2 = 10, z1 = 0
x1x2 = 00, y1y2 = 00, z1 = 0
x1x2 = 10, y1y2 = 00, z1 = 0
x1x2 = 00, y1y2 = 01, z1 = 0
x1x2 = 10, y1y2 = 01, z1 = 0
x1x2 = 00, y1y2 = 11, z1 = 0

x1x2 = 01, y1y2 = 11, z1 = 1

x1x2 = 00, y1y2 = 00, z1 = 0
```

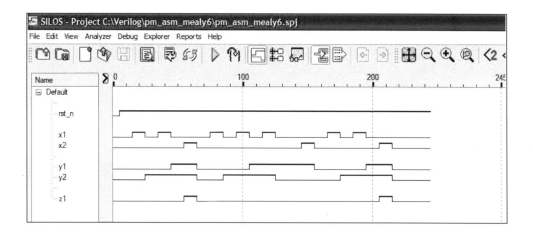

INDEX

Symbols

#0 and #10 4
#5 68
$finish 66
(%) modulus operator
$monitor 7, 66
$random 235
$stop 6
$time 11, 66, 212
<< (Left-shift amount)
>> (Right-shift amount)

A

always statement 66, 275, 499, 619
assign 36

B

behavioral modeling 63, 235, 372, 498
    always statement 66
    case statement 85
    conditional statements (if, else) 81
    event control list 67
    initial statement 64
    loop statements 89
        for loop 89
        forever loop 90
        repeat loop 90
        while loop 90
    sensitivity list 67
binary-to-excess-3 conversion 93
binary-to-Gray conversion 219, 228
bitwise operators 53
    AND (&) 53
    exclusive-NOR ($\wedge\sim$ or $\sim\wedge$) 53
    exclusive-OR ($\wedge$) 53
    negation ($\sim$) 53
    OR (|) 53
built-in primitives 2, 497, 617

C

case statement 85, 500

checksum character 477
clock segments t_1, t_2, t_3, t_4, 234
combinational shifter 151
    shift left algebraic 151, 161
    shift left logical 151, 156
    shift right algebraic 151, 168
    shift right logical 151, 165
comment (//) 4
comparator 38
conditional operator (? :) 45
conditional statements (if, else) 81, 499,
    619
continuous assignment statement 35
counters (synchronous) 173
    binary-to-Gray code converter
        219, 228
    Johnson counter 206
    Möbius counter 206
    modulo-8 counter 174
    modulo-10 counter 186
    ring counter 392
cyclic redundancy check codes 477

D

dataflow modeling 35, 498, 529, 618
    assign 36, 618
    bitwise operators 53
        AND (&) 53
        exclusive-NOR ($\wedge\sim$ or $\sim\wedge$) 53
        exclusive-OR ($\wedge$) 53
        negation ($\sim$) 53
        OR (|) 53
    conditional operator (? :) 45
    continuous assignment statement 35
    reduction operators 41
    reduction AND 41
    reduction exclusive-NOR 42
    reduction exclusive-OR 42
    reduction NAND 41
    reduction NOR 42
    reduction OR 41
    relational operators 48

greater than ($>$) 48
less than ($<$) 48
greater than or equal ($>=$) 48
less than or equal ($<=$) 48
decoder 15, 416
decoders for λ output logic 392
deterministic manner 630

E

endmodule 4, 7
error detection in synchronous
    sequential machines 473
    checksum character 477
    cyclic redundancy check codes 477
    examples 478
    Hamming code 474
    parity 473
    SLy_1, RLy_1, SLy_2, RLy_2 623
    two-out-of-five code 478
event control list 67

F

for loop 89
forever loop 90
full adder 32, 33
fundamental mode 501, 620

G

glitch 317, 654, 695
glitch elimination 317
    using complemented clock 322
    using delayed clock 334
    using state code assignment 317
Gray code 392
Gray-to-binary conversion 107

H

half adder 32, 33
Hamming code 474
hazards 505
    static-0 hazard 505
    static-1 hazard 505

I

initial statement 63, 64, 499, 618
instantiation 2, 18, 91

iterative networks 460

L

logic primitives 1, 2
logical operators 50
    binary logical AND (&&) 50
    binary logical OR ($\|$) 50
    unary logical negation
        operator (!) 50
loop statements 35, 89
    for loop 35

M

majority circuit 67, 102
map entries
    R, r 622, 656
    S, s 622, 656
    T, s, r 650
Mealy machines 254
    behavioral modeling 255
    structural modeling 259, 265
merged flow table 503
merger diagram 503
module 1
module instantiation 91
Moore machines 233
    behavioral modeling 235
    deterministic synchronous
        sequential machines 233
    structural modeling 239, 244, 249
Moore–Mealy equivalence 270
multiplexers 11
    linear-select multiplexers 348, 349
    multiplexers for δ next-state logic
        348
    nonlinear-select multiplexers
        110, 348, 376

N

nonlinear-select multiplexer 110

O

output assertion ($\uparrow$) 234
output deassertion ($\downarrow$) 234
output glitches 317

output symbol 234

P

parity 473
ports 1, 4, 91
postamble 477
preamble 477
primitive flow table 501
procedure 63
    always 63
    initial 63
programmable logic devices 421, 422
    product line 442
    programmable array logic
        21, 422, 426
    programmable logic array
        421, 422, 445
    programmable read-only memories
        421, 422, 423
pulse-mode asynchronous synchronous
    sequential machines 617

R

reduction operators 41
    reduction AND 41
    reduction exclusive-NOR 42
    reduction exclusive-OR 42
    reduction NAND 41
    reduction NOR 42
    reduction OR 41
reflective codes 107
reg 2
register transfer level (RTL) 35
relational operators 48
    greater than (>) 48
    greater than or equal (>=) 48
    less than (<) 48
    less than or equal (<=) 48
repeat loop 90

S

sensitivity list 67
shift operators 59
    << (Left-shift amount) 60

>> (Right-shift amount) 60
static-0 hazard 505
static-1 hazard 505
structural modeling 91, 501, 620
    design examples 93
    module instantiation 91
    ports 91
synchronous counters 173
    binary-to-Gray code converter
        219, 228
    Johnson counter 206
    Möbius counter 206
    modulo-8 counter 174
    modulo-10 counter 186
synchronous registers 122
    combinational shifter 151
        shift left algebraic 151, 161
        shift left logical 151, 156
        shift right algebraic 151, 168
        shift right logical 151, 165
    first-in, first-out queue 143
    parallel-in, serial-out 122
    serial-in, parallel-out 131
    serial-in, serial-out 143
    shift right register 122
synchronous sequential machine 121
synthesis of asynchronous sequential
    machines using Verilog 497
    merge rows into a single row 584
    merged flow table 503
    merger diagram 503
    merging process 503
    primitive flow table 501
    synthesis examples 502
    transition diagram 504
synthesis of pulse-mode asynchronous
    synchronous sequential machines
        617
    delay elements 620
    fundamental-mode model 620
    synthesis examples 620

T

T flip-flop 627, 662
    input map entry definitions 636, 650
    logic diagram 628
t_1, t_2, t_3, t_4, 234
test bench 1, 4
transition diagram 504
two-out-of-five code 478

U

unary operators 41
user-defined primitives (UDPs) 18
    combinational UDPs 19
    endprimitive 18
    primitive 18
    table 19

V

variable values 6
Verilog design module 4

W

while loop 90, 500, 619
wire 2

Printed and bound by CPI Group (UK) Ltd, Croydon, CR0 4YY

25/10/2024

01779408-0006